Sustainable Built Environment Conference (SBE 19)

Emerging Concepts for Sustainable Built Environment

IOP Conference Series: Earth and Environmental Science
Volume 297

Helsinki, Finland
22 – 24 May 2019

ISBN: 978-1-5108-9645-1
ISSN: 1755-1307

Printed from e-media with permission by:

Curran Associates, Inc.
57 Morehouse Lane
Red Hook, NY 12571

Some format issues inherent in the e-media version may also appear in this print version.

Institute of Physics
Dirac House, Temple Back
Bristol BS1 6BE UK

Phone: 44 1 17 929 7481
Fax: 44 1 17 920 0979

techtracking@iop.org

Additional copies of this publication are available from:

Curran Associates, Inc.
57 Morehouse Lane
Red Hook, NY 12571 USA
Phone: 845-758-0400
Fax: 845-758-2633
Email: curran@proceedings.com
Web: www.proceedings.com

TABLE OF CONTENTS

CLEAN PLANET

CLIMATE CHANGE MITIGATION

MATERIAL LIFE CYCLES AND CIRCULARITY

ADAPTATION AND RESILIENCE TO CLIMATE CHANGE

HAPPY PEOPLE

HEALTHY AND COMFORTABLE INDOOR AND OUTDOOR SPACES

IOP Conf. Series: Earth and Environmental Science **297** (2019) 011001 doi:10.1088/1755-1315/297/1/011001

Preface

Sustainable Built Environment conference 'SBE19 Helsinki' debated emerging concepts for a clean planet, happy people and shared prosperity. It was organized by the Finnish Association of Civil Engineers RIL and Sustainability Laboratory Oy SLAB 22-24 May 2019 in Helsinki. It is part of SBE conference series that were held during 2018-2019. These events organized in different regions in our planet are expected to bring their highlights to the World Sustainable Built Environment conference that will be held in Gothenburg, Sweden in June 2020.

SBE19 Helsinki conference brought together 120 experts from 23 countries to share latest knowledge in sustainable built environments. The objective was to introduce also novel approaches to climate change mitigation, adaptation and resilience, circular procurement, social license to operate and nature-based solutions for healthy indoor and outdoor environments in exciting and inclusive cities.

SBE19 Helsinki proceedings contains 50 scientific papers structured under the conference topics and sub-topics as follows

- Emerging concepts
- Shared prosperity
- Positive stimulation and well-being
- Clean planet
- Climate change mitigation
- Material life cycles and circularity
- Adaptation and resilience to climate change
- Happy people
- Healthy and comfortable indoor and outdoor spaces.

The organizers want to take the opportunity to thank the authors of the published articles, the Scientific Committee members for their improving comments to the submitted papers in the review process and all active conference participants for interesting discussions during the sessions and breaks that contributed to the SBE19 Helsinki success.

Pekka Huovila
Scientific Committee Chair

Peer review statement

All papers published in this volume of *IOP Conference Series: Earth and Environmental Science* have been peer reviewed through processes administered by the proceedings Editors. Reviews were conducted by expert referees to the professional and scientific standards expected of a proceedings journal published by IOP Publishing.

Nature Based Solutions (NBS) for sustainable and resilient cities: experiences from Europe and Brazil

C. Antuña-Rozado[1,6], C. P Herzog[2], T. Freitas[3], C. Cagnin[4], G. Wiedman[5]

[1]VTT Technical Research Centre of Finland Ltd., PO Box 1000, FI-02044 VTT, Finland
[2] Pontifical Catholic University of Rio de Janeiro, Department of Architecture and Urbanism, R. Marquês de São Vicente, 225 - Gávea, Rio de Janeiro - RJ, 22451-900, Brazil
[3] European Commission, DG Research and Innovation, CDMA 03/022, Rue du Champ de Mars 21, 1050 Brussels, Belgium
[4] Center for Strategic Studies and Management (CGEE), SCS Qd 9, Lote C, Torre C, 4º andar, Salas 401 A 405, Ed. Parque Cidade Corporate, Brasília-DF, CEP 70308-200, Brazil
[5] MCTIC – Ministry of Science, Technology, Innovation and Communications, Brazil
[6] Author to whom the correspondence should be addressed, carmen.antuna@vtt.fi

Abstract. Although Nature-Based Solutions (NBS) is a relatively recent concept, it can provide effective means for cities worldwide to become more sustainable and to enhance urban resilience, thus improving the quality of life of their citizens. Moreover, cost-effective NBS can help improve the urban environment in multiple scales, from buildings to neighbourhoods, from public spaces to peri-urban areas, providing all kinds of co-benefits positively influencing the above-mentioned challenges. Depending on the context and the need, many different stakeholders may be involved in the co-creation of systemic, inclusive and locally attuned NBS supporting social cohesion and urban equality.

The learning alliance fostered by the EU-Brazil Sector Dialogue on NBS, is currently working together to identify common research and innovation agendas for cooperation, as well as mapping relevant examples of successful NBS that have been recently implemented or are under development in both Europe and Brazil. The cases selected have been studied and compared with a particular emphasis on their potential for replicability, scalability and adaptation to the local context. In addition, they should address the three pillars of sustainability: environmental, socio-cultural and economic.

Therefore, this paper discusses the opportunities offered by NBS (even in periods of economic downturn), but also the limiting factors hindering their wider dissemination. In addition, it will present through a few inspiring case studies from both regions, clear examples of successful implementation of NBS, which could be replicated and scaled in other locations in response to specific local conditions.

1. Introduction

As the world's urban population is expected to double by 2050 according to the New Urban Agenda [1], cities will face enormous challenges related to climate threats, environmental quality, adequate services and infrastructures, mobility, affordable housing, employment, food security, etc. Even more in the case of those cities particularly affected by rapid urbanization [2].

NBS components such as parks, built wetlands, food gardens, among many others are key to reconnect people to nature and can be used as classrooms for environmental education [3]. NBS are essential do adapt cities to climate change impacts and contribute to mitigate GHG emissions [4].

The Sector Dialogue on NBS between EU and Brazil enabled the first broad research on the current scenario in the largest country in Latin America. The first Dialogue completed in 2016 "Innovating cities with nature-based solutions: co-creating knowledge on nature-based solutions with EU-Brazil Sectoral Dialogues", provided a framework for identifying and assessing NBS as well as defining future topics for collaboration. The second Dialogue recently concluded, and of which this paper can be considered an epilogue, has deepened the analysis focusing on how to harness the Brazilian potential for NBS and how the EU's experience can contribute to accelerate the process. It also maps relevant examples of successful NBS either already implemented or under development in both Europe and Brazil, with a special emphasis on their potential for replicability, scalability and adaptation to the local context, while addressing the three pillars of sustainability.

Therefore, this paper discusses briefly the opportunities offered by NBS (even in periods of economic downturn), through a few inspiring case studies from both regions, which could be replicated and scaled in other locations in response to specific local conditions.

2. NBS for sustainable and resilient cities

The EU defines NBS as *"solutions that are inspired and supported by nature, which are cost-effective, simultaneously provide environmental, social and economic benefits and help build resilience. Such solutions bring more and more diverse nature and natural features and processes into cities, landscapes and seascapes, through locally adapted, resource-efficient and systemic interventions"*[1]. Therefore, in recognition of the role that NBS can play in providing sustainable, cost-effective, multi-purpose options for more resilient cities and societies, the EU Research and Innovation Policy Agenda for NBS and Re-Naturing Cities aims at positioning the EU as a leader in "innovating with nature". In practice, the Agenda is implemented through Horizon 2020 Programme [5].

In Brazil, the Ministry of Science, Technology, Innovations and Communications (MCTIC) created the thematic programme of Technologies for Sustainable Cities that aims to support cities to transition to a more sustainable development with the incentive to innovation and technologies in several fields. The programme includes as focal themes NBS, biomimicry and urban metabolism.

3. Experiences on NBS from Europe and Brazil. What can we learn?

In Brazil, the results have shown that there are several NBS planned, designed and implemented at different scales, from local to regional. At regional and city scales, the NBS are planned, designed and implemented by local authorities, with some participation of civil society representatives and NGOs. However, local NBS are usually bottom-up initiatives promoted by citizens (grassroots) and articulated with other residents and institutions, public and private, to transform sites that they see as opportunities to create a better place to live in. Also, there are institutional initiatives that are carried out due to legal requirements to protect and enhance the local environment. Even though there is certain academic involvement in NBS at different stages, with very few research labs working in this area, the funds in Brazil are limited in this field.

Being at the forefront of investing in a green economy, the EU features a large number of examples covering a wide array of NBS (storm water retention ponds, urban food production, green façades and roofs, green corridors, etc.) for different scales in very different contexts across Europe. NBS are used for different purposes like the regeneration of degraded urban areas, storm water management, flooding risk reduction, etc. Even though most of the European examples involve multi-stakeholder engagement, the ones chosen for this paper are a bottom-up initiative and a highly participatory NBS co-design process in order to show how such approaches can contribute to social inclusion and

[1] https://ec.europa.eu/research/environment/index.cfm?pg=nbs

cohesion. They also illustrate how cities can improve their environment and their citizens' wellbeing with small-scale measures relatively cheap to implement even in times of economic downturn.

In short, NBS offer plenty of opportunities for cities to address contemporary local and global challenges such as: climate change impacts; urban heat island effect; floods and landslides; air, water and soil pollution; or human health improvement (among others). The case studies documented indicate that there is great potential for scaling up solutions, especially through international partnerships, which can contribute knowledge that can be adapted to different contexts, as well as financial support to develop innovations for climate change adaptation in tropical cities.

4. Relevant examples of bottom-up NBS initiatives and citizens participation

Bottom-up NBS initiatives offer many opportunities for improving the urban environment while enhancing social engagement and inclusion. More in particular, they allow the implementation of NBS when the economic conditions or the lack of institutional commitment are not favorable for top-down NBS development driven by the public administration, either local or regional. Similarly, citizens' participation in the decision-making process, not only before, but also during the implementation of NBS projects, particularly large and complex ones, and even after their completion, can lead to: 1) new innovative ideas better tailored to the needs and preferences of the citizens; and 2) long-term appropriation and commitment to further development with multiple benefits for all stakeholders.

4.1. The Power of Food Festival (Edinburgh, UK) and citizens' participation in co-creation of NBS (Tampere, Finland)

The Power of Food Festival[2], takes place annually in Edinburgh since 2015. Founded by Marie-Amélie Viatte, who was inspired by the community food growing activities in her city, and also by other similar initiatives in the country related to urban food production like Incredible Edible[3], the Festival was built on the idea of a "festival for the gardens and the gardeners". There are a lot of community gardens in Edinburgh. Community gardens sometimes happen on a piece of grass that belongs to the local social housing blocks. Or in schools and offices, in a library courtyard, etc.

Figure 1. Live music at the Royal Edinburgh Community Garden during the Power of Food Festival 2018. (Picture: courtesy of The Power of Food Festival)

[2] https://poweroffoodfestival.wordpress.com/
[3] https://www.incredibleedible.org.uk/

Reaching out to active groups like Granton Community Gardeners[4], and supported by a small but dedicated group of volunteers, the Festival wanted to offer people the opportunity to travel to those neighbourhoods, many of them deprived, and to walk around the gardens and know the work of the gardeners. There is no entrance fee for the visitors to see the gardens. Since the beginning, the Festival is about growing, harvesting and sharing the food together, and some gardens even offer food for the visitors to enjoy.

Figure 2. Sustrans bikes. (Picture: courtesy of the Power of Food Festival)

Figure 3. New Haven Heritage Community Garden (left). Learning about bees in Starbank Community Garden (right). (Pictures: courtesy of the Power of Food Festival)

[4] https://www.communityfoodandhealth.org.uk/community-based-activity/case-studies/granton-community-gardeners/

The gardens contribute to the promotion of the event, and therefore those that put more effort on communications typically receive more visits. Artists and artisans also participate on voluntary basis offering live performances to entertain the public. Clearly, music and the arts draw people in. After one edition ends and they thank everyone, the organizing team reflects on the results and starts planning the next edition, always looking for new ways to establish mutually beneficial partnerships in order to provide novel activities. E.g. the collaboration with Sustrans[5], a British charity focusing on making it easier for people to walk and cycle, allows the Festival to offer bicycle rides for visitors around the gardens. In the future, it is envisioned that the gardens will take more ownership of the Festival and organize it by themselves, and together they will create their own identity. In addition, new activities could be offered, like walking tours.

Since some authors argue that NBS have not always been able to involve citizens adequately and have failed to properly address food security, poverty alleviation and urban inequality, the proponents of the Edible City Solutions (ECS) conceptual framework offer a strategic approach *"towards the development of sustainable, livable, and healthy cities"* [6]. The Power of Food Festival along with Edinburgh's community gardens illustrate well the potential of edible urban green space to successfully involve citizens in the provision of socio-cultural ecosystem services.

Figure 4. Visioning Vuores session for residents and other stakeholders (left). Children watching water insects under the microscope (right). 101 participants, pupils and teachers. (Pictures: courtesy of the City of Tampere)

Located in Southern Finland, Tampere is a city of 225,000 inhabitants with an important industrial heritage. At present, Tampere is a centre for high-tech companies, technology, research, education, culture, sports and business. The city wants to become a global frontrunner demonstrator of NBS and in line with that objective, is currently a partner in UNaLab (Urban Nature Labs) project funded under the EU Horizon 2020 programme (2017-2022). Within UNaLab, Tampere is implementing an urban living lab demonstration area in Vuores eco-efficient district to address identified climate and water related challenges, more specifically flooding and storm water management, by co-creating NBS with local stakeholders and end users.

Vuores is a green district to be completed by 2030, providing residence for 13,000 people and 3,000 to 5,000 jobs. The NBS solutions co-created and demonstrated in Vuores will be scaled up and further developed in another area of the city called Hiedanranta. NBS demonstrated in Vuores include public green spaces and storm water ponds, green roofs and walls for buildings, and permeable

[5] https://www.sustrans.org.uk/

pavements. A series of workshops have been organized in Vuores in the context of UNaLab project covering "Visioning", "Ideating" and "Testing", with 139 active participants, many of them children.

Visioning Vuores workshops:

- Urban Nature - Blue and Green Vuores Workshop;
- Session for residents and other stakeholders;
- Participatory design game - end users received information on NBS while experts received information on user experience and willingness to take part in the co-creation and maintenance of NBS.

Ideating Vuores workshops:

- Storm Water Day at Vuores School;
- Session for kindergarten children and Elementary School pupils (101 participants);
- Design Thinking: participatory design game and LEGO serious play.

Testing Vuores workshops:

- The Nature Trail and the Blue and Green Treasures of Vuores;
- NBS info desk during Vuores Day Festival.

The best ideas collected from the workshops will be further elaborated and implemented. The results of the workshops can be summarized as follows: in terms of recreation, the participants proposed more accessible green areas to encourage and support physical activities; more city green and wilder parks are needed to protect and increase biodiversity; whereas in relation to water management, more information on NBS along with novel NBS to protect water quality were demanded.

In the workshops that took place in the schools, teachers and pupils were given tools to monitor water quality and macroinvertebrates. The monitoring started in May 2018 and this activity is planned to be conducted twice a year. The pupils will have water monitoring backpacks, which include the equipment needed to monitor water pH, temperature, turbidity and oxygen. Experts will give advice on where to begin and how to report the results, which will be used to evaluate the impact of NBS.

4.2. Springs Square and "Corujas" Food Garden (São Paulo, Brazil)

São Paulo is the largest city in South America, with 21.2 million inhabitants in the Metropolitan Region. During the urbanization process ecosystems were eliminated, rivers disappeared from the landscape. The waters are very polluted by point (mainly sewage discharge) and diffuse contamination by storm water run-off. Food comes from distant places, and most of the fresh produce sold in the supermarkets is treated with chemicals and pesticides. Thus, people are neither connected to nature, nor to natural processes and flows of biodiversity, waters and food production. Urban green areas are mostly covered by lawn with ornamental vegetation that offers reduced ecosystem services.

In São Paulo the concept of park is a fenced green area, whereas squares are not. There is no relation to its size nor to the vegetation cover. In this paper, two Brazilian case studies located in squares in central areas of the city are presented: Owls' Food Garden and Spring's Square. Both were transformed by the action and co-creation of citizens that wanted to transform the public green areas in their neighborhoods to generate multiple ecosystem services to enhance their quality of life and well-being.

The first is located at the Dolores Ibarruri Square, its official name, but it is known as the Owls' Square after the Owls' river that flows in a canal in the lower area. It has a total area of 48,000m^2. In 2008 the square was renovated with rain gardens and bio-swales. Most of the slopes were covered by lawn that require high maintenance and could not prevent erosion caused by frequent strong storms. It is a very popular destination for community members in search of relaxation, exercise, birthday celebrations and other social events in contact with nature.

In 2011, two residents got informal authorization from the District Deputy Mayor to start a food garden in a lawn covered 800m^2 plot of land inside the square. Both had attended a permaculture course, which turned around their lives and the history of food gardening in São Paulo, and later expanded beyond city borders with the Facebook group Hortelões Urbanos (Urban Food Gardeners).

The group has more than eighty thousand members and a list of thousands waiting to be admitted. Their initial cultivation site was at Owls' Square, now named the Owls' Food Garden. The area produces all kinds of edible plants, including non-conventional food plants, plus flowers that attract pollinators and enhance the soil biota. The applied agroforestry techniques restored the water springs that restarted to flow, conveying the waters through bio-swales into small basins, where they store the needed water to irrigate the garden even during severe droughts that happen recurrently.

The food garden relies on the voluntary work of residents. They promote collective actions to plant, manage and clean the planted site. The biodiversity thrives, with a total transformation from the high maintenance grass covered slope to a high-performance landscape that offers a myriad of benefits for people and nature. They still need to educate those neighbours that don't understand the concept of "messy" garden. The Owls' Food Garden is a social-ecological experience that has to be developed daily. The site is visited by schools and college students. The movement has gone further with events and publications that raise awareness about consumerism and the broad impacts caused by unsustainable lifestyles, organic food production compared to the conventional pesticide-based mono-cultural agriculture, and the value of nature in the city.

Figure 5. Owls' Food Garden: Before (left) and after (right) the planting of the community garden. The change in the landscape is remarkable with rich biodiversity and increased ecosystem services due to residents' participation. (Pictures: Cecilia P. Herzog)

The second case in São Paulo is located in the Sumaré neighborhood on the other side of the hill, where the waters flow to the main river named Tietê. Sumaré is a mainly residential district where there are not many green areas for people to socialize. The 12,000m² Homero Silva square is the largest green space in the neighborhood. It was a derelict place where people used to throw garbage. It was insecure, nobody used to go there. In 2013 a grassroots movement called Ocupe&Abrace (Occupy and Embrace) started to take care of the place to recover the springs of the Água Preta (Black Water) creek and the biodiversity, aiming to have direct contact with nature and natural processes and flows.

After five years they achieved more than an oasis in the hard cityscape. Ocupe&Abrace restored eight springs building two ponds to receive the waters before they disappear underground, creating new ecological niches, and planted more than 600 native and food species in collective events. The city is vulnerable to mosquitoes that transmit dengue, chikungunya and other tropical diseases. To avoid the reproduction of the vector, one hundred aquatic species were introduced for biological control. It was so successful that in January 2017, when the city suffered a burst of mosquitoes with human health paying a high toll in most of its urbanized areas, the neighborhood was the only place that had avoided the proliferation of contagious viruses.

In the slope that was eroding, an independent project named "Cerrado Infinito" (Infinite Savannah) introduced Cerrado (Brazilian Savannah) vegetation – the native dry ecosystem that disappeared from the local landscape. This project was led by Daniel Caballero who is an artist passionate for this ecosystem. He has been collecting, planting and educating residents about the importance of having back the vegetation that has deep roots, thus enhancing the water flows of the springs. So, the project "Cerrado Infinito" is an educational project that intends to raise awareness about the importance and the impact that has been caused to the water producer biome, where most of the important Brazilian rivers' springs are located, in Central Brazil.

Figure 6. Spring's Square: Built pond where native aquatic species were introduced (top left); Cerrado planting on the slope (top right); and interpretative signage to educate visitors (bottom left). (Pictures: Cecilia P. Herzog)

5. Discussion

In São Paulo, besides the ecological benefits, the grassroots movement also promotes festivals to educate about nature, and celebrate life and culture, with collective plantings, music, "Drums Circle", among other activities. Social cohesion has surpassed the limits of the square and the group is now fighting in court to avoid the construction of a high rise building in its border that will impact the springs. Similarly, in Edinburgh the Power of Food Festival attracts an increasing number of participants due to a creative and varied programme that includes arts and crafts, but also educational activities, e.g. music, storytelling, pottery making, insect safaris, bicycle tours, etc., providing lots of fun and entertainment for visitors of all ages. By celebrating the contribution of the people that breathe life into the gardens and the neighbourhoods, the Festival wishes to promote greater social inclusion and wellbeing, environmental awareness and sustainability through community food growing.

NBS have intuitively been introduced in many places, some have had the contribution of professionals that give technical support and academia that enhances the benefits with scientific knowledge. There are people who know that we all depend on nature and natural processes and flows to survive, have a good quality of life and well-being, and many are working to transform the urban landscape at the local scale. It is more sustainable when residents lead the process, since top-down practices and actions are subject to political changes, especially in countries were policies may be changed without real public participation, which is the case of Brazil. The workshops organized in Vuores by the City of Tampere, Ramboll and VTT within UNaLab project, with a high participation of different stakeholders including citizens, and even children, have provided the opportunity for mapping different viewpoints for further development of existing NBS. In addition, cooperation with local universities and a SME producing bio-char is leading to the development of new bio-filtration solutions to treat contaminated waters which also creates new business opportunities around NBS.

References

[1] United Nations General Assembly. The New Urban Agenda Adopted by the United Nations Conference on Housing and Sustainable Urban Development (Habitat III). 21 November 2016. Available online: http://habitat3.org/wp-content/uploads/N1639668-English.pdf (Accessed on 2 November 2018)

[2] Antuña-Rozado C, García-Navarro J and Mariño-Drews J. 2018. *Facilitation Processes and Skills Supporting EcoCity Development*. Energies 11, 777 (Basel: MDPI)

[3] Cole L, McPhearson T, Herzog C and Russ A. 2017. Green Infrastructure. *Urban Environmental Education Review*. Russ A, Krasny M (Eds.). (Ithaca: Cornell University) pp 261-270

[4] McPhearson T, Karki M, Herzog C, Santiago Fink H, Abbadie L, Kremer P, Clark C, Palmer M and Perini K. 2018. Urban ecosystems and biodiversity. *Climate Change and Cities: Second Assessment Report of the Urban Climate Change Research Network*. Rosenzweig C, Solecki W, Romero-Lankao P, Mehrotra S, Dhakal S and Ali Ibrahim S (Eds.). (Cambridge, MA: Cambridge University Press) pp 257-318

[5] European Union. 2015. *Towards an EU Research and Innovation policy agenda for Nature-Based Solutions & Re-Naturing Cities*. (Luxemburg: Publications Office of the EU)

[6] Säumel I, Reddy S and Wachtel T. 2019. *Edible City Solutions - One Step Further to Foster Social Resilience through Enhanced Socio-Cultural Ecosystem Services in Cities*. Sustainability 11, 972 (Basel: MDPI)

Acknowledgments

The authors wish to thank the following persons for the information they have kindly provided for this paper, as well as for their valuable feedback: Marie-Amélie Viatte, founder of the Power of Food Festival (Edinburgh, UK); Salla Leppänen from the City of Tampere (Finland); Claudia Visoni (Owls' Square), and Andrea Pesek and Luciana Cury (Springs' Square), from São Paulo (Brazil).

Smart and sustainable urban development in Egypt: the case of Nabta Smart Town

C Antuña-Rozado[1,3], **F Reda**[1], **Y El Mahgary**[2]

[1] VTT Technical Research Centre of Finland Ltd., PO Box 1000, FI-02044 VTT, Finland
[2] Nabta University of Egypt (NUE), Borg El Arab, Egypt
[3] Author to whom the correspondence should be addressed, carmen.antuna@vtt.fi

Abstract. Nabta Smart Town (NST), in New Borg el Arab, close to Alexandria, is an ambitious project under planning that includes residential and commercial facilities, in combination with high quality "Educational Magnets" capable of creating a dynamic community and generating all sorts of interesting activities. Therefore, the uniqueness of NST is building a fully integrated and inclusive neighbourhood - Educational Magnet, Residential, Commercial (shopping, leisure and entertainment), with Administrative Offices (including high-end security). The initial conceptualization of NST by the local promoters envisions a culture that protects, maintains and recycles within the community. The collaboration of VTT Technical Research Centre of Finland Ltd. (VTT) was sought to ensure that the Master Plan (MP) and the buildings of the University, as well as the Residential and Commercial facilities of NST comply with the standards of a modern EcoCity adapted to the Egyptian context. The first phase of the work focuses mainly on carrying an EcoCity Energy Efficiency Feasibility Assessment. Specific methodologies and tools developed by VTT have been used to ensure adaptation of the solutions proposed to the local conditions. This paper discusses the results obtained using the methodology developed by VTT, its applicability at different scales, as well as its replicability potential - not only in other locations in Egypt, but also worldwide.

1. Introduction

According to the World Bank, cities have a key role in climate change, accounting for roughly 80 percent of the global greenhouse gas emissions [1]. Rapid urbanization in developing countries is bringing about numerous risks in relation to social instability, poor infrastructure, potential water crises, environmental degradation, etc. Moreover, the recently adopted New Urban Agenda expects the world's population to double by 2050, which will place even more pressure on cities and exacerbate not only the demand for energy, but also the greenhouse gas emissions [2].

In this critical situation, sustainable urban planning and development can substantially contribute to decreasing the cities' adverse effects on climate change and to improve the living conditions of millions worldwide. Hence the relevance of Nabta Smart Town (NST) in New Borg El Arab, close to Alexandria, an ambitious project aiming at setting an example of sustainable urban design and construction adapted to the local conditions in Egypt. The pertinence of the positive transformation envisioned by NST project should be understood in the context of the enormous challenges faced by Egyptian cities [3], struggling with high unemployment, lack of decent housing, water scarcity, chaotic traffic, air pollution, noise, etc. Unfortunately, Alexandria and even New Borg El Arab (even though to a much smaller extent), are not an exception.

Known as "The Pearl of the Mediterranean", Alexandria was the capital of Egypt for almost a thousand years, and currently is the second largest city in the country with a population of 5.3 million as of January 2019 [4]. Between the Mediterranean Sea to the north and Lake Mariout to the south, its contemporary urban development began in the 19th century with a series of emblematic projects like El Mahmoudia Canal (1819), Egypt's first railway (1854), or the tramway (1863). Early in the 20th century, an extensive planning scheme was developed for the city (1920), and the Corniche sea boulevard was built (1934). Alexandria still has a number of beautiful buildings from this period designed by Greek and Italian architects. However, since 1952 Egypt has undergone many political, social and economic changes, and its cities, particularly Cairo and Alexandria, have experienced a constant expansion, sadly not following sustainable urban development patterns. As a consequence, Alexandria is nowadays facing numerous challenges like a very high population growth rate, unemployment -particularly among the youth-, proliferation of informal settlements, large number of unsafe buildings, inefficient mobility, or high pollution levels in El Mahmoudia Canal and Lake Mariout [5].

Figure 1. Two views of the Corniche, Alexandria´s waterfront promenade originally designed by architect Pietro Avoscani in 1870, and nowadays a major traffic corridor. It runs for over ten miles from the Citadel of Qaitbay to Al Montaza. (Pictures: Pixabay)

Figure 2. Rapid urbanization in Alexandria (Picture: Pixabay). Cases like that of the so called "Leaning tower", a symbol of the city's corruption, are sadly not an exception in a country where 400 residential buildings collapse every year [6].

New Borg El Arab City (NBC), 60 km to the south-west of Alexandria and 7 km away from the Mediterranean coast, is a new city established by presidential decree 506/1079 with a population of 150 000 inhabitants. NBC is expected to reach 570 000 inhabitants by 2022 [7]. At present, the city has around 1270 production factories and most of the workers are commuting daily from Alexandria [8].

A few years ago VTT Technical Research Centre of Finland Ltd. (VTT) and the Egypt-Japan University of Science and Technology (E-JUST) located in NBC, carried out a joint project called "EcoNBC, EcoCity Capacity Building in NBC" (2013-2015) funded by the Institutional Cooperation Instrument (ICI) under the Ministry for Foreign Affairs of Finland. The project's main objective was to increase the know-how of E-JUST on sustainable urban planning according to VTT's EcoCity concept.

Raising the awareness of local stakeholders was another important objective. To that end, EcoNBC project developed a number of activities that provided a fairly good understanding not only of the city's most pressing problems, but also of how to provide solutions that respond to the local conditions.

Figure 3. Two views of New Borg El Arab. (Pictures: Pekka Huovila)

2. Nabta Smart Town in NBC

Nabta Smart Town (NST), located in NBC, is an ambitious project under planning that includes residential and commercial facilities, in combination with high quality "Educational Magnets" capable of creating a dynamic community and generating all sorts of interesting activities. Therefore, the uniqueness of NST is building a fully integrated and inclusive neighborhood - Educational Magnet, Residential, Commercial (shopping, leisure and entertainment), with Administrative Offices (including high-end security). Nabta's mission could be summarized as follows:

- To contribute to the improvement of human conditions in the region.
- To use science, technology and culture to achieve the goals of solving society's problems through sustainable improvements of human resources in the region by providing superior education to its students, and to offer pragmatic and innovative solutions to address human needs through outstanding products and services.
- To promote and support the establishment of strong businesses, as well as technical and commercial ties between international industries and organizations and their counterparts in countries and regions covered by Nabta University.
- To capitalize on "Science & Technology Diplomacy" to promote understanding between the nations.

The initial conceptualization of NST by the local promoters of the project envisions a culture that protects, maintains and recycles within the community. Everything from regenerative urban agriculture, edible landscaping, and energy-producing building designs, will be considered when designing and creating a sustainable living ecosystem, for this generation and generations to come. Nothing goes to waste, renewable energy sources will be deployed as much as possible, other natural resources and living features will be used, reused, and regenerated efficiently and effectively. Sustainable lifestyles will be promoted among the local community, and participation encouraged.

2.1. Collaboration with VTT

In this context, the collaboration of VTT was sought to ensure that the Master Plan and the buildings of Nabta University, as well as the Residential and Commercial facilities of NST comply with the standards of a modern smart EcoCity adapted to the local conditions of Egypt. Additionally, this work should capitalize on the recommendations and findings of "EcoNBC Feasibility Study. Transforming New Borg El Arab into an EcoCity" [9], one of the main outcomes of the abovementioned EcoNBC project, and thus become the first development pilot project that makes direct use of those.

Figure 4. Concept Master Plan of Nabta Smart Town commissioned by the local promoters and developed by Atkins, a well-known global design, engineering and project management consultancy. (Image: UpScale Development Llc and Atkins)

Therefore, together with the local partner, the Khairy Foundation for Human and Social Development (KFHSD), VTT prepared a project proposal under the name of "Sustainability Assessment of Nabta Smart Town (NST) and Actions for Improvement: an Overview" that got funded by the 10YFP Trust Fund (One Planet Network SBC Programme). From its inception, the project tried to establish direct links with the UN Climate Change Conference, 21COP held in Paris in 2015, and the international campaign to decrease CO2 emissions and increase the use of renewable energy. Also linked to the 2030 Agenda for Sustainable Development, the project will significantly contribute to the following Sustainable Development Goals (SDGs) [10]:

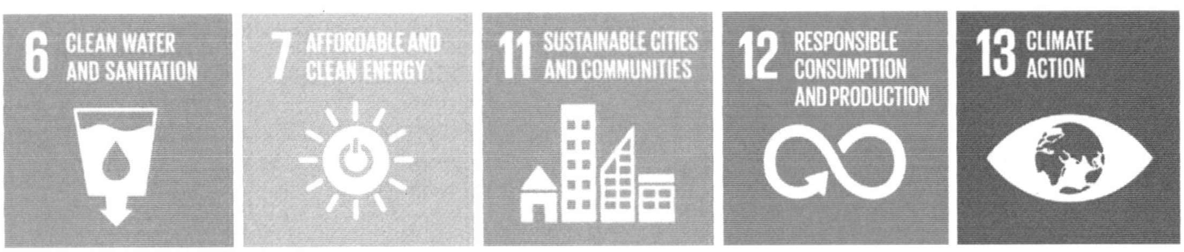

Figure 5. SDGs directly related to the main objectives of NST Project.

The first phase of the work focused mainly on carrying an EcoCity Energy Efficiency Feasibility Assessment. Therefore, special attention was paid to energy and resource efficiency when assessing the MP (including passive features, use of indigenous vegetation and local materials, comprehensive sustainable mobility, user behaviour and participation, etc.). Specific methodologies and tools developed by VTT have been used, and even improved, to ensure adaptation to the local conditions in Egypt of the solutions proposed [11].

3. Energy Assessment of NST Master Plan. Main findings

3.1. Approach and assessed cases

As mentioned above, the methodology developed by some of the authors has been used for carrying out the energy assessment of NST Master Plan. Several meetings with relevant stakeholders have been held in order to collect relevant information about the buildings to be assessed. Preliminary buildings information:

1. Shape and pre-architectural design of buildings according to the main zones in the Master Plan;
2. Construction techniques (wall, roof and floor stratigraphy) and relevant building envelope solutions;
3. Construction materials, including local materials such as white stone;
4. Occupancy patterns & Opening hours (in case of non-residential buildings).

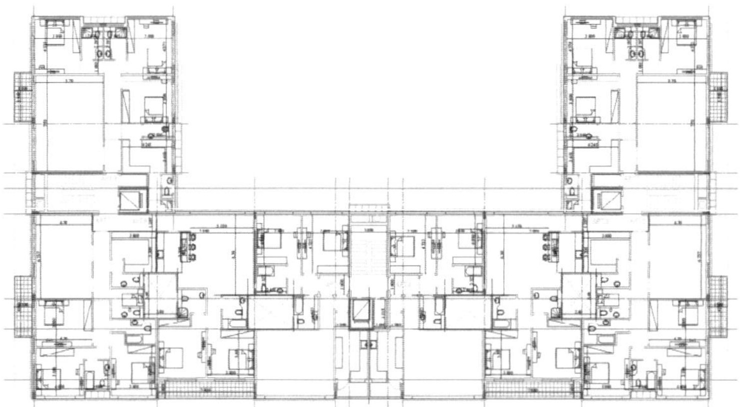

Figure 6. Example of residential building floor layout.

Building models have been generated using Transient System Simulation Tool (TRNSYS) software [12] for estimating the energy demand of the town. In particular, three different building models have been created to emulate educational, office and residential buildings. Each of these building models has been associated to a building typology present in the Master Plan. Table 1 shows the total area of the different building typologies, in accordance with the Master Plan, and the building model paired to the building typology and used to emulate the energy behaviors. Table 2 shows the thermal properties of the construction materials that have been considered in the building models.

Table 1. Total area of building typologies according to the Master Plan.

Master Plan		
Total area (m^2)	**Building typology**	***Paired building model***
101465	University Neighbourhood	*Educational*
51970	Business	*Office*
46565	Cultural	*Educational*
180471	Family	*Residential*
23490	Hotel	*Residential*
20223	Commercial	*Office*
16240	Schools	*Educational*

Table 2. Thermal properties of the construction materials used in the model.

Material	Specific Heat [J/ kg. °C]	Density [kg/m^3]	Conductivity [W/m. °C]	Width [mm]
Cement plaster	1000	2000	1.4	25
Red bricks	1000	1200	0.5	125
Insulation	900	80	0.04	50
Red bricks	1000	1200	0.5	125
Cement plaster	1000	2000	1.4	25
			[W/m^2°C] U (Transmittance)	**0.526**

Low-e windows (Tinted double glazed 4/12/4 mm Glass/air/Glass, U=1.69 W/m^2°C) have been assumed as provided by the local partner. 20° and 26°C have been set as heating and cooling indoor temperature set points. Use of energy efficient practices, such as natural ventilation (free cooling) and external shading devices, have been also considered. Different internal gains and energy consumption of electrical appliances have been estimated according to the provided occupancy patterns for each of the abovementioned building models. To assess different configuration of RES for supplying the energy to the town, a system model has also been created. A district heat pump has been assumed to be in use.

Regarding renewable energy generation, both PV and wind have been considered. Authors have estimated different RES configurations, considering PV integrated on building roofs and decentralized wind turbines. Particularly, the distribution of PV in different buildings has been optimized to minimize the energy export and, therefore, maximizing the self-consumption. Five scenarios have been considered for the total RES capacity installed in the district:

1. 25% of the district power peak (2.40 MW)
2. 50% of the district power peak (4.81 MW)
3. 75% of the district power peak (7.21 MW)
4. 100% of the district power peak (9.61 MW)
5. Enough RES capacity to achieve Net Zero Energy District (NZED) - a district that produces the same energy that it consumes

3.2. Main results

Figure 7 shows the main results of the energy Assessment of NST Master Plan, depicting imported and exported energy from and to the grid; the on-site energy consumption from PV (energy produced and consumed in the same building); the building to building energy transaction (District,B2B voice in the chart; renewable PV energy produced in a building and consumed in another); the consumed wind energy (District,Wind); the total wind and PV capacity; and the total renewable energy system capacity (sum of wind and PV capacity).

The results show that using only PV generates more energy export than using a combination of PV and wind RES. On the other hand, using the right combination of PV and wind generates less export, almost keeping the import from the grid at the same level. This can be noted comparing the cases with and without asterisk. High energy export may lead to excessive stress on the electricity grid outside the town. In order to avoid this situation, as anticipated before, PV capacity integrated in building roofs has been optimized. The results also show that the building to building energy transaction component is considerably less than the on-site consumption. It can also be seen that this is possible to achieve with conventional RES NZED. To obtain a NZED with low energy export 6.7 MW of PV and 3.6 MW of wind should be adopted. The PV distributions on building roofs associated is 25% of residential buildings, 34% of office buildings and 58% of educational buildings. Interesting to notice is that the roof of the buildings should not be fully covered by PV.

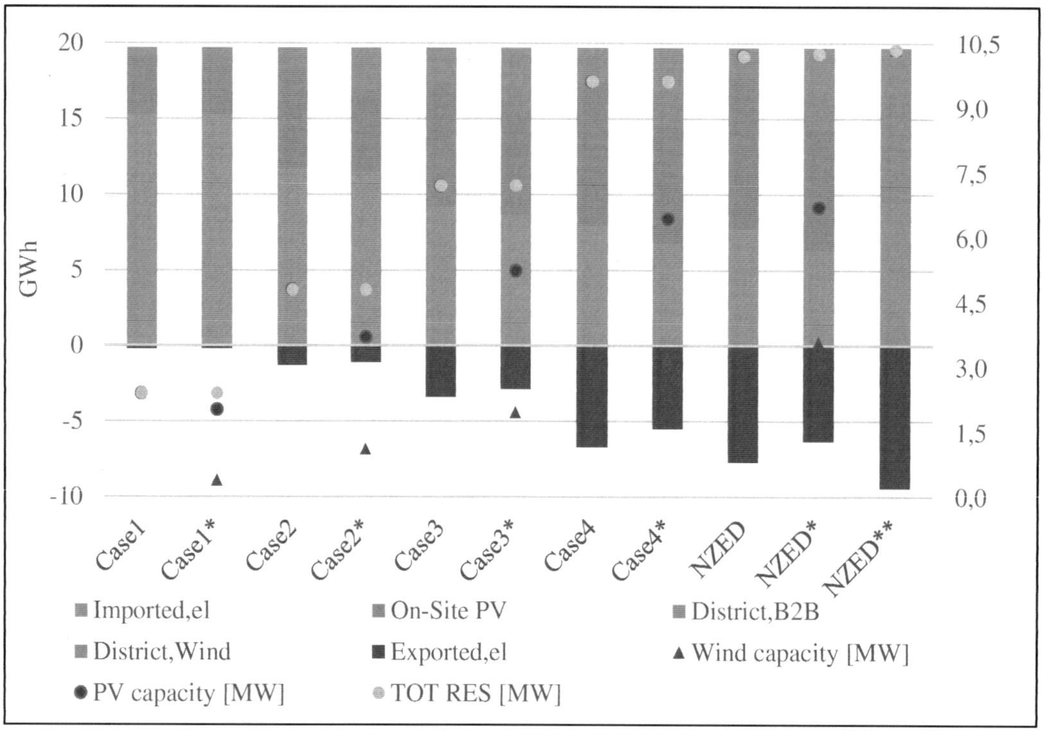

Figure 7. Main results of Nabta Town energy assessment.

4. Discussion

The main conclusions drawn from the results of the energy assessment presented above can be summarized as follows:

- Before acting on the system, buildings energy demand has to be reduced
- Different energy savings measures should be adopted for different building types
- Optimizing on-site consumption is necessary for reducing the cost on the energy infra
- NZED or Positive Energy District is possible to achieve with conventional RES

Since it has become clear that creating an adequate baseline using the most accurate possible local data and the correct assumptions is of crucial importance, to complete the study it would be necessary to cover also other aspects apart from energy like water, waste, material resources, etc., taking the same approach. Similarly, this energy study should be completed by an in-depth cost analysis.

There is also potential for scalability, and for increasing the efficiency and decreasing CO2 emissions when moving from building scale to district scale while considering the impact of features like vegetation to reduce the heat island effect, or street shading, etc.

Going beyond energy, sustainable building technologies reduce water consumption, which is becoming increasingly scarce in Egypt. Therefore, a follow-up study should include the sustainability assessment of the water system planned for NST covering the following aspects:

- Potable water management and possibility of desalinating underground water for drinking
- Recycling grey and black water for irrigation
- Use of modern irrigation systems
- Potential and economics of using underground water for irrigation

- Optimum selection of water system parameters
- Smart solutions for water management
- Consider the possibility of aquaponics systems

Similarly, the sustainability of NST solid waste management plan should be assessed as solid waste represents one of Egypt's most urgent problems, both in cities and the countryside. An array of solutions from waste separation at the source, considering transport and logistics, recycling, all the way to waste-to-energy or production of organic fertilizers should be considered for implementation in NST.

References

[1] The World Bank. 2010. *Cities and climate change: an urgent agenda*. Urban Development Series Knowledge Papers. Vol. 10

[2] United Nations General Assembly. 2016. *The New Urban Agenda*. Adopted by the United Nations Conference on Housing and Sustainable Urban Development (Habitat III)

[3] Goell E, El-Lahham N, Hussen W, El-Khishin S and Soliman S. 2009. Learning from Experience: Potentials and Preconditions for New Cities in Desert Areas. *Sustainable Cities in Egypt.* Vol. 77. The Egyptian Cabinet Information and Decision Support Center. (Cairo: Center for Future Studies)

[4] Central Agency for Public Mobilization And Statistics (CAPMAS), Egypt. 2019. Available online: http://www.capmas.gov.eg/HomePage.aspx (accessed on 5 January 2019)

[5] Sirry A. 2018. Alexandria: development challenges of a coastal second city. *"Wise Cities" in the Mediterranean? Challenges of Urban Sustainability*. Colección Monografías 06/2018. Woertz E (Ed.). (Barcelona: CIDOB) pp 145-158

[6] Jankowicz M. 2017. How Alexandria's 'leaning tower' became an emblem of the city's corruption. *Cities*. (London: The Guardian) https://www.theguardian.com/cities/2017/jun/26/alexandrias-leaning-tower-became-emblem-corruption (accessed on 5 January 2019)

[7] Hedman Å, Antuña-Rozado C, Jantunen J, Tuominen P, Balbaa O, ElMahgary Y, ElNashar A, ElShazly A, GamalEldin M, Hamza A, Kamel A, Negm A, Saeed H, Salem B, Shahin M, Tawfik A, Yousry A and Youssef W. 2014. *Readiness for EcoCities in Egypt. Insights into the current state of EcoCity systems, technologies and concepts.* VTT Technology. Vol. 161 (Espoo: VTT)

[8] New Urban Communities Authority. Available online: http://www.newcities.gov.eg/english/New_Communities/Borg_Arab/default.aspx (accessed on 5 January 2019)

[9] Antuña Rozado C, Hedman Å, Tuominen P, Reda F, ElMahgary Y, ElShazly A, GamalEldin M, Kamel A, Negm A, Tawfik A et al. 2015. *EcoNBC Feasibility Study. Transforming New Bord El Arab into an EcoCity*. VTT Technology. Vol. 220 (Espoo: VTT)

[10] United Nations Sustainable Development Knowledge Platform. 2016. *Sustainable Development Goals*. Available online: https://www.un.org/sustainabledevelopment/sustainable-development-goals/ (accessed on 5 January 2019)

[11] Antuña-Rozado C, García-Navarro J, Reda F and Tuominen P. 2016. *Methodologies Developed for EcoCity Related Projects: New Borg El Arab, an Egyptian Case Study*. Energies 9, 631 (Basel: MDPI)

[12] TRNSYS software. Available online: http://www.trnsys.com/ (accessed on 5 January 2019)

Acknowledgments

The authors wish to acknowledge: the Khairy Foundation for Human and Social Development, the Egyptian partner in NST Project, for their support and assistance during the development of the work; Mr. Pekka Huovila, Coordinator of the One Planet Network SBC Programme for his supervision; and the One Planet Network Secretariat for their guidance during the project.

Circular Economy in the Built Environment: Supporting Emerging Concepts

P Huovila[1, 4], U Iyer-Raniga[2] and S Maity[3]

[1] Green Building Council Finland, Lapinrinne 3, 00100 Helsinki, Finland
[2] School of Property, Construction and Project Management, RMIT University, 124 Latrobe Street, Melbourne VIC 3000, Australia
[3] Development Alternatives, B-32, Qutab Institutional Area, New Delhi 110016, India
[4] pekka.huovila@figbc.fi

Abstract. This paper presents circular built environment case studies from Europe, Asia and Africa covering circular housing concepts, circular office refurbishment, circular construction and demolition waste management and circular building material production. One Planet Sustainable Buildings and Construction programme is developing a data base on pilots addressing different life cycle stages of circular buildings. Those cases can be exploited when developing sustainable solutions that meet local conditions. The intent of reporting the case studies is to support the building and construction sector to move towards a circular economy.

1. Introduction

One Planet Network of 10-year framework on sustainable consumption and production consists of six programmes that operate in buildings, procurement, tourism, lifestyles and education, consumer information and food systems in both developed and developing countries. The network has chosen circular economy as a common means to implement Sustainable Development Goal 12 (Target 12.1) to accelerate the shift towards sustainable consumption and production patterns. Circular economy refers to the potential to change our production and consumption patterns and associated business models to ensure that waste is reduced and waste itself becomes a resource such that virgin materials are no longer used.

The building sector has a huge potential to mitigate climate change and practice resource efficiency through applying circularity principles providing at the same time important economic and social benefits. Ellen MacArthur Foundation [1] has identified 115 billion euros investment opportunities in the built environment in Europe for designing and constructing buildings based on circular principles, closing the loop on building construction and demolition materials, and building circular cities. Not just creating jobs, application of circular principles supports resilience, reduces resource use and lowers overall emissions in the process while also creating new and profitable business opportunities.

By 2050, the estimate of material economics [2] for only cement, steel, aluminium and plastic used for construction in the EU would result in emissions of 230 Mt CO2 if they were made with today's production processes. Demand-side measures could reduce this by 80 Mt CO2 per year, or 30% by simple measures of designing for longevity and disassembly, floor space sharing, reuse of existing buildings, materials efficiency, efficient construction practices, reduced waste in construction and building material reuse and recycling.

This paper presents examples of circular construction from Europe, Asia and Africa to show the value and success of circularity in the built environment already achieved worldwide. Policymakers, investors and construction clients have control over the business models and contractual frameworks in which built environment assets are created and operated. Their actions can maximise impacts at all scales along the value chain.

The selected case studies demonstrate different approaches applying circularity in sustainable buildings that can help in reducing CO2 emissions, lowering maintenance costs, and which can be re-used at the end of life without considerably reducing their value. The case study approach provides richness in the approaches taken, underlying drivers, value-add and key approaches.

2. The approach

Sustainable Buildings and Construction Programme (SBC) in the One Planet Network looks at circularity in buildings throughout their service life. Circularity aspects are relevant and should be considered in every phase during the service life, not only towards the end of life in renovation or demolition phases. It is important to take circularity as a starting point already in the manufacturing and design phases.

Figure 1. Life cycle phases of buildings in circular economy.

SBC has appointed circular built environment nodal points from its global network in Europe (Finland and the Netherlands), Asia and the Southern Pacific (India and Australia), Africa (Kenya and South Africa) and in Latin America and the Caribbean (Ecuador and Mexico). SBC experts in these countries collect local needs, analyse case studies and disseminate information about good practices in the region.

SBC has developed a simple framework and template to collect key information from case studies in a common format. The intention is to develop a database with a simple access to information about circular buildings in different parts of the world. Collected information contains a short summary with keywords, the life cycle phases that are relevant to the case study followed by a key message and an explanation what is unique in the project. The other aspects include the time of operation, stages of the building life cycle, stakeholders included, replicability and scalability of the case and main challenges encountered.

The next section presents successful case studies already undertaken that highlight different features of circular buildings. Circle House in Denmark is a scalable housing unit where 90% of the construction can be disassembled and reused without losing significant value during or end of life. It is intended to be scaled up to 60 units in varying shapes of row houses or apartment buildings in a new neighbourhood.

The first Dutch case, Alliander HQ, is an extension of an existing office building where 90% of its materials are reused or remain in the building site which became energy positive for the first time in the Netherlands. The other Dutch case in Brummen presents commissioning of a temporary town hall for 20 years. That building received the first materials passport turning it into a raw materials depot where approximately 90 per cent of the materials in the newly added space can be dismantled and reused.

3. Case studies

3.1. Europe

Circle House in Aarhus, Denmark, consists of 60 new social housing estates designed according to the principles of circular economy, which is expected to be completed in 2020. The goal is that 90% of the construction can be disassembled and reused without losing significant value during or end of life. These and other such projects demonstrate how sustainability and financial gain can go hand in hand to inspire future construction that has little impact on the environment. In addition to serving as housing, Circle House is a scalable demonstration project that can provide the building industry new knowledge about circular construction. Circle House consists of a range of building systems that can be assembled, disassembled and reassembled into other buildings while keeping their economic and aesthetic values intact [3].

The residential typologies are a mix of two- and three-storey terraced houses and five-storey tower blocks. The superstructure of three different typologies are built from the same six concrete elements, optimized for a quick construction, disassembly and reuse. In order to transform the building industry to a circular building practice the entire value chain of the industry needs to be engaged. The Circle House project involves more than 30 Danish companies across the entire industry. The Circle House project aims to develop, propagate and anchor knowledge about circular construction in the industry [3].

The increasing attention on resource consumption and the responsibility for future generations poses new demands and requirements for client consultancies and new constructions to reflect an integration of social responsibility and financial circumstances. Circle House is an example of this and provides insight into experiences in the circular housing development. The project demonstrates how sustainability and financial gain can go hand in hand to inspire future construction, so that we get build housing that reduce CO2 emissions, have lower maintenance costs, and can be recycled [4].

Figure 2. Circle House in Denmark, ©Søren Nielsen, Vandkunst.

Alliander HQ in the Netherlands could be perceived as a new build even though it is an extension of an existing building, where 92% of the materials are labelled as 'circular'. Originally built 30 years ago to accommodate 600 people, the expansion in 2015 has allowed for a further 900 occupants. 90% of the building materials from the existing building are being reused or remain on-site. The existing buildings, which were composed of different blocks, are respected and integrated into the new design: up to 83% of the existing constructions remain. Circularity has been an integral part of the design in many ways: respecting the majority of the existing constructions, using waste wood for the facades, reusing the concrete from the parts which were demolished, reusing the steel construction for the extensions of the buildings, recycling the asphalt from the existing roofs, reusing the existing toilets and ceiling plates and converting the existing doors into new furniture, amongst other things. A "raw material passport" is made to assure the reuse of all the materials in the future. This passport document contains information on all the raw materials, which are added during the renovation, the new installations, and the existing materials which were already there. It includes information about who has handled the materials, where they were temporarily stored and ways in which they can be reused [5].

The building complex provides more energy than needed, which can be used by neighbours in the surroundings through the Smart Grid, starting a Green Alliance in the community. The design of the building optimizes the use of energy, the design of the roof stimulates natural ventilation, and the atrium becomes a "second skin" for the closed volumes, creating a tempered 'in-between' climate. The use of solar panels and underground water for thermal storage makes it also CO2 neutral. The solar panels are placed in the parking area, acting also as shading and protection for the cars, giving them a double function. Since these were placed first, it was possible to achieve an energy positive building site for the first time in the Netherlands.

Brummen Town Hall is another circular Dutch example where due to concerns over frequently shifting municipality borders, the municipality commissioned a building for a service life of only 20 years. The answer to this request was a design made for disassembly and consistent use of reusable and renewable, high quality construction materials. Approximately 90 per cent of the materials in the newly added space can be dismantled and reused at the end of its service life. After this period, only the original 1890 building will remain on the site. The building also received the first materials passport in 2013 turning it into a raw materials depot in which the details of every piece are known, including their destination in a second life for some elements [6].

Rather than using cheap materials, the building minimises the use of concrete and incorporates a variety of high-quality reusable materials, which will be dismantled and returned to their manufacturers at the end of the building's life. The systems were developed in co-operation with manufacturers to enable easy disassembly in conditions that maximize their inherent material, component and product value after the 20-year period. Involving suppliers at very early stages in the design phase resulted in a very high degree of circularity of the building. All details of the building are known and documented, including their destination in a second life. It was the first building in the world conceived as a raw materials depot and equipped with a materials passport. The project led to a completely new way of looking at buildings and the way the temporary function they offer can be facilitated through a modular and circular building. It introduced the concept of products and buildings as a materials depot. It also made the team realise that current models of linear depreciation are not in line with circular value preservation [7].

3.2. Asia

In an increasingly resource constrained world, the main challenge for India is to find a balance between its developmental goals and constraints on minimizing the environmental impacts from increasing resource use [8]. Over the last couple of decades most of the Indian cities have experienced dynamic transformation due to rapid urbanization and infrastructural growth. This has led to increased use of construction and therefore the need for material resources. While the Indian Government is formulating effective policies to encourage use of secondary resource materials, the city of Ahmedabad has demonstrated replicable and profitable examples of utilization of wastes to replace virgin resources through a practice-to-policy connect.

Ahmedabad is the capital of the state of Gujarat and the 7th largest city in India having a population of around 5.5 million. It is one of the most important industrial and economic hubs in the Western part of India. The development and maintenance of the city is controlled by the Ahmedabad Municipal Corporation (AMC) having a control over 464 km^2. As with other metro cities the spurt of construction activities has and is generating quite a substantial amount of construction and demolition waste (C&D) within the city boundaries. Thus it was imperative to have an effective eco-system to manage this waste.

To design and implement a well-managed waste management system, the first and foremost step was to quantify the amount of C&D waste being generated and their location. Thus AMC designed and implemented a well laid out system of mandatory use of permits for demolition and new construction. This helped in quantification of waste being produced which was estimated to be around 700 metric tonnes (MT) per day. However, the actual generation was estimated to be much higher since there was construction being carried out in fringe areas without mandatory permits. The C&D waste management in Ahmedabad was based on a Public Private Partnership between Amdavad Enviro Projects Private Ltd. (AEP) and AMC. AEP was responsible for managing and processing the waste. AMC supported by designing waste collection systems and policies. 16 designated waste dumping sites helped in collection of the C&D waste from point sources. AEP transported the waste from the collection points to the processing facility and was paid Rs. 160/tonne (about $2.5 USD) as a tipping fee. The amount varied as per distance of collection points also. The collected waste was processed at a centralized processing facility into coarse and fine aggregates and used for producing standard building materials.

To understand the baseline of various building material product in and around Ahmedabad, finished products were tested which formed the minimum threshold criteria for developing products with C&D waste. The mix design was modified since the C&D waste aggregates had different properties compared to natural stone aggregates. The existing portfolio of standard building products of AEP, e.g. paving blocks, concrete bricks were augmented with high value concrete products through extensive lab and pilot scale research. Thus, heavy duty paving blocks (M80) for specialized use were also developed (Figures 3, 4). Products were also developed from the unusable fine aggregates. To increase the acceptability of the products, they were certified by external agencies (GRIHA: Green Rating for Integrated Habitat Assessment, India) which enabled users to earn green rating points. Apart from quality, product development also ensured that it fulfils all the criteria set forth by the Bureau of Indian Standards [9]. In addition to technical parameters the developed products were cheaper or at-par cost with existing products available in the market. This was the most non-negotiable criteria followed by the product development team since Indian markets are price sensitive especially when waste products are used to make finished goods.

Figure 3: C&D waste and processed aggregates at Amdavad Enviro Projects Pvt. Ltd. (AEP), ©Soumen Maity.

Figure 4: Processing facility and finished products, ©Soumen Maity.

Once the supply side of the value chain was ensured; demand of products was accelerated through public procurement policies. This was necessary since majority of the concrete products were procured by AMC at a city level and by various Government Departments at the State level. Various consultations and policy notes were prepared for preferential procurement of waste based products by AMC. They were verified through scientific and market based evidence and field visits. It was also worked out with AMC that circularity principles not only ensure sustainable use of waste materials but is also profitable for AMC since more expenses are being incurred by collection, management and dumping of wastes.

City level policies enacted by AMC, ensured that 75% procurement of non-structural products for public construction were compulsory from waste-based products. Additionally, a 5% procurement incentive was allowed for C&D waste based products compared to those produced from virgin raw materials. These public policies not only ensured an accelerated uptake of C&D waste based products in public use but also created a large market in private commercial construction. The demonstrated projects by the Government formed excellent marketable examples to ensure that waste are not dumped but put into profitable use.

3.3. Africa

In Africa, Malawi is a land locked country situated in Southern Africa surrounded by Tanzania, Mozambique and Zambia. It has a population of around 17 million growing at a rate of 2.8% per annum. As per UN reports [10], Malawi has one of the highest urbanization rates at 5.22% and is also one of the most vulnerable countries in sub-Saharan Africa from the negative effects on climate change. This is already being evident from the floods and droughts affecting the country and the loss of lives and infrastructure [11].

With the present rate of urbanization, it has been estimated that a minimum of 21,000 housing units are required over the next 10 years to meet the urban housing demand. In Malawi, the main and only affordable building material are "burnt clay bricks". Use of alternate materials is sparse, limited to subsidized construction activities. If an average of 85,000 burnt bricks are required to construct a typical urban house usually brick-fenced, approximately 1.7 billion burnt bricks are needed each year to meet the housing demand only (demand of burnt bricks due to infrastructural needs will be additional). This estimate should be seen as conservative as the requirements of public infrastructure or rural housing are not considered.

Figure 5: A Malawian clamp using fuelwood for burning bricks, ©Soumen Maity.

Currently, brick production in Malawi is highly decentralized and unorganized. The entire Malawian brick industry uses open clamps for firing (Figure 5). Due to poor quality of green bricks, clamps are not stacked high enough. There is no control over the firing process in the clamps. Fuel used is only fuelwood. It is estimated that around 850,000 tonnes of fuelwood will be required each year to produce around 1.7 billion bricks if alternate technologies are not adopted. Besides fuelwood, fuel in the form of leafy biomass is also used to provide additional energy. However, the quantity and quality is not suitable to provide additional heat to uniformly fire the upper layers of the clamp. No waste materials are used in green brick making for use as body fuel.

This high rate of urbanization puts tremendous pressure on the entire building material sector. With constraints in supply of material both the quality of product (brick) and the application (house) has degraded to an alarming extent resulting in poor quality and increasing construction costs. The current state of housing has also put an immense pressure on the low cost housing sector. Most often it has reached beyond the means of common beneficiaries. Approximately 20% more bricks are required to be constructed with traditional bricks compared to regular shaped and sized improved products. The use of quality bricks and associated eco-concrete products can drastically bring down the cost of construction of housing. Thus, the two major concerns from the Malawi brick sector are increased deforestation due to use of fuelwood and poor brick quality resulting in poor construction quality.

To solve the issues of deforestation, climate change and housing quality, it was decided to demonstrate the EcoKiln technology for producing burnt clay bricks. Some of the major advantages of EcoKiln were use of coal for the burning process, modular structure with an option to expand as and when needed, round the year production, consistent and better quality and energy efficiency with very low levels of emission. An EcoKiln was established with private investment at Mthokya, Lilongwe, Malawi. Within 2 years of production, the EcoKiln has established itself to be the World's most energy efficient and environment friendly clay brick firing technology having a Specific Energy Consumption of around 0.6MJ/kg. In Malawi the technology uses industrial wastes e.g. tobacco dust to improve energy efficiency and reduce the consumption of external coal. Experiments have proven that any carbonaceous wastes e.g. boiler ash from sugar industries, coal waste from oil industries etc., can also be used as an internal fuel. Thus, the wastes that was previously discarded and dumped creating health hazards have now found a profitable use. The burnt clay brick wastage has been reduced from 40-50% to almost 5%. Thus, quite a large amount of raw materials are being profitably utilized. Various market assessment studies have been undertaken and have found that all the raw materials required to build, operate and maintain the technology are available in Malawi. This reduces the dependence of any import or external support and demonstrates profitable examples of circular economy in the construction sector.

Figure 6: An EcoKiln in Malawi under operation, ©Soumen Maity.

This technology and the associated value chain have demonstrated that circularity is not only about profitable utilization of wastes in a large scale but a means to create sustainable green jobs round the year for both men and women. In Malawi this has helped in creating economic empowerment for women who now have become a decision maker in the family. Use of waste coal has opened up an entirely new business stream and has the possibility of arresting total deforestation.

The successful demonstration of the EcoKiln has enabled the Government of Malawi to decide on policy changes. Thus, as an incentive to entrepreneurs adopting EcoKilns, it has now been made mandatory to use non-wood fired bricks and building materials in public and donor aided construction. This will ensure a large demand of Eco Bricks thereby creating a market for EcoKilns.

4. Discussions and Conclusions

Key findings from the case studies are summarized below. The Danish Circle House is an example of a sustainable modular housing unit that can be scaled up in different architectural forms and components of which can be dismantled and re-assembled without losing their value. This offers opportunities from beyond this one case study to more modular housing opportunities, particularly in areas where materials may not be readily available. Alliander HQ shows how an extension of an office building can be undertaken using circular materials, forming a raw material passport for future reuse of all its materials. Good design supports regenerative practices. Brummen town hall presents a business model where the market delivers a temporary building removing it after 20 years. Thinking in the longer term is required, not just in current construction of operational practices.

The Ahmedabad example shows how waste-based products are profitable and their use can support the development of policies. Such business models can support a circular approach and governments can pave the way, leading by example.

In Malawi, EcoKiln has proved to be energy efficient and environment friendly way to produce burnt clay bricks using a waste by-product e.g. tobacco waste. The quality and efficiency of brick making processes has been improved; from 40-50% waste down to almost 5%. A whole industry can be supported based on this technology reducing deforestation and providing employment for both genders and the youth.

Next steps in the SBC programme are to disseminate information about good circular practices worldwide, to develop new initiatives to mainstream circularity in the built environment based on information available and to support development of policies and tools to support such policies.

IOP Conf. Series: Earth and Environmental Science **297** (2019) 012003 doi:10.1088/1755-1315/297/1/012003

References

[1] Ellen MacArthur Foundation, SUN Institute & SYSTEMIQ 2017 *Achieving "Growth Within"* January 2017, 149 p.
[2] SITRA, European Climate Foundation, Climate-KIC, Energy Transitions Commission, Ellen MacArthur Foundation, MAVA and ClimateWorks Foundation 2018 Material Economics: *The Circular Economy: A Powerful Force for Climate Mitigation, Transformative Innovation for Prosperous and Low-carbon Industry*, June 2018, 176 p.
[3] Jensen K G and Sommer J 2016 *Building a Circular Future*. Denmark 2016, 248 p.
[4] Jensen K G 2018 https://soundcloud.com/build-with-care
[5] Arup and Ellen MacArthur Foundation 2018 *From Principles to Practices: First Steps towards a Circular Built Environment*, July 2018, 26 p.
[6] Ellen MacArthur Foundation 2016 *Circularity in the Built Environment: Case Studies. A compilation of Case Studies from the CE100*. April 2016, 72 p.
[7] UN Environment 2018 *Guidance Document on Procuring Sustainable Buildings and Construction.* 59 p. Retrieved on January 14th 2019 from http://www.oneplanetnetwork.org/sites/default/files/guidance_document_on_procuring_sustainable_buildings_and_construction_final.pdf
[8] NITI Aayog 2017 *Strategy Paper on Resource Efficiency.* www.niti.gov.in.
[9] Bureau of Indian Standards 2006 *Precast Concrete Blocks for Paving – Specification.* IS 15658: 2006
[10] UNDP 2007 *Human Development Report (HDR) of Malawi.*
[11] Republic of Malawi 2014 *Intended Nationally Determined Contribution.*

Green Urban Development: A methodology to calculate site and infrastructure related GHG emissions

Prashant Kapoor[1], Ommid Saberi[2] and Neil Oliver[3]
[1] Chief Industry Specialist, IFC/World Bank Group, 2121 Pennsylvania Ave, Washington DC, USA
[2] Senior Industry Specialist, IFC/World Bank Group, 2121 Pennsylvania Ave, Washington DC, USA
[3] Director, Inani Limited, Greenhithe, Kent, United Kingdom

Email: Pkapoor1@ifc.org

Abstract The built urban infrastructure which includes the various types of residential and commercial buildings, the parking and public transport network, water and electricity networks, and delivery mechanisms determine the energy use and carbon emissions of our city.

At Building level, controlling resultant GHG emissions could be done through measures such as better insulation, more efficient heating and cooling systems, and the installation of rooftops solar photovoltaics. IFC's EDGE (Excellence in Design for Greater Efficiencies) certification has addressed the need for a mass market tool to identify such opportunities and provides third-party validation of achievement of resource efficiency in buildings. EDGE certification rewards developers who implement strategies to reduce energy and water use in their buildings and embodied energy in materials.

At Precinct level, dense, mixed use urban developments that are close to public transit nodes give people convenient, less polluting transportation options while maximising resource use and improving their access to more job opportunities and higher quality of life. To encourage such "Green Urban Developments," IFC is piloting a new concept to quantify urban strategies such as location, transportation options, achievement of optimum density, overall configuration, planning, and layout to reduce waste all the while ensuring comfort and affordability. The aim is to influence private and public-sector developers to choose "Green Urban Development" design options early in the project planning process.

This paper presents a methodology for calculating GHG impacts of the site and horizontal infrastructure options for an Urban Development and explains the benefits and challenges of putting such a concept into practice.

1. The context

1.1 Introduction to EDGE

EDGE, a green building certification system, is available in more than 140 countries. An innovation of IFC (International Finance Corporation), a member of the World Bank Group, EDGE has been used to assess and certify more than four million square meters of building floor space since October 2014 [1]. EDGE is a simple, smart and inclusive certification system applicable to residential, retail, hospital, hotel, school and office buildings across the price spectrum. The EDGE application ties into IFC's broader programme on Green Buildings, which includes investments and broader market development [2].

Figure 1. Examples of EDGE certified projects, with countries where EDGE certification is available marked in green.

Using a free access online application (www.edgebuildings.com) incorporating a locally contextualised simulation engine, EDGE empowers the discovery of technical solutions at the early design stage to reduce operational expenses and environmental impact. Based on the user's basic information inputs and then selection of green measures, EDGE provides projected operational savings and reduced carbon emissions within minutes. This overall picture of the financial and environmental performance of a building helps to articulate a compelling business case for building green.

The suite of EDGE tools includes building type specific design simulation engines connected to user guides providing both technical explanations and procedural guidance on certification.

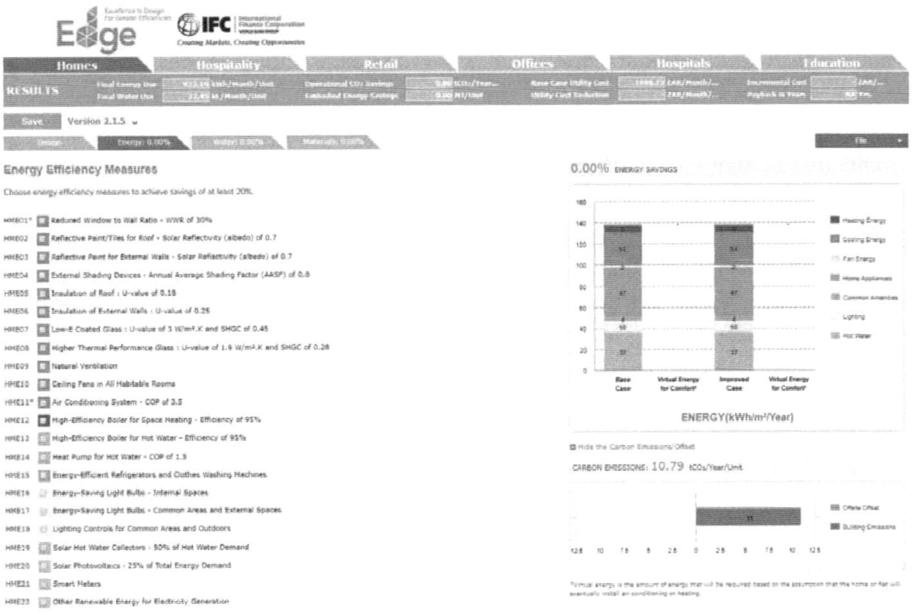

Figure 2. Screenshot of EDGE online application showing the energy section for homes

The EDGE Application is a key assessment platform that aims to accomplish the following objectives [1, 3]:

- A global solution for green buildings: more than 140 countries across the world would be able to use the app. this app is not only a rating tool but also provides the users with benchmarking and simplified energy simulation capability with minimal complexity and training requirements.
- An investment-planning tool for building owners and developers: edge app enables quick but well-informed decision-making. building owners and developers can understand the nuances of green investment and the returns on investment that they would be able to expect.
- An assessment tool for financial institutions: financial institutions currently lack a simple and effective tool by which they can judge the financial viability of a green building investment. Moreover, there is no global standard available by which they can assess risk reduction from lower utility bills. The EDGE Certification system provides the basis on which financial institutions can benchmark a building as 'green.'

While EDGE certification tackles the resource efficiency of buildings, the question is how can the program be made to accommodate and solve large scale urban issues? How can the success in quantifying efficiencies in buildings be leveraged to improve the sustainability of larger, more complex urban developments?

1.2 Need to tackle carbon intensive urban sprawl

The urban infrastructure of our cities, the types of residential and commercial buildings, the parking and public transport network, water and electricity networks, and delivery mechanisms determine the energy use and carbon emissions of a city. Research suggests that roughly 30% of "committed" future greenhouse-gas (GHG) emissions will occur as a result of new buildings and transport systems in a city. Energy-inefficient and carbon intensive urban developments can, therefore, lock us in on a high emissions trajectory.

Controlling greenhouse-gas emissions from standalone buildings can be achieved through better insulation, more efficient heating and cooling systems, and the installation of small-scale renewable energy sources, particularly on rooftops. Broader strategies, such as achieving optimum density, overall configuration or massing, and planning layouts to reduce wastage while ensuring comfort and affordability, require careful analysis and consideration.

Car dependent urban growth at the fringes of the city, known as urban sprawl, is increasing GHG emissions and private transportation energy use [4]. Sprawling, relatively low-density urban settlements make public transportation investments untenable, as there are fewer people to the bus or metro service for each transportation node. Research has also shown that low density urban forms compared to high density urban forms could increase global energy use by 2200 to 2500 TWh in 2050, equivalent of to around 10% of electricity use today [5]. A Canadian study that compared high and low-rise residential density showed that low- density suburban developments were twice as carbon intensive as compared to high-density developments on a per capita basis [6].

On the other hand, research has shown that dense, walkable, mixed-use urban developments that are close to mass transit nodes can reduce private car usage by 85% [7]. Many cities have mass transit options in place (or are in the planning stage) [8] but are yet to fully adjust urban planning controls to maximise the benefits. It is therefore important to assess opportunities beyond the boundary of a building such as benefits of mixed-use, higher density, and location (access to public transportation and amenities). Developers, investors, planners, and designers need a way to weigh the pros and cons of various site location options for their development. They also need tools for the masterplanning process, before embarking on detailed building design and procurement. These tools need to identify design and technology options to improve the environmental impacts and quantify the cost savings and investment paybacks associated with such options.

1.3 Available tools to quantify urban developments

Currently, there are two main types of tools available to quantify the environmental impacts and resource efficiency for the broader urban contexts. These are green rating systems or complex multi-domain dynamic modelling software.

1.3.1 Green rating systems

LEED for Neighbourhood Development or BREEAM Communities are well-known rating systems that take the green certification concept beyond individual buildings and apply it to the neighbourhood context as well as large-scale development plans [9, 10]. Local indigenous systems are also emerging, but most use the basic platform outlined in LEED or BREEAM. These systems have the following inherent limitations:

a. They award 'credits' or points for environmentally beneficial choices under diffrent siloed categories and do not attempt to measure a wholistic calculated value (such as CO2 emission per year or total resource consumption) for the overall performance.
b. They make subjective judgments [weighting] of different environmental issues [smart location, energy consumption, Neighbourhood Pattern] in the name of contextual economic and social concerns or priorities.
c. They do not provide efficiency improvements and related capital cost and savings of technology solutions.
d. Investors, banks and financial institutions need to be able to assess the financial viability of a Green investment through reduced risks from lower utility bills, which is not directly possible with the present rating systems [3].

1.3.2 Complex multi-domain dynamic modelling software

Urban scale models analyse a wide variety of aspects (buildings, transportation, water, waste, etc). This requires that these simulation models contain an assembly of inter-linked sub-models [11].

The analytical models that forecast resource use or GHG emissions of an urban scale rely on dynamic simulation software such as EQUEST [12], DesignBuilder [13] and EnergyPlus [14] for buildings energy demand simulation and MatSIM [15], CUBE Dynasim [16], EMME [17] , SimTraffic [18], and Paramics [19] for traffic simulation as well as GIS based systems for urban developments. These software packages are complex to use and require a large amount of data to calibrate the energy use/transport/land-use models, which impedes their wider adoption [20]. The need for a large amount of information upfront means that they are rarely used for masterplanning design and specification stage in the initial conceptual stage. This results in many iterations to the design and time and cost overhead for the urban development project.

The above limitations restrict large-scale adoption of the existing tools, especially in developing countries. Most existing green urban development modeling tools are complex and time-intensive, as they require considerable investment in training and a laborious data entry effort, making it unnecessarily difficult to use. This underscores the need for a simple, quick, and affordable tool that focuses on the efficient use of land, transport, energy, water, and materials while exposing investment costs and length of payback time. As models grow complex, there is a point at which there are diminishing returns. A simple 'quasi-steady state' model [21] can offer a useful mental tool that can inform effective decision-making rather than a faithful representation of the truth. This is especially true for complex systems such as models for urban lifestyles as there is a high degree of stochastic uncertainty in the behavior and material working of urban environments [22]. In spite of the complexity, however, empirical investigations show that behavior confirms to discrete patterns following broad universal organizing principles (for example, [23,24 and 25])

With these current circumstances in mind, the IFC team is expanding the EDGE platform which currently models the vertical development (i.e., buildings) to include the horizontal development (location, site, people density, layout, bulk infrastructure) under a pilot tool named EDGE "Green Urban Developments" (GUD).

The aim is to influence private and public-sector developers to choose "Green Urban Development" design options early in the project planning process.

1.4 Green Urban Developments (GUD)

Green Urban Developments (GUD) can range from projects built around or together with transit stations, i.e., Transit Oriented Developments (TOD), to mixed-use 'in-fill' development and 'green townships' where transit is incorporated into the masterplan. The typical features of these GUDs are (adapted from [26,27,28 and 29])

a) Mixed-use development that includes shops, schools and other public services, and a variety of housing types and prices.
b) High density and compact development to maximise land use and improve affordability.
c) More energy, water, and waste efficient design of buildings and infrastructure
d) Public transit stations or transit corridors that are easy to access, reliable, and secure, as well as street and road planning for non-motorized transport.
e) Pedestrian and bicycle friendly street and road planning. Streets should have good traffic calming features to control vehicular speeds.
f) Restriction on car parking to discourage private vehicle ownership.
g) Public and private sector participation.

GUDs can provide substantial cost savings for both city governments and citizens. A recent study concludes that China could save $1.4 trillion in urban infrastructure costs if its urban development plans prioritise optimum density over sprawl [30].

Incorporating transit into urban design allows developers to benefit from stable or higher property values. Mixed-use settings also create commercial opportunities that benefit from increased foot traffic in the area. A developer may also be able to access government incentives for such projects. Hong Kong, for example, has created a financially successful rail and property funding model, encouraging the Mass Transit Railway Corporation to partner with individual private developers to build along new and existing rail lines [31].

Planning and policy innovations have also been critical for green urban development. In Barcelona, super-blocks [32] that combine city spaces into pedestrian friendly, car-free mini grids have been designed.

Copenhagen has used a "finger plan," an urban plan, composed of five well-defined linear corridors (or "fingers"), separated by green wedges with open spaces, water sheds, and ecological preserves, to drive green growth [33]. Only compact, mixed-use developments are permitted around train stations to ensure a sustainable urban form.

Curitiba created structural corridors to promote job creation and activities away from downtown and inner-city locations and included housing and commercial density around mass transit lanes. As a result, the city has Brazil's lowest relative car use per resident [34]. The commercial success of the urban design has allowed the government to integrate low-income social housing using cross subsidies from market-priced private sector housing.

Developing a reliable transportation system requires large capital expenditure, but a substantial pool of users will offset the costs over time. Decreased pollution and fuel consumption will provide economic benefits to users, governments, and companies while enhancing a city's ability to maintain its competitiveness and environmental sustainability.

Reducing residents' dependence on private vehicles could help decrease their overall living expenses while improving health outcomes by increasing physical activity and improving air quality. Living in such developments can increase a sense of community and wellbeing for residents and help improve living standards.

2. EDGE application for Green Urban Developments

As with EDGE for Buildings, the EDGE for GUD is intended to be a web-based application, usable across geographies and pitched mainly to private sector developers, as a scenario evaluation and investment planning tool. It is envisaged that EDGE certification for GUD will allow identification of projects by investors and end-users.

The EDGE GUD will calculate environmental impacts (including, GHGs and resource used), costs (capital and operational) for design options generated by the system which will be ballpark estimates only but will be sufficient to illustrate the value of going green and to commit to the next step.

2.1 The EDGE GUD Standard

To meet the minimum requirements of the EDGE GUD standard, a project must achieve 20% GHG savings (resulting from measures to reduce passenger transportation, energy use of buildings and solid waste management measures) and 20% water savings, compared to a city-specific base case [1].

EDGE focuses on resource efficiency for a streamlined approach that enables market adoption of green buildings at scale because most actions reduce operating costs. The process is simple, with only a handful of measures required to reach the minimum requirements. The results include reduced emissions, lower utility costs, extended equipment service life and reduced pressure on natural resources.

2.2 The EDGE modelling process

2.2.1 The Base Case

The first step in assessing the efficiency of a design is to establish a project specific Base Case. A designer enters basic project data (e.g., size, location, building types, etc.) into the EDGE software, which calculates the project "Base Case," by drawing upon location-specific databases for parameters such as climate, business as usual practices and local usage patterns.

To encourage early-stage planning of green options, the EDGE software suggests many default options to a designer who has not yet decided on the details of a project. These defaults are accessible to the user to modify as required.

Defaults include city, country and regional data describing:

- Population and population density, income levels;
- Transportation data, including modal split, average trip distances, vehicle load factors and fuel efficiencies;
- Employment;
- Planning requirements, including parking provision, open space provision, block sizes, etc.;
- Utility tariffs, carbon intensities, infrastructure level losses;
- Solid waste and wastewater;
- Building energy performance for selected building types.

2.2.2 The Improved Case

An "Improved Case" is then created, when the user selects technical measures for inclusion in the design. If Improved Case GHG emissions are at least 20% less than the Base Case, a development meets the EDGE standard. In addition to energy savings, the EDGE software also reports water savings, GHG emission avoidance, and operational cost reductions. The incremental costs for the selected technical measures and the payback period from savings are also presented.

IOP Conf. Series: Earth and Environmental Science **297** (2019) 012004 doi:10.1088/1755-1315/297/1/012004

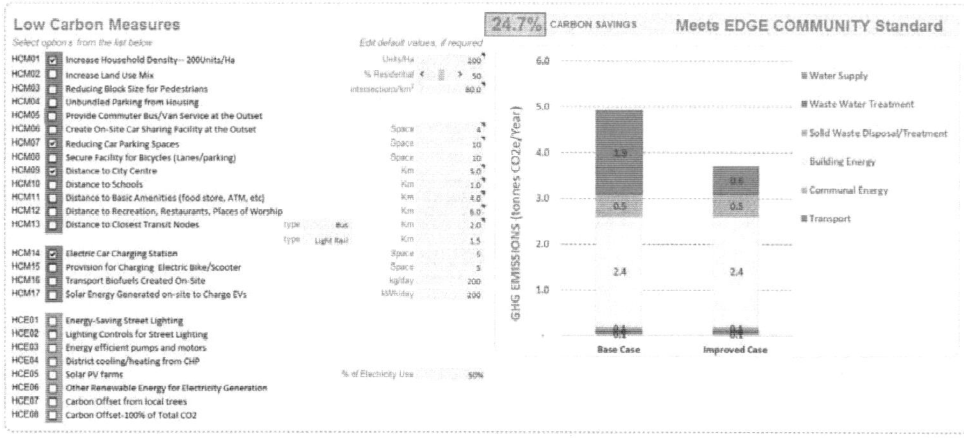

Figure 3: Screenshot of prototype EDGE GUD application

Modelling of the impacts of selected technical measures engages a number of 'engines', within the software tool, which calculate the interrelated impacts of these measures on specific key performance indicators. These indicators currently include:

- Operational GHG emissions;
- Transportation passenger and vehicle kilometers travelled;
- Building and Infrastructure related energy consumption;
- Water consumption, per source;
- Solid waste arisings, treatment and disposal, per fraction;
- Wastewater treatment and disposal;
- Employment within the proposed development.

2.3 The modelling 'Engines'

2.3.1 Employment Engine

The Employment Engine utilises the user's description of the proposed development to derive an estimate of employment opportunities, by broad job category[1], using a notional proportion of category and density of employment within each building type in the development. These categories are mapped within the tool to provide an assessment of local housing demand for these employees, for comparison to the developer's proposed residential housing provision.

The Employment Engine also calculates estimated working population within the proposed development, as a basis for calculating commuter transportation demands, linked into the Transportation Engine.

2.3.2 The Transportation Engine

The Transportation Engine estimates the impact of selected measures on person and vehicle kilometre travelled (PKT and VKT) and the resultant carbon emissions. The engine assesses these impacts based on the following location and project characteristics:

a. Increase population density

[1] Job categories are based on U.S. Equal Opportunity Commission - https://www.eeoc.gov/eeoc/statistics/employment/jobpat-eeo1/glossary.cfm

Litman, 2010 [36] suggests that an increase of 10% in people, or jobs, per unit of land typically reduces VKT between 0.5% and 1%, as an isolated factor, increasing to around 4% where associated with greater regional accessibility, transportation mix, etc.

b. Residents working within the development (job creation and mixed-use)
Commuter travel calculated based on employment creation within the development and estimated commuting out of the development into the nearest city.

c. Reducing block size for pedestrians
Rowe, et al. [37] found that per household vehicle ownership rates decline significantly with smaller block size and increased roadway connectivity.

d. Unbundling parking from housing
UCLA's California Center for Sustainable Communities [38] suggests that unbundling of parking provision from residential provision results in an overall decrease in vehicle kilometers travelled (not person kilometers travelled), due to a resultant reduction in vehicle ownership within the new development.

e. Provide shuttle bus/van service at the outset
Assuming a proportional uptake of this service by commuters, with an estimated reduction in VKT for private vehicles.

f. Create an on-site car sharing facility at the outset
Martin and Shaheen [39] provide insight into the impacts of carsharing on VKT, including the impacts of use of public transport modes. Within the Transportation engine, the provision of carshare parking spaces, and car share club members is used to estimate an overall reduction in VKT within the development.

g. Reducing car parking spaces
Litman, 2010 [36] suggests that the supply, pricing and management of available parking can reduce private automobile trips between 10% and 30%.

h. Secure facility for bicycles (lanes/parking)
The Transportation engine estimates the reduction in private vehicle usage, resulting from increased facilities for cycling within the new development.

i. Person and Vehicle kilometers travelled are modelled for each mode of transport, derived from user input and default values for:
   - Distance to city centre
   - Distance to schools
   - Distance to basic amenities (food store, ATM, etc.)
   - Distance to recreation, restaurants, places of worship
   - Distance to closest transit nodes
   - Distance to other amenities

The engine also considers measures to switch the fuel used for transportation, including:

j. Electric car charging stations
k. Provision for charging electric bike/scooters
l. Transport biofuels created on-site
m. Solar energy generated on-site to charge EVs

2.3.3 Energy engine
The energy use across the site is considered by the Energy Engine. The predicted energy use of each building type (using default building energy performance data for the region / country / city, derived from related EDGE models).

The engine also calculates site wide infrastructure energy uses, such as street lighting [40, 41], water pumping and distribution [42], district heating and cooling and solar farms. The emission factors for different energy uses are used to arrive at energy related carbon emissions.

2.3.4 Water engine
Calculation of water use across the site, by the Water Engine, considers the predicted water usage of each building type, using default water consumption data from related EDGE models, and site wide water uses such as open space landscaping and infrastructure distribution losses. Electricity consumption relating to water abstraction, transportation and treatment is used to calculate water related carbon emissions within the energy engine.

Wastewater demands for the Base Case and Improved Case are calculated, together with estimated resultant GHG from distribution, treatment and disposal activities.

2.3.5 Solid waste engine
The solid waste engine estimates tonnages of municipal and commercial solid waste arisings, using the Base Case and Improved Case development details. The engine uses waste profile data for the region, country and city (where available), to determine approximate waste fractions and prevailing local treatment and disposal practices for the Base Case.

Solid waste arisings and fractions are calculated using regional, country and city specific data, where available, coupled with the income level associated with the development [43, 44, 45, 46, 47].

Users can influence their Improved Case carbon emissions, relating to solid waste management, through a set of measures including:

a. Introduction of municipal waste management systems;
b. Material recovery and recycling facilities;
c. Bio-digesters for food waste treatment;
d. Other waste to energy opportunities;
e. Industrial/commercial waste exchanges.

2.4 Innovative Features of the EDGE GUD Tool
Some of the innovative features of EDGE GUD are:

- The EDGE GUD is the first tool to quantify the benefits of a variety of land use, urban planning and transportation interventions, along with the benefits of improved efficiencies in infrastructure and built-environment. For example, the tool enables users to quantify the GHG benefits of developing in city centre locations, creating mixed used developments, reducing parking and providing infrastructure for electric-vehicles.
- Weighting indicators to suit different climate zones: the data from individual cities have been incorporated into the GUD tool, and therefore the predicted performance reflects that of the real-case scenario.
- Translation of green practices into economic savings: The GUD is the first tool to quantify the benefits of a variety of land use, urban planning and transportation interventions, along with the benefits of improved efficiencies in infrastructure and built-environment. For example, the tool enables users to quantify the GHG benefits of developing in city centre locations, creating mixed used developments, reducing parking and providing infrastructure for electric-vehicles.

3. Making it happen

The EDGE GUD software is an entry point; an enabler; a point of entry. To effectively scale green urban developments both market and policy mechanisms will be needed, such as:

- Developing a GUD or TOD policy framework for integrated infrastructure planning that develops sustainable urban growth centres with high-density mixed land use.
- Reforming land-use regulations such as single-use zoning, low-density limits, and high parking fees to encourage mixed-used, transit-oriented developments with reduced car parking. These should be combined with effective policy changes such as mandating reduced maximum car parking for homes rather than a minimum.
- Redefining policy definitions of affordable housing to include the combined cost of housing and transport, e.g. Housing +Transport Index [35].
- Creating incentives for the private sector. Cross-subsidization using tax abatement or value capture methods could help make the higher costs of transit-accessible locations less prohibitive for private developers and investors.
- Exploring alternative implementation options, including Public-Private Partnerships (PPPs), to leverage private sector skills and financing to develop projects.
- Encouraging corporations to locate offices within green urban developments as an anchor tenant or employer through financial and non-financial policy incentives.

By using EDGE GUD at the start of a project, the intent is to draw the user in, so that the software affects the way the developer does business. Due to the current speed of industrialized development, most consideration of sustainable building elements happens at the end of the project when its implementation is already too late no longer practical or cost effective. As with most green certification systems, there is always the question of whether these efforts yield any real efficiency benefits.

Longer-term evaluation is challenging, of course. Future modifications to the software and a communication strategy aimed at occupants will consider these ever-present issues. EDGE GUD will continue to be test piloted repeatedly across various markets to ensure the accuracy of results and will remain open source, allowing other professionals to finesse it over time to ensure the highest level of integrity.

EDGE GUD will be rolled out on a pilot basis across projects. The feedback gained will be used to improve and debug the product in preparation for its first launch.

4. Conclusions

Learning from EDGE buildings certification, EDGE GUD can expand the reach to communities and neighbourhoods with wider application. The inclusion of transport and urban elements such as parking and waste management brings a fresh perspective on how best to create greener/cleaner cities.

There are a variety of reasons why developing countries would benefit from the wide-scale adoption of Green Urban Development practices. Not only would there be a reduction in built-environment-related GHG emissions and better use of land, but there would be positive economic benefits from less fossil fuel imports, fewer infrastructure investments, reduced utility costs, and the birth of a new sustainability industry. Since most of the urbanisation in the coming decades will be in developing countries, it is essential that a classification system exists that correlates directly to the unique drivers of these marketplaces. One that Improves the quality of life with better places to live work and play.

By lessening environmental impacts while simultaneously providing cost incentives, EDGE provides the financial traction that has long been missing to gain momentum from investors, buildings owners, and developers, by providing a strong catalyst for green buildings and communities in emerging markets.

5. References

[1] Saberi O, Kapoor P. Virtual Energy for Comfort: Measuring Comfort through EDGE. Proceedings of 9th Windsor Conference: Making Comfort Relevant Cumberland Lodge, Windsor, UK, 7-10 April 2016 Available from: http://nceub.org.uk.

[2] Edgebuildings.com [Internet]. USA: Excellence in Design for Greater Efficiencies, International Finance Corporation, Available from: https://www.edgebuildings.com/

[3] Kapoor P. Green Building Classification System for Developing Countries, International Journal of Sustainable Construction, Volume 1 Number 1-2012 [Online] Available from: http://ijsc.greenlines-institute.org/volumes/2012/IJSC_2012_V01_01.pdf

[4] Bart IL. Urban sprawl and climate change: A statistical exploration of cause and effect, with policy options for the EU. Land Use Policy [Internet]. Elsevier BV; 2010 Apr;27(2):283–92. Available from: http://dx.doi.org/10.1016/j.landusepol.2009.03.003

[5] Güneralp B, Zhou Y, Ürge-Vorsatz D, Gupta M, Yu S, Patel PL, et al. Global scenarios of urban density and its impacts on building energy use through 2050. Proceedings of the National Academy of Sciences [Internet]. Proceedings of the National Academy of Sciences; 2017 Jan 9;114(34):8945–50. Available from: http://dx.doi.org/10.1073/pnas.1606035114

[6] Norman J, MacLean HL, Kennedy CA. Comparing High and Low Residential Density: Life-Cycle Analysis of Energy Use and Greenhouse Gas Emissions. Journal of Urban Planning and Development [Internet]. American Society of Civil Engineers (ASCE); 2006 Mar;132(1):10–21. Available from: http://dx.doi.org/10.1061/(asce)0733-9488(2006)132:1(10)

[7] Ngai WC, Hidefumi I, Akihiro N, Masazumi A, Sustainable Urban Development Textbook, Publisher: Water Watch Penang & Yokohama City University, September 2016 Edition: 1, Chapter: 27, pp.177-183 Available from: https://www.researchgate.net/publication/308984324_CHAPTER_27_TRANSIT-ORIENTED_DEVELOPMENT_TOD

[8] Pojani D, Stead D. Sustainable Urban Transport in the Developing World: Beyond Megacities. Sustainability [Online]. MDPI AG; 2015 Jun 17;7(6):7784–805. Available from: http://dx.doi.org/10.3390/su7067784

[9] Breeam.com. BREEAM. [Online]. BREEAM Communities International Technical Standard, Building Research Establishment Ltd Available from: https://www.breeam.com/discover/technical-standards/communities/ [Accessed 27 January 2019].

[10] NDRC [Online]. USA: A Citizen's Guide to LEED for Neighborhood Development: How to Tell if Development is Smart and Green. Available from: https://www.nrdc.org/sites/default/files/citizens_guide_LEED-ND.pdf

[11] Sola A, Corchero C, Salom J, Sanmarti M. Simulation Tools to Build Urban-Scale Energy Models: A Review. Energies [Internet]. MDPI AG; 2018 Nov 23;11(12):3269. Available from: http://dx.doi.org/10.3390/en11123269

[12] Doe2com [Online]. eQUEST -The Quick Energy Simulation Tool. Available from: http://www.doe2.com/equest/ [Accessed 27 January 2019].

[13] Designbuilder.com [Online]. DesignBuilder - Simulation Made Easy. Thermal Energy System Specialists, LLC Available from: https://designbuilder.com//

[14] Energyplus.net [Online]. EnergyPlus. U.S. Department of Energy's (DOE) Building Technologies Office (BTO) Available from: https://energyplus.net/ [Accessed 27 January 2019].

[15] Matsim.org [Online]. MATSim. Multi-Agent-Based Transport Simulations Available from: https://matsim.org/about-matsim [Accessed 27 January 2019].

[16] Citilabs.com [Online]. Cube-Dynasim, Citilabs Available from: http://www.citilabs.com/software/cube/cube-dynasim/ [Accessed 27 January 2019].

[17] Inrosoftware.com [Online]. Emme- Transportation Planning Software, Inro Available from: https://www.inrosoftware.com/en/products/emme/[Accessed 27 January 2019

[18] Trafficware.com [Online]. Simtraffic, Trafficware. Available from: https://www.trafficware.com/blog/category/simtraffic [Accessed 27 January 2019].

[19] Paramics.co.uk. [Online]. Paramics Microsimulation, SYSTRA Ltd. Available from: https://www.paramics.co.uk/en/ [Accessed 27 January 2019].

[20] Boydell S & Giurco D & Rickwood P, Glazebrook, G, Using Integrated Urban Models to Respond to Climate Change in Cities. [Online] University of Technology, Sydney January 2009 Available from ResearchGate:

https://www.researchgate.net/publication/241285289_using_integrated_urban_models_to_respond_to_climate_change_in_cities [Accessed 27 January 2019].

[21] Ballarini I, Primo E, Corrado V. On the limits of the quasi-steady-state method to predict the energy performance of low-energy buildings. Thermal Science [Online]. National Library of Serbia; 2018;22(Suppl. 4):1117–27. Available from: http://dx.doi.org/10.2298/tsci170724133b

[22] Baynes TM. Complexity in Urban Development and Management. Journal of Industrial Ecology [Internet]. Wiley; 2009 Apr;13(2):214–27. Available from: http://dx.doi.org/10.1111/j.1530-9290.2009.00123.x

[23] Kenworthy JR, Laube FB. Patterns of automobile dependence in cities: an international overview of key physical and economic dimensions with some implications for urban policy. Transportation Research Part A: Policy and Practice [Internet]. Elsevier BV; 1999 Sep;33(7-8):691–723. Available from: http://dx.doi.org/10.1016/s0965-8564(99)00006-3

[24] Ewing R, Cervero R, Travel and the Built Environment. Journal of the American Planning Association [Internet]. Informa UK Limited; 2010 Jun 21;76(3):265–94. Available from: http://dx.doi.org/10.1080/01944361003766766

[25] Cervero R, Arrington GB. Effects of TOD on Housing, Parking, and Travel. National Academies Press; 2008 Oct 2; Available from: http://dx.doi.org/10.17226/14179

[26] Joshi R, Joseph, Y & Patel K, Darji V, Transit-Oriented Development: Lessons from International Experiences. Working Paper, May 2017. Available from ResearchGate: https://www.researchgate.net/publication/317580038_Transit-Oriented_Development_Lessons_from_International_Experiences

[27] Tod.org [Online]. TOD -Transit Oriented Development Institute Available from: http://www.tod.org/ [Accessed 27 January 2019].

[28] Lund HM, & Cervero R, Willson, R. Travel Characteristics of Transit-Oriented Development in California. January 2004. ResearchGate Available from: https://www.bart.gov/sites/default/files/docs/Travel_of_TOD.pdf

[29] Carlton I. Histories of Transit-Oriented Development: Perspectives on the Development of the TOD Concept. UC Berkeley: Institute of Urban and Regional Development. Retrieved from https://escholarship.org/uc/item/7wm9t8r6

[30] The World Bank. Urban China: Toward Efficient, Inclusive, and Sustainable Urbanization. World Bank Publications (July 29, 2014) Available from: https://elibrary.worldbank.org/doi/pdf/10.1596/978-1-4648-0206-5

[31] Mtr.com.hk [Online]. Hongkong: MTR Corporation Limited. Available from: https://www.mtr.com.hk/en/corporate/overview/profile_index.html[Accessed 27 January 2019].

[32] Ajuntament.barcelona.cat [Online] Let's Fill Streets with Life: Establishing Superblocks in Barcelona. Ajuntament de Barcelona May 2016. Available from: http://ajuntament.barcelona.cat/ecologiaurbana/sites/default/files/en_gb_mesura%20govern%20superilles.pdf

[33] Danishbusinessauthority.dk [Online] The Finger Plan A Strategy for the Development of the Greater Copenhagen Area. Ministry of the Environment, Denmark. Available from: https://danishbusinessauthority.dk/sites/default/files/fp-eng_31_13052015.pdf

[34] Suzuki H, Dastur A, Moffatt S, Yabuki N, Maruyama H. Eco2 Cities. The World Bank; 2010 May 4; Available from: http://dx.doi.org/10.1596/978-0-8213-8046-8

[35] Guerra E, Kirschen M. Housing Plus Transportation Affordability Indices. International Transport Forum Discussion Papers [Internet]. Organisation for Economic Co-Operation and Development (OECD); 2016 Oct 13; Available from: http://dx.doi.org/10.1787/a1fc9b79-en

[36] Litman T. Land use Impacts on Transport. Victoria Transport Policy Institute. 2018 July 24. Available from: http://www.vtpi.org/landtravel.pdf

[37] Daniel Rowe, et al. Do Land Use, Transit and Walk Access Affect Residential Parking Demand? ITE Journal, Vol. 83. No. 2, February 2013, pp. 24-28; at http://1.usa.gov/1UgmpEE. Summarizes King County Right Size Parking Project results (http://metro.kingcounty.gov/up/projects/right-size-parking).

[38] UCLA: California Center for Sustainable Communities, Unravelling Petroleum: Bundling of Residential Parking in High-Quality Transit Areas. Available at http://next10.org/sites/default/files/3.%20Bundling%20of%20Residential%20Parking%20in%20High-Quality%20Transit%20Areas.pdf

[39] Martin E. and Shaheen S. The Impact of Carsharing on Public Transit and Non-Motorised Travel: An exploration of North American Carsharing Survey Data. November 2011. Available at https://www.mdpi.com/1996-1073/4/11/2094

[40] Flagstaff Zoning Code 10-50.70.020: Outdoor Lighting Levels. Available at https://www.flagstaff.az.gov/DocumentCenter/View/14707/Chap10-50_SupplementaltoZones_Part2_Nov1

[41] Soledad Escolar, Jesús Carretero, Maria-Cristina Marinescu, Stefano Chessa. Estimating Energy Savings in Smart Street Lighting by Using an Adaptive Control System. May 8, 2014 Available from http://journals.sagepub.com/doi/full/10.1155/2014/971587

[42] World Bank: Energy Sector Management Assistance Program (ESMAP). A Primer on Energy Efficiency for Municipal Water and Wastewater Utilities. February 2012. Available at https://www.ifc.org/wps/wcm/connect/da52df004aabaace9784d79e0dc67fc6/ESMAP+EE+WASTEWATER.pdf?MOD=AJPERES

[43] World Bank. What A Waste: A Global Review of Solid Waste Management (2012). Available at https://siteresources.worldbank.org/INTURBANDEVELOPMENT/Resources/336387-1334852610766/What_a_Waste2012_Final.pdf

[44] Simone Manfredi, Davide Tonini, Thomas H. Christensen (Department of Environmental Engineering, Technical University of Denmark, Kongens Lyngby, Denmark) Heijo Scharff NV (Afvalzorg Holding, Nauerna 1, P.O. Box 2, 1566 ZG Assendelft, The Netherlands). Landfilling of waste: accounting of greenhouse gases and global warming contributions. Available from https://pdfs.semanticscholar.org/5abe/2b7e0be8ebf9874220428e064837718aba90.pdf

[45] Hwang KL, Choi SM, Kim MK, Heo JB, Zoh KD. Emission of greenhouse gases from waste incineration in Korea. March 31, 2017. Available from https://www.ncbi.nlm.nih.gov/pubmed/28371748

[46] Zero Waste Scotland. The carbon impacts of recycling and the effects of transport. Available at https://www.zerowastescotland.org.uk/sites/default/files/Carbon%20impacts%20of%20recycling%20and%20transport.pdf

[47] David A.TurnerIan; D.Williams; Simon Kemp. Greenhouse gas emission factors for recycling of source-segregated waste materials. December 2015. Available at https://www.sciencedirect.com/science/article/pii/S0921344915301245

IOP Conf. Series: Earth and Environmental Science **297** (2019) 012005 doi:10.1088/1755-1315/297/1/012005

Passive Solar design basics, a way for low emission buildings?

Mr. Søren Dietz - Bjerg Arkitektur

sd@bjerg.nu

Abstract. Danish Building Regulation allows a market that chooses non-compact and non-passive Solar optimized design. This leads to balance higher heat demand with surplus active solar gains in calculating a yearly energy balance. Expectations like „Soon we will have district heating with CO2 neutral biomass and batteries to save so is there a need for solar passive design?" further encourages to assume, that summer solar gains might be saved for winter heating demand in a not so far future. As a passive house architect, I am often confronted with this question and this way of thinking by politicians and architecture students. They do care about climate change, but do not seem to understand why we ought to change the way we plan building space and energy systems. The contribution architects can make to reduce climate impact by building spaces that ensure a passive exploit of the suns energy, not just surfaces to mount active solar devices, is not commonly understood.

1. 4 Case studies, without and with solar compact strategy

To make my case I will present 4 case studies. A row social housing [1], a single family house[2] and two kindergartens [3,4] show how different strategies lead to different climate emissions.

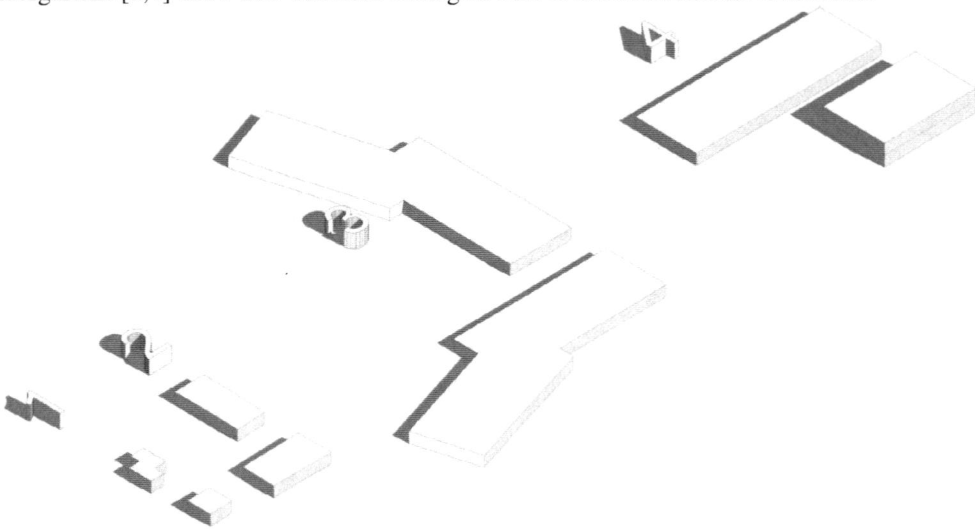

Figure 1 A schematic axonometric of the four Cases. Yellow areas are south oriented vertical elevations. The upper variants are the not solar or compact optimized versions. The lower volumes represent the same building area with an optimized south elevation and in case 1,2 and 4 a more compact building volume.

2. Case 1, a small apartment unit of a row social housing complex.

The project is a competition.

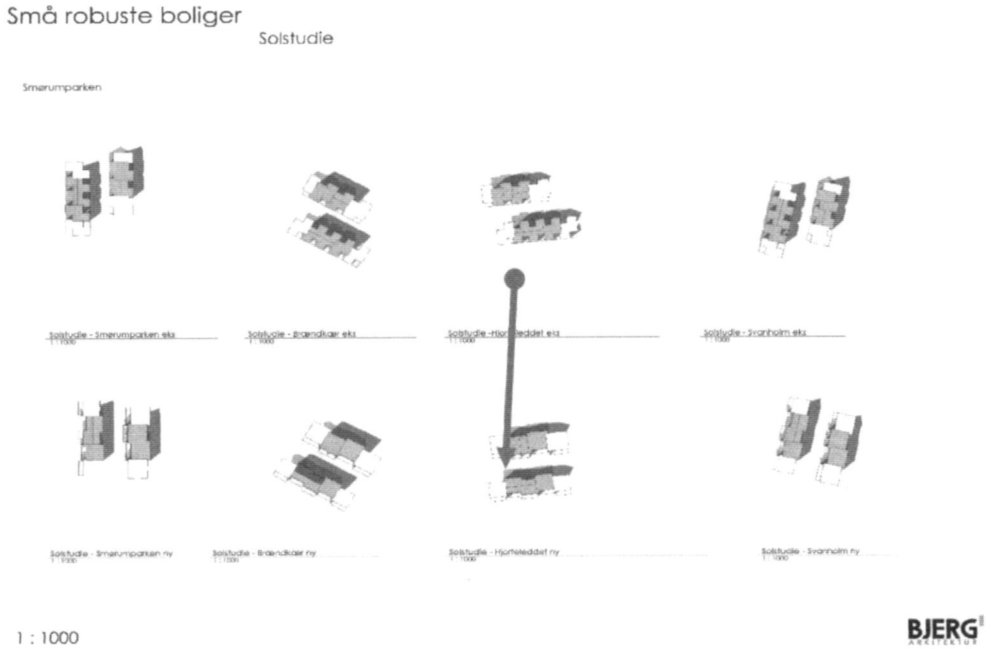

Figure 2 Layout for the apartments was given with 4 different orientations shown on the sun/shading diagram. The upper 4 different orientations are given in the competition. The lower 4 show the shading if the apartment layout were made compact instead of the proposed shifted plan.

One of four orientations were south oriented. Yet this south oriented glazing is shaded because of a split layout proposal of the apartment design were half of the apartment is shifted in front to create space. This proposed split of a small south oriented apartment unit creates shading on the south oriented glazing, thus reducing potential passive solar gains from the building itself and increasing the heat loss surface with higher building costs as a consequence. The given layouts lead to shading on potential solar gain surfaces due to built landscape, orientation and the layout of the apartment itself. In this competition only the layout of the 1-plan solution can be optimized. The other ¾ of the given apartment layouts without south orientation have no solar passive potential net gains due to the layout of the buildings decided before the competition.

According to the Danish Building Regulation 2018 and 2020 regulation it is possible to calculate up to 25 kWh/m2 and year of generated energy. It is not transparent in the "Be18" calculation software whether the yearly generated active gains cover a demand at the time of generating.

In this analysis a PHPP calculation has been used with monthly calculated demand and generated energy. Two different building designs are compared:

- Given shifted plan (called not compact).
- Compact design proposal.
- Both variants are calculated with two different levels of active systems: 5,4 m2 solar collector, 27 m2 solar collector (full south-oriented roof area).

IOP Conf. Series: Earth and Environmental Science **297** (2019) 012005 doi:10.1088/1755-1315/297/1/012005

Solar total contribution for the not compact 27 m2 solar collector is calculated in PHPP to cover 59% of the heat and hot water energy demand. The solar total contribution of the compact design is 73%.

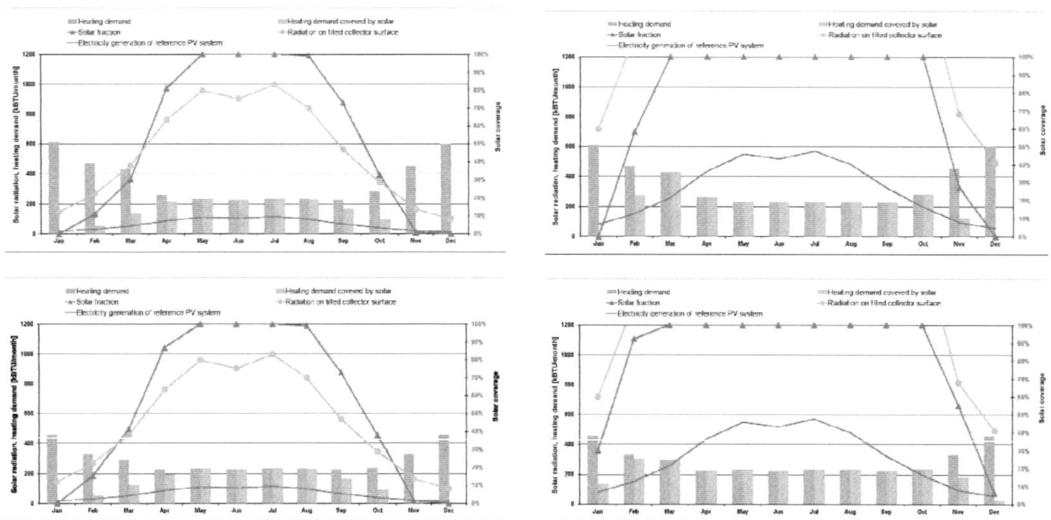

Figure 3 The upper diagrams show the shifted not compact design. On the left with 4,5 m2 solar collector and on the right with 27 m2 solar collector. The eq. area of PV coverage is calculated and shown. On the two lower diagrams the compact design with results for lower heat demand and a higher monthly coverage in the winter month is shown.

If the yearly results are listed, it becomes clear that the active potential coverage is limited. The winter gap between heat demand and active generated energy becomes a "performance gap". The U values of both variants are the same.

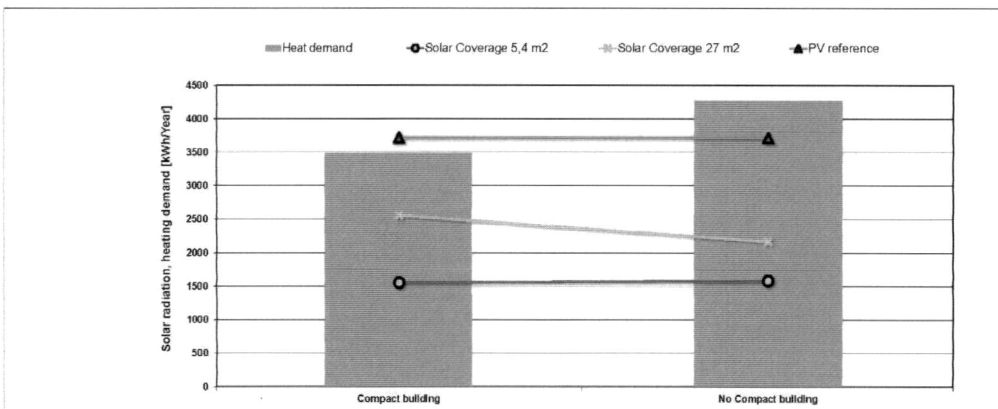

Figure 4 PHPP calculated results for a compact apartment design on the left and the proposed shifted layout of the competition project on the right. The resulting higher total energy demand is not made up for by a full roof solar collector. The winter demand cannot be covered. The compact design allows a higher coverage of the hot water energy demand and in total this results in a smaller winter performance gap.

3. Case 2, a single family house.

The second case to be analyzed is a design proposal for a south oriented villa. The client came with a reference plan – a prefabricated standard house.

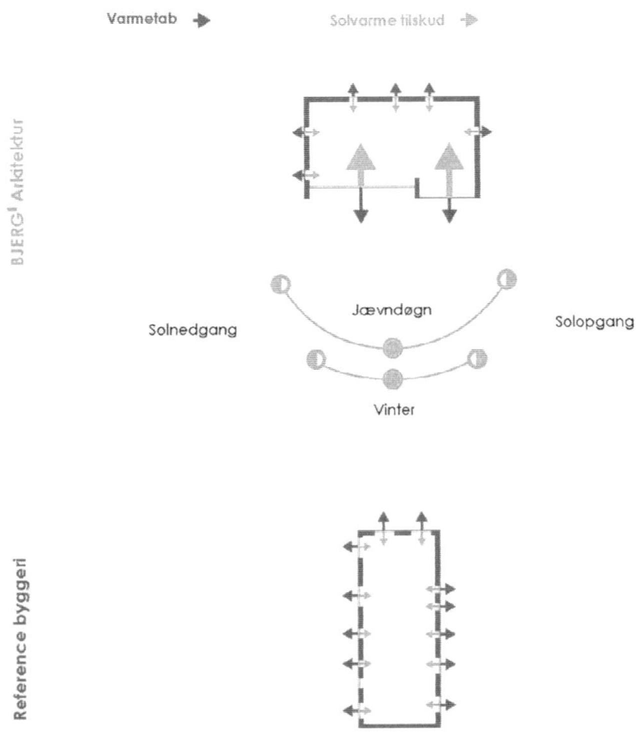

Figure 5 The upper diagram shows the solar design concept, the lower plan is the plan of the prefabricated house with mainly east- and west oriented glazing. In the middle the horizontal solar angel is shown for the 3 winter months and the winter year half. As a solar angle of more than 60 degrees to the glazing reflects the solar heat away on the glazing, it becomes clear why south oriented glazing has a high potential gain and other orientations results in net potential solar passive losses.

In this case the total economy for the client is important. The design for a prefabricated house with solar net losses is analyzed and compared with a design for the site that has a high solar net gain – but also higher costs due to the better U values and architect passive house design fee. The prefab wall's U value is 0,22 W/m²K. Compared to the PH U value of 0,08 W/m²K, the transmissions losses from the wall alone could be reduced by 36 %.

The question is however the performance of a standard prefabricated housing is compared with a solar design optimized for the site, when the higher costs for insulation and design are considered. As the standard prefabricated house with it's higher heat load would be connected to the district heating, the PH solar design is calculated with district heating in order to compare the two designs. The total energy demand and costs depends not only on the basic design but also on the energy costs. Future energy costs will in all likelihood depend on greenhouse emissions. Until now the structure of the CO_2 tax favors district heating. Electricity to be used on heat pumps has a 4,5 higher CO2 tax than heat produced by fossil energy.

District heating plays a central role in the Danish energy strategy to reach 2050 low emission society. There are discussions how this is going to be implemented. "Klima rådet" [Energi-Forsynings- og Klimaudvalget 2017][i] propose a lowering of the green CO_2 tax. In order to analyze how a CO_2 emission tax structure may influence the total costs in the future, 3 different emission tax levels are calculated.

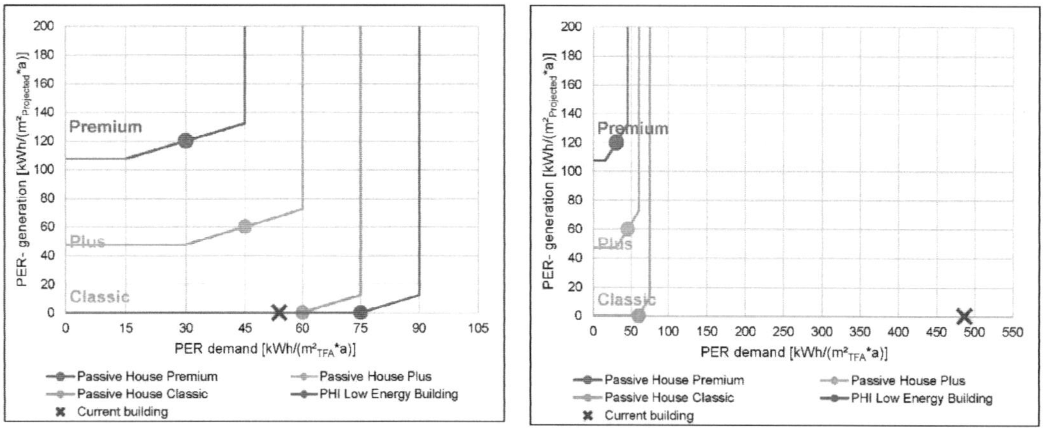

Figure 6 and 7 PHPP calculated Primary Energy Renewable demand. On the left the design for a solar south oriented passive house. On the right a standard prefab house.

Space heat demand of the PH house was calculated to be 14 kWh/m2 and year. Space heat of the standard prefab house was calculated to be 78 kWh/m2 and year, more than 5,5 times higher than the passive house. A calculation of the total primary energy with PHPP calculated primary energy factors leads to even higher difference between a PH house concept and a standard prefab house concept based on standard district heating. As mentioned before, the future factor for district heating in the 2020 regulation is 0,6. The PHPP calculates a mix of energy for district heating depending on the district plant and the resources used to produce heat. The factor calculated for a big power/district heat plant using hard coal is calculated to be 4,35 or more than 7 times higher than the primary factor used in the Danish regulation. The primary energy factor for electricity in PHPP is 1,3. The energy factor for a compact heat pump/ventilation system using electricity results in a PER energy demand for the PH house of 53,7 $kWh/_{TFA}m^2$ and year. The PER result for the standard prefab house is 486,4 $kWh/_{TFA}m^2$ and year, a factor 9 lower PER energy demand for the passive house with compact heat pump!

A significant difference in heat demand leads to a similar difference in heat load. High heat demand and heat load limits choices of heating systems based on heat pump. This can result in big differences of CO_2 emissions. The emissions for all heating, cooling, hot water and lighting energy is calculated to be 5,10 tons pr. year in the standard house and 2,69 tons pr. year in the passive house. If the passive house is equipped with a 10 m2 solar thermal collector and 38 PV panel (62 m2) the passive house house becomes a PLUS house and CO_2 "active house". The Passive PLUS house PER result is – 1,76 tons per year. A PER demand of 2,22 tons per year is made up for with the generated energy that substitute 3,98 tons of CO_2. 1,8 times more CO_2 can be taken out than used in a PH PLUS house.

To understand CO_2 emission per person we can reduce the calculated yearly emission with the 2,7 eq. number of persons calculated in the PHPP. This leads to calculated CO_2 per person results listed and compared with the approx. sustainable CO_2 mission for energy related to energy in buildings.

Emissions Tons $CO_{2eq.}$ per $Person_{eq}$ per year.

IOP Conf. Series: Earth and Environmental Science **297** (2019) 012005 doi:10.1088/1755-1315/297/1/012005

Standard prefab house District heat (hard coal heat (Power plant)	**PH Classic house** Compact heat pump/Ventilation system	**PH PLUS house** Compact heat pump/Ventilation system 10 m2 solar collector 62 m2 PV collector	**Sustainable limit max.**
1,9	1,0	0,8 Generated energy -1,5 Emission balance: ___ -0,7	≤0,8

Table 1 PHPP calculated CO_2 emissions divided with number of persons living in the house. Results are compared with the sustainable max.CO_2 per person pr. year. Approx. 40% of energy and resources are related to buildings. The sustainable CO_2 emissions per person is 2 tons per year in total for all emissions related to living. 40% or 0,8 tons of the 2 tons per year is therefore a sustainable max. emission CO_2eq. per person per year.

When the plus energy is calculated and related to the energy balance it is important to analyze the "winter gap" situation. Energy demand that leads to emission consists of space heating, hot water and lighting energy.

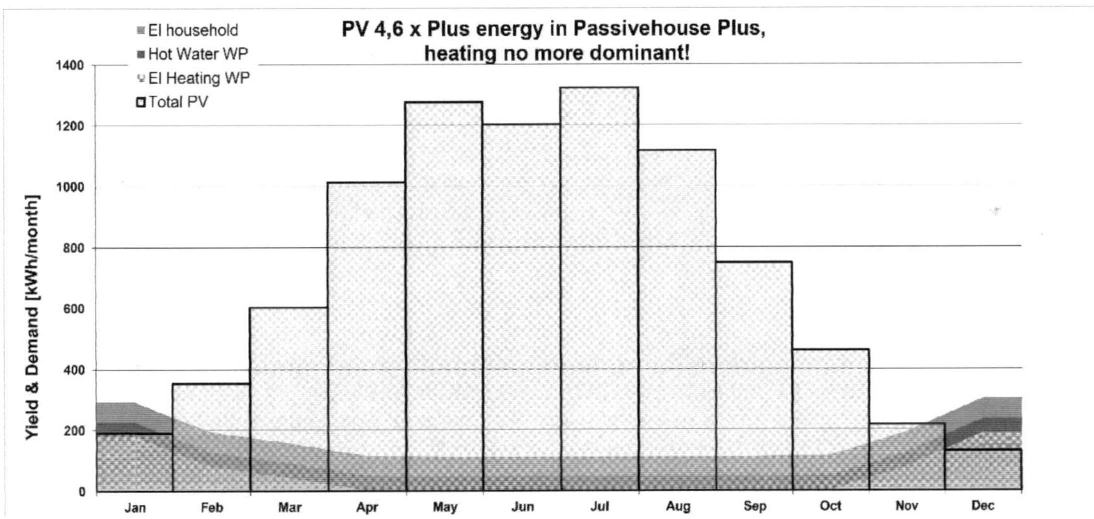

Figure 8 Monthly results for energy demand and generated energy of a PH PLUS house show that heating demand is no more dominant as the energy generated covers the electricity demand for heating almost completely. To cover the electricity demand in the winter months a plus in the electric energy balance of a factor 4,6 is the result.

This passive plus house strategy is 4,6 times "plus house", but this is not what the client is asking for. Total economy of the reference standard prefab house and a passive house classic are compared. Costs are calculated in 30 years as middle costs pr. year for construction and energy.

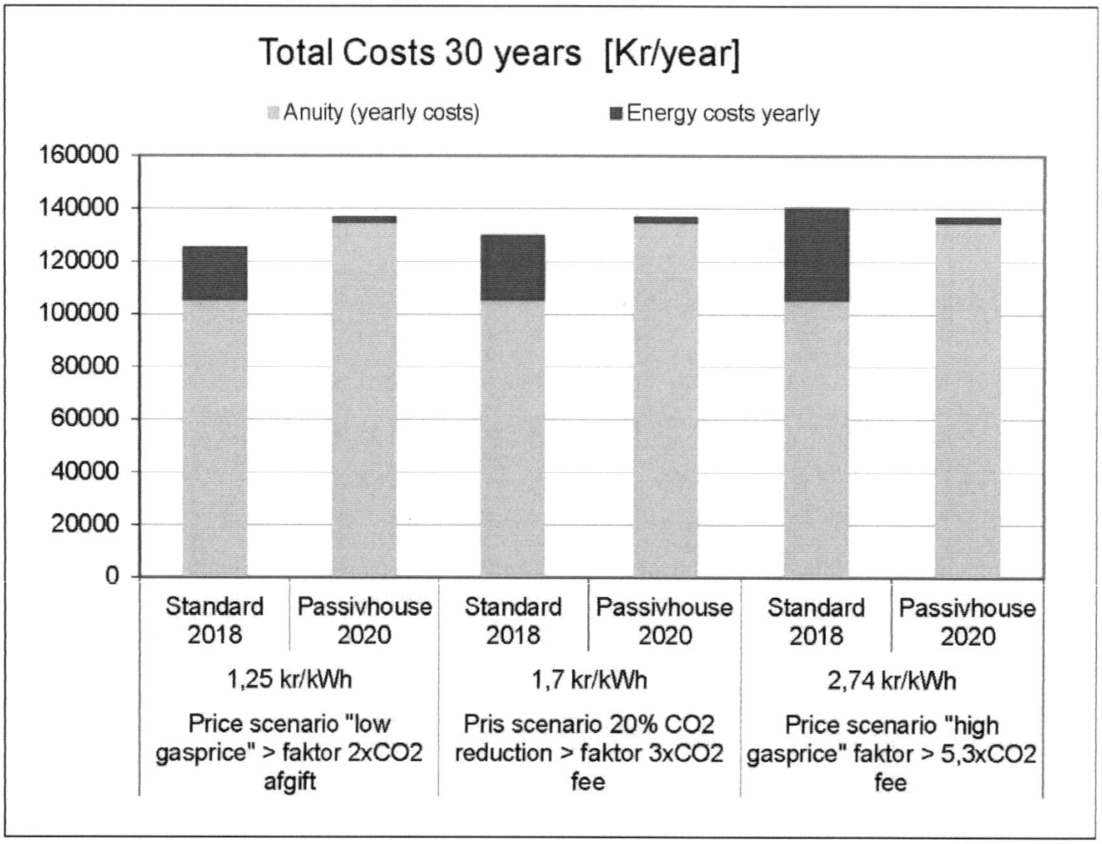

Figure 9 Result of yearly costs for 3 different energy price scenarios. With the two low-cost scenarios the total cost for construction and energy is lower for the standard prefab house using district heating. A price scenario with a sustainable CO_2 tax results in a better total economy for the PH energy standard.

When total costs are calculated for 30 years, short life investments are favored. Long life investments with a sustainable long life perspective are not taken into the result. If the client - after some years - chooses to sell the house, the value of the house is generated with the long life investments. The "rest value" of the investment costs, that is the investment costs with longer than 30 years lifetime, result in a value that can be put on the market and sold. The market "rest-value" of the long term investments of a passive house is approx. 25% more than the standard house.

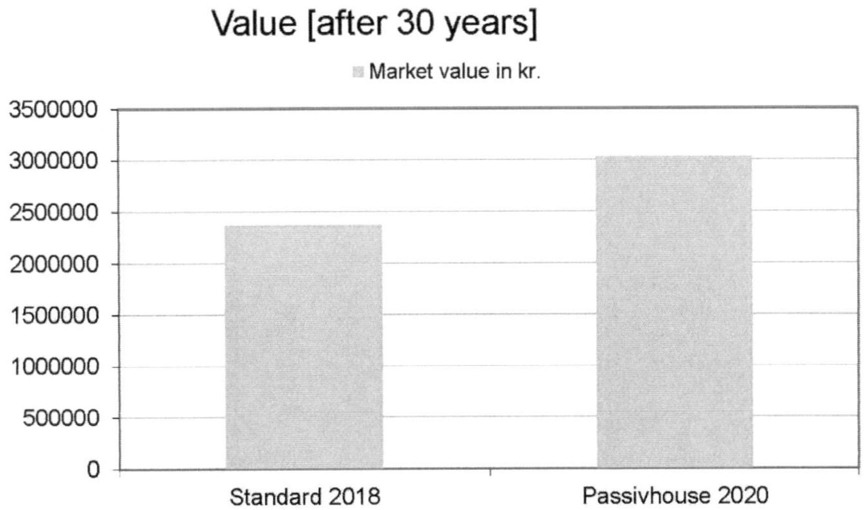

Figure 10 The "Rest-value" of long term sustainable investments. The result show that if calculated and documented – it can be a needed result to show to the bank. Often the bank sets the limit for the client – and this limit leads to investments in standard housing with lower energy related investments, short life time and potential high CO_2 emissions as shown above.

4. Case study number 3, a 1 level kindergarten competition, east-west orientation.

Early design decisions normally include form and orientation of glazing. Sometimes form decisions are limited when a competition defines a building plot or limits the number of building levels to one. In this case a kindergarten competition with ground level only and mainly east and west orientation of the vertical surfaces is given.

On the plot it was not possible to obtain south orientation of the main rooms for the children. The south oriented variant explores the potential of changing the site but keeping the building plan – so that south orientation would be possible. Results of PHPP calculated final energy demand show, that the same building turned 90 degree has a 14,7 % lower energy demand. South orientation of the vertical surfaces is critical if the building volume is limited to one level so that compact design is limited.

A higher potential of solar passive gain with a resulting lower final energy demand is not possible in the given competition used as case, as the plot is east-west oriented. 15% CO_2 emissions could be avoided if more vertical surfaces with south oriented glazing were possible.

IOP Conf. Series: Earth and Environmental Science **297** (2019) 012005 doi:10.1088/1755-1315/297/1/012005

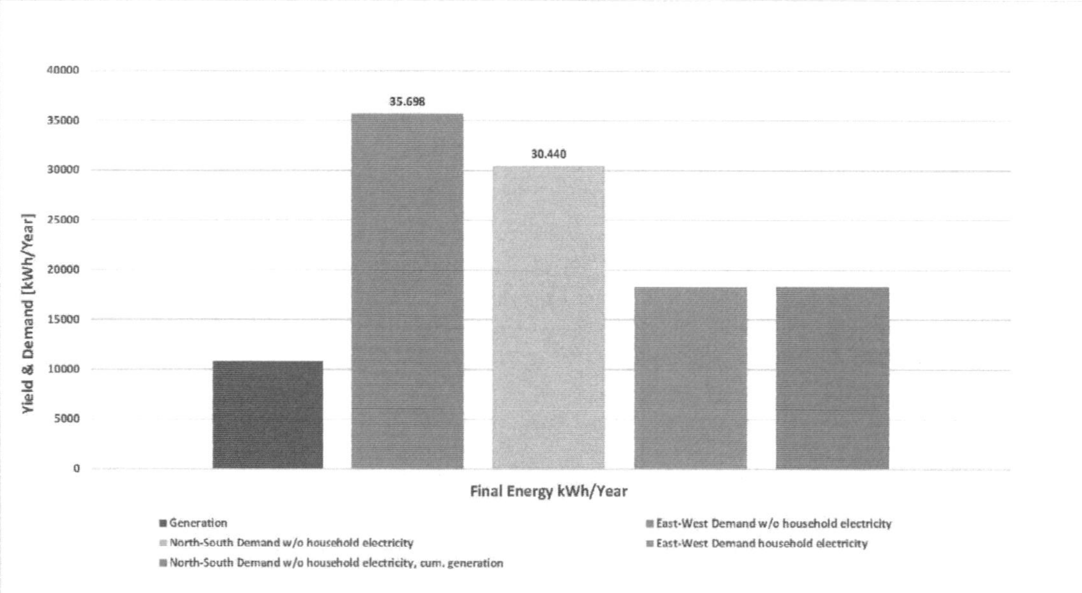

Figure 11 Case 3 PHPP results for 2 different building orientations of a kindergarten in one level. The glazing design is not optimized, only the orientation is changed. Higher passive solar gains result in a 15% lower energy demand. Energy generated by roof PV panels is the same in both variants as the panels are south oriented in both variants.

5. Case 4 is also a Kindergarten competition.

In this case it was possible to make a compact south oriented building volume on the site in one level. The optimized variant is the same proportions but in 2 levels – so a reduction of roof and floor to ground by 50%.

Results of energy demand show a 15,9% lower energy demand if the kindergarten were in 2 levels instead of one level. The same glazing area were calculated. The difference calculated is a 50% reduction of floor to ground and roof surfaces. The cost reduction of the 50% reduced roof area and ground floor cannot be calculated fully as extra costs for the 1 level floor have to be considered. But It can be estimated that potentially a considerable reduction in investment cost could be reached if the kindergarten design were possible in 2 levels.

Emissions from a south oriented plot but limited to one level results in 16% higher final energy demand and emissions.

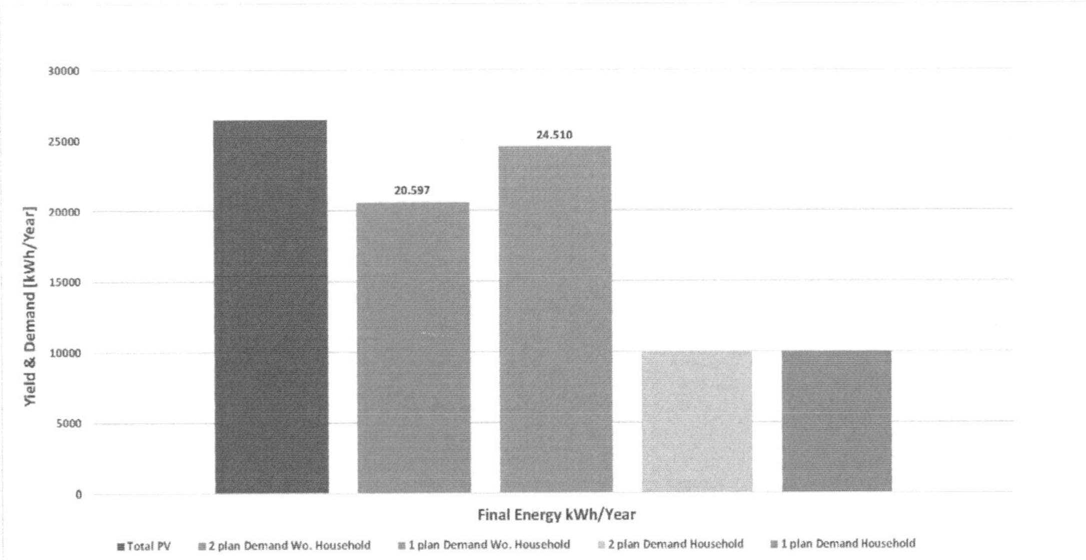

Figure 12 PHPP calculated results for 2 variants of case 4, a south oriented kindergarten competiton in 1 level and a compact variant in 2 levels. Yearly final energy demand and generated energy yield of PV panels on the roof. The generated energy on the roof is higher than the energy demand.

It can be argued that a design with a big roof surface can be used for active energy elements to generate energy and in this way "balance" the energy design. A compact variant results in a smaller roof area. To analyze how active generated energy can balance the energy demand and what the effect might be to the resulting energy balance, a monthly yield and demand balance is calculated in PHPP of the 2 design variant in case 4.

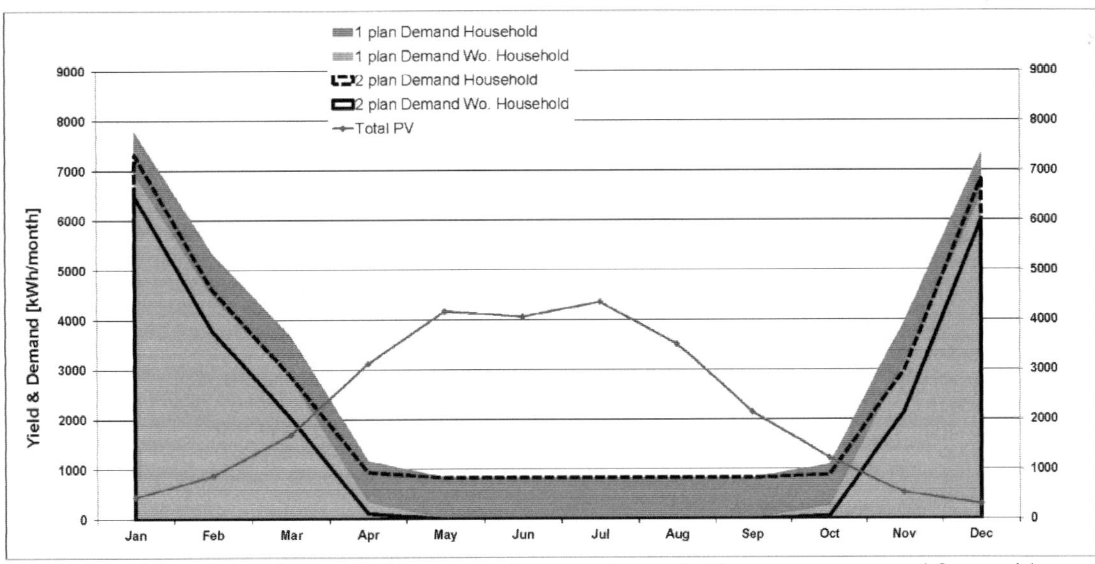

Figure 13 Monthly results of generated energy and energy demand. The energy generated from mid March to November balance the energy demand for heating and hot water of the compact 2 level design. The balance period is a little shorter when the building is in one level, resulting in a higher emission. A bigger area would mainly increase the summer peak, potentially resulting in more unused generated energy.

6. Results

The four different cases analyzed here are all real cases. The potential variants are not real possible cases – they were not possible due to the given conditions of the competitions. In order to illustrate what the potential savings might be, this paper analyzes what the results could have been – if the competition did not limit potential savings. In other words the results show how big the impact can be on the climate emission if initial primary design is limited. Often orientation and compact design do not result in higher costs. On the contrary – higher solar passive gains results in lower heat demand and lower climate emission. Approx. 15 % reduction can be expected if a design can be made solar passive optimized, show results in case 3. This reduction could be doubled if the design were not only solar passive but also compact in shape, shows results in case 4.

Case 1 and 2 analyze layout impact on climate emissions in smaller building volumes like social housing and standard houses. Layout without optimal solar passive gains and energy "performance gap", can result in higher climate emissions which could have been avoided with no extra costs or even cost savings.

References

[i] Energi- Forsynings- og Klimaudvalget 2017, -18 EFK Alm.del Bilag 214, https://www.ft.dk/samling/20171/almdel/EFK/bilag/214/1882023.pdf

Mineral wool waste-based geopolymers

Juho Yliniemi[1]*, Tero Luukkonen[1], Anne Kaiser[2], Mirja Illikainen[1]

[1] Fiber and Particle Engineering Research Unit, University of Oulu, P.O. Box 4300, 90014 Oulu, Finland
[2] Saint-Gobain Finland Oy, Strömberginkuja 2, 00380 Helsinki, Finland

*Corresponding author: juho.yliniemi@oulu.fi

Abstract. Mineral wools –a general term for stone wool and glass wool– are the most common building insulation materials in the world. The annual amount of mineral wool waste generated in Europe totaled 2.3 Mt in 2010 – including wastes from the mineral wool production and from the construction and demolition industry. Unfortunately, mineral wools are often considered unrecyclable due to their fibrous nature and low density. Thus, the utilisation of post-consumer mineral wool waste in different applications remains low.
Mineral wools have a great potential as geopolymer precursors as they have suitable chemical and mineralogical compositions. As geopolymers can provide significant CO_2 emission reductions compared to traditional Portland cement concretes, using mineral wool waste as geopolymer precursor would be an attractive utilisation path. Here, we show that mineral wool waste can be geopolymerised to form sustainable cements with good mechanical properties. Geopolymerisation of mineral wool waste therefore offers an attractive route for waste valorisation and production of low-CO_2 cements.

1. Introduction

Mineral wools, including stone and glass wools, are the most common insulating materials in the world. The production volumes are estimated to be 2/3 of stone wool (SW) and 1/3 of glass wool (GW), whereas slag wool is no longer produced in significant volume [1]. Although mineral wool production creates waste and by-products, including mineral wool offcuts, the main mineral wool waste stream is generated by building construction and demolition (C&D) industry. In 2010, the total volume of mineral wool waste produced in the 27 European Union countries was 2.3 million tons, and it is expected to rise to 2.5 million tons annually by 2020 [2].

Unfortunately, mineral wool waste is often considered unrecyclable. The main current practice is to landfill this waste which causes high costs for the C&D industry as well as for mineral wool producers. Utilising mineral wool waste as a source material for alkali-activated binders, often called as geopolymers, prevents the landfilling costs and can provide an avenue for an additional profits for C&D and mineral wool manufacturing companies.

Geopolymers can be produced from a wide range of aluminosilicate precursors, the most common ones being ground granulated blast furnace slag (GGBFS) and coal fly ash. As many industrial side streams can be used as geopolymer precursor, and geopolymer production can provide CO_2 emission reductions as much as 80% compared to Portland cement concretes [3], geopolymers are considered to be important part of the current and future toolkit of sustainable cementing binder systems.

Mineral wools have favourable chemical composition and mineralogy for geopolymer precursor. They are completely X-ray amorphous which increases their reactivity and they have a high Si content which is the main structural component of the binder. Further appealing characteristic of mineral wool waste is its high specific surface area which further increases the reactivity. [4,5]

The present study investigates the suitability of mineral wool waste for geopolymer precursor. The objective was to identify if both old mineral wool waste collected from building demolition sites and new mineral wool collected from building construction sites can be used as geopolymer precursor. The development of compressive strength is measured during the first 28 days of hardening. Furthermore, the freeze-thaw durability of the developed binders are determined. This will provide valuable information to improve the utilisation potential of mineral wool waste as geopolymer precursor.

2. Experimentations

2.1. Materials

Mineral wool waste samples were separated from C&D waste piles manually (Figure 1). In total, four samples were collected: 1) stone wool from construction site (SW_new); 2) stone wool from demolition site (SW_old); 3) glass wool from construction site (GW_new); and 4) glass wool from demolition site (GW_old). The installation dates of the old mineral wools from demolition sites are not known exactly, but they are expected to be from the 1970s or 1980s. The chemical compositions of the samples are presented in Table 1. In addition to the major oxides given in Table 1, glass wool contains typically ~2-4 wt.% B_2O_3 which is not quantifiable with the XRF technique used here. Both stone wools and both glass wool have chemical composition close to each other, which is important as the chemistry dictates the formation of the binder gels. Stone wool samples have a lower content of organic resin than glass wool samples, depicted by the loss-on-ignition at 550°C.

Figure 1. Mineral wool samples were collected manually from demolition and construction sites.

Table 1. Chemical composition and loss-on-ignition (LOI) of the mineral wool samples.

	SW_new	SW_old	GW_new	GW_old
CaO, [%]	18.2	16.6	7.9	7.3
SiO_2, [%]	39.4	44.1	61.3	61.9
Al_2O_3, [%]	15.9	14.3	2.0	3.3
Fe_2O_3, [%]	9.8	5.5	1.4	1.2
Na_2O, [%]	1.3	1.2	16.3	16.0
K_2O, [%]	0.5	0.3	1.0	0.8
MgO, [%]	11.4	14.7	2.0	2.9
P_2O_5, [%]	0.1	0.0	0.2	0.0
TiO_2, [%]	1.0	0.2	0.1	0.0
SO_3, [%]	0.1	0.0	2.0	0.3
Cl, [%]	0.0	0.0	0.1	0.1
LOI 550°C	4.3	2.6	9.4	8.8

2.2. Methods

Prior to the alkali activation mineral wool waste samples were milled with a ball mill for 3 hours. Milled mineral wool was added continuously to potassium silicate solution (SiO_2/K_2O molar modulus of 3.4 and water content of 55.3 wt.%) according to Table 2. The amount of mineral wool was adjusted so that suitable workability was achieved. Mixing was done with an IKA Eurostar 20N-mixer at 200 rpm with 8 minutes mixing time. Samples were cast into 2x2x8 cm^3 plywood molds and cured at 23°C until testing.

Compressive strengths of the samples were determined using a Zwick/Roell Z100 testing machine with loading rate of 1.2 kN/s.

Freeze-thaw durability test was conducted for 4x4x16 cm^3 samples that were cured for 28 days and half immersed in water. Environmental chamber was set to cycle between -15 and +15°C in 6 hour intervals. The duration of the test was 116 cycles in total.

Table 2. Mix designs of the alkali-activated mineral wool samples.

Sample name	SW_new (wt.%)	SW_old (wt.%)	GW_new (wt.%)	GW_old (wt.%)	Alkali activator (wt.%)
SW_new	66				34
SW_old		66			34
GW_new			46		54
GW_old				56	44
SW+GW	16.5	16.5	16.5	16.5	33

3. Results

3.1. Compressive strength measurements

All alkali-activated mineral wool samples achieved compressive strength between 25-45 MPa which is acceptable for many construction applications (Figure 2). Stone wool showed faster hardening compared to glass wool. The compressive strength of alkali-activated stone wool samples were 15-20 MPa already after 1 day curing and resulted in 44 MPa and 36 MPa for new and old stone wool, respectively, after 28 days. Glass wool samples were still soft after 3 day curing, but achieved 27-34 MPa compressive strength after 28 days. Compressive strength development of the mix of all mineral wools (SW+GW) was slower than that of stone wool samples, but faster than that of glass wool samples, thus both precursors affect the hardening process.

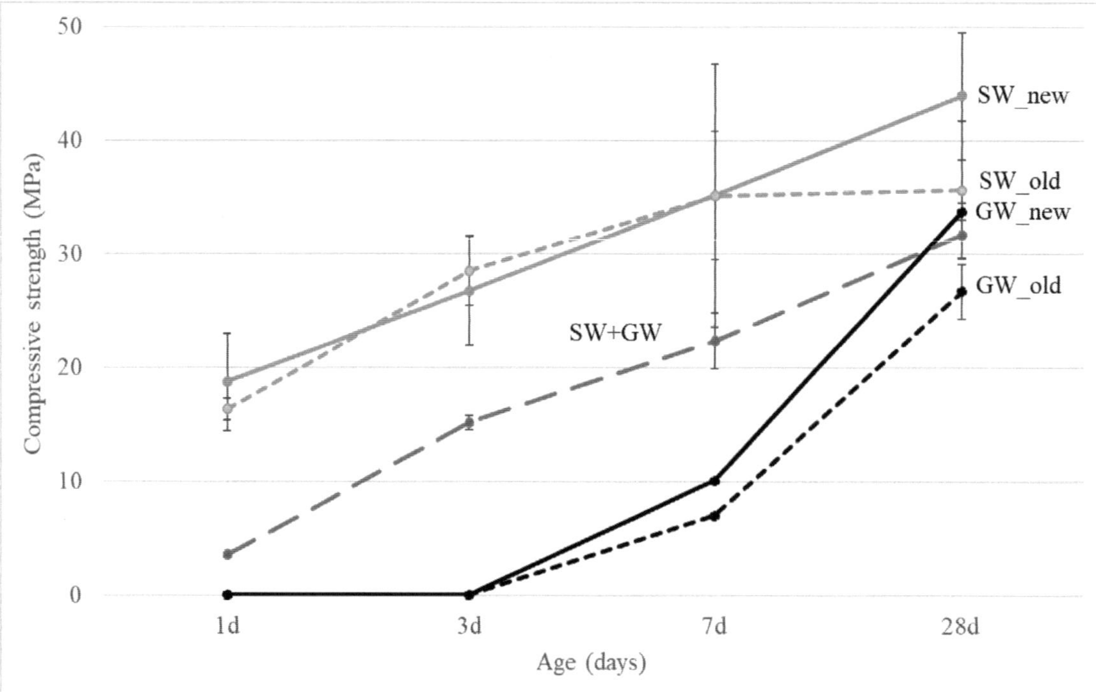

Figure 2. Compressive strengths of alkali-activated mineral wool samples.

3.2. Freeze-thaw-durability

The samples that contained stone wool showed excellent durability under the aggressive freeze-thaw test conditions (Figure 3). In contrast, the sample containing only glass wool (GW_new) disintegrated during the test. This is possibly due to formation of water-soluble binder gel in alkali-activated glass wool, which is observed in the geopolymerization studies with everyday waste glasses that have similar chemistry as glass wool [6,7]. Binder gel lacking hydrolytic stability can disintegrate during the freeze-thaw test as the samples are half immersed under water. However, co-binder that provides additional Ca or Al to the binder system can create water-insoluble gels, thus improving the hydrolytic stability of the binder [6,8–10]. This is demonstrated by the sample containing both stone wool and glass wool (SW+GW) as this sample showed greatly improved durability compared to glass wool-only sample. Stone wool contains both Ca and Al (Table 1) which can take part in the binder gel formation. However, confirmation of this would require further investigation.

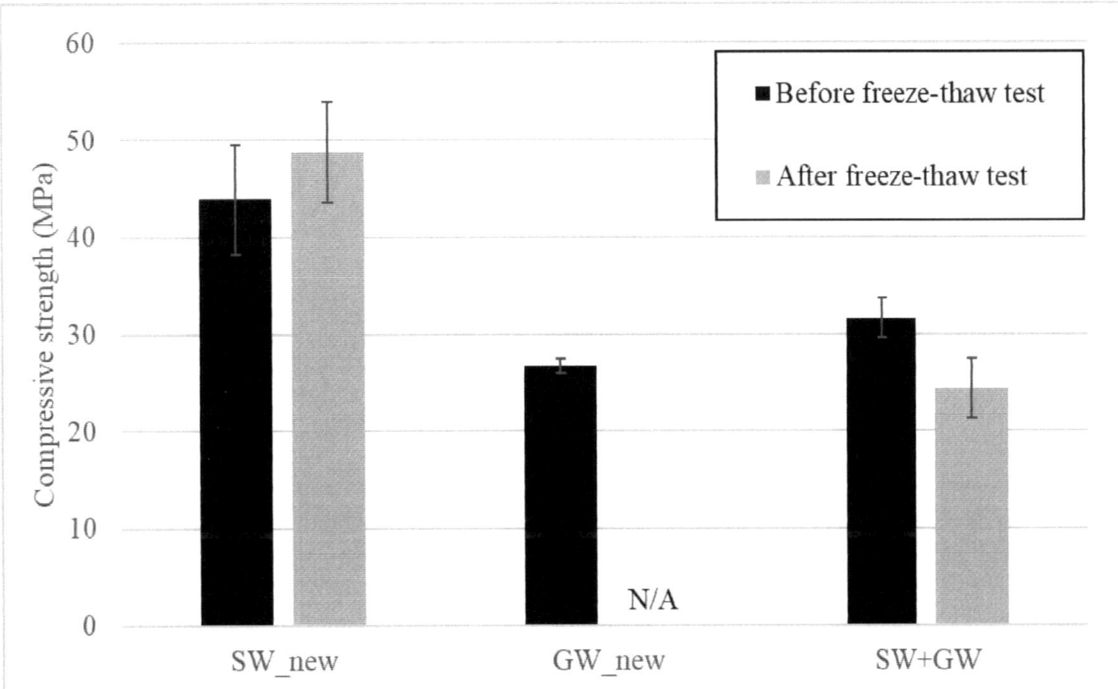

Figure 3. Compressive strengths of alkali-activated mineral wool samples before and after the freeze-thaw durability test.

4. Summary

Alkali-activated materials, also known as geopolymers, can provide significant CO_2 emission reductions for the construction industry. The key point is to use local source raw materials in order to minimise the environmental burden of transportation of raw materials. One currently underutilised inorganic mineral side stream is mineral wool waste.

Based on the results presented, mineral wool waste is suitable as geopolymer precursor despite its age and small amount of impurities. Particularly, stone wool and mix of stone wool and glass wool showed good durability even under aggressive freeze-thaw conditions.

Further research should focus on dissolution mechanism of mineral wools under alkaline conditions, optimization of mix design with particular consideration on choosing alkali-activator, and optimization of mineral wool fibre length.

Acknowledgements

This research was done under the auspices of Geodesign-project funded by the Business Finland and various companies (Boliden Harjavalta Oy, Destaclean Oy, Fortum Power and Heat Oy, Paroc Group Oy, Saint-Gobain Finland Oy, and Suomen Erityisjäte Oy). The authors would like to thank Tuomas Hirvijoki for performing the experimental work as part of his master thesis. Jarno Karvonen and Elisa Wirkkala are acknowledged for their help in the laboratory work. The authors would like to thank Destaclean Oy for providing the samples for the experiments.

References

[1] Müller A, Leydolph B and Stanelle K 2009 Recycling mineral wool waste—Technologies for the conversion of the fiber structure Part 1 *Interceram* **2009** 378–81

[2] Väntsi O and Kärki T 2013 Mineral wool waste in Europe: a review of mineral wool waste quantity, quality, and current recycling methods *J Mater Cycles Waste Manag* **2013** 1–11

[3] Luukkonen T, Abdollahnejad Z, Yliniemi J, Kinnunen P and Illikainen M 2018 One-part alkali-activated materials: A review *Cem. Concr. Res.* **103** 21–3

[4] Yliniemi J, Kinnunen P, Karinkanta P and Illikainen M 2016 Utilization of mineral wools as alkali-activated material precursor *Materials* **9** 312

[5] Yliniemi J, Laitinen O, Kinnunen P and Illikainen M 2017 Pulverization of fibrous mineral wool waste *Journal of Material Cycles and Waste Management* 1–9

[6] Redden R and Neithalath N 2014 Microstructure, strength, and moisture stability of alkali activated glass powder-based binders *Cement and Concrete Composites* **45** 46–56

[7] Giannopoulou I and Panias D 2010 Hydrolytic stability of sodium silicate gels in the presence of aluminum *Journal of Materials Science* **45** 5370–7

[8] Pascual A B, Tognonvi M T and Tagnit-Hamou A 2014 Waste glass powder-based alkali-activated mortar *nternational Journal of Research in Engineering and Technology* **3** 15–9

[9] Al Saadi T H A, Badanoiu A I, Nicoara A I, Stoleriu S and Voicu G 2017 Synthesis and properties of alkali activated borosilicate inorganic polymers based on waste glass *Construction and Building Materials* **136** 298–306

[10] Zhang S, Keulen A, Arbi K and Ye G 2017 Waste glass as partial mineral precursor in alkali-activated slag/fly ash system *Cement and Concrete Research* **102** 29–40

A Finnish Environmental Classification for Building Projects: Experiences and New Developments

Laura Sariola
The Building Information Foundation RTS sr, Po.Box 1004, 00100 Helsinki, Finland, Europe

Jorma Säteri, Helsinki Metropolia University of Applied Sciences, Finland, jorma.sateri@metropolia.fi

Timo Rintala, Green Building Partners Oy, Kutomotie 16, 00380 Helsinki, Finland, timo.rintala@gbp.fi

Abstract. There are a lot of international or national environmental classification systems for building projects. The Building Information Foundation RTS noticed a need for national classification system in Finland, where the target was to improve indoor air quality, moisture management, carbon footprint, energy efficiency, material choices and worksite technology and of course to save money in construction processes and maintenance of the premises. A Finnish Environmental Classification system called Green Leadership Tool for construction processes has been developed for managing the actual building process, environmental, economic and sociocultural aspects as described in the CEN/TC 350-standards. The criteria were made with construction industry and published by The Building Information Foundation RTS in the year 2017.

The major part of the criteria utilizes existing classifications and certifications developed in Finland as EPD's, Building Performance Indicators of GBC Finland, the emission classification of building materials and cleanliness classification of HVAC devices (M1-label), Classification of Indoor Environment (CIE2018), Energy Audit (E-Value), Green Factor Tool, Drychain10 and other relevant information provided into market.

The tool and criteria are suitable for all property owner's despite of the background and business model. There are four construction project types included: new building, renovation and partial renovation, enlargement/change of the space and two main buildings types: office/service and residential buildings. The Finnish legislation has been updated on the year 2017, the new LEVEL's certification criteria have been developed and the new Classification of Indoor Environment 2018 has been updated and consequently there was a demand to update the RTS criteria as well. The classification was launched in the spring 2017 and there are 83 active projects and 140 users in the tool on April 2019. The environmental classification of building processes follows the changes, developments and needs of the market, therefore the criteria will be updated continuously. The new criteria have been used in several projects and experiments of the criteria and projects will be presented.

1. Introduction

Responsible development of residences, offices and service buildings is valued highly. Making better use of natural resources and realizing lower property usage costs incentivizes investments in environmental factors when buildings are designed. This approach is valued by residents, property developers, and owners, as well as societal decision-makers.

RTS Environmental Classification is a system for managing construction process in a sustainable way. The system recognizes best national (Finnish) building strategies and practices. While there are globally accepted certification schemes, a need for national tool with local perspective and integration to local best practices has been recognized especially in non-profit real estate sector as municipalities renovating and building schools, day care centers, health care centers and other social property. The environmental classification tool makes it easier to agree upon matters such as the energy efficiency targets of buildings and solution development.

The classification was launched in the spring 2017, it follows the changes, developments and needs of the market, therefore the criteria will be updated continuously.

The tool has been developed primarily for the Finnish public real estate owners as municipalities, but it is suitable for all property owner's despite of the background and business model. The purpose of the classification system is to manage, monitor and lead the construction project efficiency and achieve all the way better premises to avoid problems in the future. The expected outcome of the tool is an enhanced environmental performance of the final built building. The real outcome and constructed premise will be certified. The certification after construction phase is mandatory if the owner of the project like to use certification label. The third phase, Use, is not mandatory if the highest certification level is not on purpose to obtain, but possible to show real situation after few years use.

There are four construction project types included: new building, renovation and partial renovation, enlargement/change of the space and two main buildings types: office/service buildings and residential buildings. To receive a certification, the whole construction process must pass a quality check and the outcome must be verified by a third-party auditor which is trained by The Building Information Foundation RTS sr. The auditor must pass a test arranged by the RTS. The Committee PT20 RTS Environmental Classification is responsible for future developments.

2. Best practises used in the tool

High indoor air quality is recognized as a crucial issue of both national health and economy in Finland. There are several projects and programs going on to ensure and excite to improve good indoor air quality, moisture management and use of EPD's as a source of carbon footprint calculations in Finland. When the project started, the work group named three main targets for the classification:

1. To improve energy efficiency
2. To improve indoor air quality
3. To improve moisture management

The Finnish legislation is ground for the tool, every target is above the basic requirements. In addition, there are few key classifications and certifications utilized to create the criteria for the Green Leadership Tool. These key factors are the Building Performance Indicators of GBC Finland, the Emission Classification of Building Materials and Cleanliness classification of HVAC devices (M1-label), Classification of Indoor Environment, Energy Audit (E-Value), Green Factor Tool (GF) and Drychain10.

2.1. Classification of Indoor Environment 2018

The Finnish Society for Indoor Air Quality, FISIAQ, has a leading role in indoor air improvement and especially in developing the Finnish Classification of Indoor Environment 2018, even though the publication is a result of co-operation of many organizations. The Classification of Indoor Environment has three parts:

1. The classification of indoor climate gives target and design values for thermal conditions, odour intensity, noise levels, ventilation and indoor air pollutants.
2. Guidelines for design and construction including classification of cleanliness in construction, principles and procedures for the main stages of construction works.

3. The Emission Classification of Building Materials contains target values for odours and emissions of the materials, fixtures and furniture and recommended maximum surface area of materials based on their emissions.

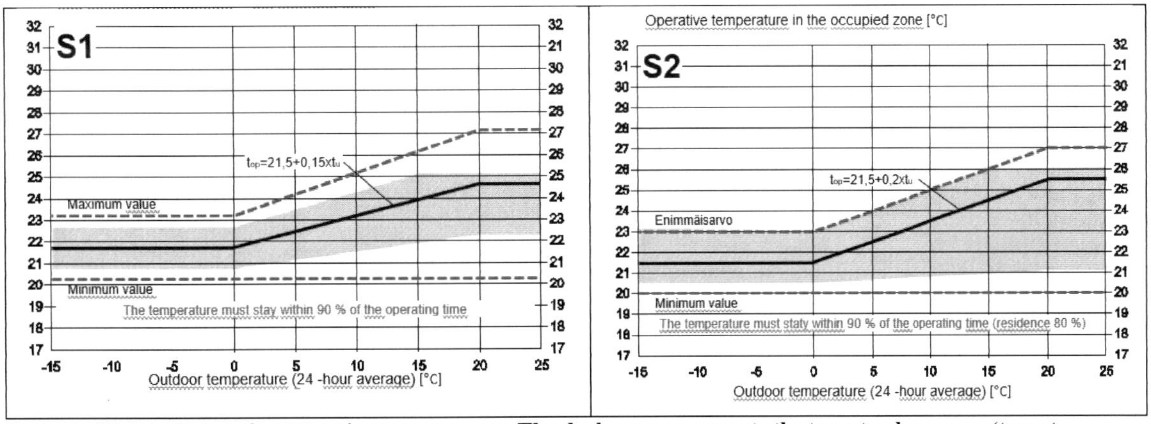

Figure 1 Target values for operative temperature: The dark area represents the target value range (target temperature+ allowed deviation).

A temperature control is essential for the indoor environment. In the updated classes S1 and S2, the temperature must stay within target value range 90% of the operating time on the dark area. There is an exception in the class S2 for residences where the temperature must stay 80% of the time within target value range within dark area. The content must be considered when setting targets in the Green Leadership Tool. One third of the points in the tool can be gathered from indoor air quality related criteria.

2.2. The Emission Classification of Building Materials

The Finnish Society of Indoor Air Quality and Climate (FiSIAQ) introduced over twenty years ago in 1995 a Classification of Indoor Climate, Construction Cleanliness, and Finishing Materials. The first revised edition, Classification of Indoor Climate 2000, was published in 2001 and the second edition Classification of Indoor Environment 2008 was published in 2008 and the third edition was published on May 2018. Based on the criteria set in the classifications, The Building Information Foundation RTS sr started M1-labelling of building products in 1996, named Emission Classification of Building Materials and Cleanliness Classification of Air-handling Components.

The new criteria for Emission Classification of Building Materials mentioned in the third part has been set in the classes M1 and M2. The content must be considered when setting targets in the Green Leadership Tool. All the standards behind the Classification of Building Materials can be found from the protocol: M1 Emission Classification of Building Materials: Protocol for Chemical and Sensory Testing of Building Materials. One of the accepted testing laboratories is in the USA and other in Europe.

Figure 2. Classification of Building Materials and Classification of Air-Handling Components

2.3. *The Building Performance Indicators*

The Building Performance Indicators were developed to react to the demand of key indicators that assesses environmental efficiency of properties and to ensure principles of sustainable development. The project was implemented with support of SITRA and published by the Green Building Council Finland in the year 2013. The eight indicators are based on the European CEN/TC 350 standards including E-Value, Life-Cycle Carbon Footprint, Life-Cycle Cost, Indoor Air Class, Measured Energy Consumption, Operational Carbon Footprint, Base-load Demand, User Satisfaction.

2.4. *The Green Factor Tool and iWater*

The Green Factor tool (GF) was developed to find out environmental impacts of green infra-structure and to measure actions for climate change adaptation, mitigation and for urban green compensation in ILKKA-project coordinated by Aalto University and City of Helsinki in the year 2014. The tool is put in place and implemented by several Finnish, Swedish, Norwegian and German municipalities. Green Factor tool idea was implemented to the Green Leadership tool to help preserve green infrastructure and nature elements characteristic to the area.

iWater – Integrated storm water management projects on the years 2015-2018 purpose was to develop Integrated Stormwater Management concept to support urban planning and to create high quality and resilient urban environments in the Baltic Sea region. The targets are integrated to the RTS Environmental Classification.

2.5. *Environmental Product Declarations (RTS EPD)*

The Building Information Foundation RTS sr launched a national EPD program first time in 1998, second time in 2004 and third time in the year 2016 and after that there have been produced several RTS EPD's published by the RTS. Environmental performance of construction products is compiled according to the standard EN 15804+A1 and (A2) and with a guidance's and methodology processed in the committee PT18 RT EPD's. The Information gathered to the Environmental Product Declaration (EPD) is verified by a third party and certified by the PT18 RT EPDs work group. It is possible to gain points in the Green Leadership Tool, if over 10 verified EPDs according to EN 15804 have been used in the project despite of the publisher of the EPDs.

IOP Conf. Series: Earth and Environmental Science **297** (2019) 012007 doi:10.1088/1755-1315/297/1/012007

Figure 3. Label of the RTS EPD

2.6. Moisture management with Drychain10

Drychain10 was developed to control moisture management by considering ten critical points as moisture outside the building, rain, roof and walls, ventilation, condition of water pipes, moisture proof in bathrooms and other spaces, moisture content of the concrete and surface coating, wetting of the building materials and poor maintenance. The project was implemented by City of Oulu in the year 2016, the operator is RALA. Part of the Drychain10 requirements have been adapted to the Finnish legislation on December 2017 and therefore excluded from the new criteria of RTS Environmental Classification system.

3. The new requirements of the RTS GLT Environmental Classification

The table below (Table 1) will show a list of the criteria to be achieved at minimum in office buildings and demonstrate the scale of the star classes. There are five classes, 1-5 stars, where the highest number of points a project earns the most precious label of five stars. The whole process has three phases: Design, Construction and Use.

Table 1 Classification scale and new requirements

Criterion	1 star	2 stars	3 stars	4 stars	5 stars
Points	25	40	55	70	85
P1.2 HVAC Commissioning			50%	50%	50%
P1.3 Userguide and training				100%	100%
P2.1 Moisture safe design			75% (DC10)	75% (DC10)	75% (DC10)
P2.2 Moisture safe construction		75%	75%	75%	75%
Y1.1 Life cycle carbon footprint			15%	30%	30%
Y2.1 Energy efficiency		20%	30%	40%	40%
S1.1 Indoor temperature			25%	50% (S2)	50% (S2)
S1.2 Air quality		50%(S2)	50% (S2)	50% (S2)	50% (S2)
S1.4 Material emissions			50%(M1)	50% (M1)	50%(M1)
In-use audit after 1or 2 years					yes

The content of the criteria set in the RTS Environmental Classification is demonstrated in the table below and new criteria set in the version 1.1 has been bolded and highlighted in grey colour:

Table 2 Criteria of the Green Leadership Tool in the version 1.1

Main group	Groups	Criteria		New RTS GLT Criteria
Process	**Commissioning**	P1.1	Design phase audit	No changes
		P1.2	Commissioning	
		P1.3	Userguide and training	
	Moisture safety	P2.1	Moisture safe design	Adapted requirements in Finnish legislation and Drychain10
		P2.2	Moisture safe construction	
	Site sustainability	P3.1	Sustainable construction	Adapted requirements in CIE2018
		P3.2	Site dust management	
Economy	**Life cycle costs**	T1.1	Life cycle costs	Adapted LEVEL's requirements
	Maintenance	T2.1	Durability	Adapted LEVEL's requirements
		T2.2	Maintainability	
		T2.3	Adaptability	
Environment	**Carbon footprint**	Y1.1	Life cycle carbon footprint	Adapted LEVEL's requirements
		Y1.2	Resource efficiency	New criterion, adapted LEVEL's requirements
	Energy	Y2.1	Energy efficiency	E-value (class A-30%), adapted requirements in Finnish legislation
		Y2.2	Energy metering	
		Y2.3	Consumption targets	
		Y2.4	Core system efficiency	
	Water	Y3.1	Water efficiency	
	Local impacts	Y4.1	Ecological value	Adapted LEVEL's requirements for resilience, iWater
		Y4.2	Cyclist and strollers spaces	
Health and wellbeing	**Indoor Air Quality**	S1.1	Indoor temperature	Adapted CIE2018 requirements and LEVEL's requirements for resilience (S1+ 3°C). Included low emitting fixture in S1.4.
		S1.2	Air quality	
		S1.3	User controls	
		S1.4	Material emissions	
	Visual comfort	S2.1	Daylight	No changes
		S2.2	Lighting quality	
	Acoustics	S3.1	Space acoustics	Adapted requirements in Finnish legislation (STI-INDEX)
		S3.2	Sound insulation	
Innovation	**Innovation**	I	Innovation	No changes

3.1. Criterion P2.1 Moisture safe design- an example of the content

This criterion is a compulsory criterion in three, four- and five-star building projects. Moisture management has been considered also in the criterion P2.2 Moisture safe construction, which is compulsory in two to five-star building projects.

Content of the criterion P2.1 Moisture safe design

Design phase includes identification of moisture risks and managing the risks by ensuring that the design solutions will work. High-quality management of moisture risks (75% of the weighted value) can also be fulfilled by acquiring the Drychain10 status.

Management of moisture-related risks during the design phase, 75% of the weighted value.

1: A qualified impartial Moisture Safety Expert, whose supervisory responsibilities cover the phases of general design and implementation planning, has been named for the project.

- Designer: Contract on appointing a person in charge and a list of tasks.
- Developer:
- User:

2: The site has been subjected to a moisture risk assessment, which was prepared by all designers. Based on the risk assessment, a specific risk category for each building section has been defined.

- Designer: Moisture risk management report and memo of the procedure.

- Developer: a report of reviewing the moisture risk assessment with the contractor.
- User:

3: Drying times for critical structures have already been calculated in the design phase, thus ensuring a realistic construction schedule in which the drying times are being considered by adding the critical times of the project to the schedule.

- Designer: A general schedule completed with critical times of the project.
- Developer:
- User:

4: The moisture risk management of the building project have been made based on description of the risk management in a separate document or risk list.

- Designer: Moisture risk management report and memo of the procedure.
- Developer: a report of reviewing the moisture risk assessment with the contractor.
- User:

5: Renovation: input parameters for renovation purposes has been obtained from a condition assessment of the building. If indoor air problems have been detected before the renovation, a qualified building health expert must be included in the project.

- Designer: Summary report of the condition assessment of the building, audit report of the building health expert AND report on the risks that have been considered in the plans for the design phase moisture safety plan.
- Developer: a report of reviewing the moisture risk assessment with the contractor.
- User:

Separate inspections for moisture risk structures, 25% of the weighted value

6: For high (category R2) and very high (category R3) moisture risk structures (according to moisture risk assessment) a separate building physics audit has been performed by an impartial building physics engineer OR the building has only normal structures considering moisture risks.

- Designer: building physics audit report OR moisture risk assessment for R1 category.
- Developer: a report of reviewing the moisture risk assessment with the contractor.
- User:

7: The project task force has dealt with the risks identified during the building physics audit, risk management plan has been decided and remarks are corrected in the plans.

- Designer: a report of the risks identified during the audit AND confirmation of settled repairs.
- Developer: Audit reports of moisture risk assessment of constructions (risk categories R2 and R3) on construction site by a qualified moisture management supervisor.
- User:

4. Audit and Certification procedure

Today there are 50 committees in the Building Information Foundation RTS sr. Twenty of these are principal committees, like PT20 RTS GLT Environmental Classification. There can be workgroups under the principal committees like classification work groups.

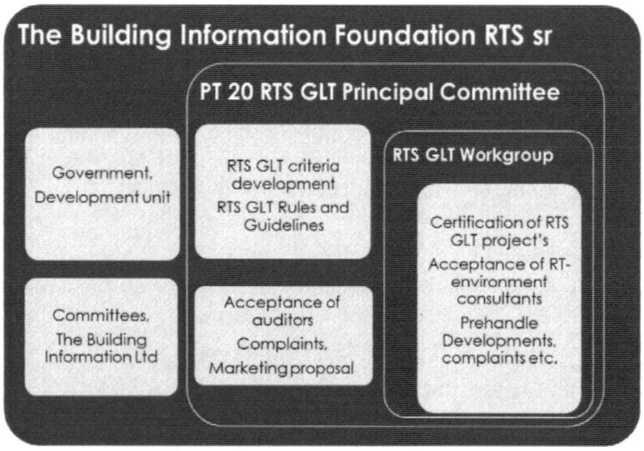

Figure 4. Organization chart

The auditors, who are trained experts authorized by the Building Information Foundation RTS, inspect the contents and scores of the reports saved on the tool and decide whether to grant approval. RTS invites auditors to audit projects; auditors cannot select projects independently. RTSs' classification working group processes the auditing reports, confirms the classification, and certifies the project. After that the RTS sends a certificate of approval to the project owner and publish basic information of the projects on the web.

Companies that have chosen to use RTS Environmental Classification for construction processes have noted, that most important thing is that the tool focus on the most relevant issues: it considers Finnish legislation and northern climate and it will guide the process and help with material choices. The RTS Environmental Classification has been used for construction processes of new buildings as hospitals, day-care centres, schools, accommodation services, restaurants, public real estate, dwellings and renovations.

5. Conclusions

RTS Environmental Classification will help construction industry in setting environmentally acceptable targets for construction projects when operating in Finland. The Finnish legislation and both indoor air and outdoor air circumstances have been considered and therefore the constructor have no reason to find out the information from other sources. This will help when setting targets to the project. The Classification of Indoor Environment (CIE) has been used in multiple projects from the beginning. By including all the CIE requirements to the RTS the process and outcome can be certified after the project. There are 90 projects going on in May 2019, most of the projects are office and residence buildings, in addition there have been started four hospital projects.

With continuous update of the existing classification systems and with wide cooperation of the construction industry it is possibly to create and maintain successful methods and tools in striving for healthier and more comfortable buildings and in addition to improve construction processes and maintenance of the premises. There is a lot of discussion about carbon footprint calculation and it will be among energy efficiency and indoor air the most relevant subject to improve in the future. To respond to the demand of life cycle approach, the Building Information Foundation started a project to create criteria for in use -period. The main targets are HVAC, safe and security and environment. There will be created "a toolbox" for real estate management team for continuous improvement in the project. The project will be concluded in the end of the summer 2019.

6. Acknowledgement

The writers like to thank all the experts and users of the RTS Environmental Classification for participating criteria development.

7. References (some of them only in Finnish)

[1] CEN/TC 350 standards
[2] Classification for Indoor Environment 2018, Finnish Society of Indoor Air Quality and Climate, FISIAQ. Espoo, Finland (will be published on May 2019 in English)
[3] Cleanliness Classification of Air-handling components, The Building Information Foundation RTS sr, http://m1.rts.fi
[4] Drychain10, Rakentamisen Laatu RALA ry, http://kuivaketju10.fi/
[5] Emission Classification of Building Materials, The Building Information Foundation RTS sr, http://m1.rts.fi
[6] Green Leadership Tool Criteria 2018, The Building Information Foundation RTS sr, http://glt.rts.fi
[7] HTP Values 2016: CONCENTRATIONS KNOWN TO BE HARMFUL, Publications of the Ministry of Social Affairs and Health 2016:8
[8] LEVEL's criteria, http://www.buildup.eu/en/news/introducing-levels-common-language-sustainable-buildings-0
[9] M1 Emission Classification of Building Materials: Protocol for Chemical and Sensory Testing of Building Materials, The Building Information Foundation RTS sr, http://m1.rts.fi
[10] The Building Performance Indicators, Green Building Council Finland ry, http://figbc.fi/en/building-performance-indicators/
[11] The Green Factor tool (GF), http://ilmastotyokalut.fi/en/
[12] iWater, http://www.stadinilmasto.fi/en/adapting-to-climate-change/iwater-integrated-storm-water-management/
Finnish legislation: (can be searched from www.finlex.fi by reference code (e.g 465/2014), some of them are described only in Finnish or Swedish)
Decree of the Ministry of Social Affairs and Health on Health-related Conditions of Housing and Other Residential Buildings and Qualification Requirements for Third-party Experts (545/2015)
Decree of the Ministry of the Environment on Foundation Structures 465/2014
Land Use and Building Act (amendment 222/2003 included)132/1999.
Land Use and Building Decree 895/1999.
Ministry of the Environment Decree on the fire safety of buildings 2017
Terveydensuojelulaki. Suomen säädöskokoelma 763/1994.
The National Building Code of Finland, Ministry of Environment, http://www.ym.fi/en-US/Land_use_and_building
Ympäristöministeriön asetus rakennuksen ääniympäristöstä 2017
Ympäristöministeriön asetus rakennusten kosteusteknisestä toimivuudesta 782/2017
Ympäristöministeriön asetus uuden rakennuksen energiatehokkuudesta 1010/2017
Ympäristöministeriön asetus uuden rakennuksen sisäilmastosta ja ilmanvaihdosta 2017

Housing and service structure concepts for areas with decreasing and ageing population

I Verma[1] and J Taegen[1]

[1]Sotera Institute, Department of Architecture, Aalto University, Po Box 3100, 00076 Aalto, Finland

Abstract. Urbanisation has often a negative impact on the population structure in remote areas. Municipalities have to find strategies to anticipate the demographic change and steer territorial shrinkage. Moreover, the current service structure is not optimal for areas with decreasing population. The aim of this ongoing research and development project is to promote resilient service structure and health prevention as well as housing services in these areas. The project comprises five case studies in municipalities with less than 20 000 inhabitants. Existing urban structures and the potentials for re-use of buildings will be identified in collaboration with local stakeholders. The comparative analyses of the municipal structures will be done at Sotera Institute at Aalto University. For each municipality, one student in architecture will provide a solution for urban design, housing design or renovation of existing building. The results of the study will help municipalities to outline a local development strategy to promote a lively and attractive municipal centre and a resilient service structure to improve living environment in remote areas. Shared use of buildings will improve the local service production and service delivery models. Efficient service delivery and housing will keep these areas attractive to live in also in the future.

1. Introduction

This paper is describing an ongoing research and development project focusing on shrinking rural municipalities in Finland. It is the continuation of a series of projects carried out between 2015 and 2018 at the Research Institute for Health Care Facilities (Sotera) at Aalto University. These development projects have been related to the national reform of social affairs and health care in Finland. The political consensus about the reform has not yet been reached. Nonetheless, it will have important impact on the role of the municipalities. The need for development of resilient service structure and housing for older people in small municipalities came first into discussion during the project *Changing society - changing services* (2015 – 2017). The consortium gathered 20 experts from public and private service providers as well as non-profit organizations. The project focused on age-friendly urban environment as well as service delivery and housing services for older population [1], [2]. Furthermore, the Government key project *Täydentäen toimivaa* (2017 – 2018) was promoting networking of local stakeholders to enhance aging-in-place policy and care at home [3]. The shared use of existing service buildings, walking friendly environment and a resilient network of local service providers was promoted in three municipalities.

The ongoing research and development project *Vital municipalities* is targeted to small municipalities with less than 20000 inhabitants. The project is carried out in collaboration with municipalities or joint municipalities of Harjavalta, Keuruu, Kurikka, Pertunmaa and Suupohja with support of the Ministry of the Environment, the Housing Finance and Development Centre of Finland (ARA) and the Ministry of Social Affairs and Health. The municipalities face different challenges regarding their population and urban structures. The loss of population and demographic development have a negative impact on local tax revenues and possibilities to maintain the services and quality of the urban environment. Careful long-term planning is needed in order to maintain and to improve the quality of living environment in these remote municipalities. The focus of the project is on housing development, densification and quality of the center of the municipality. The actions against the degradation of the built environment may include refurbishment or demolishment of existing buildings. Furthermore, the planning in the shrinking-cities often means attempts to make them greener, more compact and sustainable [4]. The municipalities may reinforce their services accepting a decline of population and recognizing the local strengths as an opportunity.

2. Background

Urbanization has often a negative impact on the population structure in remote municipalities and influence the municipalities' capacities to maintain the services. Currently, more than half of the Finnish population is living within 200 km from Helsinki [5]. The demographic dependency ratio (share of population under 15 years old and 65 years old and older compared to working age population) is increasing especially in small remote municipalities. By 2040 in shrinking municipalities, the percentage of population 75 years old and older is projected to increase by 10% [6]. Therefore, the possibility to deliver local services in remote areas is becoming increasingly difficult. Scott et al. point out that "*without a cross-section of people living in the countryside, rural economies cannot function and key services cannot be provided*" [7]. In Finland many educational facilities in remote areas have been closed down or moved to bigger municipalities, public services and bank services are going online. At the same time, the network of local grocery shops has decreased [8]. The current way of delivering services would need a larger customer base [9]. Moreover, in rural areas, the mobility of residents is based on the use of private cars. Long distances and the lack of public transportation is affecting the accessibility of local services and possibilities to live in remote areas in old age. A survey about the best living area in Finland revealed that the residents consider local services and transport connections, active community, strong identity and nature as most important features of their living environment.

Municipalities have to find new strategies to anticipate demographic change and combat loss of population. They need to steer controlled territorial shrinkage and to identify and strengthen the local potential. The decrease of population may lead to the degradation of the built environment and general infrastructure in remote municipalities. The European DART project concluded, that the "*declining regions can and should generate no growth*" [10]. However, the local long-term plans need to ensure sustainable development and combat environmental degradation to ensure good living conditions in shrinking municipalities even in the future. Furthermore, the international examples show that successful local actions may stop the migration from small municipalities, improve the quality of life of residents and generate business opportunities. Currently, in the majority of Finnish municipalities, second homes outnumber permanent dwellings [11]. Therefore, the potential of second home owners for the local development need to be studied. This project is exploring the challenges and opportunities of shrinking municipalities in the Finnish context. The project is focusing on the built environment and urban form. The framework of the project is in shrinking cities, age-friendly communities and sustainable development.

2.1. Actions in shrinking municipalities

The challenge of shrinking municipalities is global. In some of the European countries, national policies and local actions have been targeting to stop the loss of population in remote areas. For example, Germany and Ireland have undertaken concrete actions to prevent the loss of population in these areas. In Germany, the actions were focusing on the urban renewal, improvement of infrastructure and transportation networks, refurbishment and demolishment of existing buildings as well as on the management of the green landscape [12]. In Ireland the actions were focusing on the employment, promoting local services, leisure and cultural activities [13]. Furthermore, the national action plan was encouraging increased residential occupancy in town and village centres. The aim was to densify these local centres and make them attractive to live in. Moreover, the importance of mobile services, digital services and new innovative ways of delivering services was recognized [14]. The development both in Germany and in Ireland have been targeting to empower local communities.

2.1.1. Actions related to services
In Aschersleben, in Germany the strategy of the municipality was focusing on the reinforcement of the role of the municipal centre. The public services and schools were transferred from remote areas to the centre of the municipality and at the same time the buildings on the outskirts of the municipality were demolished [15]. The cultural services and art were used to improve the quality of life of residents. Small local businesses related to food production and tourism may also strengthen local economy. In Australia, for example, the improved economic situation of the small municipality of Beechworth is due to a small local bakery that became nationally known. The shrinking municipality with a population of 3000 inhabitants has become the place of the biggest bakery in Australia serving over one million customers per year [16].

2.1.2. Actions related to housing
The number of single person households has increased in all age groups. The local authorities need to be prepared to deal with the increasing number of older people and people living alone. Scott et al. [7] point out that young people who would be committed to remain in a rural areas may not be able to find an apartment to their needs. The private rented apartments and single family homes may be out of their reach, if affordable and government subsidised alternatives are unavailable. On the other hand, in Finland, the loss of population has led to the underutilization of the Government subsidised ARA-apartments in remote areas. These apartments built in the 1970's would need major renovations. Moreover, currently, the rents in these apartments are higher than in privately financed apartments [17].

2.1.3. Actions related to community
The social cohesion within the local community is one of the potential strengths of small municipalities. The challenge is to ensure participation and inclusion in society both at extremely low population density and an aging society [10]. Artistic and cultural activities may have important role in the community building. They are also building up the identity of the place. In Limburg area in the Netherlands, the Zachte G initiative has made visible the creative and artistic activity in the area. It has improved the visibility and attractiveness of the area [18]. Moreover, shared spaces for social activities are needed in the communities. In Erlau in Germany, the old railway station was transferred to a common living room for all local residents. It has enhanced the vitality and social activity in the municipal centre.

3. Aim and research questions
The aim of this research and development project was to support wellbeing and services in small municipalities with decreasing population and economy. The target was to promote resilient and inclusive service structure for remote areas and at the same time improve health promotion, mobility and housing for older population living in these areas.

The project will propose alternative development models to improve living conditions in remote areas. The local strategies in each of the municipalities participating in the project will promote a lively and attractive municipal centre and a resilient service structure. The solutions can combine new ways of delivering services and various service providers in shared premises. Shared use of buildings and collaboration with local stakeholders, public institutions and private sector will improve the local service production and service delivery models.

3.1. Research questions
The research questions are how can urban design help to increase and maintain the vitality of areas with decreasing population, and how to develop sustainable built environment, shared use of existing resources in small municipalities with shrinking economy?

4. Methods
The research and development project contains five case studies in the shrinking municipalities with less than 20 000 inhabitants. The multiple case study method enables to explore and describe the urban

form and situation in each of the municipalities. In horizontal comparison of the cases it was important to choose sites that were of the same scale, and to use units that were equivalent [20]. The sample of municipalities was chosen according to their size and population structure. The existing buildings and structures were analysed on-site and the potentials for re-use will be identified in collaboration with local stakeholders. The statistical analyses of local population structure, demographic development and population projections are bases for the development project. Moreover, workshops with local residents and other stakeholders are organized in each of the municipalities.

For each municipality, a student in architecture will provide a design solution for urban area, housing design or renovation of existing building. The designs for a specific need are based on local context and identified together with local actors. Five students in architecture will do field work on-site and participate in common workshops monthly in the Department of Architecture. The comparative analyses of the municipalities and the evaluation of population projections are carried out by the researchers at the Sotera Institute. The outcome of the project will be alternative development models for shrinking municipalities. The models will help municipalities to outline local long-term objectives for the strategic urban development plan in order to consolidate the municipal structure.

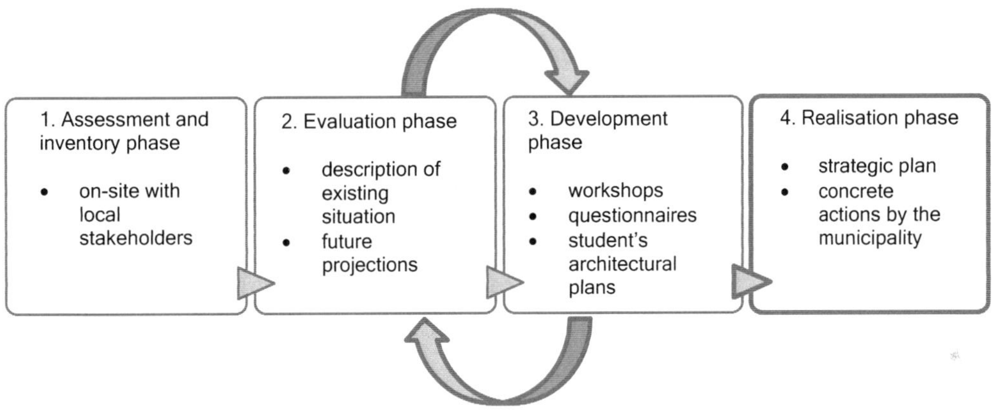

Figure 1. The development process in the project

An online questionnaire targeted to residents and workshops held in each of the municipality will provide information on the use of local services and the access to the services. Moreover, the residents are asked to report the most important places for social activities and meaningful spaces in their own community. Furthermore, they are asked to report the most important future development tasks in the municipality. The survey was launched during the spring 2019 and the final results will be published in the autumn 2019.

5. Preliminary results

All municipalities participating in the research and development project have an urbanized centre or several small centres. Population in each of the municipalities is projected to decline and get older (table 1). The challenge is to unify and to densify the urban structure in these shrinking municipalities in order to provide a good quality living environment for the residents also in the future. The preliminary results of the analyses target to identify the major development task in each municipality. The final result of project will provide a model and a process for small municipalities to improve their services. The model will help municipalities to outline an urban planning strategy based on local development actions.

Table 1. Total population and population 74 years old and older for 2018 and projections for 2030.

	total population (2018)	total population (2030)	population 74 years old and older (2018)	population 74 years old and older (2030)
Harjavalta	7 125	6 880	980	1 450
Keuruu	9 740	8 970	1 390	2 030
Kurikka	21 139	20 040	2 610	3 870
Pertunmaa	1 750	1 580	280	360
Suupohja	22 240	20 880	2 810	4 020

5.1. Harjavalta

Harjavalta municipality centre is divided between two distinct areas; a) the commercial area near the railway station and b) the public service area near the river. Some of the buildings in the commercial centre need renovation. Moreover, some of the commercial spaces are currently unoccupied. The need for good quality dwellings in the centre of the municipality has been recognised. The main challenge is to densify the commercial part of the municipality and replace old buildings with new housing development. However, the land and property ownership is one of the factors slowing down the urban development process. Moreover, the aim is to connect the areas of public and commercial services together. The project targets both short-term and long-term plans for the improvement of the urban environment that can be realised in phases. Prompt actions to improve the walking environment and green areas are possible. Moreover, the connection from the railway station to the centre would enhance the access to local services and pedestrian and bicycle lanes would connect the recreational area by the river.

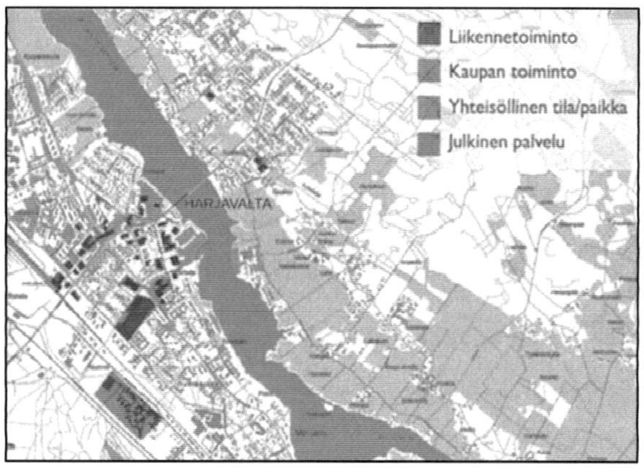

Figure 2. Commercial services and public services are located in different parts of the centre (Linnea Löytönen, 2018).

5.2. Keuruu

The historical centre of Keuruu as well as the cultural services are located near the railway station. The commercial small retail services are situated along the main street, which connects also to the public services in the north. However, a new large shopping centre has been built on the west of the old municipal centre. The challenge is to reinforce the quality and the role of the main street. This can be

IOP Conf. Series: Earth and Environmental Science **297** (2019) 012008 doi:10.1088/1755-1315/297/1/012008

done by renovating the existing buildings and, for example, by creating a new market square or a shared space near the town hall. The potential of the lake and open views from the main street to the lake can be reinforced. Moreover, the municipality is planning new housing services for the older population near the existing health care centre in the proximity of the town hall. Accessible pedestrian network along the main street and by the lake with street lighting may enhance mobility and wellbeing of all residents.

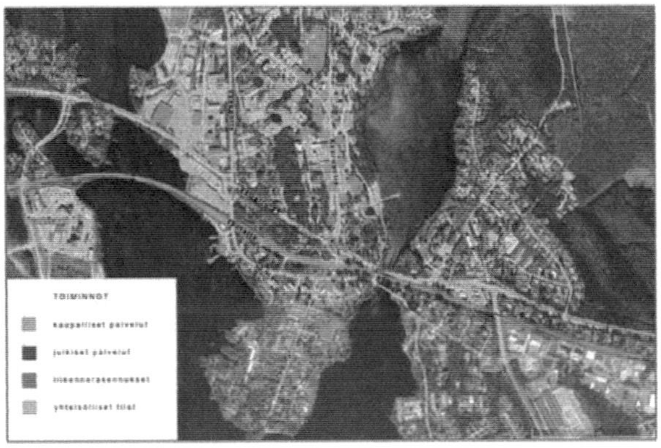

Figure 3. Retail shops and spaces for social activities are located along the main street (Susanna Leppänen, 2018).

5.3. Kurikka

Kurikka has recently gone through fusion of three small municipalities Kurikka and Jurva in 2009, Jalasjärvi in 2016. The new municipality has currently three urban centres with overlapping services in each. The student design project is focusing on the main municipal centre, Kurikka. The aim is to outline a comprehensive development plan for the municipal centre. In 2010, the municipality made an urban development plan in collaboration with the consulting company SITO. The proposal contained several concrete actions which were targeting to improve the quality and safety of streets and public spaces. As result, the municipality realized a Shared Space zone where pedestrians, bicyclists and cars share the same urban space. However, the Shared Space and other public outdoor spaces in the centre of the municipality would profit of some improvements. Moreover, the project will target to enhance the green environment and walking friendly environment in the municipal centre. The project includes densification of the centre as well as housing development suitable also for older population. Moreover, the project is aiming to find the reasons why the previous urban development plan has not been implemented.

Figure 4. The main municipal centre (Nikolay Bobrov, 2018)

5.4. Pertunmaa

The municipality of Pertunmaa has currently two centres. The public administration, health care and cultural services as well as housing area are located in the old municipal centre by a lake. However, the commercial services and pharmacy as well as industrial companies with employment opportunities are situated near the highway at 10 km distance from the old centre. To access to the commercial services a private car is needed as there are no local public transportation services available. The municipality has many second homes. In the summer, the number of resident doubles from 1750 to approximately 3700 inhabitants. Since 2018 in the summer, an open market situated by the beautiful lake has been opened once a week. It has been an attraction for the local residents as well as for the second home owners. The aim is to strengthen the local community by creating a common living room for social activities in the municipal centre. The student will provide three alternatives, how to implement the project in a) a new building, b) in a refurbished building or c) by building an extension to an existing facility. A multi-use flexible building with an open design will be planned near the open market. Amongst other things, the building will include spaces for remote work for local residents and second-home-owners.

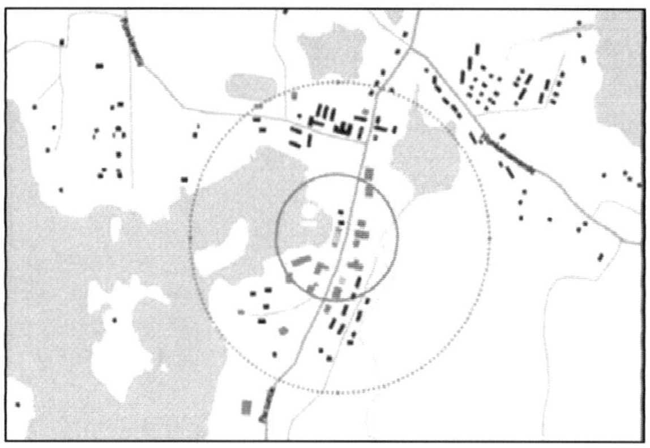

Figure 5. The public services and the market place are located in the centre of the municipality on a beautiful spot by the lake (Pirita Nykänen, 2018)

5.5. Suupohja

Suupohja is the service provider of the joint municipal authority formed by four municipalities Isojoki, Karijoki, Kauhajoki and Teuva. The overall aim is to enhance the vitality of each of the small centres but provide service jointly. Three of the municipalities are small: Isojoki and Karijoki both have approximately 2000 inhabitants and Teuva more than 5000. Kauhajoki is the biggest of the municipalities and has less than 14 000 inhabitants. The distances between these centres vary from 20km to 40km. The current public transportation services are poor and the connection between the municipalities are few. The service provider is targeting to create a network of services between the four centres, where each municipality will provide a service to all the four municipalities. The network will also include mobile and remote services. Moreover, a local transportation service based on private people may be promoted. This development work will be done together with local stakeholders and residents.

Figure 6. The public and commercial services in Kauhajoki are located along the main street (Rosa Väisänen, 2018)

6. Conclusions

The collaboration with municipal actors, local stakeholders and academic researchers may enable to enhance a resilient urban structure, with services and housing in the remote areas. The collaboration will highlight the strengths in each municipality and promote sustainable built environment in the small municipalities. The aim is to keep the remote areas attractive to live in also in the future. Many remote municipalities have a dispersed urban structure, they lack a clear municipal centre, and have no public transportation. Moreover, missing sidewalks and streetlights may decrease residents' physical and social activity opportunities, especially in winter. Densification of the municipal centre with housing, green environment and commercial and leisure activities may enhance quality of living in small municipalities. Access to local services and transportation increase mobility and daily coping. Moreover, an attractive space for social and cultural life, a common living room for residents may promote the vitality of these areas. Each of the municipalities have taken some isolated actions to improve the environment, however a vision and a comprehensive long term urban plan of the municipal centre is missing.

Second home owners are potential strength for small municipalities. However, according to Adamiak and al. [11], they do not use a great variety of local services and only rarely take part in local community activities. The local economy may benefit of the second home owners. On the other hand, they also bring instability in economic activity due to seasonality. The nature and local culture attract the second home owners and tourists. The potential growth of small municipalities may be in small businesses related, for example, to organic food provision and capacity to purchase food directly from producers. The project enables municipalities to generate a plan for sustainable change and knowledge how to implement the plan. The contrast of remote areas from busy urban environment and activities related to the nature are potential strengths for municipalities. However, further research is needed to better understand the community resources and potentials to enhance residents' wellbeing.

References

[1] Verma I (edit.), Kurkela T, Sanaksenaho P, Suominen J, Taegen J and Vauramo E 2017 *Palvelukortteli, Konseptin kuvaus ja soveltaminen erilaisiin taajamiin* (Helsinki: Ympäristöministeriö)

[2] Taegen J (edit.), Sanaksenaho P and Verma I 2018 *Taajamasta palvelutaajamaksi Muuttuva yhteiskunta – Muuttuvat palvelut –tutkimushanke* (Tampere: Asumisen rahoitus- ja kehittämiskeskus, ARA)

[3] Lehtovaara M (edit.) 2019 *Täydentäen toimivaa – TÄYTYY. Loppuraportti* (Porvoo: Porvoon kaupunki) (in publication)

[4] Hollander J B 2009 *Planning shrinking cities* 72(2009)223 (DocId: 03059006)

[5] Aro T 2016 *Elinvoimaiset kunnat nyt ja tulevaisuudessa* Data & Analytics (https://www.slideshare.net/TimoAro/elinvoimaiset-ja-vetovoimaiset-kunnat-nyt-ja-lhitulevaisuudessa)

[6] Helminen V, Vesala S, Rehunen A, Strandell A, Reimi P, and Priha A 2017 *Ikääntyneiden asuinpaikat nyt ja tulevaisuudessa* (Helsinki: Suomen Ympäristökeskus)

[7] Scott M, Gallent N and Gkartzios M 2019 *The Routledge Companion to Rural Planning* (London: Routledge) p 265

[8] Lehtola I 2007 *Maaseudun toimintaympäristön muutos ja liikkuminen* (Helsinki: Tiehallinnon selvityksiä 18/2007)

[9] Maaseutupolitiikan yhteistyöryhmä 2004

[10] DART 2012 *Declining, Ageing and Regional Transformation. Final Report* DART project p 196

[11] Adamiak C, Vepsäläinen M, Strandell A, Hiltunen M, Pitkänen K, Hall C.M, Rinne J, Hannonen O, Paloniemi R and Åkerlund U 2015 *Second home tourism. Perceptions of citizens and municipalities on the state and development of second home tourism* (Helsinki: Finnish Environment Institute) p 34

[12] Stadtbaum Ost (http://www.stadtumbauinfo.de/)

[13] Irish Government 2017 *Realising our Rural Potential. Action Plan for Rural Development* (https://actionplan.ruralireland.ie/Rural-Ireland-Action-Plan.pdf)

[14] Sepp V and Veemaa J 2017 *Shrinking regions and innovative solutions: entrepreneurship, employment and the accessibility of services* (Tartu: Ülikool Rake)

[15] IBA Stadtumbau. https://open-iba.de/en/geschichte/2002-2010-iba-stadtumbau/aschersleben/

[16] Friedman A 2018 *Fundamentals of Sustainable Urban Renewal in Small and Mid-Sized Towns* (Springer international publishing AG)

[17] Ahola H 2018 Asuntomarkkinakatsaus Vuokra-asunnot 1/2018 (The Housing Finance and Development Centre of Finland (ARA)

[18] Haase A, Hospers G-J, Pekelsma S and Rink D 2012 *Shrinking areas front-runners in innovative citizen participation* (European Urban Knowledge Network)

IOP Conf. Series: Earth and Environmental Science **297** (2019) 012009 doi:10.1088/1755-1315/297/1/012009

Processing of use cases for the development of an open platform to support the smart urban development

Alice Schweigkofler[1], Camilla Follini[1], Dieter Steiner[1], Katrien Romagnoli[1], Michael Riedl[1] and Dominik T Matt[1,2]

[1] Fraunhofer Italia Research, Bolzano, Italy
[2] Free University of Bozen-Bolzano, Italy

alice.schweigkofler@fraunhofer.it

Abstract. The topic of smart cities is assuming an increasing relevance, both locally and nationally. The key challenge for smart cities and communities is to significantly increase the overall quality of life in urban areas, for the benefit of their inhabitants and economic activities. This process occurs both through the improvement of traditional utility-grids and services, and through the implementation of digital sensor- and communication technologies in the urban ecosystem. In this scenario, the adoption of enabling tools to support the smart urban development, appears crucial to manage processes and data information flows, as well as to permit better communication and collaboration among the involved stakeholders.
This paper describes the development methodology of an open digital platform for smart cities. This platform will be an enabling tool for the smart urban development of a city i.e. will be able to support and encourage the growth of new services for - and with - citizens, companies, tourists, and public administrations. The concept derives from the ongoing research project OPENIoT4SmartCities, funded by the Operational Programme European Regional Development Fund ERDF 2014-2020 (CUP: B11B17000720008) Südtirol/Alto Adige. The main objective is to realise a product-prototype software, for the management of smart services in cities, using IoT.
The platform has been designed based on the specific service needs of Merano, a city in South Tyrol, Italy. The activities regard the identification of concrete use cases in the field of public illumination and water management. In particular, this paper will present the results of previous analysis, consultation meetings and workshops that have been conducted by the research team together with the public administration and local service providers of the city of Merano.

1. Introduction

The smart city concept finds itself in the center of an intense debate already for several years and is giving life to many initiatives. In fact, cities represent the greatest scenarios of industrial and economic activity, moreover, the growth of urban centers is constantly increasing, bringing out the urgent need of complex management tools. The smart use of information and communication technologies, enabling the connection of buildings, devices and people with each other, is the essential starting point for systems of urban innovation within all areas of urban life [1]. Considering this, the question about the most effective way to lead the urban transformation towards a smart city has not been answered yet. To give an answer to this problem, the fundamental idea of the OPENIoT project is to develop an open platform for smart cities to support the smart urban development. This platform should be a flexible and scalable operating tool, able to offer and handle different smart services based on well-defined use cases as described in the methodology section. The adoption of the platform should boost the development of new services for and with citizens, businesses, tourists and public administrations.

The aim of this paper is to present the methodology adopted for the processing of use cases for the development of the IoT platform. Since the definition of a methodology cannot be realized without the analysis of the specific reality in which it will be applied, a real city has been considered. Therefore, this work gives a look at Merano, a city located in the mountainous region of South Tyrol, Italy, which presents unique challenges and opportunities. Section 2 presents case studies of framework for the definition of a clear roadmap to the smart city, especially in the field of public illumination and water management. In Section 3, the methodology used for the development of the use cases is described. Section 4 presents the first two steps of the methodology applied to public illumination and water management in Merano and the two final implementation steps are described in Section 5. Results are presented in section 6. Section 7 concludes the paper with the main recommendations and the future avenues of the work.

2. Related works/Case studies

Although the concept of a "smart city", which generally foresees the vision of cities going through a technology-driven evolution, has been researched thoroughly in the past decade, its term remains vaguely defined causing confusion for all stakeholders. In particular, the buzzword "smart" is used not only on the urban scale, but also interchangeably for elements that are part of the city or connected to it (i.e. smart buildings, smart meters, etc.), thus creating possibilities of misinterpretation [2]. Moreover, technology may address multiple urban challenges and customers, in the spatial, social, and economic dimension, increasing the number of solutions that can be included in this field. Therefore, it is not always evident in which way solutions contribute to make the city "smarter" and concretely improve the citizens' quality of life [2]. For these reasons, research has been done to develop bottom-up frameworks to guide the digital evolution in a city. These frameworks usually start from the actual needs derived from stakeholders' requirements, which are based on the assessment of the city as-is. The latter is the starting point of several methodologies whose final aim is the creation of a smart city roadmap. One of the most notable examples is the "City Lab" model, part of Fraunhofer's project Morgenstadt [3], which builds on the evaluation of the current state of the city on several sectors with different kinds of interdependencies and interrelations. The evaluation is elaborated with the help of relevant stakeholders, as the basis to prioritize the fields which are more urgent or relevant to intervene [4]. The assessment of the city as-is may vary in the level of detail. An interesting example is offered in [5], which not only defines city sectors, but also city layers, from the physical infrastructure to the digital interface oriented to the end users. Finally, other frameworks include the creation of future visions, made in collaboration with stakeholders, from the beginning, in order to consider not only the present, but also the future, in the early stages of the roadmap [6].

The responsible use of natural resources seems to be one of the most relevant challenges of smart cities. This is especially true for limited resources, such as water. Therefore, several smart city projects focus on water management, with the aim of evaluating consumption, demand, and of monitoring possible leakages. The diffusion of IoT-enabled technology, such as smart meters, made it possible to drastically change the water management market, enabling a real time collection of data, thus granting a finer granularity of information. Moreover, wireless technologies facilitate a unified open data architecture, overcoming the logic, diffused in today's market, of point solutions to point problems, offered by different actors (such as management of water pressure, leakages, etc.) [6]. New installation of smart meters requires a long-term, holistic strategy, as real-time monitoring generates a huge amount of data. For these reasons, smart metering is often associated with the use of advanced data processing techniques, such as machine learning and data mining, to disaggregate data and provide them to customers [9]. Most importantly, data must be meaningfully delivered to end-users in a readable way; this is usually done through platforms, which may be dedicated to selected users (f. i. Autoflow© [9]) or linked to a city and open to the general public (f. i. Bristol is Open [10]). However, the latter alternative can be only used in solutions concerning information that does not carry any sensitive data and is interesting for a broad public, such as the monitoring of the water quality rather than personal use [10].

For the same reasons as water management, public illumination is a similarly recurring theme in smart city projects. Urban lighting poles offer a very convenient infrastructure which is both homogeneously scattered across the city and connected to electricity. Therefore, they are ideal carriers of IoT sensors to both monitor their surrounding streets and environment. Given this, an interesting research trend concerns the reduction of energy consumption by reacting to the collected data. This has been enabled by the advancements in LED technology, which not only remarkably reduces consumes, but also permits faster switching and better intensity control [11]. Depending on the system architecture, communication protocol, and usage scenario, an intelligent lighting system based on motion sensors can help reduce power consumption up to 33,33 % [12]. Apart from energy savings, thanks to the characteristic listed above, public lighting has been, and can be, exploited in many ways to create a smart city environment. To this end, experiments involving citizens and stakeholders from the city administrations in a co-creation smart city project have been conducted, to promote a participated vision of possible technologies and services [13]. Collected ideas ranged from light-based information for drivers to neighbourhood bonding and communication.

3. Methodology

Given the issue of ambiguity of the concept of a smart city [2] presented in the previous section, which forces to consider various aspects of a city holistically, it is clear that the successful transformation of an urban area needs the operational support of a systematic, flexible and complex tool. For the development of this tool it is necessary to define a specific methodology.

After the evaluation of successful examples from the state of the art in methodologies for smart city development, both service- [4][6] and technology-driven [5][7], it has been chosen to follow a bottom-up approach. This approach follows a specific process:

- starting from consideration of the end customer;
- definition of the service characteristics,
- specification of the data workflow management;
- choice and configuration of the required technology and infrastructure.

The next sections describe the elaboration and implementation of the first two use cases for the city of Merano, which are public street lighting and water management. In this context, use case means the precise and unambiguous specification of an urban service such as public illumination or water management, through the identification of different smart sub-services. Each use case constitutes a service development scenario and his elementary utilization by users provides the explicit requirements for the development of the platform.

4. Processing of use cases

Two workshops were conducted based on the results obtained by Fraunhofer Italia from an analysis of the city's digitization needs in the project MeranoSmart [14]. These workshops were the first of a series, which will provide, at the end of the ongoing project, a broad overview of use cases for the entire city.

4.1. Workshops

The stakeholders that took part in both working groups were mainly representatives of the public administration or belonged to the service provider companies of the city of Merano. They were selected for their knowledge of the specific features of the territory in order to realistically identify priorities of interventions. Regardless of the subject, the workshops lasted three hours, and had a unique structure, as described below. About 7-8 people were involved to each round. A first introductory phase addressed the theme of the workshop, the mutual presentation of the participants and the general description of the methodology. The results already obtained in the project 'MeranoSmart', in relation to the subject matter, were deliberately omitted to avoid influencing the participants with content information.

4.1.1. Workshop Part A: definition of use cases. Participants were asked to brainstorm about new smart services connected to the subject matter, i.e. illumination for the first and water for the second workshop. The brainstorming was driven by filling canvases (Figure 1) that suggest the identification of different end customers. The main objective was to collect a high quantity of use cases on a theoretical and strategic level. For the workshop, it was fundamental to focus on the city of Merano, without taking into account the feasibility of the idea, that could also have been visionary.

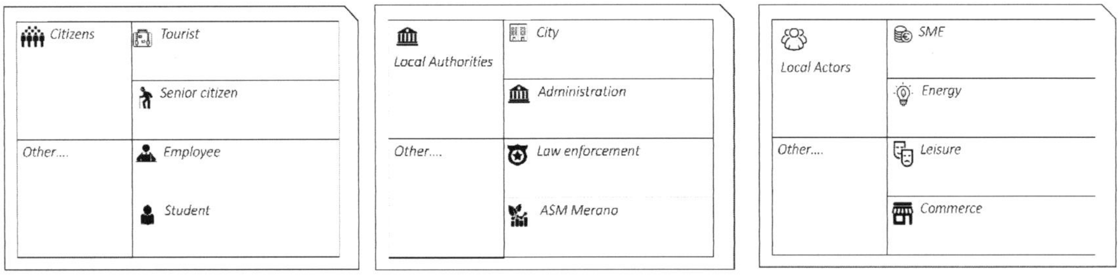

Figure 1. Workshop Part A. Canvas

4.1.2. Workshop Part B: use cases mapping. After a brief reorganization of the first results, the multiple contributions were grouped in a defined number of services. In order to classify them, participants were asked to express an evaluation and assign an indicative value of "impact" and "effort", which have been rounded to a number from 0 (none) to 4 (extreme). Another canvas was used to direct the discussion to the consideration of different aspects such as the impact on citizens, on the local economy, on city authorities, as well as on the technical, financial and maintenance effort.

4.2. Outputs of the workshops

The final output consists in two cost/benefit analysis, which classifies the 13 use cases for the public street lighting (Figure 2) and 9 use cases for the water management (Figure 3), divided in four different categories.

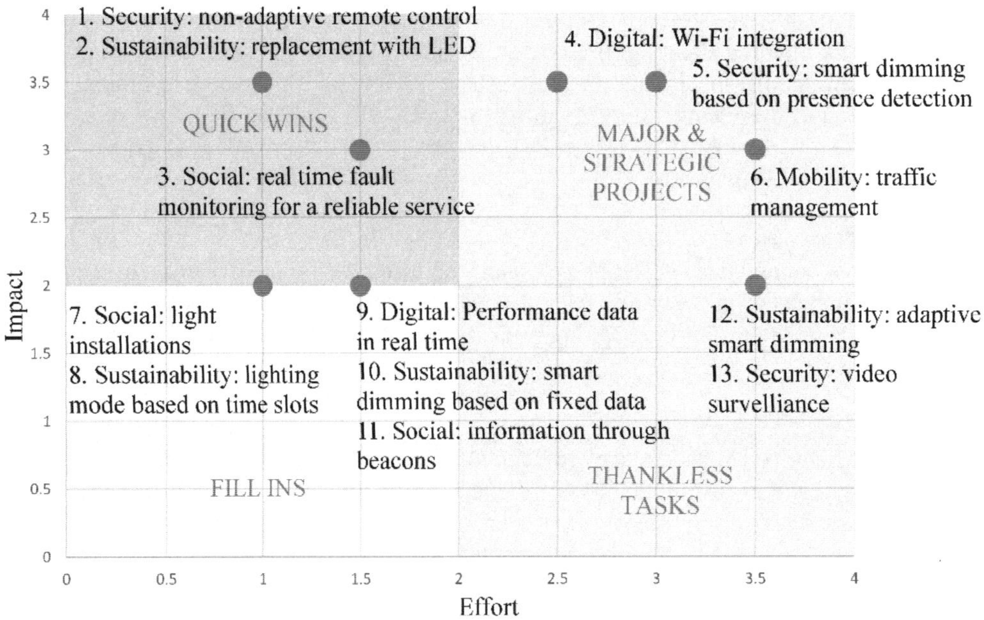

Figure 2. Workshop about public street lighting, 13 outputs.

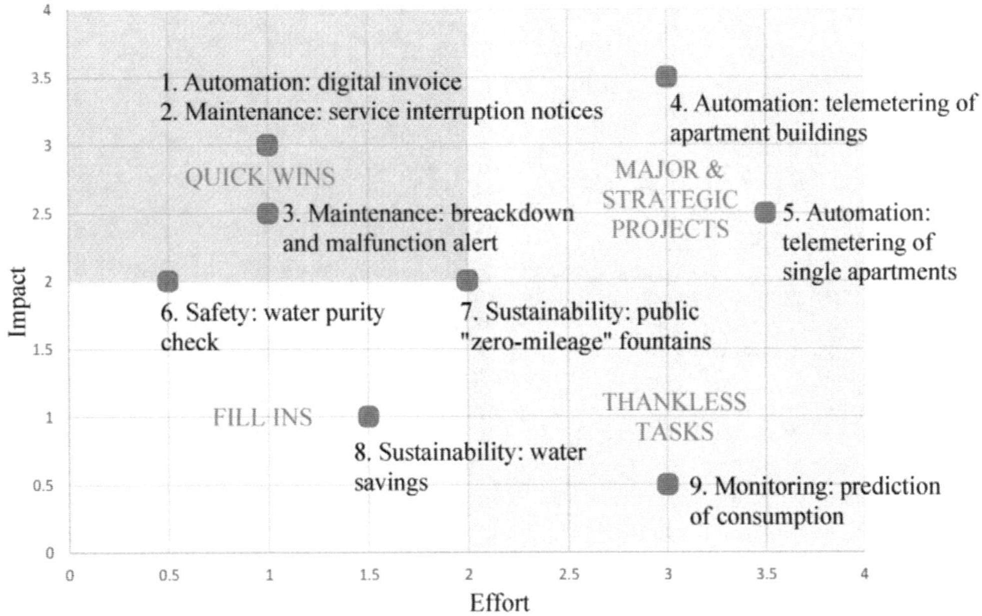

Figure 3. Workshop about water management, 9 outputs

The categories "quick wins", "major and strategic projects", "fill ins", and "thankless tasks" identify the feasibility of the various solutions. The outputs that correspond to "quick wins" are the most suitable for the platform implementation and will be considered first.

5. Implementation

After evaluating the outcome of the workshops, the following actions were implemented.

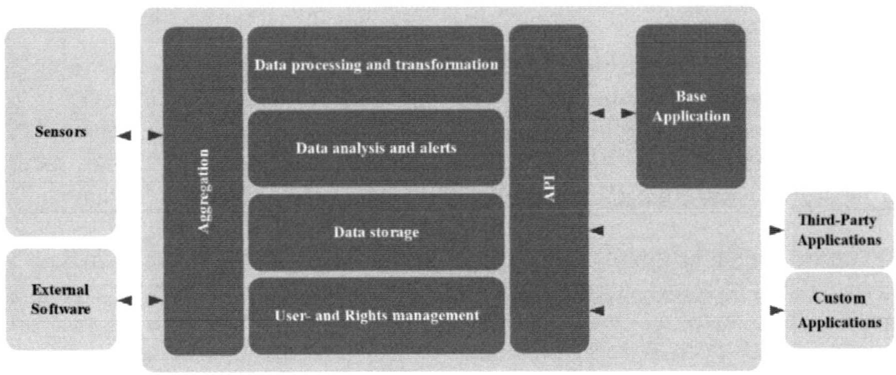

Figure 4. The OpenIoT4SmartCities conceptual platform design.

5.1. Data collection and processing

As core of the project, an open data processing platform has been designed. This platform integrates different sensor systems, wireless protocols and networks and external data providers. The aggregated data can then be stored, processed, analysed, and the results again provided using an open API to other applications, including third-party ones. (cp. Figure 4). The platform was implemented using C# and python for the backend, running on the Azure cloud, which offers a rich environment for IoT applications. The LoRaWAN network is integrated by the use of the Loriot Network Server, the leading enterprise capable tool integrated with the Azure IoT Hub.

Figure 5. LoRaWAN gateway sites. Green denotes a permanent gateway, red current test sites, and blue future permanent gateways.

Figure 6. Smart light installation sites. This figure shows where the LED light points were upgraded with a connected controller.

5.2. Communication Infrastucture

Based on the requirements elaborated in the workshops and the experience provided by the stakeholders, it was chosen to implement a LoRaWAN-based communication system. This allows each sensor to be wireless and low-power. For testing purposes, three gateways connecting the LoRa-network to an IP based network were installed on premises belonging to the municipality of Merano (cp. Figure 5). These gateways cover the current test sites. Nevertheless, due to the mountainous landscape in the city, they will not cover 100% of the city's territory. For a more permanent installation, two of these temporary gateways will be replaced by four permanently installed ones, in- and outside of Merano, providing complete coverage.

5.3. Public street lighting

The public lighting provider in the city of Merano is currently replacing over 6000 light points with modern LED lighting systems. Having already replaced about 40% of the installed lights, the work is expected to be completed in 2023 [4]. The public lighting provider decided to install sensor-ready light fixtures, which include Zhaga Book 18 Luminaire Extension connectors [16], allowing external sensor / controller modules to access a 24V power supply and the luminaires Digital Addressable Lighting Interface (DALI).

For the test site, 100 luminaires in the city (cp. Figure 6) were fitted with a controller interfacing the lamps driver with the LoRaWAN network, allowing monitoring (defect detection, power monitoring, etc.) and control (dimming) of the attached light.

5.4. Water management

For the monitoring of water consumption, residential multi-jet water meters, fitted with static pulsers and a LoRaWAN-based wireless communication module, were installed at five selected test-sites. The installed communication module is external to the meter itself. This allows for a) installation of the communication module up to a 10m distance from the meter, thereby allowing flexible positioning of the antenna and b) independence of the installed meter, as the installed meter can be replaced by a different make/model without replacing the communication module. The downsides of the chosen approach are: a) increased complexity, due to one additional component and wiring effort and b) susceptibility of wires to be damaged by rodents.

6. Results

As shown in Figure 7, a dashboard application was implemented in Microsoft Power BI on top of the OpenIoT4SmartCities platform stack. This dashboard application allowed in real-time monitoring of all connected sensors during the testing phase. The graphical user interface runs in a state-of-the-art web browser and is thereby accessible independently from the used operating system or hardware platform, enabling user access from every kind of computer or smartphone.

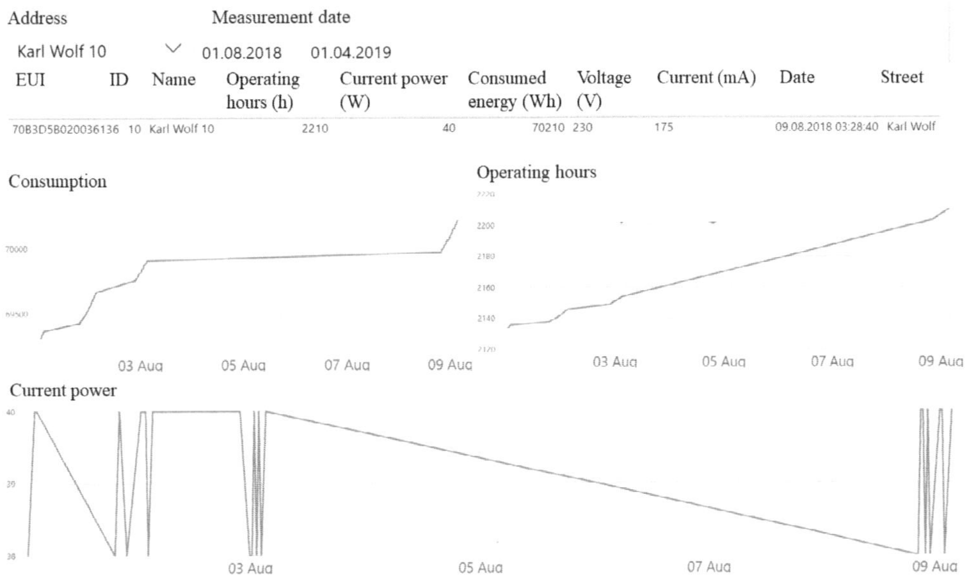

Figure 7. Detail of a specific light point, showing accumulated energy usage, power profile, and hours of operation.

The current version of the application allows the monitoring of operating hours, current power consumption, total energy use, voltage and dimming level of the installed lights, as well as current and total water use for the water meters.

At the moment, the implemented application is merely a showcase of the possibilities offered by the platform. As the implemented system is open by design, the collected data, analysis results and alerts are also available for future third party applications and services.

7. Conclusion and future work

In this research paper, a methodology for the implementation of use cases for a smart city was discussed. The methodology is scalable, meaning that it can be used for all urban areas (e.g. water, lighting, waste, etc.) and for the definition of a wide range of smart services, which allow the development of the OPENIoT platform. What turned out to be the most promising aspect is the bottom-up approach, which allows to collect possible solutions from citizens for the short, medium or long term. The limitation of this approach is its comparability between the different service areas, i.e. the interconnection of services that may have elements in common. Moreover, a future testing phase should include a greater number of elements in order to simulate realistic data analysis and management.

The discussed use cases will be implemented again with more alerts (i.e. for unusual consumption spikes or signalling of faults), and the current dashboard will be replaced by a fully custom-written application in the near future, allowing for even more flexibility and customizability. The city of Merano has the objective of becoming one of the best regional examples of smart city technology. Once a certain degree of robustness is achieved, the public administration of Merano has expressed interest in adopting a pilot version of the platform for validation.

8. Acknowledgements
This paper has been supported by funding from the European Regional Development Fund ERDF 2014-2020 Südtirol/Alto Adige – Axis 1, project OPENIoT4SmartCities FESR1069 CUP B11B17000720008. Thanks go to the City Meran, to the City Councilor for Innovation Diego Zanella, as well as, to the company Systems (Gustav Rechenmacher and Clemens Kuen).

References
[1] Matt D T, Spath D, Braun S, Schlund S and Krause D 2014 Morgenstadt – Urban Production in the City of the Future *Enabling Manufacturing Competitiveness and Economic Sustainability* ed M F Zaeh (Springer International Publishing) pp 13–6
[2] Husár M, Ondrejička V and Varış S C 2017 Smart Cities and the Idea of Smartness in Urban Development – A Critical Review *IOP Conference Series: Materials Science and Engineering* **245** 082008
[3] Fraunhofer Gesellschaft, Morgenstadt - City of the future Initative. URL https://www.morgenstadt.de/en.html (accessed 01.25.19).
[4] Padilla M, Hawxwell T and Wendt W 2016 City Lab Lisbon – Development of a Smart Roadmap for the City of the Future *REAL CORP 2016 Proceedings*
[5] Ernst & Young, Rapporto Smart City Index 2016. URL http://www.comune.bologna.it/iperbole/piancont/archivionov/tabelle_grafici/EYsmartindex/2016-EY-smart-city-index.pdf (accessed 10.10.18).
[6] The British Standards Institution, Smart city framework – Guide to establishing strategies for smart cities and communities. URL https://www.bsigroup.com/en-GB/smart-cities/Smart-Cities-Standards-and-Publication/PAS-181-smart-cities-framework/ (accessed 09.10.18)
[7] Silva B N, Khan M and Han K 2018 Towards sustainable smart cities: A review of trends, architectures, components, and open challenges in smart cities Sustainable Cities and Society 38 697–713
[8] Kulkarni P and Farnham T 2016 Smart City Wireless Connectivity Considerations and Cost Analysis: Lessons Learnt From Smart Water Case Studies *IEEE Access* **4** 660–72
[9] Nguyen K A, Stewart R A, Zhang H, Sahin O and Siriwardene N 2018 Re-engineering traditional urban water management practices with smart metering and informatics *Environmental Modelling & Software* **101** 256–67
[10] Chen Y and Han D 2018 Water quality monitoring in smart city: A pilot project *Automation in Construction* **89** 307–16
[11] Kumar N 2013 Smart and intelligent energy efficient public illumination system with ubiquitous communication for smart city *International Conference on Smart Structures and Systems - ICSSS'13* 2013 IEEE International Conference on Smart Structures and Systems (ICSSS) (Chennai: IEEE) pp 152–7
[12] Daely P T, Reda H T, Satrya G B, Kim J W and Shin S Y 2017 Design of Smart LED Streetlight System for Smart City With Web-Based Management System *IEEE Sensors Journal* **17** 6100–10
[13] Heiskanen O and Acharya K 2017 Envisioning the future of public lighting with citizens for upcoming technologies *The Design Journal* **20** S1782–93
[14] Fraunhofer Italia, MeranSmart – An inclusive and digital city. URL https://www.fraunhofer.it/en/focus/process-engineering-in-construction/meransmart.html (accessed 01.30.19).
[15] City of Merano 12.12.2018 *Öffentliche Beleuchtung: Umrüstung auf LED-Lampen wird fortgesetzt*(https://www.gemeinde.meran.bz.it/de/Oeffentliche_Beleuchtung_Umruestung_auf_LED-Lampen_wird_fortgesetzt)
[16] Zhaga Consortium 2018 *Zhaga Interface Specification* Book 18 Edition 1.0 (www.zhagastandard.org)

IOP Conf. Series: Earth and Environmental Science **297** (2019) 012010 doi:10.1088/1755-1315/297/1/012010

Promoting the energy refurbishment of social housing: the KlimaKit model

P Penna[1], A Schweigkofler[1], R Brozzi[1], C Marcher[1,2], D T Matt[1,2]
[1]Fraunhofer Italia Research, via Volta 13/A, 39100 Bolzano (BZ)
[2]Free University of Bozen-Bolzano, piazza Università 5, 39100 Bolzano (BZ)

paola.penna@fraunhofer.it

Abstract. Despite the well-known need of retrofitting the existing building stock and the availability of a wide range of energy efficient technologies, financial products and public incentives, the private as well as the public sector are struggling to invest in energy efficient solutions for buildings. Designers, companies, financers, and energy suppliers follow their own profit, instead of creating synergies to fulfil what should be the final goal of a retrofitting project: the client's satisfaction. The aim of this paper is to present the methodology developed, within the KlimaKit project [1], to support both the refurbishment process of the buildings belonging to the social housing association of the Autonomous Province of Bolzano and the energy refurbishment market of South Tyrol. The results presented in this paper are part of the ongoing research project "KlimaKit", founded by operational programme European Fund for Regional Development of the Autonomous Province of Bolzano EFRD 2014-2020 – Investments in Growth and Employment.

1. Introduction

The reduction of the energy consumptions related to the existing building stock is an undeniable need. Even if this need is well-known and despite the availability of a wide range of energy efficient technologies and public incentives, the private as well as the public sector are struggling to invest in energy efficient solutions. Some of the main barriers identified are the high refurbishment costs, the lack of information [2], the uncertainty related to the investment payback period [3], the mistrust of stakeholders and the overwhelming possibilities offered by the market [4].

Given that the major part of the social housing was built before 1990, promoting the renovation of those buildings represents an opportunity for the public administration to reduce the energy consumption of building sector and to be an inspiring example for the private sector. For this reasons, several studies [5, 6, 7, 8] have been focused on the energy refurbishment of the social housing stock. The social housing association of the Autonomous Province of Bolzano (IPES – Instituto per l'Edilizia Sociale della Provincia Autonoma di Bolzano) is already involved in some important renovation processes, such as the European Sinfonia Project [9], which aims at testing urban-scale strategies for improving the energy efficiency of social housing stock [10, 11]. However, according with the data received from IPES (figure 1), most of their buildings still present high energy consumptions and need to be renovated. Since social housing buildings are characterised by a typological and technological homogeneity [5, 6], developing standardized solution seems to be a promising strategy to promote the energy refurbishment.

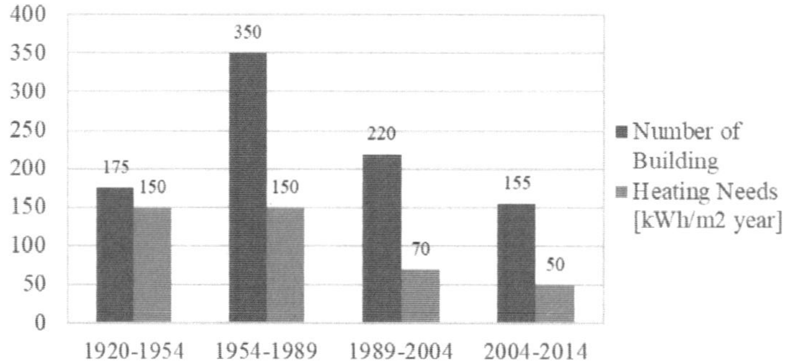

Figure 1 – Energy consumption and number of buildings from the stock of IPES, Source: IPES

This approach can drive and foster the AEC and energy sector to collaborate in the development of integrated solutions for energy retrofit of residential buildings, simplifying the decision-making process and creating synergies among stakeholders. Furthermore, the creation of these packages of Energy Efficiency Measures (EEMs), developed basing on representative buildings of the social housing stock, can take advantage of synergies between different EEMs [2, 12]. Another aspect that should be considered is the cost-effectiveness of the packages of EEMs. It should be calculated using cost/benefit analysis [2, 13], evaluating the investment costs for retrofit and the expected energy savings [7]. Especially in the early stage of renovation projects, decision support tools can help in managing this complexity, improving the awareness of decision makers owners. The growing relevance of this tools is evident in literature from the increasing development of new tools [14, 15, 16, 17, 18, 19].

The aim of this paper is to present the methodology developed within the KlimaKit project [1], to support both the refurbishment process of the buildings belonging to IPES and the energy refurbishment market of South Tyrol. The paper is organized as follows. Section 2 draws the framework of South Tyrolean AEC sector, defining strengths and weakness of the regional market. In Section 3 it is described the methodology used to identify the possible solutions for promoting the retrofit project. Section 4 presents the standardized retrofit packages, i.e. integrated out-of-the-box solutions optimized for selected representative buildings, and the development of the decision support tool. The main results and the outputs of the decision support tool are described in Section 5. Section 6 concludes the paper with the main recommendations and the future avenues of the work.

2. State of the art

At European level, buildings and in particularly the housing sector accounts for 24.8% of the total energy consumption [20]. This percentage is equal to 26.1% in Italy, constituting approximately 25% of total CO_2 emissions [21]. The remaining components of energy consumption are related to transport (35.4%), industry (23.2%), services (13%), and agriculture (2.3%). The European Commission [22] has identified the construction sector as the one with the highest potential in terms of energy efficiency while the Intergovernmental Panel on Climate Change [23] has identified it as the sector with the greatest potential for reducing pollution, although several barriers persist [23]. In the EU 2020 strategy, ambitious targets have been set for energy consumption and climate change mitigation. These include the 20% reduction in greenhouse gas emissions compared to the 1990 levels. The current political measures of the 28 countries of the EU should have an energy renovation rate of the existing building stock of about 2.5% by 2020 and 5%-5.5% by 2030 [24]. However, the renovation rates of residential buildings still remain relatively low in several EU countries, about 0.5% and 2.5% per year according to the European Institute for Building Performance and to Fraunhofer Institute [25, 26]. A more recent study [27] confirms that the annual percentage of renovations is around 0.7% in Italy, at least 0.5% in Spain, Poland and Belgium, about 1% in the Netherlands and Lithuania, above 1.5% in Germany, France and Austria, with a peak close to 2.5% in Norway. At regional level in South Tyrol, the current renovation rate is approximately 1% (550 buildings per year) [28]. The objective for 2020 aims at increasing the overall amount of renovated buildings to 1,500 per year, namely equal to 2.5% of the existing housing stock.

Table 1. Residential buildings by construction period (Source: ASTAT).

-1918	1919-45	1946-60	1961-70	1971-80	1981-90	1991-00	2001-05	2006-	Total
15.666	4.914	9.030	11.695	11.403	9.492	10.970	6.155	6.319	85.644

The analysis on local residential buildings, carried out in the framework of KlimaKit project, indicates that approximately 50% of existing buildings have been built before 1970 (table 1), confirming the large potential in energy renovation. Considering this background, further research showed additional challenges and opportunities for stimulating the construction, financial and energy local sectors towards integrated high-quality fair-price solution. On the one hand, it emerges the need of stronger collaboration among stakeholders during the retrofit processes to improve the quality of the intervention (figure 2). On the other hand, the use of standardized out-of-the-box solutions for the building stock are perceived as more desirable to reduce the range of payback times (figure 3).

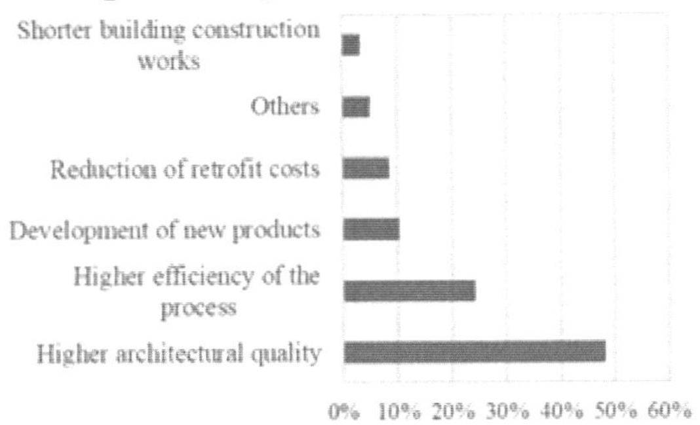

Figure 2 – Benefits of improved collaboration among stakeholders, Source: KlimaKit survey.

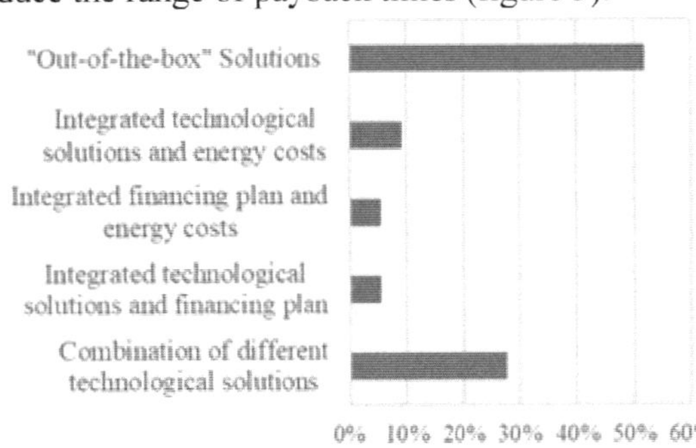

Figure 3 – Definition of "integrated retrofit package", Source: KlimaKit survey.

3. Methodology to define suitable solutions for supporting the retrofit process

Starting from these findings, a series of four workshops were organized, in collaboration with Eurac Research, in order to define solutions supported and validated by the main stakeholders, such as local social housing associations (IPES), professionals, energy suppliers, Energy Service Companies and the local agency for public tender. The workshops aim at defining preliminary concepts, solutions' requirements, potential applications in the regional context and validation of the proposed solutions. The outcomes of the workshops make it clear that to support the retrofit process is necessary to:

- Analyse the social housing building stock to define the most representative buildings;
- Define standardised out-of-the-box packages of EEMs that presents specific energy savings and certain level of indoor comfort;
- Evaluate the cost/benefit analysis of the defined EEMs packages to guarantee a pay-back time as close as possible to 10 years;
- Use the public incentives in order to reduce the payback period and increase the cost-effectiveness of the retrofit solutions;
- Support the decision makers in choosing the optimal out-of-the-box solutions for a building that has to be retrofitted.

4. Development

4.1. Reference buildings

IPES owns about 3500 accommodations in Merano that present similar building typology and construction technologies according to the construction period. Therefore, it was decided to identify a series of "reference buildings", representative of the building heritage. The results of this analysis were detailed in one of the working phases of the project by Eurac Research [30].

Based on this assumption, five reference buildings were selected within the entire social housing stock of Merano for developing standardized out-of-the-box packages composed by different EEMs. The characteristics of the selected buildings were summarized on table 2. The reference buildings considered for the definition of retrofit packages are the ones with poorest energy performance level (labelled E, F or G according to the CasaClima protocol [31]), in fact they should be the first ones to be renovated since they are the main responsible for high energy consumptions.

Table 2. Main characteristics of the selected reference buildings.

ID	Building Typology		Climatic zone [a]	HDD [b]	Construct. period	Nr. Flats
1	Small multi-family building	Terraced house	E	2894	1976-91	8
2	Large multi-family building	Apartment block	E	2894	1976-91	24
3	Small multi-family building	Apartment block	E	2894	1946-75	33
4	Small multi-family building	Multi-family house	E	2921	1976-91	16
5	Small multi-family building	Mixed	F	3223	1976-91	10

[a] According to the Italian Classification.
[b] Heating Degree Day.

4.2. *The retrofit packages: standardized out-of-the-box solutions*

Based on the reference buildings, as well as the IPES' needs, a set of five different standardized retrofit packages has been defined by Eurac Research [32]. For optimizing the synergies among different EEMs, the proposed solutions concern the improvement of the thermal envelope performance and the replacement of the technical system. The goal was not to develop new technologies, but to encourage companies and industries of the South Tyrolean building sector to identify, integrate and improve existing solutions. Each of the package is structured in three levels of energy performance, each of which are designed for accessing tax deductions or financial incentives at a national or regional level. These levels differ mostly in insulation thicknesses and characteristics of the windows' system (double or triple glazing system with improved PVC-frame) and they present a thermal transmittance of building components and envelope air tightness set according to minimum requirements defined:

- Level 1: at national level for having access to tax deduction;
- Level 2: at regional level for accessing financial incentives;
- Level 3: for achieving Nearly Zero Energy Building (NZEB) target.

While the retrofit of the building envelope is the same, according to the performance level, for all the five packages, different solutions are considered for the technical system. As it is summarized below, the existing heat generator is replaced:

- "BASIC" – with a condensing boiler (only if the existing ones is obsolete);
- "FULL" – with a heat pump coupled with a radiant floor;
- "FLEXI" – with the connection to the district heating network;
- "NZEB" – with a heat pump coupled with a radiant floor and a photovoltaic system.

The fifth package refers to the case where it is not possible to retrofit the entire building, but just a single apartment with a non-centralized technical system:

- "SINGLE" – Replacement of windows, mechanical ventilation system, replacement of the existing boiler with a new ones coupled with a radiant floor.

4.3. *Energy simulations and buildings' database*

The energy analysis, reported in this section, has been carried out by Eurac Research. The retrofit packages have been applied to the five selected reference building, for defining the impact of different

EEMs on the overall energy consumptions. The calculation of the energy performance has been evaluated by means of a dynamic simulation software, TRNSYS © [33]. The results have been collected in a database, which contains the pre- and post-retrofit information. For each building, most of the pre-retrofit information, such as year of construction, climate zone, floor area, roof typology, heating generation system typology etc., was collected from the architectural drawings provided by IPES and through the photo documentation. The energy consumption pre-retrofit has been calculated with TRNSYS, then energy model has been validated with the data from real energy consumptions. The impact of the retrofit packages has been evaluated in terms of energy savings and CO_2 emissions.

4.4. Retrofit investment costs

The costs for each EEM have been evaluated based on a regional costs' database (Regional Price List of the autonomous Province of Bolzano). To automatically provide the investment costs for each package, an analysis tool was developed with MS Excel. This tool connects the necessary information of the building's database, such as surfaces (i.e. windows, roof, floor, walls) or power of technical system, with the costs and, according to the reference building, it automatically provides the cost per square meter for each retrofit package, according to the performance level. The costs for retrofitting the envelope and the technical systems have been evaluated separately. Regarding the envelope, the following costs have been considered:

- External insulation of the walls with expanded polystyrene, scaffolding, removal and disposal of existing insulation (if older than 30 years);
- External insulation of the roof, according to the roof typologies (pitched roof, flat roof or ceiling toward unheated space), removal and disposal of existing insulation (if older than 30 years);
- External insulation of first deck toward unheated space with mineral wool, removal and disposal of existing insulation (if older than 30 years);
- Substitution of the windows system with double or triple glazing system with improved PVC-frame, including removal and disposal of the old ones.

The mechanical ventilations costs are evaluated based on the number of building units, considering an application of two mechanical ventilation for each ones. The costs for renovating the technical system has been evaluated according to power need for heating and domestic hot water and according to the chosen technical system, i.e. condensing boiler, heat pump coupled with radiant floor or connection to the district heating network. The technical system costs include the removal of the existing system, the mounting and putting into service of the new ones. The costs for the photovoltaic system have been estimated based on the installed peak power evaluated based on the reference building.

4.5. Economic analysis

The economic analysis of the defined retrofit packages has been calculated based on Net Present Value (NPV), according to the comparative framework methodology of cost-optimal level, proposed by the EU 244/2012 [34]. The NPV allows to evaluate different cash flows, occurring at a different time, by means of the discount rate. For the calculation of the NPV, the investment cost for retrofit, the annual energy costs (considering different energy carriers and rise rates of the energy prize across the time) and different public incentives for retrofits have been considered. Table 3 summarizes the data considered for the economic analysis. The maintenance and replacement costs and the residual value were not taken into account, because the purpose of the analysis was just the evaluation of the possibility to pay back the cost of retrofit interventions with the energy savings. The NPV has been calculated considering the investment costs as an outflow (with a negative value) and considering the amount of energy costs savings as an income (with a positive value). Therefore, the year when the NPV assumes a value equal to zero, corresponds to the time when the retrofit costs will be paid back with the energy savings. The economic analysis has been also integrated into the Excel tool for defining the investment costs. Therefore, for each reference building, according to the cost of each retrofit package and the energy savings reported in the database, it is possible to automatically evaluate the payback time.

Table 3. Data considered for the economic analysis.

Natural gas cost	0,08 €/kWh [a]
Electricity cost	0,20 €/kWh [b]
District heating cost	0,11 €/kWh [a]
Annual rate of increase natural gas price	2,0% [c]
Annual rate of increase electricity pice	3,2% [c]
Real interest rate	3,0% [d]

[a] Centro Tutela consumatori utenti Alto Adige (Situazione 31 marzo 2018).
[b] ARERA Autorità di Regolazione per Energia Reti e Ambienti.
[c] European Commission, [35].
[d] European Commission [34].

5. Results

Through the analysis of the impact of each standardized retrofit package on each reference building, it was possible to create an information database with the calculation results. Table 4 reports the data for the reference building nr. 2 (Large multifamily building). Looking at the payback time of the packages "Full" and "Nzeb", it emerges the importance of using public incentives as means to increase the affordability of the retrofit. Answering a series of questions related to a specific building belonging to the local housing stock, the tool identifies the most similar reference building, recommending the most desirable retrofit packages. The identification of similar reference building is operated according to a series of drivers such as building typology, year of construction and climatic context. Finally, the tool provides a technical estimation of the intervention costs in the very first planning phases.

Table 4. Impact of the standardized retrofit packages on the reference building nr.2.

Package	Level of performance	Reduction of energy costs [%]	Reduction of CO_2 equiv. emissions [%]	Retrofit costs for envelope [€/m^2]	Retrofit costs for technical system [€/m^2]	Payback time without incentives [years]
BASIC	L1	50%	50%	€ 146,16	€ 2,90	15
	L2	72%	72%	€ 166,79	€ 2,69	14
	L3	72%	78%	€ 198,92	€ 35,72	18
FULL	L1	61%	94%	€ 146,16	€ 172,35	26
	L2	82%	97%	€ 166,79	€ 162,79	22
	L3	77%	97%	€ 198,92	€ 181,48	26
FLEXI	L1	59%	89%	€ 146,16	€ 3,61	15
	L2	77%	94%	€ 166,79	€ 2,81	14
	L3	75%	94%	€ 198,92	€ 35,28	18
NZEB	L1	73%	96%	€ 146,16	€ 181,02	25
	L2	88%	98%	€ 166,79	€ 168,96	22
	L3	82%	97%	€ 198,92	€ 185,32	26

The questions concern:

- The general features of the building, such as location, year of construction, number of flats and floor, roof typology etc., which allows to identify the most similar reference building;
- The retrofit drivers (i.e. if the roof has been already insulated in the last 15 years or if the existing boiler has been already substituted) which allows to identify the EEMs that should be applied to the building envelope;

- Some information on technical system, such as energy carrier, efficiency label of the building envelope, surfaces available for PV panels etc.

This Analysis indicates for a generic building, which should be the most suitable retrofit package and the EEMs that should be applied. Furthermore, it provides a valid range of information regarding the expected energy savings (in terms of energy costs reduction), the CO_2 equivalent emissions, the retrofit costs for the envelope and for the technical system and, then, the payback time.

6. Conclusion

The paper presents some outcomes of the KlimaKit project. It describes the methodology developed for supporting the social housing association of the Autonomous Province of Bolzano in promoting a refurbishment project. What seems to be a promising strategy to support the refurbishment market of South Tyrol and to improve the collaboration among stakeholders involved in retrofit projects, is the identification of standardized out-of-the-box solutions. Those standardized package, consisting of a combination of different EEMs, are optimized for the social housing stock of Merano and they guarantee specific energy savings, access to the public incentives and, considering the investment costs, a certain payback time. For supporting the decision makers in choosing the optimal package for a building that has to be retrofitted, a decision-making tool has been developed. The tool has been realized with Excel so far, but it is planned to be developed as a free online tool. Considering that the combination of EEMs have been optimized for a specific context and the economic analysis is based on regional costs' database, the results cannot be extended on a national or European level, but the developed approach can be used for other contexts and be applied on a larger scale.

7. Acknowledgements

This paper has been funded by the European Regional Development Fund ERDF 2014- 2020 Südtirol/Alto Adige – Axis 1 Research and Innovation, with the project KlimaKit FESR1018 (CUP: B56J16001740001). The authors would like to thank IPES, the West Technical Office and the Arch. Othmar Neulichedl, as well as the Institute for Renewable Energy of Eurac Research.

References

[1] KlimaKit Project "Drive the change of the energy refurbishment market in South Tyrol. A strategy for social housing associations and public administration", www.KlimaKit.it
[2] Almeida M, Ferreira M, 2018, *Building and Environment* **143** 15-23
[3] International Energy Agency. 2010, Energy Efficiency Governance Handbook. France.
[4] Monteiro C S, Causone F, Cunha S, Pina A, Erba S, 2017, *Energy Procedia*, **134** 442-451
[5] Davoli P, Belpoliti V, Boarin P, Calzolari M, 2014, Innovative methods for a sustainable retrofit of the existing building stock. A cross-path from social housing to the listed heritage, *Techne Journal of technology for architecture and environment*, **15073** 181-198
[6] Belpoliti V, Bizzarri G, 2015, *Energy and Buildings*, **96** 261-271
[7] Agliardi E, Cattani E, Ferrante A, 2018, *Energy and Buildings*, **161** 1-9
[8] Koch C, Lutteman A, Small, Beautiful, Yet Difficult: Energy Plus Renovation in Small Social Housing Companies, *proceedings of World Sustainable Built Environment Conference 2017* Hong Kong 5-7 June, 2017.
[9] European project "Sinfonia - Low Carbon Cities for Better Living" http://www.sinfonia-smartcities.eu/
[10] Toni M, Coccagna M, 2018, The role of IPES social housing in the EU Sinfonia Project for a "Bolzano Smart City", *Techne Journal of technology for architecture and environment*, **Special issue 01** 134-140
[11] DellaValle N, Bisello A, Balest J, 2018, *Energy and Buildings*, **172** 517-524
[12] Bolliger R, Ott W, Almeida M, 2017, Investigation Based on Calculations with Generic Buildings and Case Studies (Annex 56), University of Minho, Portugal, Available on line in http://www.iea-annex56.org/

[13] Jensen P A, Maslesa E, Berg B J, Thuesen G, 2018, *Building and Environment* **143** 130-137
[14] Nielsen A N, Jensen R L, Larsen T S, Nissen S B, 2016, *Building and Environment* **143** 165–181
[15] Malmgren L, Elfborg S, Mjörnell K, 2016, Development of a decision support tool for sustainable renovation – a case study, *Struct. Surv.* **34** (1) 3–11
[16] Jensen P A, Maslesa E, 2015, *Building and Environment* **92** 1–9
[17] Nielsen A N, Larsen S T, Jensen L R, Niessen S B, Decision Making in the Pre-design Stage of Building Renovation Projects, *proceedings of World Sustainable Built Environment Conference 2017* Hong Kong 5-7 June, 2017.
[18] Dalla Mora T, Peron F, Romagnoni P, Almeida M, Ferreira M, 2018, *Energy and Buildings*, **167** 200-215
[19] Gade N A, Larsen S T, Nissen B S, JensenL R, 2018, *Energy and Buildings*, **142** 107-118
[20] Eurostat, 2014, Statistics explained. Finale Energy Consumption EU-28.
[21] Ministero dello Sviluppo Economico, 2016, Relazione annuale sull'efficienza energetica. Risultati conseguiti e obiettivi 2020.
[22] European Commission, 2006, Action Plan for Energy Efficiency: Realising the Potential, *COM(2006)* 545 final.
[23] Intergovernmental Panel on Climate Change, 2007, Climate Change 2007: Mitigation of climate change. *Summary for Policy-Makers*, **10** (5.4).
[24] D'Agostino D, Zangheri P, Castellazzi L, 2017. *Energies*, **10**(1), 117
[25] Buildings Performance Institute Europe, 2011, Europe's buildings under the microscope: A country-by-country review of the energy performance of buildings.
[26] Fraunhofer Institute, 2009, Study on Energy Savings Potentials in EU Member States, Candidate Countries and EEA Countries. *Final Report for the European Commission Directorate - General Energy and Transport*. Karlsruhe, Germany: Fraunhofer Institute for Systems and Innovation Research ISI.
[27] Zebra 2020, 2016, *Nearly Zero Energy Building Strategy 2020 - Strategies for a Nearly Zero-Energy Building Market Transition in the European Union*. D6. 2: Strategies for nZEB market transition on national level.
[28] Autonomous Province of Bolzano, Assessorato per ambiente ed energia, 2017, *Contributi per Favorire l'efficienza Energetica. Con impegno verso Klimaland* [conferenza stampa: 15 Febbraio 2017].
[29] Belleri A M, Di Pasquale C, Maturi L, 2017, Deliverable 3.1.1, Caratterizzazione di edifici di riferimento, driver per il miglioramento prestazionale e requisiti tecnici, *KlimaKit Project*
[30] Casaclima Energy Agency, 2017, Direttiva tecnica edifici esistenti e risanamento, settembre 2017, Bolzano
[31] Belleri A M, Di Pasquale C, Adami J, 2019, A framework for the technical evaluation of residential buildings' energy retrofit, *to be pubblished* in *proceedings of Rehva 13th HVAC World Congress*, 26-29 May 2019, Bucharest, Romania
[32] Solar Energy Laboratory, TRNSYS 17, TRNSYS. A transient system simulationprogram, 2012, http://sel.me.wisc.edu/trnsys
[33] European Commission, 2012, Commission Delegated Regulation (EU) No 244/2012 of 16 January 2012 supplementing Directive 2010/31/EU.
[34] European Commission, 2016, Report from the Commission to the European parliament, the council, the European economic and social committee and the committee of the regions - Energy prices and costs in Europe - *Com (2016)* 769 – 30.11.2016, Brussels

IOP Conf. Series: Earth and Environmental Science **297** (2019) 012011 doi:10.1088/1755-1315/297/1/012011

Advancement of highly environmentally sustainable business management concepts with contexts embedded in construction markets

P Huovinen[1]

[1] Business Management PJH, Light Entrepreneurship with SLP Group Oy, UKKO.fi Service, Kavallintie 1 A 8, 02700 Kauniainen; and Aalto University, Espoo, Finland

E-mail: huovinen.pekka.j@outlook.com

Abstract. Sustainable is defined to be capable of, relating to, or designating forms of human economic activity and culture that do not lead to environmental degradation, especially avoiding the long-term depletion of natural resources. Herein, the aim is to advance the design of business management (BM) concepts with contexts embedded in construction markets along the environmental sustainability dimension. The paper is part of the on-going reviewing of research on construction-related BM. The reviewing has resulted in the identification of 79 construction-related BM concepts published between 1990 and 2017. Thereof, only 16 (20%) authors have designed their concepts along the environmental sustainability dimension. There are 3 (4%) high-sustainability BM concepts including sustainable business goals, offerings, advantages, synergies, competitiveness, processes, organizations, and/or project portfolios, followed by 1 (1%) medium-sustainability and 12 (15%) low-sustainability BM concepts. In the future, the design of high-sustainability BM concepts could be advanced by (i) creating and distributing value in sustainability for all stakeholders, (ii) re-designing strategies and leveraging knowledge about and competences in sustainability through value chains, (iii) making sustainability a priority for decision makers, (iv) empowering middle managers, experts, and employee teams, and (v) redefining businesses and public services as well as collaborating inside and across borders.

1. Introduction

This paper is part of the on-going reviewing of research on construction-related business management (BM). Readily, this reviewer has conducted the two focused reviews on the environmental sustainability as part of 62 construction-related BM concepts published between 1990 and 2009 [1] and as part of 71 such concepts published between 1990 and 2013 [2].

In the same vein, the aim of the paper is to advance the design of BM concepts with contexts embedded in construction markets along the environmental sustainability dimension. The sub-aims include (i) the introduction of the pioneering reviewing of construction-related BM concepts and the overview of 79 construction-related BM concepts published between 1990 and 2017, (ii) the reporting on the conduct and results of the third focused review of the design of these 79 construction-related BM concepts along the environmental sustainability dimension, and (iii) the suggestions for the advancement of the design of high-sustainability, construction-related BM concepts in the future among various concept designers.

2. Reviewing of 79 construction-related BM concepts published between 1990 and 2017

The six review rounds have been carried out in 1999-2003, 2006, 2010-2012, 2014, 2017, and 2018. Cooper's approach and the original limitations have been re-adopted to protect the review validity [3]. The **coherent nature** has been maintained by focusing on research on firms that are based in the OECD countries. Exceptionally, references originating from Singapore and Hong Kong have been included due to these authors' British Commonwealth heritage and interests in international construction. Hart's guidelines have been relied upon [4]. The synthesis and use of the **method for the reviewing of conceptual research**, i.e. ways of searching, browsing, in-/excluding, retrieving, coding, describing, analyzing, and interfering have been reported upon in [5], [6]. The search for eligible BM concepts has been conducted comprehensively within the volumes of 28 construction-related journals published between 1990 and 2017 and those of 47 business administration journals published between 1990 and 2013. Concerning the other formal channels, the degrees of the comprehensiveness of the search have varied markedly [7]. This reviewer will submit the coverage of the channels including the lists of the journals on request.

Construction relatedness includes the ownership, financing, life-cycle, design, implementation, and servicing aspects of capital investments in natural resources usage, energy supply, manufacturing, telecommunications, transportation, infrastructure, real estate, and general building concerns. Herein, the terms "construction" and "construction-related" are used to encompass all kinds of built environments as well as real estate and investment concerns [2].

So far, the reviewing has resulted in the identification of 79 construction-related BM concepts published between 1990 and 2017. Discipline-wise, 37 (47%) BM concepts belong to construction management (CM), 16 (20%) BM concepts belong to industrial/international management and marketing, 15 (19%) BM concepts belong to project management (PM), and 14 (15%) BM concepts belong to real estate ownership and management. Business-wise, 25 (32%) concepts address project-based business, contracting, complex product systems, or engineering, purchasing, and construction (EPC) projects, 26 (33%) construction or building, 14 (18%) real estate management and services, 6 (8%) design and consulting services, 5 (6%) capital investments-based businesses, and 3 (4%) building products supply. Context-wise, the 79 BM concept designers have specified 81 geographically bounded contexts. There are 26 (32%) worldwide, 16 (20%) UK, 13 (16%) US, 7 (9%) Finnish, 5 (6%) Swedish, 3 (4%) generic, 3 (4%) Swiss, 2 (2%) Australian, and 2 (2%) Dutch contexts as well as 1 (1%) Austrian, 1 (1%) German, 1 (1%) Irish, and 1 (1%) Hong Kong –based context. Practice-wise, it seems that the degrees of the applicability of the 79 BM concepts are fairly low, on average. Most of these BM concepts do not contain (inter)national context-specific or business-specific elements. Only 35 (44%) BM concepts are supported with the initial, empirical, case-based evidence on a possible positive causal relationship between the adoption of the focal BM concepts and the highly successful managing of these case companies' businesses with the targeted contexts embedded in the construction markets at the time, respectively. This reviewer will submit a list of 76 references containing these 79 BM concepts on request.

3. Environmental sustainability as part of company management and business management

In general, **Oxford English Dictionary** defines sustainable "to be capable of, relating to, or designating forms of human economic activity and culture that do not lead to environmental degradation, especially avoiding the long-term depletion of natural resources" [8]. In the same vein, **Intergovernmental Panel on Climate Change's (IPCC)** pathways limiting global warming to 1.5°C with no or limited overshoot would require rapid and far-reaching transitions in energy, land, urban and infrastructure (including transport and buildings), and industrial systems. These systems transitions are unprecedented in terms of scale, but not necessarily in terms of speed, and imply deep emissions reductions in all sectors, a wide portfolio of mitigation options, and a significant upscaling of investments in those options [9]. In turn, the **World Economic Forum's** System Initiative on Shaping the Future of Environment and Natural Resource Security aims (i) to improve the effectiveness of the international community's response to environmental challenges, (ii) to support public-private action that accelerates the climate action agenda, improves ocean health, realizes deforestation-free supply chains, enables more water-secure economies and helps unlock a trillion-dollar opportunity in the circular economy, and (iii) to explore how Fourth Industrial Revolution innovations can be harnessed to address environmental issues, including redefining what effective global environmental governance looks like. Responding to these challenges requires breaking down traditional silos and building new forms of cooperation and innovation across the public and private sectors to quickly effect change at scale [10].

Typically, **Boston Consulting Group** is advocating companies to incorporate environmental sustainability into company management in terms of eight best practices, i.e., setting a sustainability vision and ambition, focusing on material issues, designing a right organization to achieve ambition, exploring business model innovation opportunities, developing a clear business case for sustainability, getting the board of directors on board, developing a compelling sustainability value-creation story for investors, and collaborating with multiple stakeholders to drive strategic change [11].

In turn, Battacharya and Polman posit that when it comes to practicing and not just preaching environmental sustainability, many companies struggle, and most flounder in developing and implementing a sustainable business model. They have identified the **six biggest stumbling blocks** and they have offered the practical advice on how to surmount them as follows. (1) Create value for all stakeholders in the ecosystem and view profits as a consequence of such value creation. Implementing a sustainable business model requires executives to engage with the entire organization as well as multiple external stakeholders (such as nongovernmental organizations, shareholders, suppliers, regulators, and competitors), and to balance multiple goals that are sometimes in conflict. To accomplish this task, a business needs to operate in a much more transparent and symbiotic way and strive to make everything it does purpose-driven. (2) Look at the entire value chain and identify the best leverage points for value creation via a sustainable business model. Performance analyses concern both the planet (carbon, water, and waste footprints) and people (taking stock of working conditions). (3) Make sustainability a priority for the board. The challenges of engaging the board involve management's inability to explain the financial impact of sustainability initiatives, a lack of sustainability expertise, other priorities, an orientation toward short-term results, and a focus on shareholder value. To overcome these barriers and engage the board, the first task is to understand and explain the board's oversight role. (4) Gain buy-in from the "undecideds." Typically, senior leadership and young recruits "get it". But there is a big middle layer of "undecideds" that leaders need to bring on board by giving permission, removing barriers, empowering employees, emphasizing economics, and investing in training. (5) Make sustainability part of every employee's day job. This is the only way the sustainable business model can be successfully implemented. In contrast, most managers continue to believe "sustainability is great, but it's someone else's job, and I have more important things to do." This is a huge hurdle that leaders need to overcome, function by function, by training, developing relationships, promoting benefits, and urging responsible consumption and proper disposal, etc. (6) Redefine the competitive space by collaborating. Typical hurdles include different cultures and goals, additional management resources, and fear that, if they collaborate with competitors, they will not be able to differentiate themselves from their peers. Instead, we collectively need to put sustainability in the precompetitive space. Industry collaborations are required to solve complex supply-chain challenges such as deforestation. Collaborations allow companies and people to access expertise and the networks of relationships as well as to benefit from partners' political influence, standard-setting authority, and ability to impact public opinion [12].

Recently, Geradts and Bocken advocate that [business] managers promote **sustainability-focused innovations** in their respective organizations, by (i) having a clear direction and articulating goals to employees, (ii) providing an adequate budget and other resources (including the space, time, and training) to pursue goals, (iii) arranging "room for collaboration" and allowing people to work with other parts of the organization or to partner with suppliers, customers, non-governmental organizations, and other third parties to address gaps in skills and resources, (iv) providing positive reinforcement to employees who get involved in sustainability-oriented innovation projects, and (v) instituting measures of accountability that establish social and environmental value creation as a priority [13].

4. Third focused review and the assessed degrees of the design of 79 construction-related BM concepts along the environmental sustainability dimension

4.1. Original focused review method

For this third focused review, the **four degrees** of environmental sustainability were pre-specified. An author may have designed a construction-related BM concept along the environmental sustainability dimension to [2]:

- a high degree, i.e., environmental sustainability is one of the primary elements of BM such as a business goal, a key attribute of offerings, a competitive advantage, a synergy among strategies, an edge of competitiveness, and/or a key performance indicator of business processes, organizations, project portfolios;
- a medium degree, i.e., environmental sustainability is one of the supportive elements of BM;
- a low degree, i.e., environmental sustainability is only mentioned as a requirement or a tendency in stakeholder behavior, a criterion in decision making, a factor in environmental analyses, and alike, or it is only an implicit part of BM;
- no degree at all, i.e., an author is silent, no single 'thing' is explicitly written along the environmental sustainability dimension in the reference.

The concept-specific assessments have been compiled in the tables. The corresponding sentences, phrases, or single terms have been quoted and the page numbers have been referred to, respectively (see Table 2). This reviewer will submit a complete set of the tables containing the 79 assessments on request.

4.2. Results of the third focused review

This review resulted in the revelation that only 16 (20%) authors have designed their construction-related BM concepts along the environmental sustainability dimension, i.e., there are 3 (4%) high-degree BM concepts, 1 (1%) medium-degree BM concept, and 12 (15%) low-degree BM concepts. In other words, there are 63 (80%) no-degree BM concepts (Tables 1 and 2).

This reviewer knew that there are at minimum the two high-degree construction-related BM concepts that he himself had designed [1], [14]. Ex ante, this reviewer erroneously assumed that some other authors had designed and published their recent high-degree construction-related BM concepts as well.

Table 1. Design of 79 construction-related BM concepts (published between 1990 and 2017) along the environmental sustainability dimension.

Degree of the design of the BM concepts along the environmental sustainability dimension	High-degree concepts		Medium-degree concepts		Low-degree concepts		No-degree concepts		All BM concepts	
	No.	(%)	No.	(%)	No.	(%)	No.	(%)	No.	(%)
Sum	3	(4%)	1	(1%)	12	(15 %)	63	(80%)	79	(100%)

4.3. Protection of the validity of the third focused review against the four biases

Concept Inclusion Bias 1 involves this reviewer perceiving that an author has designed a concept along the environmental sustainability dimension even if this author has not done so. This bias has been minimized by assessing each reference in the same way and quoting only the necessary part of each of the references that contain 16 (20%) construction-related BM concepts designed along the environmental sustainability dimension (Table 1). Future reviewers can test the inter-concept consistency of

Table 2. Three high-degree, one medium-degree, and 12 low-degree, construction-related BM concepts (published between 1990 and 2017) designed along the environmental sustainability dimension.

Author (year)	Construction-related BM concept and its focal context	Assessed degree of environmental sustainability based on key quotations (page number)
Huovinen 2011a [14]	4 environmentally sustainable businesses in 8 arenas	HIGH: 23 high-sustainability drivers into 4 businesses via 23 competitive arenas (6-12)
Huovinen 2011b [1]	High-sustainability BM concept as a 5-element system for managing businesses with contexts in construction markets	HIGH: 1 customizing into offerings and strategies, 2 leveraging into processes, 3 crafting into competitiveness, 4 fusing into a frame, 5 sustainable collaborators (11-13)
Keenan 2016 [15]	Real estate strategy continuum with relations btw sustainability resilient operations, and adaptive capacity for US corporations	HIGH: Paradigm shift (409) and a framework for managing relations btw 8 sustainability applications, climate mitigation implications, resilient operations, adaptive capacity (419).
Chinowsky with Meredith 2000 [16]	7 areas of strategic management, feedback with a competency spectrum and maps in US civil engineering organizations	MEDIUM: Environmentally sensitive core designs and solutions, support strengths (130, 142), environmental engineering as a surface characteristic (146), testing competency (150)
Flanagan 1994 [17]	Successful construction company in the year 2000 (based in the UK)	Low: Consciousness and sustainability are driving forces (312). Energy, traffic, waste, ecology impacts into designs (316).
Veshosky 1994 [18]	Analytical framework for the US design segment	Low: Environmental systems (43) and facilities (44), hazardous waste competency (45).
Lowendahl 1997 [19]	Strategies, resources, 3 phases for professional service firms	Low: Environmental protection as a market (106).
Huovinen 2001 [20]	Competitive strategy in technology-intensive contracting	Low: Solutions' environmental impacts as one of clients' decision criteria (73)
Love et al. 2002 [21]	Model for alliances founded on TQM and supply chains in the contexts of Hong Kong	Low: To take ethical consideration of the social and environmental responsibility in cooperative alliances (12).
Huovinen, Hawk 2003 [22]	Model for building product suppliers to manage their client relationships and collaboration	Low: "A building product's environmental impacts" is one of 11 criteria that clients use for choosing a winning bid (158).
Anderson, Merna 2005 [23]	Business development process in PM services in the UK	Low: Environment management is one of 11 domains of development (175).
Madritsch, Ebinger 2011 [24]	Process model for built environment lifecycle within an organization	Low: Disposal/recycling is defined as the exit from the "built environment" lifecycle (114). Facility condition index is specified (117).
Mutka, Aaltonen 2013 [25]	8-element business model framework for managing a process technology supplier	Low: To provide sustainable life-cycle solutions and guarantee best returns to customers; environment-friendly technologies (170-171).
Brege et al. 2014 [26]	Business model for industrial dwelling building in Sweden	Low: Frame system supplier's climate-proof structural offering as a business model (221).
Then et al. 2014 [27]	Business-real estate (RE)/ FM alignment model	Low: Sustainability is part of social alignment (btw. business and facilities) criterion (87)
Kämpf-Dern, Konkol 2017 [28]	Assessment concept, success factors, and menu for workplace change management	Low: Energy savings is one of performance indicators. Environmental responsibility as one of performance criteria (219, 221).

of inclusion by repeating the assessments, i.e., reading the references and confirming this reviewer's quotations or rejecting some of them and, thus, excluding the same.

Concept Exclusion Bias 2 involves this reviewer perceiving that an author has not designed a concept along the environmental sustainability dimension even if this author has done so. So, 63 "no-degree" assessments indicate that this reviewer did not identify any aspects on environmental sustainability. Future reviewers can test the inter-concept consistency of exclusion by repeating the assessments, i.e., reading the references and confirming the exclusions of 63 concepts or identifying sustainability elements in some of them and, thus, including the same.

After the inclusion, **Degree Assessment Bias 3** is related to this reviewer's reliance on a pre-specified scale of the three analytical degrees instead of a quantitative scale. The 3-degree lens corresponds to the explorative nature of the focused review. This reviewer could assign one of the three degrees to each of 16 BM concepts without hesitation. Future reviewers can request this reviewer to submit the concept-specific quotations and assessments to them and test the inter-concept consistency of degree assignments by reading the references and confirming the same degrees or assessing changes in some of them and justifying such changes with quotations. Or, they could re-specify a scale of degrees, e.g., by dividing each degree into two sub-degrees.

Concept Author-Reviewer Bias 4 is related to a fact that this reviewer has designed 13% or 10 out 79 construction-related BM concepts. Thereof, only 2 concepts have been co-designed. Further, I have assessed that I have designed 2 high-degree, 0 medium-degree, 2 low-degree, and 6 no-degree construction-related BM concepts along the environmental sustainability dimension. Future reviewers can carefully test the inter-concept consistency of the assessment outcomes versus each of the three other biases in the case of my 10 concepts. Other reviewers can come up with some explanations for this reviewer being, so far, one of the two designers of the three high-degree BM concepts. When I could not assess any high-degree BM concepts from among 62 construction-related BM concepts published between 1990 and 2009 during the first focused review [1], I perceived that I had to design these two high-degree BM concepts and, thus, to open the environmental sustainability dimension for future BM concept designers with contexts embedded in construction markets.

4.4. Briefing of three high-sustainability construction-related BM concepts

Huovinen has designed the **high-sustainability managing of four businesses** based on the multiplication of Porter's (1980) five forces framework and the design of the 8-arena framework for capturing complexity of managing businesses in competitive arenas in construction markets, i.e. 1 Uses, 2 Ownership, 3 Life-cycle servicing, 4 Capital investing, 5 Contracting as wholes, 6 Contracting as parts, 7 Prefabrication, and 8 Materials. A high degree of environmentally sustainable BM is achieved by implanting 23 drivers into the four main business types coupled with 23 specified (sub-)arenas. In life-cycle contracting and development businesses (in Arena 4), the drivers include the coupling of object development ideas with sustainability advantages. In design-build contracting businesses (in Arena 5), the drivers include the re-engineering of value chains. In design businesses (in Arena 6), the drivers include the transformations of design firms into long viewers, path dependency breakers, stock programmers, object planners, impact blockers, and impact cause tracers. In supply businesses (in Arenas 6-8), the drivers include the adoptions of cradle-to-cradle certifications, product renewals, and responsibility takings over life-cycles [14].

In addition, Huovinen has designed a **high-sustainability, 5-element business system**, i.e., environmental sustainability has been (i) customized into competitive strategies and offerings so that no/low negative impacts enable a firm to pre-empt or over-satisfy client needs, excel among competitors, and meet its business-specific, high-sustainability goals in the short term, (ii) leveraged into business processes to minimize carbon footprints and, thus, ensure and increase targeted effectiveness across contracts to be won, by managing a firm's business processes and contract tasks as a dynamic, IT-supported matrix where teams play integrative, flexible roles, (iii) crafted into a firm's core competitiveness, based on its high-sustainability foresights that involve business opportunity perceptions, technology foresights, a technology platform, a core competence architecture, a core offering portfolio, and innovation paths, (iv) fused into a business frame, governance, and the preferred ways of firm-market interactions, along an interdependent set of ownership, managerial, legal, financial, venturous, organizational, institutional, social, and sustainable dimensions, and (v) linked to an extended business frame in terms of partnerships, networking, and similar new forms of physical and virtual collaboration with like-minded stakeholders [1].

Keenan has developed the **high-sustainability, conceptual framework for designing sustainable corporate real estate (CRE) strategies** in the case of US multinational companies. Sustainable real estate is defined by investments in technology and design that reduce consumption and promote efficiencies that limit a building's consumption of resources and minimizes its waste to such an extent that the value created and/or money saved justifies the alternative investment. The framework attempts to demonstrate the potential for causal, reciprocal connections between sustainability, adaptation, and resilience. The framework has been explored through a case study of the CRE strategies that Goldman Sachs developed over the course of the consolidation and development of its corporate headquarters. The evidence supports the existence of an applied relationship between sustainability, resilience, and adaptation. Sustainability advanced the adaptive capacity and resilience of the firm. Evolving sustainable processes had transformative impacts in adapting to change in both material and organizational CRE terms. In addition, robust adaptive capacities may promote the diffusion and execution of sustainable practices. The eight environmental sustainability applications within CRE included less paper, less furniture, less power consumption, a consolidated campus plan, workplace standards and an occupancy management system, preventative maintenance and life-cycle management, the integration of design and operations planning, and the organizational integration of facilities and engineered systems [15].

5. Advancement of the future design of high-sustainability construction-related BM concepts

It is herein posited that in the future many dedicated concept designers adopt also the environmental sustainability dimension and incorporate such sustainability to high degrees into BM concepts as a criterion, an element, or an attribute of managing, or a criterion in decision making. Readily, 63 (80%) no-degree BM concepts form a potential platform for advancement, i.e., simply these authors could incorporate high-sustainability elements into their existing BM concepts, respectively. It is suggested that the design of environmentally highly sustainable BM concepts with contexts embedded in construction markets be advanced in the five potential areas (aligning with Battacharya and Polman [12]) as follows.

(1) Create and distribute value in environmental sustainability concurrently for all stakeholders chained within global and local construction markets. Novel, sustainable business models can be designed for coupling internal and external stakeholders into seamless value chains as well as balancing multiple, stakeholder-specific goals and risks. The designers of emerging forms of integrating multiple actors on equal bases can exploit accumulated knowledge about managing constellations, ecosystems, heterarchies, long-term alliances, joint ventures, networks, orchestrations, partnerships, and alike.

(2) Re-design construction-related business models with strategies for leveraging in-depth knowledge about and competences in environmental sustainability through entire value chains. High-sustainability drivers can be implanted into the chains of competitive, value adding arenas embedded in (inter)national construction markets, i.e., the eight Porterian arenas of ownership, uses, life-cycle servicing, impact investing, wholes contracting and consulting, parts engineering and contracting, building systems and products supply, and construction materials supply (applying [14]).

(3) Make environmental sustainability a priority for various decision maker groups within construction markets, i.e., shareholders, financiers, institutions, board members, top management teams, business managers, entrepreneurs, governmental and other public entities, non-governmental organizations, and alike. Internally, sustainable mindsets can be cultivated, adaptive capacities can be enhanced, and resilient assessment and choice processes can be adopted (applying [15].

(4) Empower middle managers, expert task forces, and employee teams in companies and public organizations also along the environmental sustainability dimension. Make environmental sustainability

part of every employee's job. Sustainable business (and public service) models can be successfully implemented in this way. Novel BM concepts can accommodate semi-independent behavior from back ends such as research, development & innovation, competitiveness, core systems, solutions, and services through business processes and support processes up to front ends such as offerings and competitive strategies (applying [1]).

(5) Redefine businesses and public services related to construction markets through collaboration inside and across regional and national borders. Multi-party collaboration can trigger solutions to many wicked environmental problems such as not-yet circular economies, excessive carbon footprints, scarce clean technologies, energy over-consumption, non-shared well-being, severe air pollution, and dirty waters embedded in (inter)national construction markets. Collaboration allows companies, public organizations, and people to access expertise and the networks of relationships as well as to benefit from partners' political influence, standard-setting authority, and ability to impact public opinion.

6. Conclusions

It is herein envisioned that the majority among the owners and managers of firms and businesses with contexts embedded in construction markets across the globe prioritize high environmental sustainability in the future. Among others, this is empirically indicated by Corporate Knights Inc. [29], a leading sustainable business magazine and ranking organization. The Global 100 ranking is based on a rigorous analysis of 7,500 companies with revenues of over USD 1 billion. The Global 100 companies show that doing what is good for the world can also be good for financial performance, i.e., top sustainability performers have tended to also provide better revenue for investors and have greater corporate longevity. The prime example is KONE Corporation of Finland, a global leader in the elevator and escalator industry, ranked as the 43rd most sustainable company in the world. KONE [30] is already a trusted partner for green buildings and sustainable cities. The goal is to be a leader in sustainability, which means providing energy efficient solutions and services that are smart and reliable. Sustainability is an integral element of the value that KONE creates for its customers. Sustainability is embedded into KONE's organizational culture. Sustainability embraces the entire supply chain, including suppliers and partners. KONE is a signatory of the UN Global Compact and supports the UN Sustainable Development Goals.

Likewise, it is foreseen that more and more construction-related researchers adopt the environmental sustainability dimension, integrate it with the other dimensions of BM concept design, and incorporate circular economy, carbon footprint minimization, clean technologies, energy savings, shared well-being, air cleaning, water purification, and resource reuses as key elements into their construction-related BM concepts, respectively.

In turn, this reviewer intends to continue the pioneering reviewing up to 2025 and to protect it against many biases. Especially, the reviewing has been protected only in part against the formal publishing channel bias, i.e., an increasing difference between the number of 79 identified construction-related BM concepts and the number of an existing population of such concepts published via all formal scientific channels. It is likely that in total 90-95 construction-related BM concepts have been published between 1990 and 2018, including 3-5 high-sustainability, construction-related BM concepts that just wait for identification, assessment, and exploitation.

References

[1] Huovinen P 2011 Managing of Construction-Related Businesses in Environmentally Sustainable Ways - A Focused Review of 62 Concepts *Proc. of MISBE2011 Int. Conf. on Management and Innovation for a Sustainable Built Environment* eds H Wamelink, R Geraedts and L Volker (Delft, Univ. Delft) Downloadable at http://misbe2011.fyper.com/proceedings.htm

[2] Huovinen P 2015 Leveraging Concepts for Environmentally Sustainable Business Management in Construction - A Focused Review. *Proc. CIB 2015 on Going North for Sustainability: Leveraging Knowledge and Innovation for Sustainable Construction and Development* ed C Egbu (London: IBEA Publications) pp 286-296

[3] Cooper H M 1998 *Synthesizing Research - A Guide for Literature Reviews* 3rd edn (Thousands Oaks: SAGE)
[4] Hart C 1998 *Doing a Literature Review* (London: SAGE)
[5] Huovinen P 2003 Firm Competences in Managing a Firm's Dynamic Business in Particular in Construction Markets. Unpublished Licentiate Thesis in Construction Economics and Management (Helsinki University of Technology)
[6] Huovinen P 2006 Reviewing Conceptual Research on the Targeted Area of Construction-Related Management *Proc. Symp. CIB W55, W65, W86 on Construction in the XXI Century: Local and Global Challenges* eds R Pietroforte, E De Angelis and F Polverino (Rome: Edizioni Scientifiche Italiane) pp 1-12
[7] Huovinen P 2018 Advancing Future-Oriented Business Management with Contexts in Construction Markets *Proc. 1st Int. Conf. on Construction Futures (ICCF)* (Wolverhampton: Univ. Wolverhampton) pp 1-10
[8] OED 2018 *Oxford English Dictionary* 2nd edn (Oxford: Oxford University Press)
[9] Intergovernmental Panel on Climate Change 2018 *Global Warming of 1.5°C, Summary for Policymakers* eds V Masson-Delmotte et al. (Geneva: IPCC) Available from www.ipcc.ch
[10] World Economic Forum 2019 *How to Set Up Effective Climate Governance on Corporate Boards, Guiding Principles and Questions* (Cologny/Geneva: WEF with PwC) Available from www.weforum.org
[11] Boston Consulting Group 2017 *Corporate Sustainability at a Crossroads - Progress Toward Our Common Future in Uncertain Times* Available from https://www.bcg,com
[12] Bhattacharya B E and Polman P 2017 Sustainability Lessons from the Front Lines *MIT Sloan Management Rev* **58**(2) pp 71-78
[13] Geradts H J and Bocken N M P 2019 Driving Sustainability-Oriented Innovation *MIT Sloan Management Rev* **60**(2) pp 1-9 (preview)
[14] Huovinen P 2011 Advancement of Sustainable Development, Contracting, Design, and Supply Businesses vis-a-vis Construction Markets. *Proc. of MISBE2011 Int. Conf. on Management and Innovation for a Sustainable Built Environment* eds H Wamelink, R Geraedts and L Volker (Delft, Univ. Delft) Downloadable at http://misbe2011.fyper.com/proceedings.htm
[15] Keenan J M 2016 From Sustainability to Adaptation: Goldman Sachs' Corporate Real Estate Strategy *Building Res. & Inform.* **44**(4) pp 407-422
[16] Chinowsky P S with Meredith J E 2000 *Strategic Corporate Management for Engineering* (New York: Oxford University Press)
[17] Flanagan R 1994 The Features of Successful Construction Companies in the International construction market. *Proc. Etkin International Seminar on Strategic Planning in Construction Companies* ed A Warszawski (Haifa: Tecnion) pp 304-318
[18] Veshosky D 1994 Portfolio Approach to Strategic Management of A/E Firms." *J. Management in Eng.* **10**(5) pp 41-47
[19] Løwendahl B 1997/2001 *Strategic Management of Professional Service Firms* 1st edn/2nd edn (Copenhagen: Copenhagen Business School)
[20] Huovinen P 2001 A Framework for Designing an International Competitive Strategy in the Case of Technology-Intensive Contractors *Proc. 2nd Int. Construction Marketing Conf.* ed C N Preece (Watford/Leeds: University of Leeds) pp 68-75
[21] Love P E D, Irani Z, Cheng E and Li H 2002 A Model for Supporting Inter-Organizational Relations in the Supply Chain *Eng., Constr. and Architectural Management* **9**(1) pp 2-15
[22] Huovinen P and Hawk D L 2003 Towards Collaborative Customer-Supplier Relationships in Global Building Product Businesses *Information Technology-Enabled Global Customer Service* ed T Reponen (Hershey: Idea Group Publishing) pp 143-162
[23] Anderson D K and Merna A 2005 Project Management Is a Capital Investment Process *J. Management in Eng.* **21**(4) pp 173-178
[24] Madritsch T and Ebinger M 2011 A Management Framework for the Built Environment:

BEM2/BEM3. *Built Env. Proj. and Asset Management* **1**(2) pp 111-121

[25] Mutka S and Aaltonen P 2013 The Impact of a Delivery Project's Business Model in a Project-Based Firm *Int. J. Proj. Management* **31** pp 166-176

[26] Brege S, Stehn L and Nord T 2014 Business Models in Industrialized Building of Multi-Storey Houses *Constr. Management and Econ.* **32**(1-2) pp 208-226

[27] Then D S S, Tan T H, Santovito R and Jensen P A 2014 Attributes of Alignment of Real Estate and Facilities Management to Business Needs *J. Corporate Real Estate* **16**(2) pp 80-96

[28] Kämpf-Dern A and Konkol J 2017 Performance-Oriented Office Environments – Framework for Effective Workspace Design and the Accompanying Change Processes *J. Corporate Real Estate* **19**(4) pp 208-238

[29] Corporate Knights Inc 2019 *Global 100 Most Sustainable Corporations* Available from https://www.corporateknights.com/reports/global-100

[30] KONE Corporation 2019 KONE Ranked among the World's Most Sustainable Companies by Corporate Knights, Press release (published on 23 January 2019) Available from https://www.kone.com/en

IOP Conf. Series: Earth and Environmental Science **297** (2019) 012012 doi:10.1088/1755-1315/297/1/012012

Developing a transition agenda towards a circular economy: the Dutch case of the province of Overijssel

T J N M de Bruijn[1, 2] and A G Entrop[1]

[1] Saxion University of Applied Sciences, School of Business, Building & Technology, M.H. Tromplaan 28, 7513 AB Enschede, the Netherlands

[2] IAA Architects, M.H. Tromplaan 55, 7513 AB Enschede, the Netherlands

Email corresponding author: a.g.entrop@saxion.nl

Abstract. Over the last years, it has become clear that societies need to increase the adoption rate of circular economy principles. Even though the number of circular products, services and building projects is increasing, innovators obviously experience obstacles in their quest towards a more circular economy. Moreover, relatively many governmental organisations still seem to struggle how to effectively stimulate a circular economy. Governments face the challenge to develop a playground in which a linear economy gradually is replaced by circular practices. This paper presents the case of the Dutch province of Overijssel. The key question Overijssel put forward is what the province can do to stimulate innovative practices specifically in the building sector. After a literature study, a research collective brought together highly successful and less successful local and regional examples of circularity. Document study, website analyses and interviews, provided insights on the drivers and barriers. Our analysis shows what actions the province needs to take to further stimulate or at least to maintain the drivers and what to do about the barriers. Our findings are relevant not only to the province of Overijssel, but also to other regions in Europe.

1. Introduction

The next wave in sustainability is in circular economy. Although the notion of limited stocks of raw material was already noted by the Club of Rome in 1972 by their report Limits to Growth [1], only the last couple of years we see the idea of circular economy getting a foothold. Since our economies are firmly grounded in linear patterns, this implies a major transition. In our study we focused upon the building sector. In the Netherlands this sector is responsible both for around 50% of material use as well as 50% of all waste. Circular building is, therefore, a necessary ingredient of a circular economy. This, however, is a major task. Both in content given the impact of the sector, but also with respect to the structure of the sector: many different actors from architects to scrapyards, mainly SMEs, and with a large time lap between the design and construction of buildings and the possible reuse of the material in them. Business cases of all of these actors are built upon added value in a linear think model.

This is one of the reasons why at the national level a transition agenda for circular building has been developed. Below, in Section 4, we briefly discuss its contents. Given that the building sector holds a strong position in Overijssel, the province of Overijssel decided to translate the national agenda to the regional level. A research consortium was asked to develop this regional transition agenda. The authors were part of the consortium. The regional agenda has been published in last February. Also in Section 4, we describe the contents of this agenda too. Next to identifying current practices, potential obstacles and needs, we also addressed the role of the province. The question we put forward is: what role of the province is needed to foster circular building?

Our theoretical perspective is found in policy sciences. In this literature a distinction can be found between two models of government action, see for instance De Bruijn [2]. The first, more traditional model is called the hierarchical model. Its steering concept positions the government in a subordinate place. The government issues regulations that guide private actors. The second model is the network model. Policy development is seen as an interactive negotiating process between public and private actors. The government is one of the actors, positioned equally as the other partners. This last model is especially applicable in the case of wicked problems, in which no easy solution is visible that can be captured in regulation. Working in a network model requires governments to reflect upon the question which actors are vital. Policy development takes place in close interaction with these actors. We see the transition towards a circular building sector as a wicked problem. In developing the agenda we, therefore, used the network model as inspiration.

(Policy) Development processes in a network setting follow a different path than hierarchical processes [2]. In the latter the focus is on the substance of the problem, possible solutions, and the implementation of the most optimal alternative. In a network setting the focus shifts to exploring the context of the problem and the actors involved. Below we highlight the main steps in the process.

1. Exploring the substance of the challenge
Although the main focus is on the network, it is wise to start with an exploration of what is at hand. It is difficult for an actor to engage others without some basic understanding.

2. Exploring the actors involved
The network setting is particularly useful, if there is mutual dependency. The second step, therefore, is an exploration of who is involved, in some way. As this is not easily answered often, it is wise not to limit the group of actors to engage.

3. Determining the architecture of the process
An interactive process does not run itself. Rather, it requires careful planning and management.

4. Consultation rounds on the problem definition
The first consultation round should focus on jointly exploring what the perspective of each actor is on the challenge at hand. Of course, in a network setting these perspectives can be rather diverse. Before exploring possible solutions, it is wise to get a clear view on how the challenge is perceived.

5. Consultation rounds on possible routes for meeting the challenge
A logical final step is the exploration of how possible solutions may look like. An important factor for success is the extent in which the various actors recognize their problem definition in the various routes. Also to be discussed is the contribution relevant actors can make to the solution.

In our research setup, that we discuss in the next section, we used this network model as a guidance.

2. Research setup

What makes an agenda effective? Especially in the case of wicked problems, the agenda in our viewpoint should reflect the perception of the sector itself. As explained above, with wicked problems governments have too little information and knowledge to simply regulate by themselves. The knowledge of the private partners, their innovation capacity and other resources are necessary ingredients. Therefore, the research setup is as follows.

We started with a literature review to get a good grip on what circular building is. Academic as well as professional publications have been reviewed. Special attention was paid to identifying possible obstacles towards circular economy and circular building. The results were used to feed the rest of the process. Some of the results of the literature review is presented in the next section.

Next we started exploring who should be involved in the development process of the transition agenda. More than six workshops in diverse settings and with partly different participants were organized. The participants were representatives of the building industry, but included also people working in education and research. All in all more than one hundred people were involved over a period of six months. Next to the workshops individual interviews were held with opinion leaders and innovators.

Through the workshops and interviews we learned what perspectives the different actors have on circular building. We explored current practices, innovative projects, partnerships, and last. but not least. obstacles. The second part of the workshops were used to explore possible routes for a transition towards circular building. All information was brought together in the draft agenda. This draft was presented and discussed with the province. The project ended with a presentation of the agenda to the network.

3. Theoretical framework

This section elaborates on our theoretical framework. Circular economy is a concept that many scholars try to define and try to make more tangible. It probably comes as no surprise that the Ellen MacArthur Foundation [3] seems to be regarded as an authority in the quest of scholars, as well as practitioners, i.e. entrepreneurs, to grasp the principles of circular economy and to come to actions.

3.1. Defining circular economy, its principles and actions needed

According to Kirchherr et al. [4] 114 different definitions of circular economy can be found. Considering that that paper was written in 2017, the number of definitions probably even increased in the meantime. Although it is by no means our intention to set out all definitions of circular economy that are available in literature, we would like to reflect on some to illustrate some of the different interpretations that seem to exist. The Ellen MacArthur Foundation defines a circular economy as follows [3]: *a circular economy is one that is restorative and regenerative by design and aims to keep products, components, and materials at their highest utility and value at all times, distinguishing between technical and biological cycles*. When narrowing our search for literature on circular economy down to the construction industry, Mahpour addresses the definition for circular economy of the Ellen MacArthur Foundation as follows [5]: *"an industrial system that is restorative or regenerative by intention and design. It replaces the end-of-life concept with restoration, shifts toward the use of renewable energy, eliminates the use of toxic chemicals impairing reuse, and aims at eliminating waste through the superior design of materials, products, systems, and business models."*

Korhonen et al. [6] define circular economy as follows: *"an economy constructed from societal production-consumption systems that maximizes the service produced from the linear nature-society-nature material and energy throughput flow. This is done by using cyclical materials flows, renewable energy sources and cascading-type energy flows.*

It becomes clear that internationally different points of view led to many definitions. Considering the specific situation of the province of Overijssel, which is located in the Netherlands, the efforts of the Dutch government to define this concept cannot be ignored. According to Reike et al. [7] the Dutch government defined circular economy in 2013 as an *economic system that takes the reusability of products and materials and the conservation of natural resources as starting point. It also strives for value creation for people, nature and the economy in each part of the system*" [8]. In 2018 the national Transition Agenda Circular Building of the Netherlands offered the construction industry the following translated description [9]: circular building means to develop, use and reuse buildings, areas and infrastructure without unnecessary depleting natural sources, without wasting the environment and without affecting ecosystems. It is a way of building that can be justified economically and that contributes to human and animal welfare. Here and there in the present and for the future.

When explaining what a successful circular economy is, Korhonen et al. [6] refer to three aspects, that are very familiar to those involved in sustainable development, namely people, planet and profit.

The people aspect becomes apparent in, like [6] describes it, *the sharing economy, increased employment, participative democratic decision-making and more efficient use of the existing physical material capacity through a cooperative and community user (user groups using the value, service and function) as opposed to a consumer (individuals consuming physical products) culture."*

The environmental objective of circular economy, or planet aspect, *is to reduce the production-consumption system 1a) virgin material and 1b) energy inputs and 2a) waste and 2b) emissions outputs (physical throughput) by application of material cycles and renewables-based energy cascades."*

The economic objective of circular economy is to reduce the economic production-consumption system's raw material and energy costs, waste management and emissions control costs, risks from (environmental) legislation/taxation and public image as well as to innovate new product designs and market opportunities for businesses."

Making the concept of circular economy again less abstract, the Ellen MacArthur Foundation also provides fundamental characteristics to describe what needs to be done or, preferably, already has been done in a circular economy [3]:

A. waste is designed out, because in a circular economy waste does not exist;
B. diversity builds strength, because a circular economy values diversity;
C. in a circular economy renewable energy sources power the economy;
D. systems-thinking is applied broadly in a circular economy;
E. prices or other feedback mechanisms reflect on real costs.

Besides these characteristics, the Ellen MacArthur Foundation also identifies six actions that businesses and governments can take. These are referred to in the so called ReSOLVE framework [3]. This framework encompasses six categories of actions [3]: Regenerate, Share, Optimise, Loop, Virtualise and Exchange.

3.2. Defining drivers and barriers

The Ellen MacArthur Foundation identifies four building blocks that might help to foster a more circular economy ([3], pp. 16-17):

1. Circular product design and production;
2. New business models;
3. Reverse cycle;
4. Enablers and favourable system conditions:
   a. Education;
   b. Financing;
   c. Collaborative platforms;
   d. A new economic framework.

Although, for the reason of conciseness, the complete descriptions of the building blocks are left out here, two things seem inconsistent to the stranger's eye. Firstly, it can be debated if it is wishful to give name to an individual building block "1. Circular product design and production" that addresses directly the overall target, namely that of coming to a circular economy. Secondly, as one can see, the fourth building block consists of four elements, which suggests that these four, a to d, are less important than the first three, 1 to 3.

On the basis of analysing 195 articles from the Web of Science Core Collection and Scopus databases, Galvão et al. [10] identify seven main barriers in literature to come to a circular economy. These barriers are: 1) Technological, 2) Policy and regulatory, 3) Financial and economic, 4) Managerial, 5) Performance indicators, 6) Customer, and 7) Social. However, the authors fall short in defining or describing these categories. Strangely, most examples given are not even related to these seven main barriers, but to other additional categories of barriers.

This again brings us back to Korhonen et al. [6], who do not speak of barriers. However, they do mention six categories of limits and challenges for the circular economy concept, namely 1) thermodynamic limits, 2) system boundary limits, 3) limits posed by physical scale of the economy, 4) limits posed by path-dependency, 5) lock-in limits of governance and management and 6) limits of social and cultural definitions.

4. Policy context

In this section an overview will be given of existing national and regional policies.

4.1. National policies

The Dutch national government has put down its ambitions and targets on circular economy in a memo called '*Nederland Circulair in 2050*' [9]. By 2030 the use of virgin material should have bisected. In 2050 all material should be used in a sustainable manner, without any harmful emissions and without using any virgin material at all. Products should be designed in such a way that reuse and recycling is possible without a loss of quality.

Following the memo on circular economy early in 2018, the national government published its memo on the building sector in this respect (see Figure 1) under the name of '*Transitieagenda Circulaire Bouweconomie*' [11]. This national transition agenda has its main ambition in reaching a completely circular building sector in 2050. At the same time, due to existing shortages in the Dutch residential real estate market and due to demographic developments, one million new houses need to be built. The agenda notes that reaching these goals will be far from easy. The agenda holds four priorities, namely:

- market development by, amongst others, creating 'early adopters' through stimulating 'launching customers', by developing passports for materials in buildings and by stimulating Design, Build, Finance & Maintain tenders;
- measuring and monitoring: developing an uniform method for determining levels of circularity;
- policies and regulation: creating space for experimentation in current rules;
- knowledge and awareness: the transition process requires (new) knowledge of all parties involved through research and education, educational programs and communication.

Figure 1. Cover of the national transition agenda for circular building economics.

Figure 2. Cover of the regional transition agenda for circular building sector.

4.2. Regional policies

The national ambition for a circular economy was a stimulus for the province of Overijssel to issue its own regional agendas, for six different sectors of our economy, namely the construction industry, infrastructure, manufacturing industry, plastics, biomass and food, and consumer goods. Figure 2 shows the cover of the regional transition agenda for a circular building sector [12]. As the construction industry is prominent in Overijssel, it is one of selected sectors. The resulting regional transition agenda recognizes that the transition will not only be technological. Rather, new business models and social innovation are necessary ingredients. It also recognizes that the road towards a circular building sector must be explored step by step. Along the way the actors involved will discover what roles they can play and how their practices should change and what is needed for that. That is why developing experimentation platforms are an important element on the agenda.

5. Exploring circular challenges

In coming to the regional transition agenda circular building of the province of Overijssel, interviews were conducted with six different front runners in educating circularity. Interviews at universities and at an Intermediate Vocational Education institute were taken. In the field were circularity needs to be practiced entrepreneurs were interviewed. Eleven interviews in total were taken to find out how well circularity has landed in company policies and in the daily routines. By studying literature and policy documents on the one hand and by interviewing political, entrepreneurial, research and educational leaders on the other hand, it became clear to us that nine major obstacles still exist hampering circularity becoming mainstream. These obstacles have been formulated as questions which the construction industry might want to ask herself:

1. How can the behaviour within and culture of the construction industry be transformed to a situation in which skilled and unskilled clients and contractors recognize the scarcity of virgin materials and act accordingly, placing circularity upfront?
2. How to ensure that laws and regulation in the construction industry do not obstruct a circular economy, but rather require and support this new economy without lowering quality standards, climate objectives, CO_2 and energy targets?
3. How do we come across all stages in the whole life cycle of buildings to lasting partnerships in which long term common interest and social relevance are united with short term individual financial benefits?
4. How can functionality be valued profitable by a business model that takes into account environmental impact and the circularity of products and materials offering that functionality?
5. How can financial constructions be developed that assess risks by also considering the added value of building and exploiting circularly?
6. How can a solid, comfortable and sustainable building for the user be guaranteed, while at the same time material cycles are closed?
7. How can building components, products and materials be reused offering high quality with the lowest environmental impact possible?
8. How can the construction industry come to a circular supply chain of (circular) components, products and materials for the entire life cycle of a building?
9. How can educational programs prepare future employees in and adjacent to the construction industry for (contributing to) the circular building economy?

As shown by the challenges identified above, it is clear that the transition towards a circular building economy is a major challenge. From what we have learned from the forerunners there is yet much to be experimented and learned. A central element in the agenda, therefore, are the so-called experimentation platforms. At these platforms more knowledge can be gained on how to overcome obstacles. The experimentation platforms have been spotted by the research team. We further developed them in close collaboration with the (local) partners involved.

6. Discussion

The transition process will need close collaboration between all actors involved: the sector itself, governments, research and educational institutes. A study on the role of the province of Overijssel in the energy transition showed that governments and particularly the province play an important role [13]. It concludes that there is yet not fully functioning market. Without an active role the energy transition will not likely reach its goals timely. We feel this situation is comparable with the transition towards a circular building economy. The question we address in this section is what role there is for the province of Overijssel.

In a recent study by the University of Twente/BMC a couple of action perspectives for provinces is described [14]. They distinguish four categories:

- the outstretched hand: at the start of a process a province act as equal partner, exploring the playground;
- the helping hand: as the actors involved sometimes do not oversee a challenge, a province can intervene by for instance commissioning research or appointing a process manager;
- the guiding hand: guiding all actors into the direction wanted. The role of the province is to provide a vision and to keep all actors aligned and on track;
- the strong hand: as not all actors will move automatically, a province should in the end also use its regulatory power to force change.

Of course there is some sequentiality. At the start of a transition process the outstretched hand is more fitting than the strong hand. Also, in cases where all is clear a province shouldn't be holding back from direct regulation. We feel that all 'hands' will be necessary in the end. For the coming period, however, the emphasis should be on collaborating, gaining knowledge, getting all relevant actors on board. Once there is sufficient mass, and there is more clarity on what circular economy actually means for the building sector, the focus can shift towards guiding and regulating. This development of different strategies stands at the basis of our advice to start with experimentation platforms.

7. Conclusion

Although there is much talk about circular economy, in practice still much has to be learned. Major issues, as the nine challenges we identified, remain open. That is why it is too early for scaling up already. The challenge towards a circular construction sector presents itself as a wicked problem. Not clear yet is what needs to be done, who has to do what and how governments can stimulate action. We, therefore, looked at the challenge from a network perspective. Some conclusions can be drawn: many of the actors in the construction industry know about circularity, but community-building is only in its early stages. That is why we started up our process of research iteratively. For the next round a more firm process architecture should be developed. Through our research we discovered the challenges the sector faces. As a next step projects need to be developed that address these challenges.
As for the role of the government: there are some early adopters, but the majority of the building sector is, at best, only starting up. Further experimentation and learning are necessary. That is why we recommended experimentation platforms. This calls for a government who is supportive and collaborative. When sufficient knowledge has been gained, regulatory action may be a future phase.

Acknowledgements

The Province of Overijssel is acknowledged for providing financial support to this research project. This paper builds on the agenda written by the research team. The authors would like to thank the team members for their cooperation. The full text of the agenda is available in Dutch [12].

References

[1] Meadows D H, Meadows D L, Randers J and Behrens W W 1972 *Limits to growth.* Club of Rome
[2] Bruijn T J N M de 2014 Beleid voorbereiden in netwerken in: *Overheidsbeleid. Een inleiding in de beleidswetenschap,* Kluwer
[3] Ellen MacArthur Foundation 2015 *Towards a circular economy: business rationale for an accelerated transition.*
[4] Kirchherr J, Reike D, Hekkert M 2017 Conceptualizing the circular economy: An analysis of 114 definitions *Resources, Conservation & Recycling* **127** 221-32
[5] Mahpour A 2018 Prioritizing barriers to adopt circular economy in construction and demolition waste management *Resources, Conservation & Recycling* **134** 216-27
[6] Korhonen J, Honkasalo A and Seppälä J 2018 Circular economy: the concept and its limitations. *Ecological Economics* **143** 37-46
[7] Reike D, Vermeulen W J V and Witjes S 2018 The circular economy: new or refurbished as CE 3.0? — Exploring controversies in the conceptualization of the circular economy through a focus on history and resource value retention options *Resources, Conservation & Recycling* **135** 246-64
[8] Circular economy: *some definitions*, Retrieved from http://www.circular.academy 11 Feb 2019.
[9] Ministry of Infrastructure and Environment and the Ministry of Economic Affairs 2016 *Nederland circulair in 2050; Rijksbreed programma Circulaire Economie*
[10] Galvão G D A, Nadae J de, Clemente D H, Chinen G and Carvalho M.M. de 2018 Circular Economy: Overview of Barriers *Procedia CIRP* **73** 79-85
[11] Nelissen E et al. 2018 *Transitie-agenda circulaire economie; circulaire bouweconomie*
[12] Bruijn T de, Bults J, Engelsman L, Entrop B, Smit M, Straatman J and Vrielink R 2019 *Circulair bouwen; een transitieagenda voor Overijssel.* Provincie Overijssel
[13] Entrop B and Dorée A 2015. The efficacy of policy instruments to reduce the fossil energy use of privately owned dwellings. *10th Conference on Advanced Building Skins*, 3-4 November 2015, Bern, Switzerland.
[14] Wit J de, Boogers M and Tempelaars R 2018 *Meervoudig verbonden: de provincie in opgavegerichte allianties; Evaluatieonderzoek naar de rol, samenwerkingsvaardigheid en impact van de provincie Zuid-Holland in de bestuursperiode 2015-2019: spiegelen en vooruitblikken.* BMC

IOP Conf. Series: Earth and Environmental Science **297** (2019) 012013 doi:10.1088/1755-1315/297/1/012013

Health & wellbeing in offices - A study of literature on the Nordic perspective

Melina Forooraghi, Holger Wallbaum, Nina Ryd

melinaf@chalmers.se,
holger.wallbaum@chalmers.se
nina.ryd@chalmers.se

melinaf@chalmers.se

Abstract. **Aim**: This review aims to explore the relation between the physical office environment and employee health as found in the Nordic literature. **Background**: The Nordic countries have a worldwide fame for their high living and working standards. Nevertheless, they have also been facing global challenges related to demographic changes, aging population, and a rise in long term conditions which require a creative way of thinking, and innovative approaches in the ways we deliver buildings and the built environment. The built environment is a health determinant and a resource for health and wellbeing of building users. In office sector, however, there is little known regarding the design implications of the physical environment for health outcomes. A review of the literature on the current state of the art in the Nordic context could offer an opportunity to rethink office design and support dialogue on how better design can improve employee health outcomes. **Methods**: An explorative document analysis of recent literature on workplace environment in the Nordic context was carried out. **Results**: Three main gaps in the literature were identified, including (1) limited understanding of health, (2) unrecognised role of the physical work environment and design professions in relation to health, and (3) diverse research orientations. The combination of results suggests the paucity of the studies relating the physical office environment to positive health approaches. **Conclusions**: This review highlights the need for more quantitative and qualitative methods to give relevant knowledge for complex questions regarding the physical office environment and health outcomes of employees. Future research should enable dialogue and collaboration between different actors such as managers, human resources, occupational health professionals and designers, which can benefit the users of office building.

Keywords: Built Environment, Health, Nordic, Office Design, Workplace, Wellbeing

1. Introduction

The Nordic countries have a worldwide fame for their high living and working standards. These countries are often at the top of global wellbeing and quality-of-life-related comparisons [1]. Several health outcomes, like long life expectancy and low child mortality rank among the best in the world [2, 3]. Nevertheless, there is no certainty that these countries will maintain their good rankings. Like all other countries, the Nordics have also been facing global challenges related to demographic changes, aging population, and a rise in long term conditions. These challenges require a creative way of thinking, and innovative approaches in the ways we deliver buildings and the built environment.

Most people spend 90% of their lives within buildings [4]. Everything from light [5, 6], noise levels [7-9], ventilation [10, 11], temperature [12], layout [13, 14], and colour [15-19] affect the daily health of occupants, all of which are influenced and shaped by the built environment. As a result, the workplace plays a key role in employees' health as a place where they spend almost 40% of their waking hours [20].

In the field of workplace environment, most of the research focuses on 'Psychosocial' environments. This refers to various factors, such as social relationship at work and organizational factors. These factors in turn can affect general health and sick leave [21], stress related ill health [22], depression and anxiety [23]. While healthcare building design has long dealt with the built environment and its health-related outcomes, the relationship between physical work environment and employees' health and wellbeing has gained interest only through recent years [24, 25].

Over the past decades, several topics have been brought up in studies concerning physical office environment and employee health and wellbeing. A great number has focused on ambient factors, such as air quality, light and noise [26-28]. Others studied the impact of office type on health-related outcomes [24, 25]. Several factors are associated with the layout of the office. For example, privacy can affect employee wellbeing and job satisfaction [29, 30]. Sense of control over physical work environment is also of importance and has positive effects on employees' satisfaction and wellbeing [31, 32]. Office layout can encourage physical activity as well. There is a relatively new but growing body of research that deals with design features which stimulate movement in the building. As a result, Active Building Design has been developed which promotes access to facilities and services such as height adjustable desks, gyms, bicycle parking and green space. Nevertheless, much of the research in this area has focused on stairs and elevator i.e. [33-35]. Besides the effects of the built solutions, others focused on the design process of offices in relation to employee health and wellbeing. For example, studies highlight that the participation of employees in planning and design processes increases sense of control over the work environment [4] and wellbeing among employees [36].

Employees are the most valuable resource for organizations. Their costs, such as salaries, benefits, but also sick leave accounts for 90% of business expenses [37]. As employee health and satisfaction relates to productivity, and sick leave, their health and wellbeing can have significant financial implications. This equation has made it imperative for many organizations to understand productivity loss and negative health outcomes in relation to the design of workspaces. However, the research is limited, and there is little known about design implication of the physical environment for health outcomes. A review of the literature on the current state of the art in the Nordic context could offer an opportunity to rethink office design and support dialogue on how better design can improve employee health outcome. Therefore, the goal of this review is to explore the relationship between physical office environment and health as recognized within a Nordic perspective.

2. Background

The World Health Organization (WHO) states: 'Health is a state of complete physical, mental and social well-being and not merely the absence of disease or infirmity' [38]. Wellbeing can be defined as 'one's feeling about oneself in relation to the world' [39].

Positive health approaches

Researchers have long tried to explain the positive dimensions of the health continuum by describing terms such as Self-actualization [40], Wellness [41], Psychological wellbeing [42], Flow [43], Flourishing [44], Wellbeing [45] and Salutogenics [46]. For instance, Antonovsky developed the Salutogenic view on health which focuses on the origins of health rather than determinants of disease. He argues the factors that create health are often different from those causing illness. In medicine and health-related disciples, health is defined as the absence of disease [47] what Antonovsky would consider pathogenic. In pathogenic approach, the starting point is disease and from there, we identify the causes of the problem and eliminate them. Pathogenesis is therefore reactive to disease. Antonovsky argues that, to move toward health, the elimination of risk factors is not enough. Rather he argues we should increase positive factors such as imagination, play, will and meaning would be necessary. Thus, Salutogenesis with a proactive approach, aims to generate better health.

Health perspectives in office research
The majority of previous research on the indoor (workplace) environment has tended to focus on threats, illness and negative impacts on people. For instance, studies investigated a range of irritative symptoms in office buildings such as eye and nose irritation, difficulty concentrating, skin reactions, mental fatigue and headaches. This led to the recognition of 'sick building syndrome' which refers to a situation when 20 per cent of buildings occupant complain of a similar medical condition, while in the building, because of an unknown cause over a period of at least 2 weeks [48]. Most research thus has focused on removing the harmful stimuli, while the factors which generate a positive experience in the workplace have been overlooked [49]. Only few studies have studied the positive aspects related to work environments that could promote health and well-being [32]. Bluyssen (2014) calls for a change of mindset from the current focus on single component to users and improving the quality of their lives. Positive aspects could include participation in planning and design, space personalization, building a sense of control and ability to meet changing needs and preferences.

WHO defines five keys to healthy workplaces as follows [50]: A healthy workplace is one in which workers and managers collaborate to use a continual improvement process to protect and promote the health, safety, and well-being of all workers and the sustainability of the workplace by considering the following identified needs:
- Health and safety concerns in the physical work environment;
- Health, safety and well-being concerns in the psychosocial work environment including organization of work and workplace culture;
- Personal health resources in the workplace; and
- Ways of participating in the community to improve the health of workers.

The definition reflects a shift in perspective from a major focus on physical environment to a more holistic view on the workplace factors including psychosocial and individual health factors [31].

That said, it was only until recent years that health became an integral part of many post-occupancy evaluations (POE), building certifications and standards such as WELL, LEED and BREEAM. Among those, WELL building standard specifically focuses on health and wellbeing of occupants in buildings. However, there are not may WELL certified office buildings. Moreover, many of these POE's focus on physical measurements and do not include occupant' perception [51].

3. Method

Aim and research question
The aim of the study was to explore the relation between the physical office environment and employee health as found in Nordic literature.
This review was guided by the following question:
How does the Nordic research relate employee health to office's physical environment?

Sample and setting

The study is situated in the Nordic context, including institutes from Norway, Finland, Sweden, and Denmark. All data has been collected from institutes that focus on workplaces. More literature was identified by analysing the references of the major reports in the field. From their public reports we selected those that addressed the combination of health and workplaces.

Data collection
The data was collected between December 2018 and February 2019. The identified institutes were listed in a table included the name of the institute, a summary of their aims and the country (Table 1). Their reports were collected mainly from their websites.
The texts concerning the built environment was extracted into one document. The focus was on texts sections that concerned office design in relation to employee health and well-being.

Data analysis
This study is based upon a selective document analysis using written material, developed by others [52]. This included descriptive comparisons for the document set, short enough to allow clear and simple qualitative comparisons. All the selected texts was analysed using content analysis framework [53].

Table 1. Listing of identified institution

Institutes	Summary of objectives of the institution	Country
The Nordic Institute for Advanced Training in Occupational Health (NIVA)	"NIVA gathers top experts and knowledge from the Nordic countries and transforms research findings into advanced courses and seminars within the field of occupational health and safety for researchers and practitioners from both the Nordic countries and world-wide."	Nordic countries
Nordic Innovation Centre	Nordic Innovation is a Nordic organisation working to promote cross-border trade and innovation. Nordic Innovation is a Nordic organisation working to promote cross-border trade and innovation.	Nordic countries
Forskerforbundet	The Norwegian Association of Researchers (NAR) organises employees and students in research, higher education and dissemination of knowledge. NAR's main purpose is to increase salaries and improve working conditions for our members.	Norway
Helsebiblioteket.no	Helsebiblioteket.no provides free access to subject procedures, reference books, databases, journals and other knowledge resources for health professional.	Norway
Finnish Institute of Occupational Health, FIOH	"The Finnish Institute of Occupational Health (FIOH) is a multidisciplinary research and specialist organization that focuses on well-being at work, research, advisory services and training. We operate under the Ministry of Social Affairs and Health as an independent legal entity."	Finland
Institutet för stressmedicin (ISM)	The Institute for Stress Medicine (ISM) is the Västra Götaland Region is a knowledge centre in the stress area and part of the administration Health and Stress Medicine. They carry out research and development, while also conveying knowledge and experiences about stress, health and psychosocial work environment.	Sweden

Institutet för miljömedicin (IMM). Karolinska Institutet	The Institute for Environmental Medicine (IMM) is an institution at Karolinska Institute and a national expert body within environmental medicine. Activities that span a wide area around the environmental impact on health, with a focus on chemical and physical factors and lifestyle.	Sweden
Det Nationale Forskningscenter for Arbeidsmiljo	The National Research Centre for Occupational Health and Safety (NFA) researches, communicates and educates, and thus contributes to developing a healthy and safe working environment in Denmark.	Denmark
The Danish National Centre for Social Research	The Centre conducts research and carries out commissioned projects in the area of welfare state policies and disseminates the results. Its primary task is to generate new knowledge of relevance to society.	Denmark

4. Results

The details of the publications included in the review were documented in a chart form which included source information (authors, year), title, and addressing health in relation to office environments (Table 2). The results show that the Nordic countries recognise the importance of the workplace impact on employee health. Nevertheless, there is little research relating the physical office environment to positive health approaches in the Nordic countries.

Table 2. Listing of included articles

Author(S), year	Title	Addressing health
Bakke, 2007	DEKAR – Den Nordiske Kunnskapsarbeidsplassen (The Nordic Workplace Design for Knowledge Work)	Does not mention anytime direct health outcome, however process-related factors such as employee's participation and space-related such as visibility, proximity and degree of flexibility are addressed.
Bakke & Fostervold 2017	Kontorlandskap – arbeidsmiljøfaglig veiledning	Identifies health outcomes in relation to open plan office, such as Reduced short-term memory, concentration, productivity, wellbeing, false stress alarms and work motivation; Increased fatigue, sick leave days, physiological stress reactions from the heart and vascular system Increased.
Conference proceedings	Towards better wellbeing at work	Among 8 articles from Finland, Sweden and Denmark, 0 papers addressed the physical office environment as a resource for wellbeing enhancement.
Ruohomäki et al., 2015	Salutogenic and user-centred approach for workplace design	Relates salutogenic approach to workplace design
Hultberg, 2018	Sammanfattning av forskningsläget avseende aktivitetsbaserade kontor (ABW)	Identifies risk factors in Activity Based Offices

Toivanen, 2015	Framtidens arbetsplatser: Att utveckla hållbara och friska kontor	Mentions health related outcomes such as better comfort, performance, and job satisfaction
Hultberg et al., 2018	Hälsa på arbetsplatsen	Explores the connections between work, stress and health, and how good conditions for health in the workplace are created.
Kwak et al.,2017	Hälsofrämjande insatser riktade till arbetsplatsens fysiska miljö och organisationsstruktur: effekt på arbetsrelaterade utfall. En kartläggning av forskningen.	Relates health promotion to absenteeism, presenteeism, work performance and workability as well as productivity.

4.1. Reports relating to health

Health as a term is used in 7 articles, however not all of them provide a definition. In fact, only 1 article explicitly referred to the WHO definition; 'Health is a state of complete physical, mental and social well-being and not merely the absence of disease or infirmity'. One document provided an interpretation of health; "*Health is to feel good and have sufficient resources to meet everyday demands and to be able to realize both personal and professional goals*" [54]. This report goes on and gives a comprehensive overview of studies dealing with the connection between work, stress and health and gives an interpretation of health promotion as well; "*The characteristics of health-promoting workplaces is that the workplace as a starting point should offer good working conditions, a good working climate and provide the pre-conditions for personal and professional development*".

Other articles included health-related outcomes such as healthy behaviour, mental illness, absenteeism, and work ability. Several others referred to organisational outcomes, such as productivity which is connected to health.

The analysed documents represent two different health perspective:

- a traditional biomedical and pathogenic approach (i.e. physiological stress reactions, fatigue and absenteeism) [55-57];
- a salutogenic perspective referring to presence of positive factors that create health and not the mere absence of disease also known as positive health (i.e. coping, motivation, workability, work engagement and job satisfaction) [54, 58].

The identified positive health approaches, such as workplace health promotion research, were found to focus on psychosocial work environments and simply neglect the complex interplay of the physical environment and the health outcomes. A report reviewed literature on health in the workplace [54]. This document mainly focused on psychosocial work environment. Additionally, it was targeted towards who work in any supporting role to managers, such as HR and occupational health professionals, as well as managers, safety representatives, and students while completely neglecting to address architects and designers who play a major role in the creation of the physical environments.

Similarly, in a conference proceeding 'Toward better wellbeing at work' 8 articles from Finland, Sweden and Denmark were extracted and no paper addressed the physical office environment as a resource for wellbeing enhancement [59]. Another study aimed to produce evidence base for health promotion related to eating habits and physical activity [58]. The study included analysis and proposals for how actors at national, regional and local level can participate in long-term work to promote health related to eating habits and physical activity. Nevertheless, the role of the physical environment was limited to promoting healthy behaviour, such as access to healthy food and facilities for increased activities.

4.2. Reports relating to the physical environment

In the reports with a focus on the physical office environment, the health perspective is mainly dominated by traditional risk-oriented health, such as sick leave rate. For example, Bakke & Fostervold [56] looked into open-plan offices and their impact on health, wellbeing and productivity. This study addresses a national debate which started since the Norwegian Ministry of Local Government and Regional Development introduced a maximum of 23 square meters per employee and flex office as the default office type for state buildings. Several influencing factors were identified in the literature such ventilation, acoustic, ergonomics, the need for recovery, daylight and views, privacy and control. Consequently, the outcomes were increased stress, fatigue, general health, lower motivation and physiological reactions from heart and vascular. It also highlights the need for recovery in relation to office design. Recovery is considered "*the time it takes to come back functionally to the post-stress level after a stress episode has ended*". This study concluded that society is forced to pay for the benefits of the open plan offices by increased sick leave days and productivity losses.

Two similar reports, focused on future offices and research on open plan and activity-based offices with more focus on the physical environment, but dominated with traditional pathogenic notion of health [55, 57]. There was only one article relating the physical work environment to positive aspects of health [60]. This article identified social, psychological and social dimensions of the workplace that promote wellbeing and presents a framework. They conclude a wellbeing promoting workspace:

- has a good indoor environment;
- supports work tasks and work process;
- is ergonomic and accessible to all;
- respects privacy and the personal need for space;
- strengthens sense of control;
- enhances work flow and work engagement and
- enables communication, collaboration and mutual learning.

Finally, a variety in research orientation was identified within the data. The focus of the results varied depending on the institutes' disciplinary focus. For instance, several included documents appeared to have a focus on occupational health investigating psychosocial environments in relation to health [54]. Those with starting from an architectural perspective focused on employees' health and performance in relation to the office type, function, spatial and aesthetic features, as well as factors such as light or sound conditions, indoor air quality and temperature [55, 56, 60]. One document had a management-oriented direction studying how the Nordic managers understand the workplace design and its role as a strategic tool for knowledge sharing [61].

5. Discussion

This review set out to explore how literature in the Nordic countries relate employee health to office physical environment. Three main gaps in the literature were identified, including (1) limited understanding of health, (2) unrecognised role of the physical environment and design professions in relation to health, (3) diverse research orientations. The combination of results suggests the paucity of the studies relating the physical office environment to positive health approaches.

Although healthcare building design has dealt with the influence of the built environment on patient health, there is little understanding of health in the field of workplace research. This might be due to lack of transdisciplinary studies especially when (trans)disciplinary research has been considered challenging [62]. Another reason could be due to the contradictory and inconsistent body of research in this field hence making it difficult to draw clear conclusions [63]. Without a common definition or explicit description of health, it becomes difficult to outline health related goals and desired outcomes. Moreover, how health is viewed determines research areas and thereby influences health related policies in a longer run. Our results are in accordance with a study investigated workplace health promotion research in the Nordic countries and revealed no definition of health was provided by the identified literature [64].

This review reveals the role of the physical office environment in promoting health has been overlooked. Some reports never mention the physical environment and the ones which do are mainly limited to promoting physical activity without explicitly addressing the contribution of physical environment. For example, one of the report gives a relatively comprehensive overview of the existing research on health promotion strategies, however it completely misses to mention the physical environment [54]. Our "environment" includes both social and physical environmental determinants of health. By focusing on one dimension such as the psychosocial environment, it becomes difficult to mobilise accessible resources for better health, in this case, the physical office environment. Especially when the relationship between the environment and health has already been identified as complex.

The diverse research orientation among workplace actors such as management, occupational health professions, architecture and real estate developers presents several challenges. Those challenges might lie in the nature of a discipline and how research questions are determined. For example, occupational health profession research recognizes the influence of the physical work environment on employee health. However, they do not consider the physical work environment as a resource for health promotion. This might be due to the nature of occupational health field that is closer to psychology and therefore more attention is directed towards psychosocial environments rather than the physical environment.

6. Conclusion

There is limited knowledge on the relation between the physical environment of offices and employee health. This study investigated how different Nordic institute address this relation in their documentation. The results reveal that more research is needed to design healthy offices, especially including more holistic perspective on health in relation to the physical office environment.
This review offers an opportunity to rethink office design and support dialogue on how better design can improve employee health outcomes. If we seek to incorporate health consideration in the design of the offices, we firstly need to define what health means and what it entails. The conception of health may guide different design choices. Secondly, we need to relate the physical office environment to positive health approaches such as workplace health promotion research including strategies and evaluations. Lastly, with ongoing changes in workplace development, robust terminology among different disciplines can form a basis for discussions between different workplace actors
This review highlights the need for more quantitative and qualitative methods to give relevant knowledge for complex questions regarding the physical office environment and health outcomes of employees. Future research should enable dialogue and collaboration between different actors such as managers, human resources, occupational health professionals and designers, which can benefit the users of office building.

7. Reference

[1] Gallup 2010 *Gallup Global Wellbeing: The Behavioral Economics of GDP Growth* Washington, D.C
[2] WHO 2018 *Healthy life expectancy (HALE)* Editor Volume Pages
[3] Lozano R, Fullman N, Abate D, Abay SM, Abbafati C, Abbasi N, Abbastabar H, Abd-Allah F, Abdela J, Abdelalim A, and Abdel-Rahman O 2018 Measuring progress from 1990 to 2017 and projecting attainment to 2030 of the health-related Sustainable Development Goals for 195 countries and territories: a systematic analysis for the Global Burden of Disease Study 2017 *Lancet* **392** 10159

[4] Evans GW and McCoy JM 1998 When buildings don't work: The role of architecture in human health *Journal of Environmental Psychology* **18** 1
[5] Richards K. C and Bairnsfather L 1988 A description of night sleep patterns in the critical care unit *Heart & lung: the journal of critical care* **17** 0147-9563 (Print)
[6] McColl SL and Veitch JA 2001 Full-spectrum fluorescent lighting: a review of its effects on physiology and health *Psychological Medicine* **31** 6
[7] Bayo MV, Garcia AM and Garcia A 1995 Noise levels in an urban hospital and workers' subjective responses *Archives of Environmental Health* **50** 0003-9896 (Print)
[8] Haddock J 1994 Reducing the effects of noise in hospital *Nurs Stand* **8** 43
[9] Morrison WE, Haas EC, Shaffner DH, Garrett ES and Fackler JC 2003 Noise, stress, and annoyance in a pediatric intensive care unit *Critical Care Medicine* **31** 1
[10] Chow T.T and Yang X.Y 2003 Performance of ventilation system in a non-standard operating room *Building and Environment* **38** 12
[11] Menzies D, Fanning A, Yuan L and FitzGerald JM 2000 Hospital ventilation and risk for tuberculous infection in canadian health care workers. Canadian Collaborative Group in Nosocomial Transmission of TB *Annals of Internal Medecine* **133** 10
[12] Bell P.A 1981 Physiological, Comfort, Performance, and Social Effects of Heat Stress *Journal of Social Issues* **37** 1
[13] Poulter Hair L Satisfaction by design. 'Place' component of the marketing mix takes center stage for some providers *Marketing Health Services* **18** 1094-1304 (Print)
[14] Zimring C, Joseph A, Nicoll GL and Tsepas S 2005 Influences of building design and site design on physical activity: research and intervention opportunities *American Journal of Preventive Medicine* **28** 0749-3797 (Print)
[15] Jacobs KW and Suess J.F 1975 Effects of four psychological primary colors on anxiety state *Perceptual and Motor Skills* **41** 0031-5125 (Print)
[16] Steffes R and Thralow J 1985 Do uniform colors keep patients awake? *Gerontological Nursing* **11** 0098-9134 (Print)
[17] Rabin M 1981 Medical-facility colors reduce patient stress *Contract* **23** 2
[18] Edge KJ 2003 *Wall Color of Patient's Room: Effects on Recovery* Editor University of Florida
[19] Connellan K, Gaardboe M, Riggs D, Due C, Reinschmidt A and Mustillo L 2013 Stressed Spaces: Mental Health and Architecture *HERD: Health Environments Research & Design Journal* **6** 4
[20] Duffy F 1997 *The New Office* Londin, England Conran Octopus
[21] Ganster DC, Fusilier MR and Mayes BT 1986 Role of social support in the experience of stress at work *Applied Psychology* **71** 1
[22] Frese M 1999 Social support as a moderator of the relationship between work stressors and psychological dysfunctioning: a longitudinal study with objective measures *Occupational Health Psychology* **4** 3
[23] Netterstrøm B, Conrad N, Bech P, Fink P, Olsen O, Rugulies R and Stansfeld S 2008 The relation between work-related psychosocial factors and the development of depression *Epidemiologic Reviews* **30** 1
[24] Pejtersen JH, Feveile H, Christensen KB and Burr H 2011 Sickness absence associated with shared and open-plan offices--a national cross sectional questionnaire survey *Scandinavian journal of work, environment & health* **37** 5
[25] Danielsson B.C and Bodin L 2008 Office type in relation to health, well-being, and job satisfaction among employees *Environment and Behavior* **40** 5
[26] Lahtinen M, S-D C and Reijula K 2004 Psychosocial work environment and indoor air problems: a questionnaire as a means of problem diagnosis *Occupational and Environmental Medicine* **61** 2
[27] Veitch JA 2001 Psychological Processes Influencing Lighting Quality *Journal of the Illuminating Engineering Society* **30** 1

[28] Clausen T, Christensen KB, Lund T and Kristiansen J 2009 Self-reported noise exposure as a risk factor for long-term sickness absence *Noise and Health* **11** 43
[29] De Croon E.M, Sluiter Jk, P P Kuijer and Frings-Dresen M.H 2005 The effect of office concepts on worker health and performance: a systematic review of the literature *Ergonomics* **48** 2
[30] Duvall-Early K and Benedict JO 1992 The Relationships between Privacy and Different Components of Job Satisfaction *Environment and Behavior* **24** 5
[31] Clements-Croome D 2011 Sustainable intelligent buildings for people: A review *Intelligent Buildings International* **3** 2
[32] Bluyssen PM 2014 What do we need to be able to (re)design healthy and comfortable indoor environments? *Intelligent Buildings International* **6** 2
[33] Bassett D, Browning R, Conger SA, Wolff DL and Flynn JI 2013 Architectural Design and Physical Activity: An Observational Study of Staircase and Elevator Use in Different Buildings *Physical Activity and Health* **10** 4
[34] Nicoll G and Zimring C 2009 Effect of Innovative Building Design on Physical Activity *Public Health Promotion Policy* **30** 1
[35] Nicoll G 2007 Spatial Measures Associated with Stair Use *American Journal of Health Promotion* **21** 4
[36] Myerson J and Ramster G 2017 *Architecture: Workplace health and wellbeing: Can greater design participation provide a cure?* Taylor and Francis Inc. 347-357
[37] W G B Council 2014 *Health, Wellbeing & Productivity in Offices* Council W G B
[38] WHO 1946 *Preamble to the Constitution of WHO as adopted by the International Health Conference* New York
[39] Clements-Croome D 2017 *Creating the Productive Workplace* London Routledge
[40] Maslow A.H 1987 *Motivation and Personality, 3rd edition* New York, US Harper and Row
[41] Dunn H.L 1959 High-level wellness for man and society *American journal of public health and the nation's health* **49** 6
[42] Ryff C.D 1989 Happiness Is Everything, or Is It? Explorations on the Meaning of Psychological Well-Being *Personality and Social Psychology* **57** 6
[43] Csikszentmihalyi M 1990 *Flow: The Psychology of Optimal Experience* New York, US Harper Perennial Modern Classics
[44] Keyes C.L 2002 The mental health continuum: from languishing to flourishing in life *Health and Socail Behavior* **43** 2
[45] Healey-Ogden MJ and Austin WJ 2011 Uncovering the lived experience of well-being *Qualitative Health Research* **21** 1
[46] Antonovsky A 1996 The salutogenic model as a theory to guide health promotion1 *Health Promotion International* **11** 1
[47] Boorse C 1977 Health as a theoretical concept *Philosophy of Science* **44** 4
[48] Abdul-Wahab AS 2011 *Sick Building Syndrome: in Public Buildings and Workplaces* Berling Heidelberg, Germany Springer
[49] Heerwagen J.H, Heubach J.G, Montgomery J and Weimer W.C 1995 Environmental Design, Work, and Well Being: Managing Occupational Stress through Changes in the Workplace Environment *AAOHN Journal* **43** 9
[50] Burton J 2010 *WHO Healthy Workplace Framework and Model: Background and Supporting Literature and Practices* Geneva: Organization W H
[51] Bae S, Asojo A, Guerin D and Martin C 2017 A Post-Occupancy Evaluation of the Impact of Indoor Environmental Quality on Health and Well-Being in Office Buildings *Organizational Psychology* **17** 6
[52] Bowen GA 2009 Document analysis as a qualitative research method *Qualitative research* **9** 2
[53] Stemler S 2001 *An overview of content analysis in practical assessment* 7
[54] Hultberg A, Ahlborg jr G, Jonsdottir I, Winroth J, Corin L and Heimdahl M 2018 *Hälsa på arbetsplatsen. En sammanställning av kunskap och metoder*

[55] Toivanen S 2015 *Framtidens arbetsplatser. Att utveckla hållbara och friska kontor* AB N P D
[56] Bakke J.V and Fostervold K.I 2017 *Kontorlandskap – arbeidsmiljøfaglig veiledning* Norway Helserådet
[57] Hultberg A 2018 *Sammanfattning av forskningsläget avseende aktivitetsbaserade kontor (ABW)*
[58] Kwak L, Grimani A, Aboagye E and Hagströmer M 2017 *Hälsofrämjande insatser riktade till arbetsplatsens fysiska miljö och organisationsstruktur: effekt på arbetsrelaterade utfall. En kartläggning av forskningen.* Stockholm Institutet K
[59] Anttonen H and Vainio H 2010 Towards better work and well-being: an overview *Towards Better Work and Well-being* **52** 12
[60] Ruohomaki V, Lahtinen M and Reijula K 2015 Salutogenic and user-centred approach for workplace design *Intelligent Buildings International* **7** 4
[61] Bakke J.W 2007 *DEKAR – Den Nordiske Kunnskapsarbeidsplassen (The Nordic Workplace Design for Knowledge Work)*
[62] Femenías P and Thuvander L 2018 Transdisciplinary Research in the Built Environment: A Question of Time *Technology Innovation Management Review* **8** 8
[63] Sailer K, Budgen A, Lonsdale N, Turner A and Penn P 2009 *Evidence-Based Design: Theoretical and Practical Reflections of an Emerging Approach in Office Architecture* Editor Sheffield, UK Sheffield Hallam University
[64] Torp S and Vinje H.F 2014 Is workplace health promotion research in the Nordic countries really on the right track? *Scandinavian Journal of Public Health* **42** 15 Suppl

IOP Conf. Series: Earth and Environmental Science **297** (2019) 012014 doi:10.1088/1755-1315/297/1/012014

Multidimensional evaluation of tenants' temporal relocation during a renovation: A comparison between two case studies from Sweden.

A Serrano-Jiménez[1*], P Femenías[2] and L Thuvander[2]

[1] Department of Architectural Construction I. University of Seville. Avenida Reina Mercedes 2. 41012. Seville (Spain).
[2] Department of Architecture and Civil Engineering. Chalmers University of Technology. Sven Hultins Gata 6. SE-412 96. Gothenburg (Sweden).

[*]E-mail: *aserrano5@us.es*

Abstract. This research focuses on the tenants' temporal relocation or stay in the home during renovation works and the property owners' possibility to refurbish with intense and global interventions. The study compares two renovation strategies applied in two residential neighbourhoods in Gothenburg, Sweden, built between the 1950s and 1960s. In one neighbourhood, the tenants stayed in their homes during the renovation works, and in the other neighbourhood, the tenants were relocated to other buildings for over six months. Retrofitting interventions, investment costs, rent increase, time periods and the constructive processes are considered for each renovation strategy, as well as the impressions of the tenants and property owner entities in each neighbourhood were collected. The results compare, through a report on the affordability and effectiveness, the pros and cons of a relocation or non-relocation strategy. This research offers a real and experimental comparison that reach to political, architectural, social and economic conclusions to facilitate the decision-making regarding the tenants' temporal relocation. Finally, it is also highlighted the need to involve the tenants in the design of effective renovation proposals.

1. Introduction

One of the major global and European challenges for this century is to renovate the existing building stock through urban regeneration strategies that should improve people's quality of life in the built environment [1]. Building renovation acquires a substantial role since more than 40% of the European housing stock is over 50 years old and 75% of buildings required for 2050 are already built [2]. These facts justify the need to develop new urban regeneration procedures and evaluate, from different dimensions, the effectiveness and adequacy of refurbishment actions in the housing stock [3].

During recent years, different studies have highlighted the importance of integrating the social and economic dimensions into the decision-making process of housing renovation. Mangold et al. [4] emphasised the need to consider the socioeconomic context in each renovation process in order to avoid the risk of creating social inequality due to rent increases in renovated buildings; Santangelo et al. [5] defined new challenges of building renovation through the influence of the occupant behaviour on the success and sustainable performance of different energy retrofitting measures; lastly, Femenías et al. [6] highlighted the possibility of carrying out partial renovations that achieve sustainable proposals with a social and economic responsibility.

In addition, there is a progressive aging population that is highly significant in Europe, where the number of people over 65 years old is expected to rise to one third of the European population by 2050 [7]. Thus, the consideration of an aging population is a relevant factor in urban regeneration, hence recent studies by Kovacic et al. [8] or Serrano Jiménez et al. [9] have promoted new procedures and renovation strategies specifically focused on the elderly requirements.

Regarding the construction process, one of the main influential factors in major renovation of housing is to decide to either carry out the works with the tenants staying in their apartments or relocating them, with the main restriction of the extra cost that the relocation supposes [10]. Nowadays, in different major renovation projects either of these strategies is chosen depending on the budget and the socioeconomic conditions of each region, and adapting the intervention levels depending on the investment size, the duration or the complexity of works, but without making an exhaustive comparison about the influence of the relocation or non-relocation duality [11–13]. However, all these facts usually limit the intervention levels to the point that housing renovation is reduced in order not to disturb the tenants. It can also be that deep renovation is not implemented, as temporal relocation is considered as complex and costly and challenging to manage [14]. Given these conditions, there is a need to specifically study the influence of this renovation factor in detail, and identify the advantages and inconveniences to be considered when deciding on relocation of the tenants or not during the works.

This research presents and compares the experiences of the renovation of two residential neighbourhoods in northern Europe that have followed different strategies in terms of the tenants' relocation during the process. In the first neighbourhood, the renovation proposal considered keeping the occupants in their apartments during the works. In the second neighbourhood, the tenants were relocated to other buildings for a period of more than six months. The main aim of the study is to evaluate the performance and the consequences of each project with respect to relocation but also the experiences of the elderly tenants. The paper aims at serving property owners support for decision-making based on the characteristics of the buildings, the tenants' perception of the renovation and by taking into consideration the property owners' ability to carry out a major renovation. Furthermore, the paper contributes to the knowledge building of sustainable renovation of the European housing stock, providing technical, economic and social understandings of the consequences derived from a relocation of the tenants or not.

The paper starts with a background description and justification of the topic, explains the methodology applied on both neighbourhoods, describes the two neighbourhoods, and ends with the presentation of the results reached through a multidimensional evaluation of the renovation strategies used in the two cases. The results lead to a series of conclusions about the advantages and disadvantages of relocating tenants or no by showing the impressions and opinions, from the survey results.

2. Methodology

This section presents the procedure used to evaluate and compare the two different renovation strategies, a medium renovation where the tenants stay during the renovation works (Case Study 1), and a deep renovation with relocation of tenants (Case Study 2). Figure 1 shows a graphic outline for the procedure to be performed, including the analysis factors that are taken into account for each case study in the multidimensional comparison and evaluation of the two neighbourhoods.

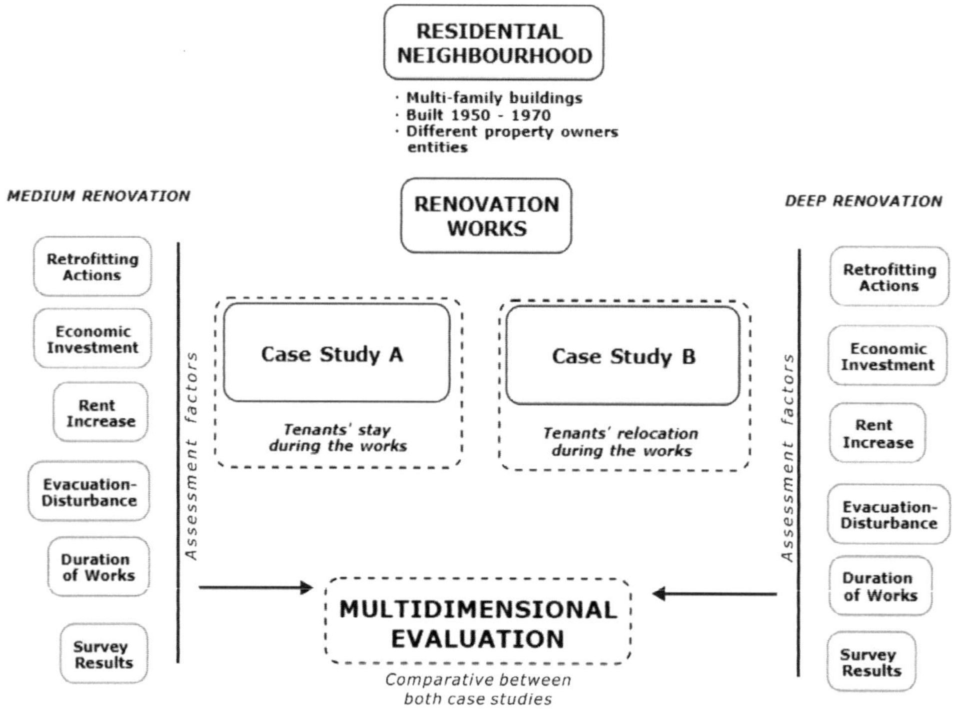

Figure 1. General outline of the procedure developed in this research.

The outline also defines the scope of application, two residential multi-family neighbourhoods, built between 1950 and 1970, and owned by two different semi-public property owners. For each neighbourhood data on retrofitting actions, investment costs, rent increase, evacuation conditions and the duration of work were collected. An analysis of these technical, economic and social aspects were accompanied by the results of a post-renovation questionnaire carried out among the tenants.

The data from the two cases were gathered through interviews, meetings, and e-mail contact with representatives of the property owner entities. These data were complemented with basic diagnosis procedures on the location, and the questionnaire. The research has been developed thanks to the collaboration of a research team made up of architects from universities in two countries in northern and southern Europe, bringing in knowledge from the two different countries in order to find common references and strategies for housing renovation. Therefore, this multidimensional evaluation procedure should be applicable in any other residential neighbourhood as a means to assess the feasibility of relocation, depending on the socioeconomic context of each property owner and the demands of the occupants.

3. Case studies

The case areas are two neighbourhoods with multi-family buildings which are located in the city of Gothenburg, Sweden. In Sweden, where around 50% of the apartments in multi-family buildings were built between 1941-1975 [15], there are many similar residential neighbourhoods facing the same needs for renovation. The two studied neighbourhoods are currently in a similar renovation phase, where some of the buildings have already been renovated and others are in the early design stage of a renovation. The two case studies areas are defined below:

A) A residential area composed of lamellar and tower block buildings that were built in 1950. The buildings that are renovated consist of several lamellar buildings of three storeys with over 200 apartments. Each building has five access portals with six apartments per staircase or nine apartments, if it is the corner. The property owner A decided to keep the tenants in their homes during the works. Figure 2 illustrates the kind of typology of buildings in area A.

B) A residential area composed of several larger lamellar buildings that were built in the early 1960s. Each building had initially 12 storeys and four entrances with independent staircases. So far, two buildings have been renovated comprising around 200 apartments. The property owner B decided to move the tenants to other residential buildings, which they own, during the renovation works. Figure 3 illustrates the kind of typology of buildings in area B.

Figure 2. Generic picture of the buildings in Case A. Photo: Authors

Figure 3. Generic picture of the buildings in Case B. Photo: Authors

4. Results

This section presents the results corresponding to each case study in two different subsections. Subsection 4.1. presents the multidimensional analysis of each renovation including several technical and economic factors, and subsection 4.2. presents the results of the questionnaire.

4.1. Multidimensional evaluation

Table 1 and 2 define and characterise the renovation works carried out in the selected neighbourhoods. Each table presents basic information about the neighbourhood and describes a sequence of the assessment factors related to the renovation strategy, the economic consequences and the development of the works.

Table 1. Renovation works in Case A.

Case A *(without tenants' relocation)*			
Location	**Year of construction**	**Buildings** (reference building)	**Apartments** (per building)
Gothenburg (Sweden)	1950	7	33
Storeys	**% of elderly** (area)	**Annual rent** (before, average)	**Floor area per apartment** (average)
3	23%	105€/m^2	64m^2
Assessment factors			
Retrofitting actions	Medium renovation with retrofitting actions mainly applied in the building envelope, the wet rooms of the apartments, new electricity, and HVAC systems. The actions were: · Building envelope: new plastered, isolated façade system adhered to the existing brick facade and replacement of the windows. · Bathroom: complete renovation with the replacement of pipes and new sanitary elements and tiles on floor and walls. · Kitchen: new electricity, new fan, the tenants can either keep the old furniture or make changes with they pay for separately. · Ventilation: new ventilation with heat recovery system (FTX) in each building.		
Investment costs	Total cost of the intervention of 360,000€ for each building, or an approximate cost of 10,800€ for each apartment or a ratio of 170€/m^2.		

Rent increase	An increase in the pre-established rent of 23% is applied after the renovation, which means an annual increase of 22€/m^2.
Time period	The total time of the renovation in the neighbourhood was one year, the period of renovation works inside the apartments while the tenants resided was 10 weeks.
Evacuation - Disturbance	The renovation process of the façade was mainly carried out from the outside, the windows were replaced from inside the apartments. The bathrooms and kitchens were renovated progressively and sequentially in each apartment, and led to a short temporal relocation of the occupants. In the most complicated situations, there was no access to the bathroom and kitchens for two weeks and there were water cuts for up to two days.

Table 2. Renovation works in Case B.

Case B *(with tenants' relocation)*			
Location	**Year of construction**	**Buildings** (reference building)	**Apartments** (per building)
Gothenburg (Sweden)	1961	2	100
Storeys	**% of elderly** (area)	**Annual rent** (before, average)	**Floor area per apartment** (average)
12	18%	95€/m^2	69m^2
Assessment factors			
Retrofitting actions	Deep renovation with retrofitting actions applied mainly in the common areas of the building and a new distribution inside the apartments. The actions were: · Building storeys: addition of a new storey with apartments in the roof to increase the economic benefits of the intervention. · Common areas of the building: replacement of elevators and a complete adaption of portal access and staircases. · Bathroom: complete renovation with the replacement of pipes and new sanitary elements and ceramic coatings. Expanded size of the room with the redistribution. · Kitchen: complete renovation with the replacement of pipes and new distribution and materials. · Ventilation and heating system: new ventilation system with heat recovery (FTX) in each building and a new heating system and new radiators and pipes.		
Investment costs	Total cost of the intervention including the tenants' relocation of 15,000,000€ for each building, or an approximate cost of 150,000€ for each apartment or a ratio of 2,150€/m^2.		
Rent increase	An increase in the pre-established rent of 45% is applied after the renovation, which is almost the double and means an annual rent increase of 43€/m^2 compared to before the renovation. The increase in rents will be progressive during the first 15 years.		
Time period	The total renovation time reached 15 months, tenants were transferred to other buildings belonging to the same property owner for a period of seven months.		
Evacuation - Disturbance	The construction process was carried out mainly from the interior of the building, generating practically a new one, except in its structural elements. One of the main reasons why the property owner decided to evacuate the tenants was to install a centralised heating and ventilation system throughout the building, which made it impossible to have a building thermally adapted throughout the whole renovation.		

4.2. Survey results

This subsection presents the results of the questionnaire among the tenants in both neighbourhoods. This survey was carried out post-renovation, with the objective of understanding the tenants' satisfaction with the renovation process, including the relocation and their satisfaction with the ready renovation.

The questionnaire was designed by a multidisciplinary team and having the double intention of informing the property owners and increasing the general knowledge about renovation. The questionnaire, with 61 questions (28 about the results of the renovation, 18 about the renovation process, and 15 about the household), to be filled in anonymously, was put in the mailboxes. The questionnaire also included a link to an on-line version. More details about the questionnaire can be read in the study developed by Femenías et al. [16].

The response rate was rather low (29 %), of a total of 392 households, 113 filled in the questionnaire. In Case A, the respond rate was 34,4 % (42 answers from 122 delivered questionnaires) and in Case B, 24% (65 out of 270). In both projects, 56% of the respondents were women and about 42% men. Figure 4 shows the age distribution among respondents in both cases. The distribution of age groups is rather equal, with a slightly higher number of +65 in Case B.

Figure 5 shows the satisfaction with the renovation (Yes or No) related to age groups. There seems to be more dissatisfaction with the renovation in Case B, and especially among elderly. On the one hand, this can be related to the deep renovation in general, with important changes and generating numerous annoyances in their usual residential environment. On the other hand, it could also be connected to the relocation or to the rent increase. According to the property owner, no more than 7 tenants in Case B actually moved back to the same building after the evacuation. 33 of the tenants moved to Case B from another building in the same property which is now about to be renovated. Furthermore, as many as 33 tenants, choose to move directly to another apartment in the area which the property owner provided instead of being evacuated and having to move twice.

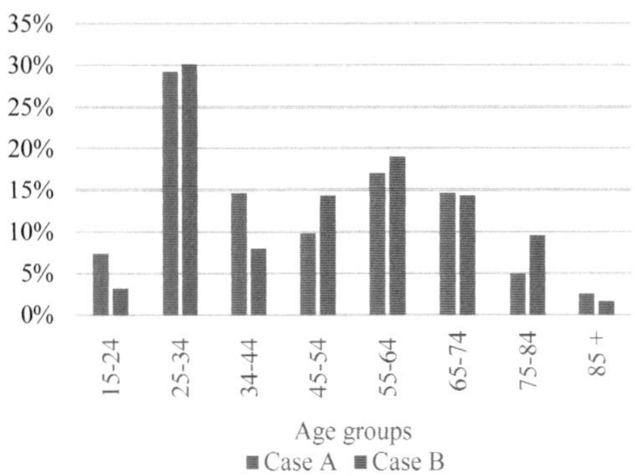

Figure 4. Age groups among respondents

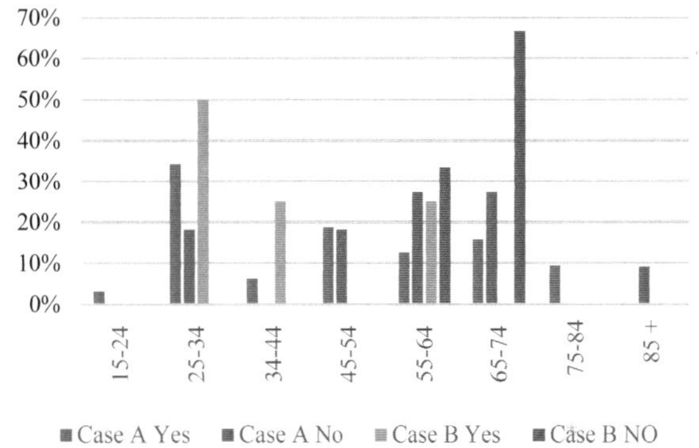

Figure 5. Satisfaction with renovation related to age group

Table 3. Main questions and responses regarding the renovation works in each neighbourhood.

Case A *(without tenants' relocation)*			
Questions	**Answers**[c]	**% Yes**	**% No**
1. Are you satisfied with the renovation?	N=57	73	27
2. Did you think you got sufficient information about the renovation?	N=56	78	22
3. Did you experience that you had influence on the renovation?	N=57	51	49
4. Did you live in a temporary dwelling during the renovation?	N=37	75[a]	25
5. Where you satisfied with the relocation during the renovation?		86	14

Case B *(with tenants' relocation)*			
Questions	**Answers**[c]	**% Yes**	**% No**
1. Are you satisfied with the renovation?	N=57	57	43
2. Did you think you got sufficient information about the renovation?	N=56	50	50
3. Did you experience that you had influence on the renovation?	N=57	80	20
4. Did you live in a temporary dwelling during the renovation?	N=37	50	50[b,c]
5. Where you satisfied with the relocation during the renovation?		100	0

[a] Of these 75%, 22% says they arranged for another place to live themselves during the renovation.
[b] This probably corresponds to tenants that have moved in after the renovation, they were never relocated.
[c] Number of answers of both case studies. It is lower because those that moved in after the renovation didn't answer.

Table 3 shows the tenants satisfaction with the renovation and the relocation. With respect to the relocation, although Case A did not have relocation as a strategy, actually 75% answers that they actually did relocate. Of these 22% arranged an alternative living themselves during the renovation, which shows the need for some people to find alternatives to avoid nuisance and discomfort during the works on their own. The free text comments testify about what the tenants' experiences as an impossibility to stay in their apartment during the works. The comments show that some tenants have approached the property owners with requests on an evacuation apartment, and some managed to get that. Those with small apartments (no separate bedroom), health conditions or night shift work were in priority for the temporal relocation arranged by the property owner. Regarding satisfaction, Table 3 also shows that residents of Case A are more satisfied (73%) with a medium renovation strategy compared to residents of Case B, with lower satisfaction (53%) despite being a deep renovation strategy.

Focusing on aspects that created nuisance for the tenants (Table 4), it is only dirt and dust that corresponds to both cases. In Case A, having no access to private bathrooms and kitchen, disorder, noise and strangers in the apartments were often noted by the respondents. In Case B, the relocation, the length of the process, and "other" scored high besides no opinion which could indicate that those respondents moved in after the renovation. Only one indicated that moving in after the renovation was this "other" reason.

Table 4. Aspects which has caused the most nuisance for the tenants in each case.

Aspects	Case A	Case B
Dirt/dust	**16%**	**10%**
No access to private bathroom	**12%**	0%
Disorder	**10%**	0%
Noise	**9%**	0%
Strangers in the apartment	**9%**	0%
No access to private kitchen	**9%**	0%
Length of the renovation process	7%	**10%**
Lack of information	5%	0%
The relocation	5%	**10%**
No opinion	5%	**50%**
Other	5%	**10%**
Strangers outside the apartment	5%	0%
Limited access to common areas	3%	0%
Light in the apartment	0%	1%

Regarding information and experienced influence. In Sweden, rents, in both public and privately rented housing, can only be increased if the original qualities or living standards of the apartments are improved. The rent increase is negotiated between the property owner and the Swedish Tenants' Association and includes the active approval by all tenants for interior changes in the individual apartment and the approval of 50% of the tenants for standard raising of common facilities.

Our results show that the almost 80% of the tenants in Case A think that they had sufficient information about the renovation. However, they are divided 50-50 about their experience of being able to influence the renovation, the level of interventions etc., or not. In the free text one tenant in area A says that there was even too much decision to be taken by the individual tenant at the same time as the real choice was very limited to pre-defined solutions. Other found the actual freedom of choice unclear. In Case B, 50% think that they had good information and 50% not, but as many as 80% think that they could influence the renovation. Something which has to be considered with the answers of the questionnaire, is that the authors had a rather low respondent rate and that and not all of the respondents did answer all questions, giving a lower response rate to some these.

Although the tenants witness about problems during the renovation process, in general, they recognise the improvement after the renovation. Counting both areas, 72%, think that the renovation was necessary.

5. Discussion

The two cases show two different renovation strategies. The strategy carried out in Case A aims for a new service expectation of around 15 or 20 years while the strategy developed in Case B is a profound renovation with a long-term perspective, with a service expectation of at least 50 years.

Regarding the financial investment, it could be understood as an excessive cost in the strategy carried out in Case B, since the total investment is around one million euros for a building with 100 apartments. It also requires an unusual availability of empty apartments to relocate so many residents. This action demands significant resources from the property owner. However, this action enables a deep renovation of the building, with changes that allow for a complete regeneration of the interior in order to meet the functional and social requirements of the 21st century. In contrary, the investment carried out in Case A has been typical for a medium renovation, taking care at all times of the excess in spending and carrying out the most important adaptation and repair operations to adapt the buildings to the current requirements. In addition, in Case B, the duration of the works has been extended to a higher number of months in order to carry out important works in the building with the guarantee of not disturbing the tenants. In Case A, the renovation duration has been adjusted to reduce the inconvenience and introduce the necessary benefits in the shortest possible time. Case A demands a much greater optimisation in the planning of the works.

Regarding the rents, both case studies have experienced an increase in the rent, but at different rates. In Case B, a progressive increase of 45% with respect to their previous rent will be much more difficult to assume for the tenants. In this sense there is a duality between the architectural benefit, in terms of modernisation of carrying out a deep renovation, and the doubling of costs for the tenants. In the initial processes, a mutual agreement needs to be reached between the property owner and the tenants.

The renovation involves a process of discomfort and inconvenience for the occupants that can be modified and reduced according to the relocation or not of the tenants during the works. However, it has been found that in both cases there are disturbances for the tenants in one way or another, and that even in Case B many of the respondents (corresponding to no more than 34% of the total tenants) were actually relocated as well but for a shorter time period. In Case A, the tenants had to endure noise, discomfort, and the lack of access to some basic services in their home for some period of time. On the positive side they could, in most cases, stay in their permanent homes, without breaking the social tie and without making the extra effort that a displacement to a new residential environment implies. In Case B, the tenants have been able to avoid the inconveniences directly linked to the works, but have had to accept, well in advance, a move to another home with the inconvenience of packing and unpacking in a lengthy process, and having to adapt to a new habitat.

Since there is an aging population that exceeds 23% in housing area where Case A is located and 18% where Case B is located, this research has also a specific focus on the tenants' age as a determining factor in the decision-making regarding the relocation [15]. Although one might think the most appropriate strategy is to relocate the tenants to reduce the impact during the work in this vulnerable population, one of the most counterproductive factors in this population is to abandon their homes for temporary permitting rehabilitation operations [17]. These circumstances generally lead the elderly to reject deep renovations, due to apathy to a long and expensive process and for refusal to break the bond and attachment to its place, which are very important psychosocial aspects highlighted in other research. In fact, the results of the questionnaire carried out have shown an increase in the percentage of non-satisfaction in this population sector for Case B, where a deep renovation has been carried out. Therefore, this research suggests that it is important to have a prior design process with the participation of tenants through meetings or surveys [18].

6. Conclusions

This research has carried out a comparison between two real renovation projects and studied the experiences of housing renovation applying two different renovation strategies, a relocation of the tenants during the renovations works and to let them stay at home. The impact of selecting one strategy or another has been analysed from the social, economic, and constructive point of view, and from the perspective of the property owners as well as the residents through a post-renovation survey. One of the main results of this research is that there is no unique and optimal strategy without having previously analysed the socioeconomic context of each case of application. It also illustrated the importance to consider the residents' views as well as their demographics, in order to create a technically, economically and socially balanced renovation proposal.

Focusing exclusively on the tenants' during the works, the research has shown advantages and disadvantages when integrating this premise in the design of a renovation. Relocating tenants allows to carry out deep renovations in buildings without exceeding the inconveniences to users and ensuring an improvement of comfort and well-being conditions inside the apartments. However, it has been shown that this is an expensive process, which implies a subsequent increase in rent to tenants and that requires a large initial investment with the availability of numerous housing units to relocate. This high rent increase also involves a risk for some of the socioeconomic vulnerable tenants not being able to move back to their former apartment and instead being forced to move to an apartment with cheaper rent. In addition, it has been expressed that the relocation itself can cause many inconveniences and nuisances in the tenants, breaking psychosocial links when moving their habitat to other environments for an extended period. Finally, this process has a higher incidence for elderly people since in this population sector the relocation generates a greater apathy and therefore a greater rejection, preferring simple and precise renovations that do not disturb the rhythm of their daily life. Therefore, this population sector deserves specific attention in the early-stage design of the renovation proposals, introducing particular plans in their relocation by reducing to the maximum the duration of relocation and without breaking the bonds with their permanent habitat.

The inclusion of the results of a post-renovation survey in both cases studies highlights the importance of integrating the occupants in the process of urban regeneration, through identifying their limitations, as well as preferences and opinions. In this case, the results have been related to satisfaction with each strategy and the demands and comments that these processes have generated. This research also demonstrates the utility of this tenants' valuation in an early design stage to develop a better structured process and taking into account the economic and social considerations in the final renovation proposal.

Finally, this research reveals the complexity of the housing renovation process and its multiple influencing factors. Some of the questions are difficult to answer in form of post-renovation survey, and therefore, more qualitative studies, like interviews or focus groups could provide complementary explanations about the tenants' experiences with the renovation.

The paper provides the basis for future research to develop a complete model for decision-making in residential renovation, according to different strategies, and to include various disciplines of analysis so that in each case it can be evaluated from a more exhaustively, assigned a value to each factor, what social, technical and economic impact each proposal has to develop a feasible and successful urban regeneration process.

References

[1] World Health Organization 2015 *World report on Ageing and Health.* http://apps.who.int/iris/bitstream/10665/186463/1/9789240694811_eng.pdf.

[2] European Commission 2018 Directive 2018/844 of the European Parliament and of the Council of 30 May 2018 amending Directive 2010/31/EU on the energy performance of buildings and Directive 2012/27/EU on energy efficiency *Official Journal of the European Union*.

[3] European Union 2015 *Sustainable regeneration in urban areas urbact ii.* http://urbact.eu/capitalisation-and-dissemination.

[4] Mangold M, Österbring M, Wallbaum H, Thuvander L and Femenias P 2016 Socio-economic impact of renovation and energy retrofitting of the Gothenburg building stock *Energy Build.* **123** 41–49. doi:10.1016/j.enbuild.2016.04.033.
[5] Santangelo A and Tondelli S 2017 Occupant behaviour and building renovation of the social housing stock : Current and future challenges *Energy Build.* **145** 276–283. doi:10.1016/j.enbuild.2017.04.019.
[6] Femenías P, Mjörnell K and Thuvander L 2018 Rethinking deep renovation: The perspective of rental housing in Sweden *J. Clean. Prod.* **195** 1457–1467. doi:10.1016/j.jclepro.2017.12.282.
[7] European Commission 2015 *Population ageing in Europe. Facts, implications and policies* https://ec.europa.eu/research/social-sciences/pdf/policy_reviews/kina26426enc.pdf.
[8] Kovacic I, Summer M and Achammer C 2015 Strategies of building stock renovation for ageing society *J. Clean. Prod.* **88** 349–357. doi:10.1016/j.jclepro.2014.04.080.
[9] Serrano-Jiménez A, Barrios-Padura A and Molina-Huelva M 2018 Sustainable building renovation for an ageing population: Decision support system through an integral assessment method of architectural interventions *Sustain. Cities Soc.* **39** 144-154 doi:10.1016/j.scs.2018.01.050.
[10] Egan M, Lawson L, Kearns A, Conway E and Neary J 2015 Neighbourhood demolition, relocation and health. A qualitative longitudinal study of housing-led urban regeneration in Glasgow, UK *Heal. Place* **33** 101–108. doi:10.1016/j.healthplace.2015.02.006.
[11] Serrano-Jimenez A, Barrios-Padura A and Molina-Huelva M 2017 Towards a feasible strategy in Mediterranean building renovation through a multidisciplinary approach *Sustain. Cities Soc.* **32** 532–546. doi:10.1016/j.scs.2017.05.002.
[12] Holopainen R, Milandru A, Ahvenniemi H and Hakkinen T 2016 Feasibility studies of energy retrofits. Case studies of nearly zero-energy building renovation *Energy Procedia* **96** 146–157. doi:10.1016/j.egypro.2016.09.116.
[13] Baek C and Park S 2012 Policy measures to overcome barriers to energy renovation of existing buildings *Renew. Sustain. Energy Rev.* **16** 3939–3947. doi:10.1016/j.rser.2012.03.046.
[14] Miltenburg E, Van de Werfhorst H, Musterd S and Tieskens K 2015 Consequences of forced residential relocation: Early impacts of urban renewal strategies on forced relocatees' housing opportunities and socioeconomic outcomes *Hous. Policy Debate* **28** 609–634. doi:10.1080/10511482.2018.1424722.
[15] SCB 2018 *Statistics Sweden's switchboard.* http://www.scb.se/en/.
[16] Femenías N, Forsemalm J, Jonsdotter L, Punzi E, Bogdanova E, Thodelius C and Granath K (Forthcoming) Residential movement patterns in connection to renovation of rented multi-residential housing: A pilot study. Submitted to SBE Helsinki 2019.
[17] Ujang N and Zakariya K 2015 The notion of place, place meaning and identity in urban regeneration *Procedia - Soc. Behav. Sci.* **170** 709–717. doi:10.1016/j.sbspro.2015.01.073.
[18] Kleit R G and Manzo L C 2006 To move or not to move: Relationships to place and relocation choices in HOPE VI *Hous. Policy Debate* **17** 271 – 308. doi:10.1080/10511482.2006.9521571.

Acknowledgments
The authors would like to thank the Spanish Ministry of Education and Research for financial support via the contract: "Formación del Profesorado Universitario" FPU2015/00070 granted to Antonio Serrano-Jiménez, and the financial support for the research stay EST17/00243 of the author at Chalmers University of Technology in Sweden. Authors also acknowledge the Swedish Research Council Formas, grant number 2013-1804, for the finances of SIRen, a National strong research environment on Sustainable, Integrated Renovation. This research is also possible due to the support of the "Institute of Architecture and Building Science" IUACC from the University of Seville (V-IUACC grants, mode B, 2018).

The role and value of local welfare institutions in multicultural areas

Authors

Kari Moseng, Inge Vestergaard, Birgitte Geert Jensen
Aarhus School of Architecture, Nørreport 20, 8000 Aarhus, Denmark

Corresponding author's e-mail address
kam@aarch.dk

Abstract. This abstract addresses the role and value of welfare institutions as social structural elements of daily life, framing the informal -but important- meeting between residents in an urban fabric.
The discussion is contextualized through the demolition of the primary school Nordgårdsskolen, a local school serving the area of Gellerup, a social housing area back then framing the life of app. 7.000 people counting more than 80 nationalities in the western part of Aarhus, Denmark. Demographics vary from the rest of the city of Aarhus as 35 %. of tenants are under 18 years of age, meaning that a very high percentage of the local population has direct or indirect relations to schoolchildren and their institution (being parents, siblings, grandparents or other relatives).

Due to integration problems and political pressure, a close-down of two kindergartens and Nordgårdsskolen was decided on in 2007 and the demolition of the school itself took place in 2013 – 42 years after its initiation. This caused distribution of children to a large number of schools not only to the surrounding school districts but also to districts in other parts of the municipality.
In a Danish context the area of Gellerup was considered a highly modern "new town" at its initiation in 1968 – the architect Knud Blach Petersen unfolding international inspirations in terms of local welfare-, cultural- and sports facilities supporting the housing structure and the quality of life of the tenants.

New Towns have always been places characterized by migration: by definition, they have no 'original' inhabitants – every resident is a migrant. This situation makes it even more important to gather residents to promote integration.

The aim of the paper is to investigate and discuss the role and value of local welfare institutions not only as centers of service, but also as creators of cohesion, trust and identity.
In the specific case of Gellerup, the paper will unfold the social effects of demolishing Nordgårdsskolen as a physical place of informal meetings in a local multicultural community as Gellerup.

1. Background

Like in many other European countries housing needs in the cities of Denmark expanded in the post-war period. Along with increasing national ambitions related to welfare this led to a Scandinavian interpretation of the international term New Town in Brabrand in the western part of Aarhus, the second biggest town in Denmark after the capital Copenhagen.

Gellerup was developed during the late sixties in a collaboration between the housing association Brabrand Boligforening and the architectural office of Knud Blach Pedersen. The masterplan unfolded comprehensive social and architectural ambitions of modern middleclass housing and additional welfare institutions, framing the lives and recreation of its app. 7000 inhabitants – segregated from work life as was the ambition of ideal modernistic housing.

Also inspired by the garden city movement the masterplan layout of Gellerup consisted of 23 housing blocks of both four and eight levels centering a park-zone only crossed by pedestrian traffic. [1]

1.1 Welfare in the context of a Danish new town

Knud Blach Petersen himself visited France during the sixties and most likely found international inspirations in the collective housing of other modernist architects among others Le Corbusier. To understand the concept of Gellerup it makes sense to look into the conceptual ideas of the building structures of Le Corbusiers five versions of Unité d'habitations, (all realized between 1946-67) As it is well-known the first "prototype" in Marseille was later realized in three other French locations (Nantes en Reze, Briey, and Firminy) and in one location in Berlin, Germany.

Based upon Le Corbusiers concept of "The house as a machine to live in" featured in "Verse un architecture" from 1923 [2] the building concept unfolded the vision of the "vertical garden city" [3].

In the Unites d'habitation, dwellings, residential services as well as collective facilities such as kindergarten and nursery school are all implemented into the building structure, leaving surrounding nature out of touch. Situated on the top floors, the school and kindergarten are only accessible through the structure, which makes its relations to the surrounding area of lesser importance underlining its belonging to the "living-machine". One could say that in the Unités the institutions mainly served the families practical need of child-care to make modernist everyday life function.

1.2 Disassembling the living machine

Like the different versions of Unites d'habitations, the New Town of Gellerup also offered residential welfare services for its inhabitants; a library, a church, swimmingpools, shopping and sports facilities as well as school and kindergartens. Unlike in the layout of Corbusiers Unités however, the plan of Gellerup implements its welfare facilities into the surrounding landscape, connecting the various buildings by a system of footpaths through the landscape. One could say that in spite of strong references to Corbusiers concepts of living-machines Knud Blach Petersen in the case of Gellerup disassembles his "machine", introducing a closer contact between man and (park-) nature.

2. Nordgårdsskolen

An important element of welfare in the masterplan was of course the 7.600 sqm. school buildings of Nordgårdsskolen, initiated in 1971 after severe attention in the local medias because of a financial overrun of more than a hundred pct. on the building costs. The building commitee was criticized for having "thought more about the project than the economy" when applying for 31 mil. DKK. and not the 15 mil. DKK that was first assumed. [4]

With its offset one-level units in three levels, sloping slightly, like the landscape, towards south, the school and the park seemed interconnected. Situated in the western part of the area the school worked as a sort of hinge between the inner park and the surrounding suburb, being accessible by car from the west, but connected to the rest of the area by the system of footpaths that characterized the non-car area of the inner park.

On the eastern side of the school access to a sports hall and outdoor sports area was offered. Off-school hours the area, which was easy accessible, would also serve as overall sport facilities for the inhabitants of Gellerup [1].

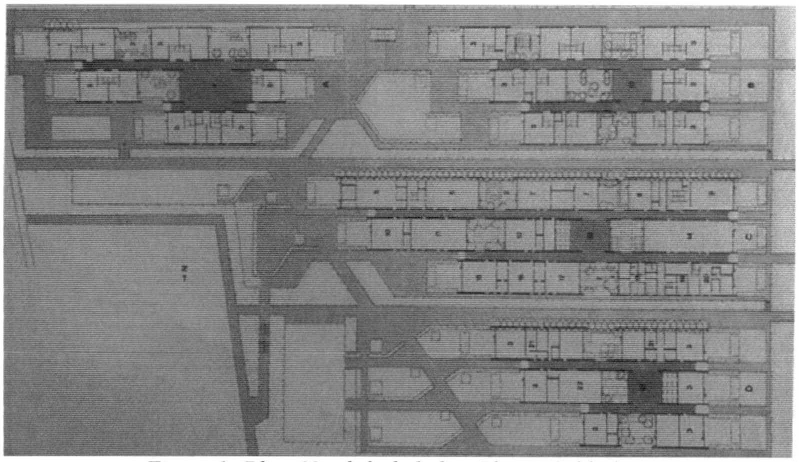

Figure 1: Plan, Nordgårdsskolen Photo: Arkitekten

2.1 The buildings

The school buildings divided facilities for first to tenth class into three units containing the infant-, junior-, and continuation-schools. The three minor units were centered around a joint building offering specialized classrooms for creativity, music, science as well as library, administration, dentist-, and medical facilities. The ambition of dividing the units into three mirroring different ages of schoolchildren was seen as untradtional and was by some inhabitants critized because a lot of families in the new area of Gellerup were young and had small children, which created a severe pressure on the infant school [5].

Each building consisted of a number of long, parallel bays of 9.60 meters width, separated by 3,60 meters wide glass-roofed corridors [1].

2.2 The various outdoor spaces

The building offered various types of outdoor spaces as all classrooms had direct exit to a smaller courtyard that took in daylight and framed outdoor teaching and breaks. A larger outdoor area between the infant and the junior school would serve as meeting point and playground and of course the park on the eastern side of the school was accessible for sports activities and other events.

2.3 Supporting community and dialogue

The school was developed under the influence of Danish school policy and the pedagogical thinking in the late sixties and early seventies abandoning "industrial" school buildings with uniform classrooms in favor of a various scheme of teaching rooms – many of which were meant to support both formal and informal dialogue and community-thinking with the aim of supporting the upbringing of democratic citizens. Seemingly, the village was an architectural metaphor for the spacial organization of Nordgårdsskolen, working as its own little township in the larger plan of Gellerup. The buildings centered all classrooms around a common square in every unit. Streets connected the classrooms to the outdoor patios of the school, making the school accessible yet shielded from the surrounding world [1].

According to a former pupil, who attended Nordgårdsskolen between his first and fifth form in 1985-90, the most valuable place of the school however was the outdoor square between the infant- and the juniorschool. This particular place would gather children in play and talk also in their free time, as the schools open structure allowed inhabitants of the school district to use the school area all day long [6].

2.4 An untapped future flexibility

A consciousness of potential future changes in school policy as well as potential changes in the needs of society was implemented in the building from the beginning. The buildings of Nordgårdsskolen were assembled in precast concrete components; columns, beams and 1.20 metres wide wall units. All building elements were assembled by bolts which actually made it possible to dismantle, replace or reuse the elements in other contexts. Even the lavatory units could be removed and replaced elsewhere and below the floors were cat walks for inspection purposes.

According to engineer Torben Sigurdsson, who was employed at Birch & Krogboe, the then engineers on the project, the buildings were designed to adapt to changed requirements and could actually be modified to suit completely different layouts [1] [7].

3. From emerging democracy to demolition

From its inauguration in 1971 Nordgårdsskolen was an attractive district school for the people of Gellerup -in the year of 1973 the school had eight first forms at the same time mirroring the predominantly young families of Gellerup [8].

However through the seventies, not long after the estate of Gellerup and its welfare institutions were commissioned, it turned out that a lot of the middle class residents the 2000 rental units were addressing, rather wanted to buy their own houses as conditions for private loans were favorable.

Gellerup, built and run by Brabrand Boligforening, one of many Danish Housing Associations that organizes rental housing built with public financial support, got through the seventies and eighties inhabited by people who were in need of affordable accommodation and by raising numbers of migrants, refugees and their descendants [9].

Mirroring the inhabitants of Gellerup the student body of Nordgårdsskolen in the mid-eighties counted twenty percent bilingual pupils and eighty percent ethnic Danes. According to a former pupil who started in first form in 1985, this was a perfect situation seen from an integration perspective, as language was learned quickly and the integration and the cultural exchange had good conditions [6].

By the end of the eighties and during the nineties a large percentage of inhabitants however, had no connection to the labour market, crime and violence rates were high, also causing a lot of children opting out of the local school to seek alternate possibilities.

In 2004 Nordgårdsskolen was the first school in Denmark to have a completely bilingual student body. Around the school smaller groups of young people caused vandalism on the school buildings and political focus on the social challenges of the area were servere, the main argument being on integration-isssues [10].

The same year, in 2004, the Danish Gouvernment publiced their first strategy for disadvantaged housing areas and has every year since 2010 published so-called ghetto-lists placing the area of Gellerup in the cathegory of so-called hard ghettos. [11]

In spite of severe resistance from the locals the political decision of a close down of the school was taken in 2007 and in 2008 the 452 pupils left Nordgårdskolen for the last time to continue schooling by commuting to 14 different schools around the Municipality of Aarhus – the longest distances causing two hours of commuting every day [12].

Between 2008 and the actual demolition in April 2013, the school buildings were used on and off, the joint building in the middle of the complex temporarily serving as a home for the manifold local associations (at the moment Gellerup is the home of 133 associations) [6].

It was decided not to demolish the unit of the infant school as the Municipality planned to use this as supplement for the new kindergarten buildings on the site of the school, but the junior school, the continuation school and the joint building were demolished as well as a lot of the trees and vegetation on the site were cut down.

As a result of the demolition the concrete elements were made into 20.000 tons of gravel. According to P. Olesen & Sons, the demolition company in charge of the task, to be recycled as land fil in road construction. Glass from windows was to be recycled as new glass or as filter sand. [13].

The demolition of the school was the very first part of a severe transformation process of Gellerup taking place in these years to unfold the ambition of creating a heterogeneous housing area and to accommodate the Danish Government vision of ghetto/non-ghetto At present 340 apartments have been demolished in order to add other types of housing with various ownerships to the area. The Housing association and the Municipality of Aarhus are at present debating whether to only demolish an additional 200 apartments or whether to demolish up to 600 apartments [14].

Figure 2: Demolition, Nordgårdsskolen Photo: Skræppebladet

Figure 3: Demolition, Nordgårdsskolen Photo: Skræppebladet

4. Methodologies

The methods of investigating the topic of this paper are various.
Literature studies on modernistic housing have been carried out and, furthermore, the newspaper articles from the initiation of Gellerup has brought knowledge into the work.
Attention has also been paid to local networks on the internet exposing the views of and discussions between tenants.
As less than 20 pct. of the former buildings of Nordgårdsskolen, the subject of this paper , remaining, researching the school's architectural qualities has been impeded. Hence the investigation relies on studies of drawing material, photos, literature and interviews with people involved in the building process.
Studies of scientific literature on the field of sociology and the value of networks in housing areas have been studied. Studies on the field of commuting have also brought knowledge to the work.

5. The overall consequences

Human mobility may seem beneficial from an acute integration point of view, when aiming to dissolve social challenges in a local area. When the political decision of closing down Nordgårdsskolen in Gellerup was taken Louise Gade, who was then councilman in the Municipality of Aarhus argued that "the close-down and following replacement of the children was meant as a very solid school - and integration policy effort" adding that the municipality offered free bustransportation and even support for moving expenses, if families chose to move closer to their new schools, the limit of transportation time being up to two hours a day [15].
According to an employee of Brabrand Boligforening there are no known cases of families that have moved away from Gellerup because they wanted to live closer to the schools their children were distributed to. As he puts it "no one would give up their community and their close relations in the area" [6].

The human consequences of commuting have been explored in recent years by, among others The University of Roskilde [16]. There are manifold signs of severe negative effects of commuting on personal, health and societal well-being. From a health point of view longer commutes are linked with increased rates of obesity, high cholesterol, high blood pressure, back and neck pain, divorce, depression and death.

At the societal level, people who commute more are less likely to vote. They are more likely to be absent from work, and they have kids who are more likely to have emotional problems. Speaking more abstractly, there is a huge pool of more or less untapped human potential locked up in long commutes [17].

6. The role and value of welfare institutions

The attitudes towards the outcome of dissolving the large group of schoolchildren from Nordgårdsskolen, distributing them over a larger area are manifold – some see benefits, some see drawbacks for the pupils and the surrounding society. But what are the costs of losing welfare institutions in an area - institution that a lot of citizens have had a relationship to?

As a welfare institution a public school is by definition for everyone – democratic in the sense of being an open, free of charge and non-religious institution.
In the first place schools are a necessity, framing the needs of society and of families to make everyday life work. Apart from framing children's lives and learning in important years of childhood it also structures the daily lives of parents and siblings going back and forth – meeting other parents or maybe teachers informally. Furthermore a lot of Danish public schools also let local associations use their facilities for activities ranging from sports to all kinds of teaching, meetings and events. This means that the school building serves as a community center to a greater extent. Before the demolition of the school locals criticized the Municipality, pointing out the fact that the platform of the manifold associations of Gellerup would also be homeless [18].

6.1 Social capital - Bonding and Bridging – the strength of the weak ties

American political scientist Robert D. Putnam describes his theories of social capital in "Bowling Alone", 2003 Furthermore he creates the theories of "bonding and bridging" as different valuable ways of networking in society. [19].

Bonding, he argues is when individuals take part in networks that amplifies homogeneity and identity by including people that have things in common – close family, ethnic minorities, Sunday schools or pokerclubs The positive aspect of "bonding" is mobilization of solidarity within the group providing social and mental support to its members, the negative aspects is that the groups can appear excluding to others.

6.2 The strength of the weak ties in the case of Gellerup

The other kind of network, according to Putnam, is *bridging*, describing networks that are open and including – for instance religious organizations or political groups. The ties between members in bridging networks are weaker, but very important in order to obtain information and possibilities of social mobility. Mark Granovetter, American sociologist, therefore describes "the strength of the weak tie", arguing that the weak ties are important because they have potential in terms of social mobility, understanding and cultural exchange [20]. Bridging may hold potential for instance when talking of citizens in multicultural areas such as Gellerup, but also in every public school in every society, because the bridging effect creates contact between social classes, religious minorities, gender etc.

7. FINDINGS

Exploring the field has caused various findings, the most important ones are unfolded in the following.

7.1 Environmental findings

Investigating drawing material and interviewing engineer Torben Sigurdsson, in the late 1960es an employee at Birch & Krogboe, it becomes clear that Nordgårdsskolen unlike many other buildings of the time was actually designed for disassembly. This means that in this particular case there was an implemented but later untapped potential of either rebuilding the school for another function or reusing building elements for other purposes. In an environmental perspective both solutions would have had lower impact than the chosen demolition of 75 % of the buildings in 2013. Demolishing 20.000 tons of concrete into gravel and transporting it to its destination as landfill in road construction has both economic and environmental costs that have to be taken into consideration when concluding on the decisions.

As mentioned above research show that commuters experience negative consequences in terms of health, their social lives and well-being. Furthermore bus transportation of up to two hours a day instead of walking or biking to a local school can hardly be considered sustainable from an environmental point of view. In 2008 the spokesman Ib Bremshøj, Municipality of Aarhus, told the local newspaper Aarhus Stiftstidende that twenty new busroutes had been established to solve the needs of bustransportation from Gellerup [12]. Danish children go to school 200 days every year, meaning that the children with the longest commute have bustransportation of up to 400 hours a year. Needless to say that the impact on the environment over the years has been considerable.

7.2 Human and social findings

Schools are a social platform for more than teaching. In spite of the social problems Nordgårdsskolen may have had, in the perspective of Putnam the demolition leaves a larger wound than a whole in the ground. When dissolving the platform for bonding and especially bridging theres a lack of potential positive social capital, coherence and trust the area could need.

A report from Department of Learning and Philosophy at Aalborg University [21] made in 2014, explored the presence of local associations and voluntary work in depopulated villages on the west coast of Denmark. According to its conclusions there is a linearity between the volume of voluntariness and cohesion in small societies – the voluntary work very often initiated and taking place in the local school. Also, it is concluded, the societies with many associations are more likely to have a vertical contact with the municipality or other authorities negotiating resources, which local areas usually benefits from.

7.3 Findings on the consequences of commuting
It has now been more than ten years since the school children started commuting from Gellerup to the fourteen schools in the municipality as a result of the close-down and following demolition. In an overall view it makes sense to look at what human, social and environmental consequences this has resulted in, and how we can conclude on the decision as a platform of an overall societal discussion. Long commutes have human consequences. In the case of Nordgårdsskolen we are, of course, dealing with schoolchildren aged between 6 and 15, not grown-ups on their way to work, which a lot of the research in the field is looking at. However, it is most likely that the effects of commutes for a school child are even more serious considering that young bodies and minds are under development, and also given the fact that, unlike adult commuters, the choice of commuting is completely out of their hands.

It is remarkable that the close-down and demolition of Nordgårdsskolen was carried out as a political ambition aimed at integration, among other things aiming to give children and society a wider range of possibilities, social mobility and future-choices and at the same time leaving the children in question with remarkably less time to socialize, make friends, exercise and do homework or read, all of which seems to lower life quality, future possibilities and happiness in other ways.

Also in relation to the human health the up to 400 hours of commuting -or 16,5 days- spent on a bus every year seemingly has severe consequences. Both the physical and the social wellbeing of the commuting children and young people ought to be in focus in further investigations.
According to an inhabitant, who is the parent of six, the commuting children have a hard time engaging in activities of the local community especially after a Danish school reform in 2014 dictated fairly longer schooldays [6].

7.4 Findings of common memory connected to places
Literature studies and interviews show a common memory of the school as a place of special qualities. Some former pupils describe their lives as being enriched by their Gellerup upbringing and remembers their teachers as involved and engaged, however it has sometimes been connected with prejudices from the outside world to have been a pupil there because of the later social problems and vandalism. [22] A Danish-Palestinian man, still residing in Gellerup, describes how he is still in contact with his former teachers and classmates and how the school served as a social landmark for the children in the neighborhood, even after he left the school to go to another school in his fifth grade. He also describes how he always tells his daughter, who goes to one of the new kindergartens built on the site of Nordgårdsskolen, where his favorite spot in the school was and how happy the place used to make him, underlining how sad he feels that a few groups of youngsters caused social problems and vandalism.
A former teacher describes Nordgårdsskolen as a "tough place to be" in the last functioning years, however, it also had a certain team spirit and feeling of coherence between teachers, pupils and parents [23]. Seemingly, the now torn down place accumulates rather strong feelings of affiliation and of a cultural-historical memory of community that is present although it is now physically absent and cannot be visited anymore.

8. Discussion
When discussing the political act of solving the social problems of an area by demolishing its school buildings there are various perspectives to unfold, some of which includes severe paradoxes. Looking into the UN Sustainable Goals [24], as a background for the discussion, the focus is -among other things- on *sustainable cities and communities* as well as on *good health and well-being.*

On a national level it is interesting that the Danish Government presented an ambitious and humanistic architecture policy [25] under the headline "People in the center" in February 2014, less than half a year after the actual demolition took place. Among other things it is stated that:

> "*It is important that renovations are made with an eye for user needs, including the need for accessibility, cultural-historical preservation values, potentials for architectural lifting and energy improvements, and therefore the government will develop methods and examples of how this is best done in practice.*" [25]

Whether looking at the UN goals or at Danish architecture policy there seems to be a lack of correspondence between the ambitions described and the fact that a large group of children from Gellerup are commuting with the all known disadvantages this entails. Furthermore their local community does not have the advantages, comfort and social potential of a local school. Seemingly, the ambition of sustainable communities, good health and well-being is downsized in order to reach a local political goal in the name of integration. Further research ought to look into the children's well-being in terms of health and social well-being and integration parameters. To some extent the aim of "user needs, accessibilty and cultural-historical preservations" as argued in the Architecture Policy contradicts the consequences of the demolition described in this paper.

The recent masterplan of Gellerup in fact – apart from massive architectural and infrastructural transformations – also involves plans for a new school in the area. Paradoxically, the argumentation is that to attract citizens to the area it will have to offer local school facilities, because no families would like to do without a local school - contradicting what the locals of Gellerup was offered back in 2007 when decisions on the school closing were taken. [26]

Also, the perspectives of cultural-historic value and common memory when transforming and demolishing housing areas and places of social value, seems under-researched. The process of commissioning and identifying oneself with urban fabrics of former, existing and new structures is of great importance when consolidating transformed areas.

9. Conclusion

This paper attempts to highlight and discuss the role and value of public welfare institutions as more than the framework of pragmatic functions.

The aim has been to explore accessible sources of knowledge to explore the situation before and after the demolition of the local school in a multiethnic- and cultural area in Denmark.

It seems however that an important perspective is - whether in the context of a depopulated village on the West Coast of Denmark or, as in this case a multicultural area with more than 80 nationalities, a school close-down has severe consequences not only for the children involved but also for local associations, existing and potential cohesion and the resulting trust in the local community.

Taking the scale of Gellerup into account and also considering that recent demographics show a local population of 35 pct. inhabitants under 18 years of age (a deviation from national average of approx. 20 %) logics would call for a local school no matter the nationality of the children [27] [28].

Therefore, the conclusion of this paper must be that if we consider welfare institutions like local schools to be more than "buildings of service", but also places of local human exchanges and voluntary activities, they are platforms of potential coherence and trust and must be seen as important social landmarks when planning or reestablishing housing areas.

Acknowledgements

The authors would like to acknowledge Chadi Ali Kayed, inhabitant of Gellerup and an employee at Brabrand Housing Association. Chadi has shared insightful knowledge of the development of Gellerup and of the life at Nordgårdsskolen and the consequences after the demolition.

Also we would like to thank Engineer Torben Sigurdsson for sharing his knowledge of the concepts of construction when developing Nordgårdsskolen, and for sharing his articles, drawing material and building-site photos.

References

[1] Arkitektur (Danish) 1974 *Gellerupplanen, Brabrand* (København: Arkitektens Forlag) p 289 –
[2] Corbusier 1927 *Towards a new architecture* (London: The Architectural Press)
[3] Cohen JL 2018 *Le Corbusier The buildings* (London:Thames & Hudson)
[4] Jyllandsposten 1970 *Så mere på projektet end økonomien* (Viby J:Jyllands Posten 9.10.1970)
[5] Jyllandsposten 1970 *Ny skole til 30 mill. uegnet til de børn som skal bruge den* (Viby J:Jyllands Posten 23.11.1970)
[6] Interview 2019 *Chadi Kayed* (Gellerup: authors sound file 07.01.2019)
[7] Interview 2018 *Torben Sigurdsson* (Gellerup: authors sound file 28.08.2018)
[8] Hansen H *Lederen Nedbrud og genrejsning (Gellerup: Skræppebladet 3.4.2013*)
[9] Bach JS 2019 *Ambivalent Hopes - Residents' Views on Architectural Transformations in Socially Disadvantaged Estate* (NAF)
[10] Politiken 2007 *Århus lukker skole for to-sprogede* (København: Politiken 20.11.2007)
[11] Danish Governments ghetto list 2018 (København:2018)
[12] Mathiesen S S 2008 *Integration giver lange skoledage* (Aarhus: Aarhus Stiftstidende 12.8.2008)
[13] Helbo F 2013 *Udtjent folkeskole bliver brugt som vejbelægning* (Maskinteknik 7.10.201)
[14] Yding H og Lichscheidt E *Boligforening om Gellerupparken: Riv kun 200 lejligheder ned i stedet for 600* (Østjylland:DR Nyheder 30. 11. 2018)
[15] Politiken 2007 *Aarhus lukker skole for to-sprogede* (København:Politiken 20.11.2007)
[16] Freudendal-Pedersen M *Pendling gør mennesker ulykkelige* (Roskilde:1.9.2014)
[17] Washington Post *How much of your life you're wasting on your commute* (Washington:2016)
[18] Kvorning J 2017 *Velfærdsinstitutionen som urbant fortætningspunkt* (København: Arkitektens Forlag) p. 112 Skræppebladet
[19] Putnam RD 1995 *Bowling Alone: America's Declining Social Capital* (Journal of Democracy, January 1995) pp 65-78
[20] Granovetter M *The Strength of Weak Ties* The American Journal of Sociology, Vol. 78, No. 6., pp. 1360-1380
[21] Lolle og Lauridsen AKE *En undersøgelse af skolers betydning for den sociale kapital i lokalsamfund* (Aalborg Universitet, 2014)
[22] Hansen H *Et sidste farvel til Nordgårdsskolen* (Aarhus 2013)
[23] TV2Østjylland 2013 *Nordgårdsskolen var et hårdt sted at være* (Aarhus 2013)
[24] UN Sustainable Goals (https://sustainabledevelopment.un.org)
[25] The Danish Gouvernment, Ministry of Culture *Architecture Policy – People in the center* (København 2014) https://english.kum.dk/services/publications/2014/danish-architectural-policy-putting-people-first/?L=0
[26] Helhedsplanen 2007 (Gellerup: https://helhedsplangellerup.dk/)
[27] Samvirket for Gellerup og Toveshøj (https://www.samvirket.dk/fakta)
[28] Danmarks Statistik Homepage https://dst.dk/da/Statistik/emner/befolkning-og-valg

Infrastructure, innovation and industry as solutions for breaking inequality vicious cycles

Irene Josa and Antonio Aguado

Civil and Environmental Engineering Department, Universitat Politècnica de Catalunya (BarcelonaTECH), Jordi Girona 1-3, 08034 Barcelona, Spain

irene.josa@upc.edu

Abstract. Currently, poverty and inequality are crucial social issues around the world. They demand objective evaluations with the purpose of defining policies and prioritizing actions. Most of the times the most important problem lies on the fact that individuals under poverty circumstances or belonging to vulnerable collectives cannot escape these conditions by themselves and need external assistance or interventions. Another added difficulty is that both poverty and inequality can be studied from different angles and require a multidimensional approach whose definition is not straightforward. There exists a vicious cycle defined by a collection of determinants or events that lead to a phenomenon in which various disadvantages work circularly so that it is impossible for individuals or households to break the cycle.

In this context, setting the background in which these vicious cycles arise can be helpful in order to conduct in-depth studies as a first step to establishing possible solutions. Even though many researchers have proposed alternative schemes for the poverty's vicious circle, none has been introduced for inequality. In this contribution, a meaningful and integrated cycle is defined for inequality at two different levels: first of all, at an individual level; secondly, at a collective level in which a certain group is made vulnerable. This cycle includes similar parameter as those comprised in the poverty cycle such as income, access to education and access to healthcare. However, it additionally places emphasis on other aspects such as participation in politics and access to leadership positions.

Besides, a conceptual and analytical framework is built, whose objective is to describe ways in which infrastructure, innovation and industry can help interrupt the continuation of the cycle. This model integrates a collection of different measures that is to be potentially helpful for policy makers. Additionally, the frame accounts for not only the fight of current existing cycles, but also of future generations ones, hence considering sustainability dimensions.

1. Introduction

Inequality and poverty are growing problems around the world and many studies have been showing how the levels of these phenomena both within and between countries have increased in the last decades. Infrastructures have been recognised as having a fundamental role to play both in the generation and in the mitigation of these problems. Actually, there exists a growing amount of literature dealing with the relationship existing between infrastructure and growth and or human development. But infrastructure development not only has effects on human development and growth, but also on inequality. Despite the fact that some researchers have analysed the relationship between infrastructure, poverty and inequality [1,2,3], this work has failed to address the multidimensional nature of these phenomena. Current analyses focus on inequality in economic terms and therefore fail to approach other dimensions such as education and health. An inherent issue to poverty is the fact that the conditions under which it occurs generate a collection of characteristics that make it difficult to escape them. In fact, many authors have dealt with the so-called vicious cycle of poverty and have proposed different versions for it [4,5,6]. Even though there exists a parallelism with the phenomenon of inequality, there is not yet literature describing how inequalities arise and propagate along time, making it difficult for individuals in vulnerable situations to escape from this vicious cycle. However, the understanding of these cycles is fundamental for the consideration of sustainability holistically.

Having said this, the objectives of this paper are two-fold. First of all, this paper aims to propose a structure for the vicious cycle of inequality with the objective of providing a better understanding of how this phenomenon arises and when and where are most useful the methodologies to tackle with it. This cycle exists both at an individual level and at a collective level and both are analysed. Secondly, we examine how these cycles can be interrupted through industry, innovation and infrastructure, which are the basis of the ninth Sustainable Development Goal.

The structure of the paper is as follows. Section 2 describes the methodology that has been followed for the analysis carried out in the development of the paper. Section 3 briefly introduces contributions dealing with the effects of infrastructure on problems such as human well-being, poverty and inequality. Then, section 4 describes the vicious cycles of poverty and of inequality. On the basis of these descriptions, section 5 discusses measures to be taken in order to break these cycles. Finally, section 6 concludes.

2. Methodology

The methodology that has been followed for this study is the one shown in Figure 1. First of all, a literature review was conducted with the objective of establishing the relationships that have been found between poverty and inequality with infrastructure, industry and innovation. Secondly, the results of the literature review were used to establish how the vicious cycle of poverty has been defined in other publications. This, together with the references dealing with inequality, was used at the same time as a source of information to describe a vicious cycle of inequality. On the basis of this cycle, several different key spots that may be helpful to break the cycles are identified and possible solutions are analysed.

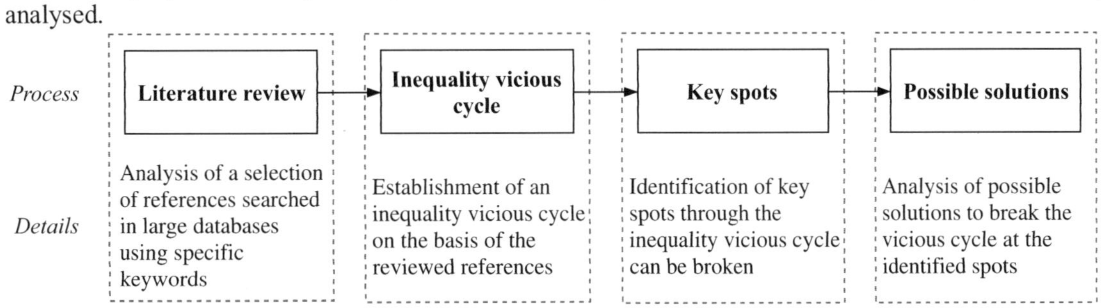

Figure 1. Diagram of the followed methodology

3. Literature review

There is general agreement on the fact that infrastructure development plays an important role both in developed and developing countries. However, authors tend to diverge with regard to the characteristics of this role. While some of them emphasise the benefits of infrastructures on economic growth [1], others point out at the fact that when the impact is positive in the sense that it leads indeed to an economic growth, it can have negative effects on the way assets are distributed among population, leading to an increase in interpersonal and interregional inequalities [2]. Besides, some analyses show that growth due to investment in infrastructure is not always positive and it depends on the kind of infrastructure, such as health, education, transport, energy and communication [7,8]. Other authors have analysed the conditions under which investment reduces inequality [3]. Infrastructures can help in the reduction of inequality by improving mobility and geographic access and therefore enhancing people's access to job opportunities or children access to schools; also, it can serve to increase the value assets of those at the bottom of the distribution. Besides, communication infrastructure can be helpful for vulnerable groups to search, find and apply for productive opportunities to which they would otherwise not have access to. It must be emphasised at this point that although many authors have traditionally focused on inequality in monetary terms, inequality is a multidimensional problem and therefore it can occur in different areas apart from economy, such as health, education or participation. This means that even though in some situations infrastructure provisioning affects only one of these dimensions, they are correlated and other dimensions can indirectly be impacted.

Apart from the effects of infrastructure on inequality, several authors have analysed and empirically shown evidence on the relationship between innovation and disparities [9,10,11]. Nevertheless, there is disagreement on the kind of effect it has. [12], for example, show a positive correlation between innovation and social mobility; but [13] emphasise that even though innovation can lead to economic development, it does not ensure this growth to be totally inclusive.

It needs to be considered that there is a strong connection between built environment and poverty or inequality. This connection is related to sustainability. Even though a considerable amount of literature has analysed infrastructures' environmental sustainability, few writers have been able to draw on any systematic research into its social pillar [14,15,16,17]. However, according to some authors, it is only when a good or a service is socially sustainable that it is possible to start considering environmental sustainability [18].

Among the Sustainable Development Goals, the ninth provides a basis for understanding and dealing with issues concerning infrastructure, such as transportation networks, water provision and sanitation and energy infrastructure.

4. The vicious cycle of inequality

The circumstances under which children that live in poor families grow up substantially affect their personal and educational development. In this context, the access of the household to resources to cover their basic needs is minimal; therefore, these children lack access to enough food, to safe water and sanitation and to healthcare. As a result, the children may be malnourished and have health problems; this has as consequence a low productivity at school due to the lack of energy of the children or due to their incapability to attend school classes. They are disadvantaged in education and skills and have difficulties finding a job when older, which will make it complicated for him to escape the vicious cycle.

The cycle can also be analysed from the point of view of an adult. If he is living in a poor household he is probably disadvantaged in education and skills. At the same time, poverty will also restrict his access to food, safe water and healthcare, which leads to a higher probability of undergoing diseases or infections, as well as to a lack of energy. Altogether has as consequence less job opportunities or a low productivity at work, leading to unemployment or a low personal income. This can have effects on the personal stability and self-esteem of the individual and in some cases can result in increased child births; together with the high probability that there are sick adults at the household there is a greater number of dependants that need to survive with the household's low family income.

These cycles, the children and the adult ones, can be seen in Figure 2. They include the factors that authors have most commonly included in their analyses, even though small variations from the diagrams presented here can exist; these differences come generally from the point of view taken and from the depth of details in the description.

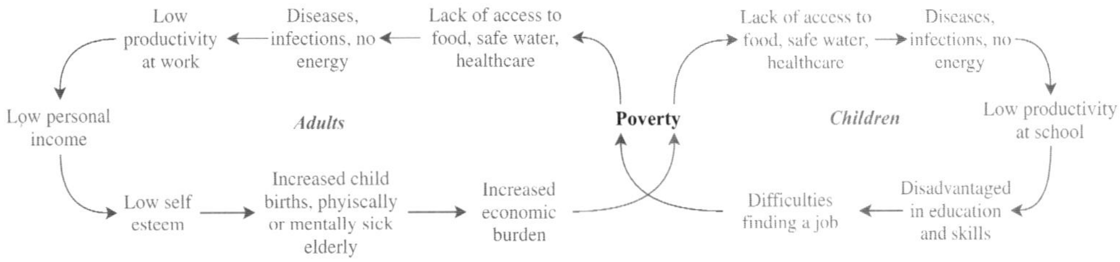

Figure 2 Poverty vicious cycles

Poverty vicious cycles have attracted greater attention and generally less attention has been paid to the vicious cycle occurring with inequality. While families living in vulnerable situations due to poverty it is possible to define the cycle for adults and for children, in the case of inequality there is a collective component that adds up to the household one. This is why in the case of inequality cycles, apart from differentiating between children and adults, we also describe the differences existing between inequality at individual and at collective levels. Figure 3 shows the inequality cycles at an individual or household level and at a collective level.

Firstly, at a household level, individuals (children or adults) that are in a vulnerable situation in terms of economy, health, education, etc. are disadvantaged in the access to basic needs (such as food, water and energy) and to basic services (such as healthcare and education). Therefore, they are more vulnerable to diseases and are prone to receiving worse education and achieving lower skills. As a consequence, there is more risk of being depleted and of getting sick; this, together with a worse level of education, leads to fewer work opportunities in comparison to other people who is more advantaged in these dimensions. This gives rise to an increase in unemployment or, in the case that they succeed in finding a job, to worse working conditions. In the end, they receive lower incomes and even though they live in worse conditions, they don't have enough resources to move to a better place, which restricts their social mobility.

Secondly, looking at inequality from a collective perspective, vulnerable groups are usually excluded or disadvantaged in social activities and environments and are often stereotyped; this affects their communicating strength within society, which has consequences on their influence and political power. In consequence, they are inadequately represented within society and their needs may be less visible among stakeholders or they may have less opportunities to impact on decisions made at different levels. Hence, their fundamental rights, such as basic needs, the right to no discrimination or the right to a decent life may be affected or even completely denied. In the end, the social mobility of the group is very limited and they continue to be in a vulnerable situation.

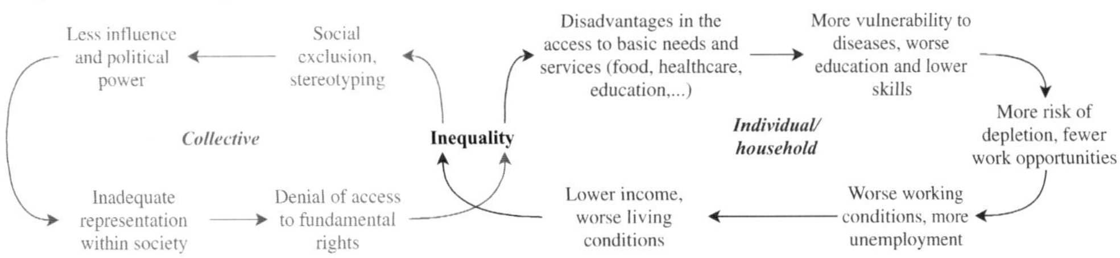

Figure 3 Inequality vicious cycles

5. Discussion: infrastructure, innovation and industry to tackle with inequality

In the previous sections we have introduced the state-of-art of the relationship between poverty and inequality and infrastructure and we have described the vicious cycles of poverty and proposed the cycles of inequality. The aim of this section is to analyse measures to tackle with the problem of inequality from three perspectives: infrastructure, innovation and industry.

On the basis of the conducted research, we have identified four domains through which the cycle can be broken at individual level, and one at collective level. For the former, there is: education, economic activity, health and quality of life, leisure; for the latter, there is participation in decision-making. We have described each domain in Table 1. In this table we also introduce the measures that we propose as key actions to break the vicious cycles. These measures are classified according to whether they correspond to infrastructure, innovation or industry.

First of all, as for education, the identified aspects relate to the lack of access to education or to the (low) levels of skills and education. [19] document the relationship between regional disparities and class sizes, number of teachers and counsellors, amount and quality of academic courses, amount of extracurricular activities, provision of books, materials, supplies and computers, quantity of libraries and special services. Education infrastructure such as schools plays an important role in education inequality. For example, [20] shows regional differences in education levels due to disparities existing in terms of infrastructures and of public education resources in Ghana. Therefore, it is necessary to invest in building and improving schools and other infrastructures with educational objectives and make them both sustainable and resilient (Aspect 1.1). Another aspect to account for is the access to education services. If there are no physical facilities or if there are barriers for children to attend school, some collectives might fail to achieve good levels of education. [21] emphasise how transport inequalities can have consequences on the capability of people to access certain activities such as education and training. If children belong to a vulnerable group, they either must rely on public transportation or they need to find their own means to get to school. Therefore, improving the public transportation system or the roads and paths through which children have to get to school is key to tackle this sort of inequalities; also, facilities need to be developed offered to children living in remote locations such as special transportation, platforms to connect families with the same problems and willing to share transportation or distance education (Aspects 1.2 and 1.3). Aspect 1.4 refers to the need for including social problems issues such as inequality in the education of children, and also of possible training of adults and of employers and employees in order to avoid the perpetuation of stereotypes and social stigmas. Finally, Aspect 1.5 proposes developing and producing new forms of school materials, to make them accessible and affordable for everybody, such as through ICTs.

Secondly, as for economic activity, [22] empirically show the relationship between income inequalities and geographical access to goods. [20] points at the fact that inequalities in access to employment can arise as an effect of transport inequalities. Therefore, ensuring that transportation can be accessed and afforded by all members of society is key, and this can be done by improving transportation infrastructure (Aspect 2.1). However, not all the problem comes from inequalities in physical access, but also from the lack of access to job opportunities. Some authors [23,24] argue that ICTs can have impacts on the quantity, quality and distribution employment opportunities as well as on income levels of vulnerable groups. However, this impact is not always positive in terms of inequality and quality of life; [24] present two different scenarios, a pessimistic and an optimistic one, and describe the effects of ICT on employment relations and employment protection, time and work autonomy and skills and careers. In Aspect 2.2 we emphasise the need of developing ICTs that are inclusive and accessible for all and that can have positive impacts on employment. Aspect 2.3 focuses on investments in science and technology research, which can generate new jobs and improve the existing ones. For instance, [17] empirically analyse the relationship between public investments in research and development with the change in poverty and inequality. Aspects 2.4 and 2.5 target the industry and claim first of all that in the working environment it is fundamental that all practices and standards are satisfied so that vulnerable people are not given a differential treatment; secondly, they advocate for the development and implementation of equity policies that can influence the industrial development positively, namely, inclusively and sustainably.

IOP Conf. Series: Earth and Environmental Science **297** (2019) 012016 doi:10.1088/1755-1315/297/1/012016

Table 1. Description of the main areas through which the vicious cycle of inequality can be broken and specific measures of action in the fields of infrastructures (INF), innovation (INN) and industry (IND)

	Key area	Description	Solution					References
				INF	INN	IND	Description	
							Individual level	
1	**Education**	Aspects related to disadvantages in the access to education, to the achievement of lower skills, to the unaffordability of education support materials and to raising awareness about the inequality problem	*1.1*	√√			Improve schools and buildings, make them more resilient.	[20,21]
			1.2	√√			Improve and make more affordable the physical access to schools (roads, bridges, etc.) from remote locations	
			1.3		√√		Develop methodologies and structures to facilitate access to education for children that are disabled or live in remote locations	
			1.4		√√	√√	Include inequality issues in education and training, raise awareness among society but also among companies	
			1.5		√√	√√	Develop and produce new forms of school materials, to make them accessible and affordable for everybody	
2	**Economic activity**	Aspects related to inequalities in the access to job opportunities, to mobility-related access problems to work positions, to the inclusivity and sustainability of employment	*2.1*	√√			Improve physical access from vulnerable locations to locations with high economic activities	[20,22,23, 24,25]
			2.2		√√		Facilitate access to ICT to vulnerable groups, which can help searching and finding jobs	
			2.3		√√		Invest in science and technology research, which can generate new businesses, create and improve jobs	
			2.4			√√	Ensure that all practices and standards related to labour are satisfied	
			2.5			√√	Develop and implement policies for more inclusive and sustainable industrial development	
3	**Health and quality of life**	Aspects related to access to healthcare services, to environmental control	*3.1*	√√			Improve physical access to healthcare centers	[26,27]
			3.2	√√			Healthcare infrastructure	
			3.3	√√			Environmental infrastructure	
			3.4		√√		Develop more affordable, accessible and sustainable: (1) methodologies to remove toxic substances from drinking water and food, (2) energy sources	
			3.5			√√	Implement and promote the removal of toxic substances from drinking water and food	
4	**Leisure**	Aspects related to activities performed during free time	*4.1*	√√			Improve urban areas and build healthy spaces	
							Collective level	
5	**Political power and participation in decision-making**	Aspects related to the relationships between society and vulnerable groups and to the inclusion of these groups in decision processes	*5.1*	√√			Improve built environment where vulnerable groups get together	[31]
			5.2	√√			Facilitate access to buildings or develop projects of buildings where vulnerable groups can meet, organise themselves, discuss, etc.	
			5.3		√√		Develop accessible forms through which disadvantaged groups can participate in decision-making processes, such as ICTs	
			5.4		√√	√√	Improve inequality-related policies and raise awareness	

Fourthly, one aspect that has been less analysed is related to how individuals spend their free time. As it was observed in the vicious cycle, social problems such as inequality can psychologically affect individuals. Actually, many authors have described and empirically examined evidence on the relationship existing between leisure and physical, social, emotional and cognitive health [28,29,31]. Therefore, promoting and engaging people in mentally and physically healthy activities can strengthen their self-confidence.

Finally, it is necessary to work towards strengthening the political power of disadvantaged groups and towards increasing their access to participation in decision-making processes that may affect them. As pointed out in [31], factors making groups more vulnerable are their living environment (such as their neighbourhood) and the type of social relationships established between them. Therefore, a first way to tackle with this problem is to fight against the effect that a degraded or negative environment has on the reinforcement of inequality (Aspect 5.1) by improving urban areas. Also, it is necessary to facilitate and reinforce the relationships between people belonging to vulnerable groups and between pressure groups and vulnerable people so that they can fight against their exclusion or stereotypes collectively (Aspect 5.2). Aspect 5.3 points out at the need for making it easier for disadvantaged groups to taking part in decision-making processes by developing new forms of participation. Finally, Aspect 5.4 emphasises the fact that one way to break the cycle is through awareness campaigns and development of inclusive social policies.

6. Conclusions

This paper has analysed the relationship between inequality and infrastructures, innovation and industry and has described the main factors that boost the perpetuation of inequality in time, generating a vicious cycle of inequality. The considerations taken in the presented cycle of inequality are: the level at which it occurs (individual/household or collective) and if it is representative of children or of adult.

The analysis carried out has been used to identify at which points the inequality vicious cycle can be broken. The four areas that are fundamental at individual level are: education, economic activities, health and quality of life and leisure. As for the collective level, it is important to take measures on political power and on participation in decision-making processes.

For each of the identified areas specific measures have been proposed, each one belonging to at least one of the categories infrastructures, innovation and industry. The number of measures that have been described are five for education, five for economic activity, five for health and quality of life, one for leisure and four for political power and participation in decision-making.

The following key characteristics have been established for each of the three areas of action:

- Infrastructure: construction of inclusive basic facilities (aimed at everyone), mobility logistics, physical accessibility, improvement of the living physical environment.
- Innovation: expansion of technologies, development of new skills.
- Industry: production of goods, production of services, satisfaction of ethical standards, promotion of good practices.

Acknowledgments

Irene Josa was supported by the Catalan Government through the grant of Agència de Gestió d'Ajuts Universitaris i de Recerca (AGAUR), with reference number 2018 FI_B 00655.

References

[1] Calderón C and Servén L 2004 *The Effects of Infrastructure Development on Growth and Income Distribution* (The World Bank)
[2] Bajar S and Rajeev M 2016 The Impact of Infrastructure Provisioning on Inequality : Evidence from India *J. Comp. Asian Dev.* **15** 122–55
[3] Klenert D, Mattauch L, Edenhofer O and Lessmann K 2016 Infrastructure and inequality: insights from incorporating key economic facts about household heterogeneity *Macroeconomic dynamics*

[4] Vorster H H 2010 The link between poverty and malnutrition: A South African perspective *Heal. SA Gesondheid* **15** 1–6

[5] Karekezi S, McDade S, Boardman B, Kimani J and Lustig N 2012 Energy, Poverty, and Development *Global Energy Assessment: Toward a Sustainable Future* ed Global Energy Assessment Writing Team (Cambridge: Cambridge University Press) pp 151–90

[6] Bain L E, Awah P K, Geraldine N, Kindong N P, Sigal Y, Bernard N and Tanjeko A T 2013 Malnutrition in Sub - Saharan Africa: Burden, causes and prospects *Pan Afr. Med. J.* **15** 1–9

[7] More I and Aye G C 2017 Effect of Social Infrastructure Investment on Economic Growth and Inequality in South Africa: A SEM Approach *Int. J. Econ. Bus. Res.* **13.2** 95–109.

[8] Simon N S and Natarajan P 2017 Nonlinearity between infrastructure inequality and growth: Evidence from India *Rev. Mark. Integr.* **9** 66–82

[9] Falkinger J 1994 An Engelian model of growth and innovation with hierarchic consumer demand and unequal incomes *Ric. Econ.* **48** 123–39

[10] Zweműller J 2000 Schumpeterian Entrepreneurs Meet Engel ' s Law : The Impact of Inequality on Innovation- Driven Growth *J. Econ. Growth* **5** 185–206

[11] Pieroni L and Pompei F 2008 Evaluating innovation and labour market relationships: the case of Italy *Cambridge J. Econ.* **32** 325–47

[12] Aghion P, Akcigit U, Bergeaud A, Blundell R and Hémous D 2015 Innovation and Top Income Inequality *Natl. Bur. Econ. Res. Work. Pap. Ser.* **No. 21247**

[13] De Palo C and Raab R 2018 *Innovation and inequality in the EU: for better or for worse?*

[14] Sahely H R, Kennedy C A and Adams B J 2005 Developing sustainability criteria for urban infrastructure systems *Can. J. Civ. Eng.* **32** 72–85

[15] Ugwu O O and Haupt T C 2007 Key performance indicators and assessment methods for infrastructure sustainability-a South African construction industry perspective *Build. Environ.* **42** 665–80

[16] Newell J P, Seymour M, Yee T, Renteria J, Longcore T, Wolch J R and Shishkovsky A 2013 Green Alley Programs: Planning for a sustainable urban infrastructure? *Cities* **31** 144–55

[17] Bocchini P, Frangopol D M, Ummenhofer T and Zinke T 2013 Resilience and Sustainability of Civil Infrastructure: Toward a Unified Approach *J. Infrastruct. Syst.* **20** 04014004

[18] Vallance S, Perkins H C and Dixon J E 2011 What is social sustainability? A clarification of concepts *Geoforum* **42** 342–8

[19] Darling-Hammond L 2007 Race, inequality and educational accountability: the irony of 'No Child Left Behind' AU *Race Ethn. Educ.* **10** 245–60

[20] Senadza B 2012 Education inequality in Ghana: gender and spatial dimensions *J. Econ. Stud.* **39** 724–39

[21] Lucas K 2004 *Running on empty* (Bristol University Press)

[22] Redding S and Venables A J 2004 Economic geography and international inequality *J. Int. Econ.* **62** 53–82

[23] Arun S, Heeks R and Morgan S 2010 ICT initiatives, women and work in developing countries: reinforcing or changing gender inequalities in South India?

[24] Rubery J and Grimshaw D 2001 ICTs and employment: The problem of job quality *Int. Labour Rev.* **140** 165–92

[25] Fan S, Zhang L and Zhang X 2002 Growth, Inequality, and Poverty in Rural China: The Role of Public Investments. Research Report 125 *IFPRI, Washington, DC* 82

[26] Stuckler D, Basu S and McKee M 2011 Health care capacity and allocations among South

Africa's provinces: Infrastructure-inequality traps after the end of apartheid *Am. J. Public Health* **101** 165–72

[27] Ele-Ojo Ataguba J, Day C and McIntyre D 2015 Explaining the role of the social determinants of health on health inequality in South Africa *Glob. Health Action* **8** 28865

[28] Cassidy T 1996 All work and no play: A focus on leisure time as a means for promoting health *Couns. Psychol. Q.* **9** 77–90

[29] Caldwell L L 2005 Leisure and health: why is leisure therapeutic? AU - *Br. J. Guid. Counc.* **33** 7–26

[30] Mannell R C 2007 Leisure, Health and Well-Being *World Leis. J.* **49** 114–28

[31] Mechanic D and Tanner J 2007 Vulnerable People, Groups, And Populations: Societal View *Health Aff.* **26** 1220–30

New housing concepts seducing young adults to settle; the case of Deventer and Enschede

A G Entrop[1] **and T J N M de Bruijn**[1]

[1] Saxion University of Applied Sciences, School of Business, Building & Technology, M.H. Tromplaan 28, 7513 AB Enschede, the Netherlands

Email corresponding author: a.g.entrop@saxion.nl

Abstract. In multiple cities with polytechnics and universities, companies have difficulty finding the proper well-educated candidates for their vacancies. These young adults do not stay within the city's proximity to reside and work after having obtained their bachelor or master degree. Apart from job opportunities, it might be that they cannot find proper housing. Two municipalities felt the urge to let us study the housing requirements of this particular group of young adults and to design housing concepts for them. Two research and design teams were formed. Without obtaining a strict delineation and with both teams having mutual interests and collaborative meetings, it can be said that one team focused slightly more on architectural design and the municipality of Enschede and a second team focused more or less on city planning and the municipality of Deventer. Within the time available, it was not possible by means of a prototype to prove that housing constructed with our specifications regarding surroundings and the building itself really helps the companies, but in this paper our research in coming to promising designs for housing concepts, that just might seduce these young adults, are disseminated.

1. Introduction

Residential buildings come in all sizes and forms, because every resident or future resident has his or her own specific demands. Not only the diversity in building designs and specifications are large, surroundings also vary tremendously of course. The popularity of a particular building, and therefore its value, is according to real estate agents even most strongly related to its location. Traditional classifications on housing can among others relate to [e.g. 1]: 1) its ownership, rental or owner-occupied, 2) its form, low-rise or high-rise, or 3) its position on a land plot, terraced, detached or semi-detached for example. However, actors in the construction industry more and more often try to offer new original and complete housing concepts, in which the user, the interior, the building and the surroundings complete each other.

The research project described in this paper has its origin in a particular problem experienced in at least three cities in the Netherlands. Even in municipalities with polytechnics and universities, it can namely occur that companies, commercial and industrial, have difficulty finding the proper well educated candidates for their vacancies. In the Netherlands this problem is being experienced in the municipality of Enschede (approximately 158.000 inhabitants), of Groningen (approximately 202.000 inhabitants), and, to a smaller extend in the municipality of Deventer (approximately 100.000 inhabitants). Many graduates leave these cities to start a career elsewhere. Of course, job opportunities play an important role here, but an additional reason for their leaving might be that these cities do not deliver proper housing concepts for this particular group. A problem that might be partially solved by a housing concept fitting properly to the designated user; young bachelors and masters.

In two of the three cities mentioned, namely Enschede and Deventer, we were able to conduct an in-depth study on the requirements and wishes of young adults that soon finish their bachelor or master programme. With these insights and by gaining input from innovations in the housing industry, housing concepts were developed to match their requirements. It is expected that not only requirements regarding e.g. the size, lay-out, energy use or environmental impact will play a role, but also the availability of amenities nearby.

For a particular location in both municipalities this foreseen favoured housing concept will be worked out in a conceptual design. Will we be able to seduce these young bachelors and masters to settle into Enschede and Deventer? Within the time available to us, we do not expect to be able to prove that housing constructed with our specifications regarding surroundings and the building itself really helps the companies, but we would like to disseminate our research setup, methods, results and analysis in coming to promising designs for housing concepts, that just might seduce these young adults.

2. Research setup

The principals of this research project are the province of Overijssel and the municipalities of Deventer and Enschede. These three organizations, and in particular the two municipalities, were looking for opportunities to become more attractive for those that local companies are currently in need to, namely young well-educated adults. Our research department Sustainable Built Environment within the School of Business, Building & Technology of the Saxion University of Applied Sciences took the challenge to design the requested housing concepts. Considering Saxion is an important employer for well-educated young adults, and is located in Enschede, as well as in Deventer, one can imagine the mutual interests. The main research question was: *how can new housing concepts seduce young well-educated adults to settle in the cities of Enschede and Deventer?*

In line with the wishes of the principals, our research department assigned two student project teams. One team consisted of five students studying building engineering, building engineering business (2x), industrial product design and small business & retail management. Considering their background and the fact that this team has its basis in the municipality of Enschede, the architectural design and the municipality of Enschede was more or less their focus [2].

The second team consisted of four students studying urban planning (3x) and regional planning. Logically this team focused more or less on city planning, the availability of amenities and the municipality of Deventer [3]. All students were and some still are in their final year, which means they have theoretical knowledge needed to become a young professional and, soon, they will belong to the target group of the principals and companies with vacancies.

In their research these two teams dealt with the following steps to tackle the main research question. Firstly, they were collecting among their fellow students the requirements and wishes these young adults have regarding housing. This means questionnaires were developed and interviews were taken. Secondly, the teams were out in the field and on the internet to spot innovative housing concepts. Thirdly, each team dove in a design process to make more explicit and visible what is needed in both cities. The final step was to check, if the developed designs really match with the interests of the foreseen user groups. Unfortunately, but for obvious reasons, it was not possible within the budget and time frame of this project to realize the designs and to check if in real life these buildings at the suggested proper locations will attract the young well-educated adults.

At the start of the academic year in September 2018 the teams were formed. After having read background documents, the teams introduced themselves by the end of that same month to the principal. During approximately twenty weeks they were occupied by this research and design project. So by the end of January 2019, they were able to hand in their reports and final designs. In total three meetings were organized together with the three principals, namely the one already mentioned in September 2018 and two others in early December 2018 and in late January 2019. In their role of managing principals, the whole process was supervised by the authors of this paper.

3. Collecting requirements

The two student project teams applied different methods to get insights into what young well educated adults require regarding housing after they finish their bachelor or master degree programme. The team that had its base in Enschede applied a rather innovative approach in the sense that they used a form of participatory photo mapping [e.g. 4], in which the respondent was offered the opportunity to provide photos to the student project team practising the role of researcher. Respondents were completely free in choosing their preferred photos. They were not only asked to provide photos on what they would like, wish for or even require in and on their new future neighbourhood and house, but also what they would dislike. In interview sessions thirty respondents provided 300 photos, that were analysed by the project team. The project team in Deventer was able to set out a questionnaire to a staggering 500 students, of which 339 were in the final stage of their study or already had their bachelor diploma. In this section the findings of these two studies are presented.

3.1. Requirements regarding surroundings and amenities

The photos and interviews of the project team in Enschede revealed that 8% demanded a supermarket to be close by and 67%, the highest score, have the wish that it is nearby. The young well educated adults like the surroundings neat and tidy, because 17% indicated that this is a requirement and 46% wished for it. Also the availability of green areas scores high, with 29% demanding them and 38% wishing for them. Another important statement regarding the neighbourhood was, that many of the respondents do not want to have loiterers in the proximity of their homes. Most surprisingly, the respondents stated quite clearly that they do not want to live in the city centre. Furthermore, living in a low class workers district or next to a busy road is unwanted.

The project team in Deventer asked in their questionnaire what distances to amenities would be acceptable. For the daily groceries 31% of the 339 respondents answered that at maximum 2 km would be acceptable, another 32% was willing to increase that figure to 5 km. The amenities that were favoured the most where: 1) supermarkets (22%), 2) bus stops & railway stations (15%), 3) shops (13%), 4) general practices & pharmacies (11%,) and 5) restaurants & bars (10%). Other answers revealed less popular amenities like indoor (9%) and outdoor sport facilities (5%), play grounds & parks (6%), police stations & hospitals (4%), cultural facilities (3%) and libraries (2%).

The suggestion for the need of having daycare for children nearby was discounted, because almost all respondents in Enschede as well as in Deventer regard parenthood as a next stage in their lifecycles. This same reason made a nursery of children's room for the time being redundant.

3.2. Requirements regarding the built object

The project team in Enschede found out that 50% of the respondents demand and 38% would like to have some form of (private) outdoor space. However, if a garden is available, then it should require little maintenance. Although most of the Dutch students do not privately own a car, in this research 54% of the respondents interviewed by the project team in Enschede indicated that parking space is a requirement and 4% indicated parking space is a wish. The project team in Deventer noticed that only 18% of their respondents to the questionnaire travels by car. For the project team in Enschede other important requests of the respondents regarding their home were spacious livings, low on maintenance, privacy and not a vintage, but a modern look. In Deventer it was found that the targeted audience wants a floor surface of 45 to 65 m^2.

3.3. Requirements regarding the housing concept

Current modern housing concepts aiming at young people often try to let residents share space, facilities and even consumer goods. However, the respondents in Deventer do not want to share living space. Furthermore, the housing concept that is going to be designed, needs to be affordable. The project team in Deventer received indications that housing rents or a mortgage of € 800,- to € 1000,- per month would be acceptable to the respondents.

4. Designing housing concepts

The former section of this paper showed insights on the requirements of young well educated adults. In this section we elaborate on the housing concepts that were designed by the two project teams. In the municipality of Enschede no particular land plot was assigned by the municipality to design the desired housing concept for. In the municipality of Deventer a land plot was selected in the proximity of the central railway station and the city centre (see Figure 1). Although this land plot currently accommodates some buildings, these are partially already in disuse and their demolition is foreseen. Considering the different backgrounds of the students in the project teams, project team Enschede focused on the interior and exterior of the to be built object and project team Deventer searched for the integration between surroundings and building.

Figure 1. The land plot in Deventer that was assigned to the project team to design the housing.

4.1. Housing concept for young adults in the municipality of Deventer

The project team in Deventer studied eight housing concepts that are on the Dutch market offering appropriate housing to (studying) young adults. These housing concepts were rated by the project team herself on bases of compactness, flexibility and the capability to share living space. The housing concept ONE of Heijmans [5] was chosen by 43% of the respondents as appealing. Another 31% chose for the so called Superloft [6]. Becoming familiar with these housing concepts by pragmatically analysing them, helped the team to work out some first sketches.

Figure 2. Result of the project team in Deventer; buildings designed for young well educated adults.

However, taking into account the land plot accommodating the housing, affordability and the requirements and wishes as addressed by the respondents, the first designs were made. After having received feedback from the target group and experts in urban and regional planning, the project team in Deventer showed the principals the designs as presented in Figure 2. A huge wide screen display forms the centre piece drawing the attention from passengers in passing trains and offering possibilities for social events in the park. Figure 3 shows a cross section of the area, in which parking space is accommodated in the slope at the left side and existing houses are shown on the right side.

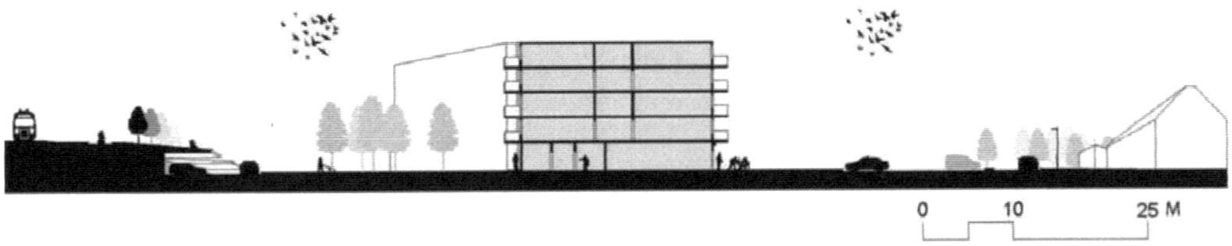

Figure 3. A cross section of the foreseen site in Deventer.

4.2. Housing concepts for young adults in the municipality of Enschede

The project team in Enschede also became familiar with different modern housing concepts. However, their design trajectory differs from the one in Deventer. After sketching and internally discussing multiple exteriors and interiors, one exterior was chosen keeping the requirements and wishes of the target group in mind. Three different preliminary internal layouts fitting in the designed exterior were presented at the Smart Solutions Festival on the 15th of January 2019; a public event visited by many senior year students [7]. By making it possible for the visitors to vote, the project team tried to ensure that their final design will be appreciated by the target group.

The three preliminary designs presented at the festival looked quite similar from the outside (see Figure 4) as apartments form the basis, but their names 'shared living', 'functional living' and 'exceptional living' leave little to the imagination. Offering four 14 m^2 bedrooms and 73 m^2 of shared living space on a single floor, the first housing concept is strongly linked to the American television program 'Friends'. The second housing concept offers 62 m^2 of private living space to one or two persons again on one single floor (see Figure 5-7). The 'exceptional' concept offers a two story apartment and a larger balcony than the second one. The total indoor gross floor area is around 56 m^2.

Figure 4. Result of the project team in Enschede; the designed exterior of the building.

Figure 5. Floor plan as designed by the project team in Enschede.

Figure 6. Impression of the interior as designed by the project team in Enschede.

Figure 7. Impression of the recessed balcony or loggia as a form of personal outdoor space.

After counting the 91 votes of the festival visitors, it became apparent that the 'functional living' housing concept was with 70% of the votes by far favourite. Least favourite with 10% of the votes was 'shared living', leaving the remaining votes for the 'exceptional living' housing concept. Supported by these convincing voting results, the project team put some extra effort in investigating the financial burden for the future users. Although in the first round of interviews and questionnaires quite many students responded that a land based family house would be their ideal, the current housing market and financial systems in the Netherlands will provide limited to no possibilities for starters to live up to this dream. With estimated construction costs of € 50.000,- (excl. land plot) per couple, the monthly costs are expected to be relatively low for the first housing concept. The second housing concept costing € 100.000,- (excl. land plot) can probably be offered for an average price. The last concept is with € 130.000,- (excl. land plot) the most expensive one.

5. Discussion

The two project teams were able to offer great insights and developed designs to visualize what is necessary in Enschede and Deventer. However, one might have some doubts about how these residential buildings could be financed and how reliable the findings are, hence this section.

Three options to finance housing for well-educated young adults seem to exist, namely:

- in the form of social housing owned by a housing association;
- as a rented house owned by a commercial organisation, or;
- via a mortgage as owner occupied property.

The first option shows long waiting lists and is accompanied by limitations regarding the maximum annual income and the savings of the tenant. Practically, the second and third option to finance this form of housing are more likely. For a mortgage a substantial amount of the total costs needs to be paid directly, so in advance money needs to be saved by the former student and student loans need to be preferably as small as possible.

It was due to obvious reasons not possible by means of a prototype to prove that housing constructed with the collected specifications really quickly will be occupied by the targeted user group and really helps the companies in the two municipalities to reduce the number of vacancies. However, the design trajectories of both project teams encompassed moments of reflection with the future user group. The received feedback strengths our believe that the findings are broadly supported by the bachelor and master students in Deventer and Enschede. At the same time, this is also the reason that, although last year students made up the project teams and the risk of bias seemed to exist, we are confident that the project teams did not let their personal opinions prevail. The data the project teams collected is arranged in appendices of their reports and in digital form provided to the supervisors of the project. However, principals, supervisors and the project teams themselves can imagine that young well-educated adults in other cities have different preferences regarding design and location. They also might have other possibilities to acquire the dwelling they prefer.

6. Conclusion

Let us recap the main research question, which was: how can new housing concepts seduce young well-educated adults to settle in the cities of Enschede and Deventer? This study started with the unspoken expectation that young well educated adults would be enthusiastic about sharing living space, open cosy indoor as well as outdoor areas and commuting in an upmost environmental friendly way. This expectation matches similar developments in cities like Amsterdam. However, these expectations of the principals and supervisors were shattered, because it seems that the ideal living concept is quite traditional. Well, at least for the students in Deventer and Enschede our project teams questioned and interviewed.

These young adults expect to be able to get a nicely paid job and they want some traditional privacy and comfort for it. Students still living at their parents' house, also relate to that particular picture. Students having already their own place during study, seem to be willing to share some space and at the same time they seem to be aware about the costs certain preferences might have. Last but not least, all students want with their future house or apartment a parking lot close by. Instead of imitating developments elsewhere, a strong case can be made to develop more traditional yet affordable housing concepts in cities like Enschede and Deventer. More speculative, one hypothesis that came to mind was that innovative housing concepts based upon sharing as found in cities like Amsterdam might be the result of lacking opportunities to find proper housing at all, at least not affordable ones.

7. Recommendations

The starting point of this research and design project was the conception that young adults that recently finished their bachelor or master program can be seduced to stay in their student cities by appropriate housing. Due to rising study and housing costs the number of students living at their parents' house has increased. Considering that more than half of the respondents, involved in the questionnaire of the project team in Deventer, was living at home, it is recommended researching how to improve the bond between students and the municipality in which they study.

A second recommendation lays in the given, that this study specifically focused on the student populations of two cities, namely Deventer and Enschede. Other student populations and cities have specific preferences and challenges. In other cities, where young well-educated adults leave, specific building designs might be necessary.

Acknowledgements

The authors would like to express their gratitude to the principals at the municipality of Enschede, municipality of Deventer and at the Provence of Overijssel for their trust and valuable comments. The same gratitude applies to the students in the project teams for their enthusiasm in conducting the research and developing the designs. Furthermore, the Province of Overijssel is acknowledged for providing financial support to this research and design project "Studenten behouden voor de woningmarkt van Enschede en Deventer; een onderzoek naar de mogelijke rol van innovatieve woonproducten."

References

[1] Henilane I 2016 Housing concept and analysis of housing classification *Balt J Real Estate Econ Constr Manag* **4** 168-79
[2] Berg B vd, Keemers L, Lodewijk J, Velthuis R and Zanden L vd 2019 *Studenten behouden voor de woningmarkt van Enschede*
[3] Borgman L, Jager S, Rood T and Rozema V 2019 *Wonen in het spoorkwartier; een onderzoek naar de randvoorwaarden van een vernieuwend woonconcept voor (bijna-)afgestudeerden*
[4] Teixeira S and Gardner R 2017 Youth-led participatory photo mapping to understand urban environments *Child Youth Serv Rev* **82** 246-53

[5] Website https://www.heijmans.nl/nl/heijmans-one/ visited on January 27th, 2019
[6] Website http://superlofts.co/ visited on January 27th, 2019
[7] Website https://www.saxion.nl/nieuws/2019/01/festivalreport-3e-editie-smart-solutions-festival visited on January 28th, 2019

IOP Conf. Series: Earth and Environmental Science **297** (2019) 012018 doi:10.1088/1755-1315/297/1/012018

Identify Optimal Renovation Packages for Residential Buildings: A State-of-the-Art Computational Model

Mohamad Kharseh, Claudio Nägeli, Holger Wallbaum

Chalmers University of Technology, Sweden

The corresponding author's e-mail: mohamad.kharseh@chalmers.se

Abstract. Renovating the existing building stock has a significant potential to achieve the goal of reducing greenhouse gas (GHG) emissions in the European Union. However, a common European renovation project focuses primarily on improving the thermal performance of the building shell by adding insulation to the opaque surfaces and improve the thermal performance of the windows. The potentially positive contribution of renewable energies (RE) in balance with energy efficiency measures is often underestimated. Consequently, a more holistic approach can contribute to a reduction in total net energy demand up to 40-45% for the entire buildings sector. Thus, in order to achieve the goal of GHG emission reduction in an economic most responsible way, the share of RE in a renovation project needs to be increased. However, building renovation projects are becoming - apparently - more complicated if more factors are considered in the planning of a renovation project. Thus, a computational tool for evaluating hundreds of different renovation options, including the implementations of renewable energy resources, to obtain an optimal or nearly optimal set of renovation options is essential. Therefore, a novel planning tool has been developed within the framework of DREEAM project, a project funded by the European Union within the Horizon 2020 research framework. The DREEAM-Tool has been designed in the way that it helps designers and other stakeholders to plan a renovation project of a single building or even on a multi-building scale. The tool was built in the way to optimize the renovation project taking into consideration the most critical factors in planning and decision-making processes, such as the economic or environmental performance. In other words, the tool combines an energy calculation model for a building or multiple building with an economic and environmental assessment to identify and optimize the most beneficial refurbishment solutions. The current study presents the concept of the DREEAM-Tool and shows examples of how the optimal renovation packages of a considered building will be determined and how this will support designers or buildings owners in decision-making processes.

1. Introduction

Because of increasing greenhouse gas emissions and its well-known impact on the climate, reducing energy consumption became an urgent issue. 40% of the total European Union energy use is consumed in buildings [1]. Knowing that about two-thirds of the energy used in the EU-15 buildings goes to space heating, the most significant potential for saving energy lies on renovation and upgrading of the old building to modern energy standard [2,3]. This leads to the fact that 30% of global carbon dioxide emissions come from buildings [4]. Heating and cooling system, which is expected to increase between the year of 2010 and 2050 by 79% in residential buildings and 84% in commercial buildings, account for the majority of energy consumption in building [5].

Unfortunately, 72% of residential dwellings in the EU were built before 1972, namely, were built to the thermal performance standards lower than the current standards which are imposed under current building regulations. In Sweden, for example, 77% of the buildings were built before 1980, while the first Swedish regulation of thermal comfort in buildings was published [3,6]. Thus, refurbishment of the old building is a curtail factor not only to fulfill the target of the EU in reducing the energy consumption but also in reducing the environmental load of the building stock. In other words, the most substantial potential for saving energy toward mitigating the climatic changes lies on renovation and upgrading the old buildings to modern energy standard. The European Commission established a long-term target of reducing the greenhouse gas (GHG) emissions levels for the building sectors by 88-91% up to 2050 compared to 1990 levels [7]. In line with this, the Swedish environmental quality objectives have established a goal of reducing energy consumption by 50% per unit heated area in the built environment in 2050 compared with 1995 [8].

To achieve an economical GHG emissions reduction, the share of RE in buildings renovation projects needs to be increased. However, building renovation projects are becoming - obviously - more complicated if more factors are considered in the planning of a renovation project. Thus, a computational tool for evaluating hundreds of different renovation options, including the implementations of renewable energy resources, to obtain an optimal or nearly optimal set of renovation options is essential.

Several tools are developed to simulate the renovation on the district level to assist designers and stakeholders before deciding on buildings refurbishment. The District ECA that is formed by the EU MODER project, for instance, is available for free as the international version. The D-ECA tool can perform energy efficient strategies and technologies for the energy efficient renovation of districts. However, the calculation is based on the use of pre-defined archetype buildings. This means the geometry and user profile of the archetype buildings specific for the individual country is fixed, which reduce the freedom of the user.

Recently, Fan et al. presented an optimized mathematical model to help decision makers to identify the best combination of retrofit options for buildings to ensure policy compliance in the most cost-effective way in the office sector [9]. The model determines optimal retrofit plans for a whole building in a systematic manner, considering both the envelope components and the indoor facilities. The model also considers the rooftop PV system to reduce the usage of electricity produced from fossil fuels. The model was built based on the dynamic method of calculating heating and cooling load of the building. However, the model can be used for a single building and cannot be used at the district level. Also, the optimization method is a genetic algorithm meaning, taking into consideration a large number of variable, a long computational time.

Wu et al. developed a model using EnergyPlus for the simulation of the energy demands and CPLEX for optimizing [10]. The aim of the model is optimizing both annual costs and GHG emissions. The model considers both retrofits in the building envelope as well as local production of energy with boilers, heat pumps, solar thermal and PV modules. The optimization is posed as an epsilon-constrained mixed integer linear program and takes into consideration a set of previously simulated retrofit scenarios, along with the rest of the local energy production measures. It does not detail how much insulation should be put; it just targets U-values that come from the Swiss regulation. The space heating simulations are thorough, but just a few scenarios are explored.

FASUDIR is another European project [11], which was conducted by industrial and academic parents, has developed FELICITY web-based decision support tool, termed FELICITY, for assessing the condition and energetic performance of large building stocks as well as setting up and comparing retrofitting concepts based on key performance indicators [12].

Finally, designPH EDU was developed specifically with students and universities in mind [13]. The model was developed to help designer and building owner to pan a refurbishment in order to achieve a passive house level. Therefore, the model cannot be used to optimize the retrofitting. Also, it cannot be used at the district level.

Hence, The DREEAM-Tool, which is developed by the Chalmers University of Technology, is the core of the approach to refurbishment concept design taken within the DREEAM-project. The tool considers the possibility of integrating the available renewable energy resource as well as the interaction between the different buildings to define optimal renovation scenario and energy demand reduction. In other words, the developed tool aims at achieving high energy reduction demands by cost-optimally balancing between energy efficiency (EE) and renewable energy supply measures (RES) to optimal energy demand reduction.

2. DREEAM-Tool

In the light of the fact that most renovation projects take place in a multi-building scale accounts for 36% of all EU's residential buildings, innovative approach in which the balances between energy saving measure and implementing of renewable energy on multi buildings scope seems to be the method to increase the saving in energy demand cost-effectively. However, many of the technologies to be considered for cost-effective solutions are available on the market or having been piloted in real-life environments, the integrated approach for the targeted market segment does not exist.
DREEAM-Tool is a computational tool to help in optimizing a building retrofitting project.

2.1. Simulation steps

By taking a multi-building approach to refurbishing their residential building stock, building owners can act more strategically than it is the case when renovating one building at a time and aim to achieve better energy efficiency results.
The overall process can be divided into three necessary steps, see Figure 1:

1) Building Portfolio: The aim here is to give an overview of the status quo of the buildings including, but is not limited to, building type, U-value of different components, occupancy and usage profile.
2) Select what to be renovated: Pre-select what retrofitting measure or technology options to be included based on preferences and pre-assessment. Namely, identify the essential hotspots to be addressed in refurbishment concepts (e.g., windows, wall, roof, install a solar system, or any other renovation option).
3) Optimization Indicators: this step specifies which indicators to be optimized. These can be energetic (i.e., energy demand reduction), economic (i.e., return on investment) or environmental (i.e., greenhouse gas emissions) indicators.

This way the model is ready to run. The energy demand reduction of the concepts is calculated and optimized, based on the indicators selected in step 2. The results visualized on a Pareto-curve allow a building owner to select the most optimal refurbishment concept in an informed way.

Figure 1. Simulation steps of DREEAM-Tool.

2.2. Concept

The concept of the tool can be broken down into three key parts (see Figure 2). First, the building portfolio and refurbishment database have to be added. In the second part, the energy performance of the building is calculated according to multiple target criteria (e.g., max. energy demand reduction vs. ROI) and then matched with suitable renovation solutions. As a result, selections of optimal renovation scenarios are presented on a Pareto-curve.

The tool combines an energy calculation model for the building with economic and environmental assessments in order to assess and optimize refurbishment concepts. The optimization is done by multidimensional optimization approach, based on an evolutionary algorithm that can automatically find the Pareto-boarder for multiple criteria selected by the user. Thereby, the DREEAM-Tool enables the development of an optimized design of renovation concepts that highlight the trade-offs between the different indicators (e.g., environmental vs. economic benefits) and allows the building owners to select the concept that is most suited to their preference. By doing so, the tool can support building owners in making strategic refurbishment decisions for their portfolio and help them translate those decisions in corresponding refurbishment concepts and select between different refurbishment approaches in line with overarching targets.

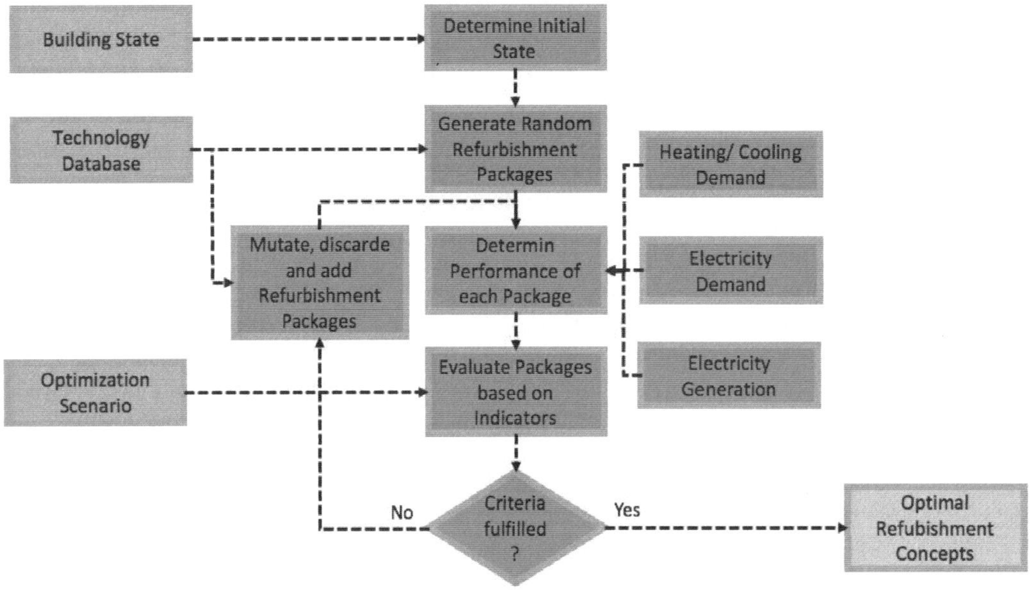

Figure 2. the concept of the DREEAM-Tool.

2.3. DREEAM-Tool user interface

The user interface of the DREEAM-Tool has been built to be user-friendly and composed of three sections as follows:

2.3.1. Building Portfolio

The user can add and manage the existing building data with an easy-to-use editor. This includes the location, building groups and individual buildings with their properties. The properties include all the technical and geometrical data of the different buildings.

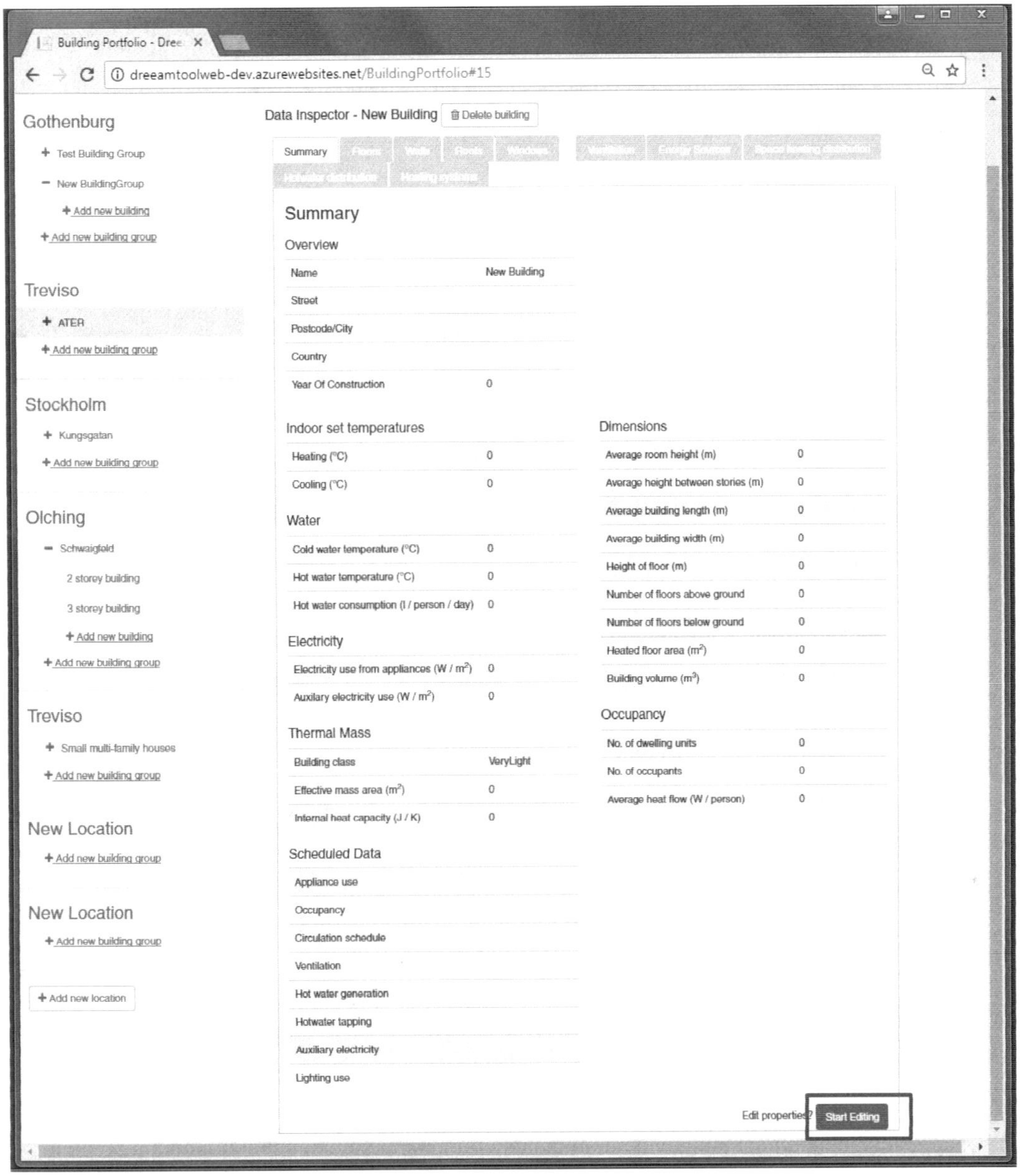

Figure 3: Editing of basic information about the building

2.3.2. *Refurbishment Database*

The user adds own custom refurbishment options and can edit properties of all the refurbishment options that were added. Those refurbishment packages will be stored in the database, and the user can select later on which are chosen for the calculations.

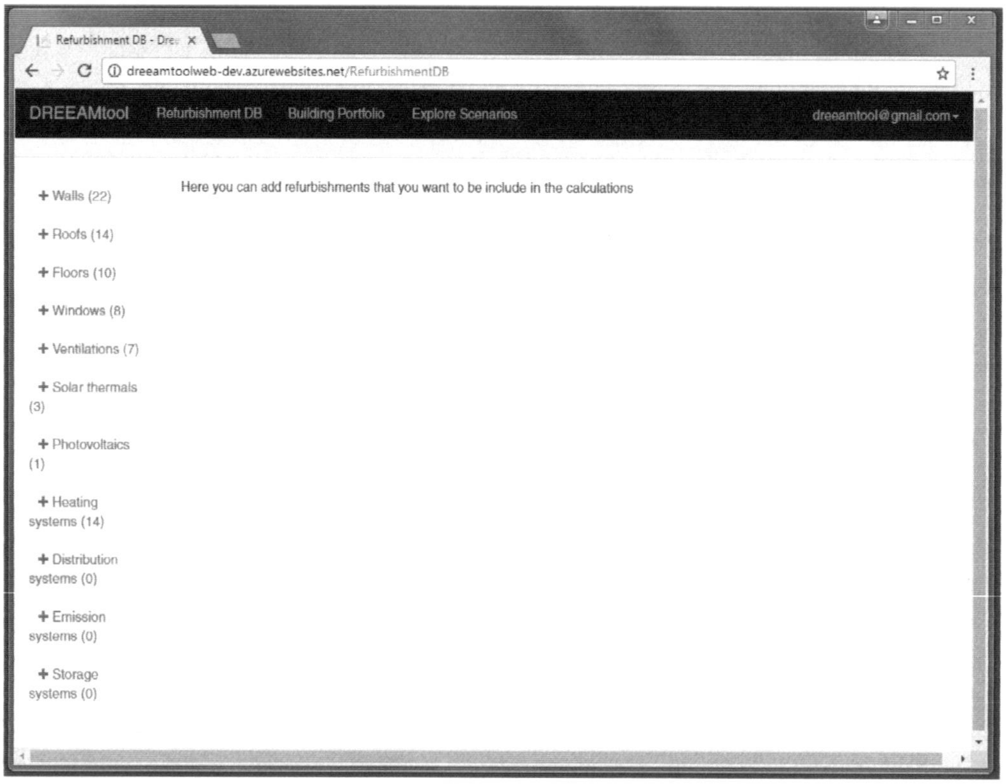

Figure 4: Refurbishment Database Section in the DREEAM-Tool

2.3.3. Explore Scenarios

Here new scenario calculations for the building portfolio can be started, and the optimal strategies can be explored when the calculations are done. The different strategies can be selected and exported for external use.

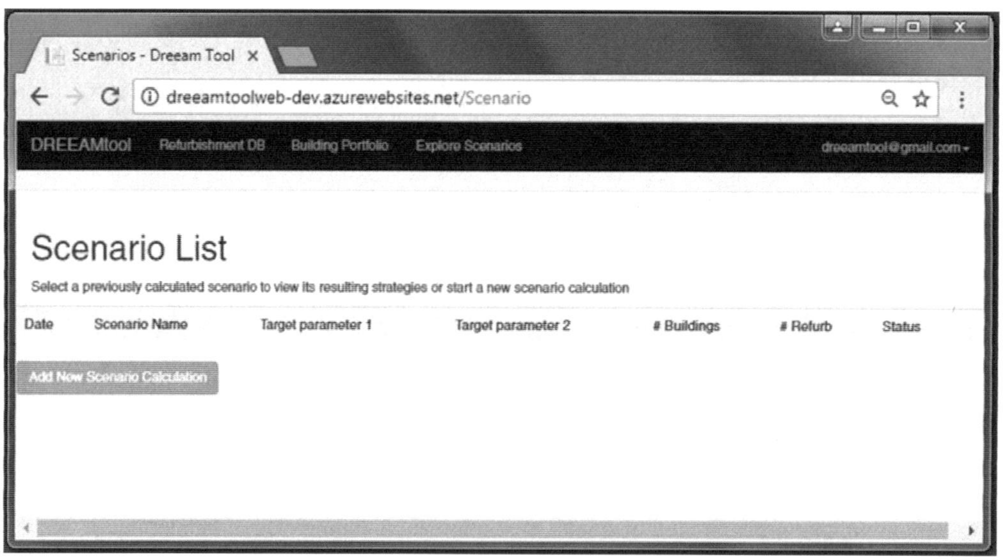

Figure 5: Explore Scenarios in the DREEAM-Tool

After starting the scenario by clicking "Explore Scenarios" it will appear in the scenario list. The status shows how far the calculation has gone. It might take up to 20 min until a calculation reaches 100%.

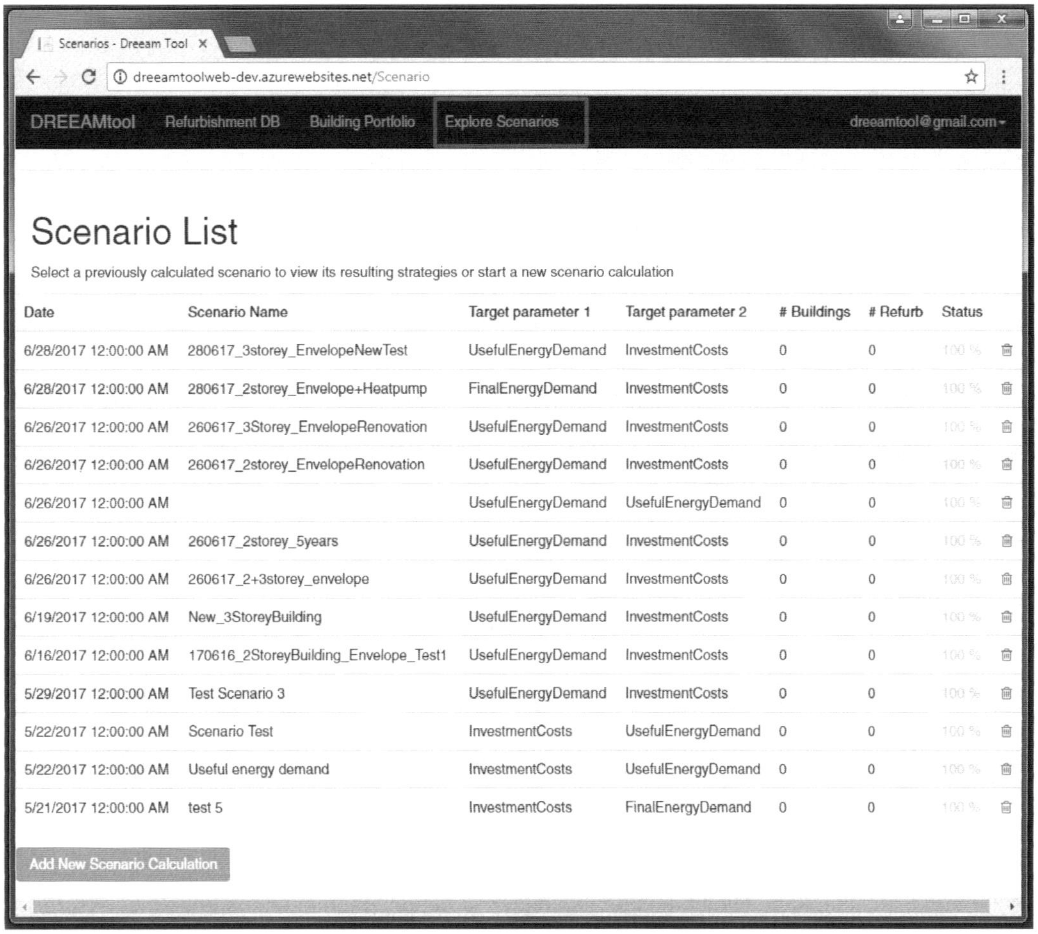

Figure 6: Calculation of the scenario

By clicking on the scenario, a new window opens which shows the results of the calculation between the two chosen target parameters in a graph and table.

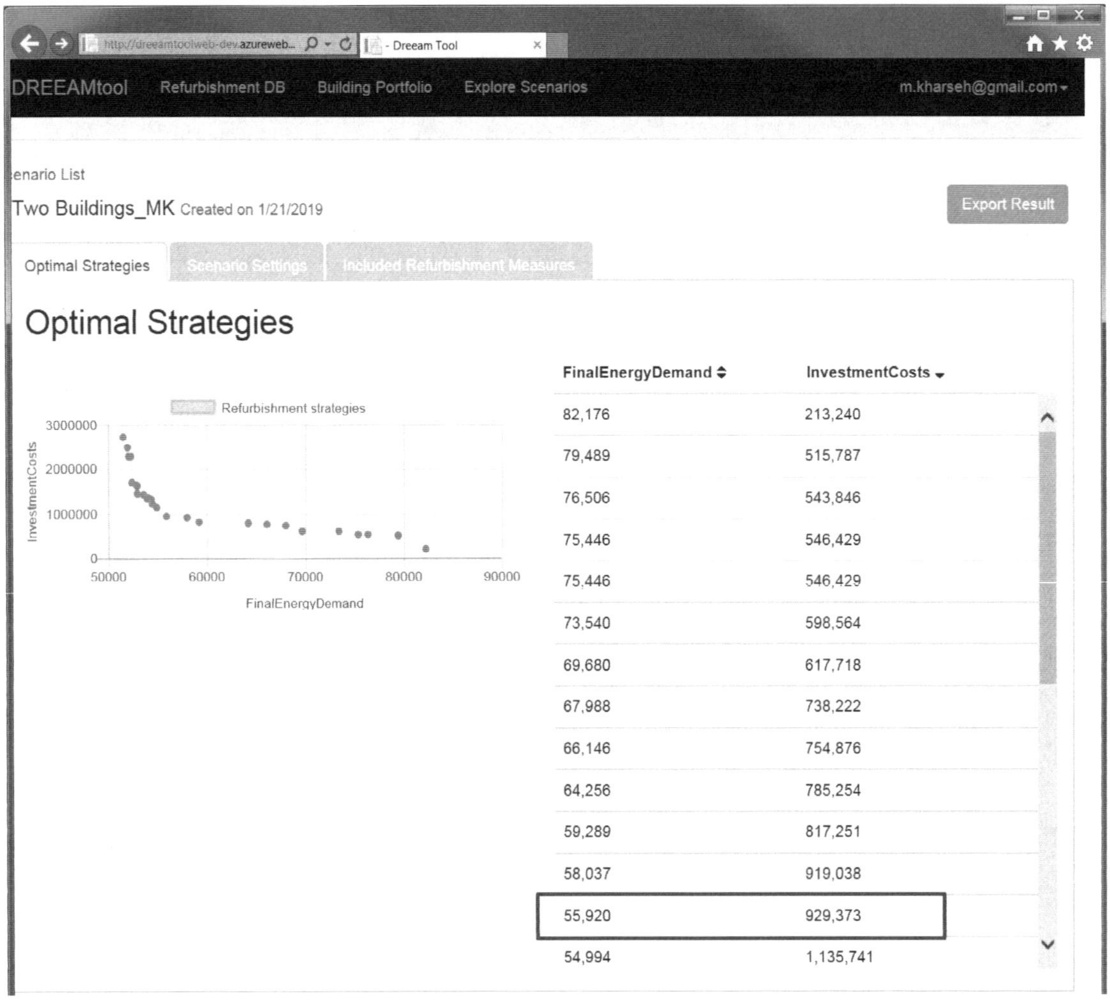

Figure 7: Scenario results

As also shown in Figure 7, one can recall the scenario settings and included refurbishment measures for the specific scenario. By clicking on a specific result in the table or graph, the strategy results are listed in more detail (Figure 8):

- Target parameters
- Costs
- Energy use by energy carrier
- Energy production
- Energy use by type:
- Useful Energy
- Final Energy
- Primary Energy
- GHG Emissions
- Economic indicators

Figure 8: Detailed scenario results

When opening the building on the left side of the interface (here: Validation) one can also see the selected refurbishment solutions for that specific strategy.

3. Conclusion

Renovating the existing building has a significant potential to achieve the goal of reducing GHG emissions in the European Union. However, the main challenges that face the designers and stockholders are the identification of the optimal renovation approach in which the GHG emission in maximized or the economic visibility of the project maximized. To achieve this goal an innovative design DREEAM-too has been developed. DREEAM toll has been developing in a way that helps designers and stockholders to plan a renovation project on a single building or multi-building. The tool was built in the way to optimize the renovation project taking into consideration different criteria, such as economic or environmental criteria. In other words, the tool combines an energy calculation model for the building with economic and environmental assessments to assess and optimize refurbishment concepts.
This work aimed to describe the approach of the DREEAM-Tool to refurbishment, which takes a larger renovation scale (multiple buildings or even portfolio level) as a starting point for renovation planning and design. Taking a multi-building approach to refurbishing the residential building stock, building owners can act more strategically than it is the case when renovating single buildings and aim to achieve better energy efficiency results. The DREEAM-Tool allows an optimized design of renovation concepts that best meet multiple objectives (e.g., energetic, environmental or economic). By doing so, the tool can support building owners in making strategic refurbishment decisions for their portfolio and help them translate those decisions into corresponding refurbishment concepts. A first online version of the tool was developed and can now be tested by potential users.

References

[1] J. Quinten, V. Feldheim, Dynamic modelling of multidimensional thermal bridges in building envelopes: Review of existing methods, application and new mixed method, Energy Build. 110 (2016) 284–293. doi:10.1016/J.ENBUILD.2015.11.003.
[2] M. Gustafsson, G. Dermentzis, J.A. Myhren, C. Bales, F. Ochs, S. Holmberg, W. Feist, Energy performance comparison of three innovative HVAC systems for renovation through dynamic simulation, Energy Build. 82 (2014) 512–519. doi:10.1016/J.ENBUILD.2014.07.059.
[3] L. La Fleur, B. Moshfegh, P. Rohdin, Measured and predicted energy use and indoor climate

before and after a major renovation of an apartment building in Sweden, Energy Build. 146 (2017) 98–110. doi:10.1016/J.ENBUILD.2017.04.042.

[4] E. Antipova, D. Boer, G. Guillén-Gosálbez, L.F. Cabeza, L. Jiménez, Multi-objective optimization coupled with life cycle assessment for retrofitting buildings, Energy Build. 82 (2014) 92–99. doi:10.1016/J.ENBUILD.2014.07.001.

[5] J. Yang, A. Chong, M. Santamouris, D. Kolokotsa, S.E. Lee, K.W. Tham, C. Sekhar, D.K.W. Cheong, Energy utilizability concept as a retrofitting solution selection criterion for buildings, J. Civ. Eng. Manag. 23 (2017) 541–552.

[6] E. Brandsma, J. Valik, Underlag till den andra nationella strategin för energieffektiviserande renovering Ett samarbete mellan Boverket och Energimyndigheten, 2016:15, Boverket, Karlskrona, 2016. https://www.boverket.se/globalassets/publikationer/dokument/2016/forslag-pa-uppdaterad-strategi-for-energieffektiv-renovering_slutversion.pdf.

[7] C. Rodrigues, F. Freire, Environmental impact trade-offs in building envelope retrofit strategies, Int. J. Life Cycle Assess. 22 (2017) 557–570. doi:10.1007/s11367-016-1064-2.

[8] N.W.O. Brown, T. Malmqvist, W. Bai, M. Molinari, Sustainability assessment of renovation packages for increased energy efficiency for multi-family buildings in Sweden, Build. Environ. 61 (2013) 140–148. doi:10.1016/J.BUILDENV.2012.11.019.

[9] Y. Fan, X. Xia, Energy-efficiency building retrofit planning for green building compliance, Build. Environ. 136 (2018) 312–321. doi:10.1016/J.BUILDENV.2018.03.044.

[10] R. Wu, G. Mavromatidis, K. Orehounig, J. Carmeliet, Multiobjective optimisation of energy systems and building envelope retrofit in a residential community, Appl. Energy. 190 (2017) 634–649. doi:10.1016/J.APENERGY.2016.12.161.

[11] FASUDIR, Project partners | FASUDIR, (n.d.). www.fasudir.eu (accessed June 11, 2018).

[12] FeliCity web-based decision support tool, (n.d.). www.felicity.tools (accessed June 11, 2018).

[13] EnerPHit, Passivhaus Institut, (n.d.).
http://www.passivehouse.com/04_phpp/04_phpp.htm (accessed June 11, 2018).

IOP Conf. Series: Earth and Environmental Science **297** (2019) 012019 doi:10.1088/1755-1315/297/1/012019

Thermal comfort analysis of a representative multi-story social housing unit with wood as an alternative construction material in Brazil

Esaiy Valdenebro[1,2], Udo Dietrich[1] and Akemi Ino[3]

[1] REAP research group (resource efficiency in architecture and planning), Hafencity University, Überseeallee 16, 20457, Hamburg, Germany
[2] Corresponding author
[3] Habis research group (housing and sustainability), Universidade de São Paulo, avenida trabalhador são-carlense 400, 13566-590, São Carlos, Brazil

Abstract. This research intends to link thermal comfort as an important part of achieving energy efficiency in the operational phase of a building with the use of wood as a low embodied energy and carbon efficient construction material. The thermal performance of a representative multi-story social housing unit with common construction materials is evaluated in contrast to wood-based construction systems in Brazil. A project located in São Carlos, Brazil was selected. The construction system for walls consists of concrete blocks while roofs are composed of a fiber cement sheet and a concrete slab ceiling. As alternative wood-based wall systems, plywood panels with an air chamber or three different insulation materials (expanded polystyrene, cellulose and mineralized wood fiber) were tested while two alternative roof systems were analyzed (plywood lining with air chamber or insulation material). The project with common construction materials as well as the wood-based alternatives were evaluated through the Brazilian standard NBR 15575 and the European Standard ISO 13786. Furthermore, a transient simulation was conducted following the procedure established by NBR 15575 and using the adaptive comfort model ASHRAE 55-2013. The analysis evidenced the neglection of the recommended passive design strategies according to the Brazilian Standard 15220 and the psychometric chart of Givoni. In regard to the wood-based constructions, results showed that insulation of mineralized wood fiber in panels performed better than cellulose and expanded polystyrene. The transient simulation also demonstrated better thermal comfort conditions with a hybrid wood-based model compared to the construction typology of the representative project.

1. Introduction

The use of wood in buildings brings many environmental benefits; it is renewable, recyclable, represents a permanent carbon deposit and requires low energy in its manufacturing process for construction. In spite of having the second largest forest reserves on the planet, wood-based building components are not commonly used in the construction sector in Brazil [1].

In less than 50 years, Brazil increased its urbanization rate from 50% to 85% [2]. As a response to this rapid urbanization, in 2009 the Brazilian government launched the program "Minha Casa, Minha Vida" - MCMV ("My house, My Life"), as a strategy to address the housing deficit in the country [3]. Previous researchers have pointed out the great financial interest of the main actors involved in the production of social housing and the low concern regarding design quality [4].

IOP Conf. Series: Earth and Environmental Science **297** (2019) 012019 doi:10.1088/1755-1315/297/1/012019

The selected project is Spazio Mont Royal (SMR) located in the city of São Carlos, Brazil. SMR presents characteristics of social housing developments being currently built in Brazil, including construction system, building arrangement, and space distribution [5]. In addition, the project takes part in MCMV program and has been used as a reference for other research [4].

SMR has 602 housing units distributed in 36 buildings of 5 stories with 20 apartments each. The building units have a symmetrical floor plan with central circulation and are arranged as symmetrical twin blocks (see figure 1).

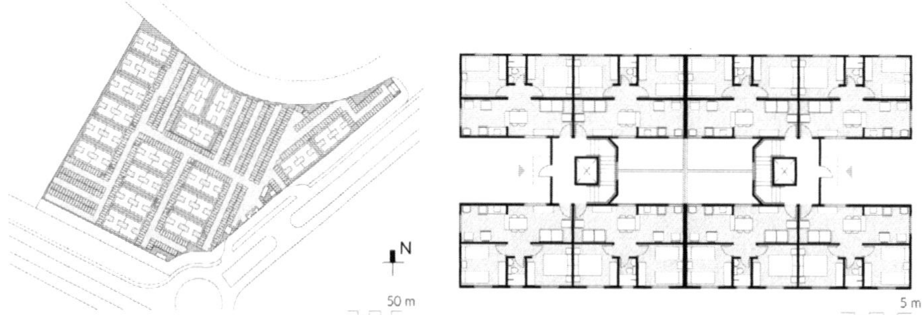

Figure 1. Distribution plan of residential complex SMR (left) and typical floor plan of building units in SMR (right).

The apartment unit has a total internal area of 37.65 m^2 with two bedrooms, a living room, a bathroom, a kitchen, and a service area. The construction system for walls consists of concrete blocks of 9 cm with external and internal plaster. The roof is composed of a fiber cement sheet with metal structure and a 10 cm concrete slab ceiling. The composition of the intermediate floors is 10 cm concrete slab with ceramic tiles.

Due to the high degree of prefabrication and benefits in terms of costs and quality, a platform wood-frame construction technique was assumed for the selection of wood-based alternative systems. The basic concept of the wall system is a sandwich panel with air chamber or insulation layer encased between two skins of plywood boards. Three insulation materials are tested; expanded polystyrene EPS, cellulose, and mineralized wood fiber. The construction system of the roof consists of repetitive wood joists sheathed internally with plywood panels and fiber cement sheets on the external surface. Two roof alternatives were tested; air cavity and insulation material (EPS). The floor system consists of repetitive wood beams at a prescribed spacing encased with wood boards or panels (see figure 2).

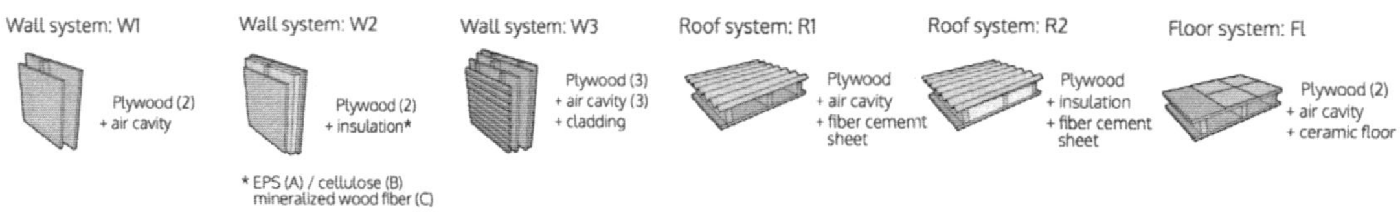

Figure 2. Alternative wood-based construction systems considered for the evaluation.

2. Methods of assessment of climate, comfort and constructions; first conclusions for passive optimized buildings

2.1. Assessment of climate and comfort

The city of São Carlos is situated in the center of the state of São Paulo in Brazil. It has a temperate-warm climate with rains throughout the year [6]. The temperature in summer ranges from 14 °C to 30 °C while in winter from 8 °C to 26 °C. Relative humidity (RH) is high for most of the year and remains around 70% on average. Prevailing winds in São Carlos come from east throughout the year with important variations ranging from northeast to southeast.

The software Climate Consultant 6.0 [7] makes use of the Psychrometric Chart developed by Givoni [8]. It relates the air temperature and the relative humidity of the environment with design strategies to achieve thermal comfort conditions. According to this chart and applying the ASHRAE 55 comfort model [9] and the EnergyPlus weather file (EPW) of São Carlos [10], the city has a very mild climate with comfort in 23.3% of the hours along the year (see figure 3). The main design strategies in buildings are internal heat gain thermal (41.7%) and natural ventilation (25.3%). With these design strategies, the predicted number of hours in discomfort is reduced from 76.7% to 9.7%, a significant reduction with only passive design strategies.

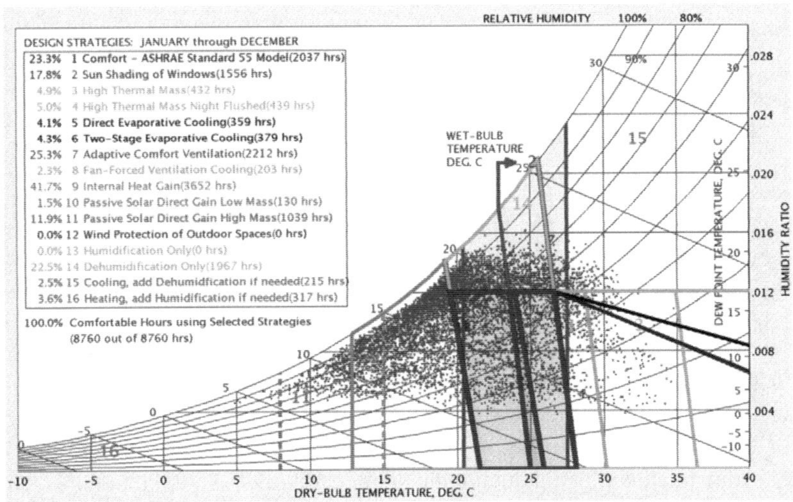

Figure 3. Psychometric diagram of Givoni with comfort model ASHRAE 55 and hourly data of weather file EPW of São Carlos, Brazil through Climate Consultant 6.0.

In respect to the assessment of climate, the Brazilian standard of thermal performance in buildings NBR 15220 [11] divides the Brazilian territory into eight relatively homogeneous zones called bioclimatic zones (BZ). It offers passive design strategies for each BZ to optimize the thermal performance of housing. São Carlos forms part of the BZ 4. For this BZ the standard recommends evaporative cooling, thermal inertia for cooling and cross ventilation. Good natural ventilation is achieved when windows are oriented to prevailing breezes and with openings located on opposite sides of building. The building units in SMR present openings in opposite facades, but prevailing winds (from east) and external obstacles were not considered to define the arrangement (see figure 1). In winter, solar heating and heavy internal walls are recommended. The purpose is to receive and store the heat from the sun during the day and to release it during the night for a comfortable temperature while sleeping. In this regard, the optimal orientation of long-stay areas (rooms where residents spend most of the time) and windows is north since this facade has a long-day sun exposure in winter. However, the layout of the representative project is symmetrical, with long-stay areas in opposite facades (see figure 1).

For the assessment of climate, the adaptive model ASHRAE 55 was considered. It is frequently used for the analysis of the thermal performance in the context of thermal comfort research in Brazil [12]. The ASHRAE 55 adaptive model assumes that occupants adapt to thermal conditions through the opening of windows and clothing. This comfort model applies only in naturally ventilated buildings without heating, ventilation and air conditioning systems (HVAC), as it is in the case of SMR. As result of the calculation proposed by ASHRAE 55-2013 adaptive model [9], figure 4 shows the comfort range for indoor operative temperatures in comparison with the outdoor temperatures considering the EPW file of São Carlos.

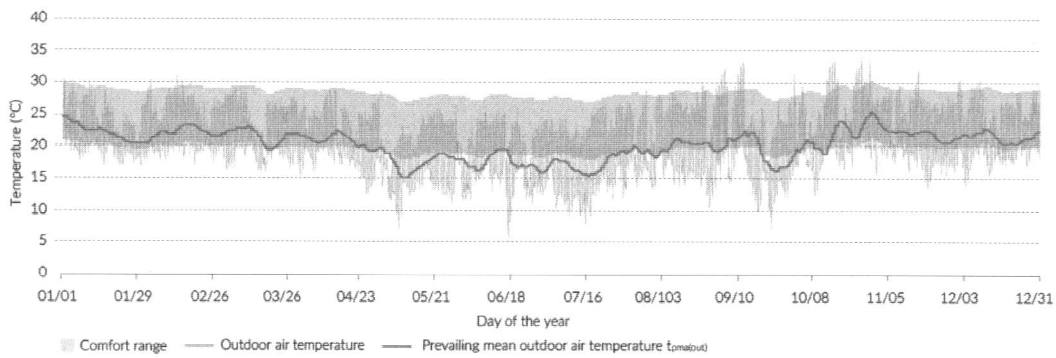

Figure 4. Prevailing mean outdoor air temperature, comfort range of adaptive model ASHRAE 55 and outdoor temperature according to EPW weather data of São Carlos, Brazil.

The quantification of discomfort of the operative temperature in the building was carried out using the degree-hours method (DH). This method is used to classify the thermal performance of residential buildings by the Brazilian Energy Labeling methodology RTQ-R [13]. The degree-hour method allows the quantification of degrees out of the thermal comfort range every hour in the period of interest. The quantification of degree-hours is divided into discomfort by heat (upper limit) and discomfort by cold (lower limit). With the assumption of the outdoor temperature as indoor temperature and the comfort model ASHRAE 55-2013, 625 DH by heat and 16,393 by cold, for a total of 17,018 DH of discomfort are found out of the comfort range. This shows the potential of passive optimization; with only few hours of discomfort by heat, comfort can be widely reached with natural ventilation while the hours of discomfort by cold can be lifted into the comfort range by using internal and solar heat gains. In addition, a medium thermal storage mass would help in both periods.

2.2. Assessment of constructions.

At present, Brazil has one mandatory standard of thermal performance of housing: NBR 15575 [14]. This standard offers minimum pre-requisites regarding thermal properties of building components to meet thermal comfort performance in different BZ. Both, construction of external walls and roof of the representative project meet the requirements of thermal transmittance U (W/m^{2}K), thermal capacity CT (kJ/m^{2}K) and solar absorptance (α) according to NBR 15575 (table 1). The solar absorptance for walls was assumed to be 0.33 ("pérola" color) and 0.514 ("amarelo-antigo" color) in roof according to the existing characteristics of SMR and the common paints used in Brazil [15].

Table 1. Compliance of SMR against minimal requirements by NBR 15575.

	Thermal Transmittance U (W/m^2 K)			**Thermal Capacity (kJ/m^2 K)**		
	Required $\alpha \leq 0.6$	SMR	Compliance	Required	SMR	Compliance
Wall	≤ 3.7	2.78	☑	≥ 130	209	☑
Roof	≤ 2.3	2.06	☑	-	233	-

Regarding the wood-based alternatives, all wall systems were considered deficient for the requirement of thermal capacity (kJ/m2K) according to NBR 15575. The highest number was obtained by the system with mineralized wood fiber W2C (see table 2). Low values of thermal capacity mean low thermal mass, reducing the thermal inertia against daily temperature fluctuations. The influence of these values will be investigated and assessed through simulation in section 3.

Table 2. Compliance of wood-based alternatives against minimal requirements of NBR 15575 in BZ 4.

		Thermal Transmittance U (W/m² K)			Thermal Capacity (kJ/m² K)		
		Required $\alpha \leq 0.6$	WBA	Compliance	Required	WBA	Compliance
Wall	W1	≤ 3.7	1.43	☑	≥ 130	40.48	☒
	W2A	≤ 3.7	0.56	☑	≥ 130	42.25	☒
	W2B	≤ 3.7	0.55	☑	≥ 130	47.48	☒
	W2C	≤ 3.7	0.76	☑	≥ 130	88.45	☒
	W3	≤ 3.7	0.76	☑	≥ 130	73.6	☒
Roof	R1	≤ 2.3	1.96	☑	-	23.98	-
	R2	≤ 2.3	0.35	☑	-	27.53	-

To predict the thermal comfort conditions in a building, the behavior of the construction system should be considered dynamical. The European standard EN ISO 13786 [16] describes the calculation method of the dynamic behavior of constructions. Important results are decrement delay and temperature-amplitude-damping TAD. The decrement delay is the time shift between the amplitudes of external and internal temperature and it is determined by the thermal properties of the material, including density, conductivity, heat capacity and thickness of the construction system. TAD refers to the temperature attenuation of the amplitudes of exterior temperature and internal surface temperature. The external surface of a wall can heat up remarkably under solar radiation. That high temperature is transferred to the interior surface of the wall. If the attenuation of the amplitude is very small, a high amount of the corresponding heat flow will be transferred to the interior and heat up the room. Values for TAD under 15 can be regarded as critical, below 10 as very critical (strong risk of overheating). When TAD is in critical range, the decrement delay plays further an important role. A short decrement delay (less than 6 hours) implies that the heat is transferred already during the day increasing the overheating effect by direct solar heat gains. When the time shift is about 8 to 12 hours, the heat transfer occurs during night hours, with the possibility to ventilate the space with cooler night air (night cooling).

Table 3 shows the TAD and decrement delay values of the wood-based alternatives. Regarding TAD, all wall systems showed a risk of overheating, while system W1 (plywood boards and air cavity) has a strong risk of overheating. The roof construction R2 (roof with insulation) was the only system with a recommended value for TAD. Regarding the decrement delay, all constructions have a short shift time. The range goes from 5 hours for the wall with mineralized wood fiber W2C to less than one hour for the roof system with air cavity R1. This means that heat transfer occurs during the day. These effects will be investigated in section 3.

Table 3. Evaluation of TAD and decrement delay of wood-based alternatives according to EN ISO 13786.

		TAD dynamical		Decrement delay (hours)	
		Value	**Assessment**	**Value**	**Assessment**
Wall	W1	5.51	Very critical, strong risk of overheating	1.39	Short shift time, heat transfer during day
	W2 A	14.37	Critical, contributes to overheating	1.92	Short shift time, heat transfer during day
	W2 B	14.90	Critical, contributes to overheating	2.46	Short shift time, heat transfer during day
	W2 C	13.65	Critical, contributes to overheating	4.94	Short shift time, heat transfer during day
	W3	12.08	Critical, contributes to overheating	4.08	Short shift time, heat transfer during day
Roof	R1	4.46	Very critical, strong risk of overheating	0.90	Short shift time, heat transfer during day
	R2	22.99	Recommended value	1.81	Short time shift, heat transfer during day

To summarize, all the wood-based alternatives for walls comply with the limit values for thermal transmittance (U-value) but were considered as deficient for the requirement of thermal capacity according to the mandatory Brazilian standard NBR 15575. Regarding temperature-amplitude-damping, all wall systems showed a risk of overheating. It is important to ensure a minimal quality in thermal insulation (U-value) to keep higher indoor temperatures in cold periods while a minimal quality for dynamic properties of the constructions to avoid quick heat transfer through constructions during hot periods.

3. Transient simulation of thermal comfort: comparison of representative model and wood-based model

3.1. General assumptions

The program adopted in this research is EnergyPlus, version 8.3.0 [17]. Since almost 80% of the building units are oriented on the axis Northwest-Southeast to maximize the number of building units (longest facades towards Northeast and Southwest), these conditions were assumed for the simulation model. The ground, intermediate and top stories were modeled with internal thermal zones per long-stay room (living room/kitchen, bedroom 1 and bedroom 2) since only these three floors will be considered for the analysis (see figure 5). In order to consider the thermal influence of the 2nd and 4th story, these apartments were considered as a single thermal zone with windows and without internal walls. For the analysis of the simulation results, the DH of discomfort per apartment represents the sum of the three long-stay areas based on the adaptive comfort model ASHRAE 55, while the DH of discomfort in the building wing is the sum of the DH of the apartments located in the ground, intermediate and top story on that side of the building unit. A-wing represents the facade facing southwest while B-wing faces northeast.

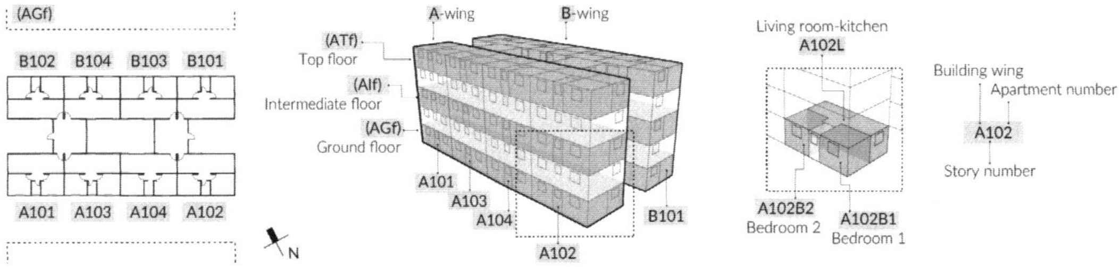

Figure 5. Simplified model of representative project for computer simulation.

A shading surface was modeled in parallel to the biggest facades to consider the shading effect of adjacent buildings. The apartment unit has two types of windows - 2 sashes horizontal slide in the living room, kitchen and bedrooms and 4 blades louver window in the bathroom. The windows in bedrooms (largest facade) have an internal aluminum screen in one of the sashes, avoiding direct solar infiltration. This condition was also modeled and considered for the simulation. The glazing type assigned to the model is single glass of 4 mm similar to the existing conditions of SMR. All doors in the three fully-modeled stories were included. As recommended in the evaluation procedure through simulations proposed by NBR 15575 [14], one air change per hour (ACPH) was assumed for the simulation during the day and year. Heat gains from internal loads (e.g., people, lights, and equipment) and schedules followed the recommended values stated on the Brazilian Standard NBR 15575.

3.2. Representative model

As showed in figure 6, the representative model with common construction systems presented a significantly higher level of discomfort by cold (147,884 DH, 97%) than by heat (4,029 DH, 3%). B-wing had slightly more DH of discomfort by heat than A-wing, while A-wing has much more DH of discomfort by cold than B-wing. This is directly related to the orientation and solar exposure of the building wings; the northeast facade (B-wing) has a day-long sun exposure in winter and half-day incidence during summer. In total, A-wing presents 19% more DH of discomfort than B-wing. This different comfort qualities shows the lack of passive design strategies addressed on the representative project.

The intermediate story showed a better thermal performance compared to the ground and top story. As expected, the top story has a higher DH of discomfort by heat, due to external exposure through the roof and the high transfer of absorbed solar radiation. The ground story has the highest total level of discomfort from the three stories, caused mostly by cold. The apartments on this level have almost no direct solar exposure due to the proximity of adjacent buildings.

According to the simulation results, apartment B501 has the highest discomfort level by heat (512 DH, 7%), while A502 has the highest by cold (8,549 DH, 95%). A502 is the least comfortable apartment with a total 8,955 DH of discomfort. The best thermal conditions are found in B304, located on the third floor of B-wing. This apartment unit showed a total of 3,311 DH of discomfort with 2% produced by heat and 98% by cold. The reduced surface exposed to outdoors and the higher sun exposure of that facade in winter offer these conditions.

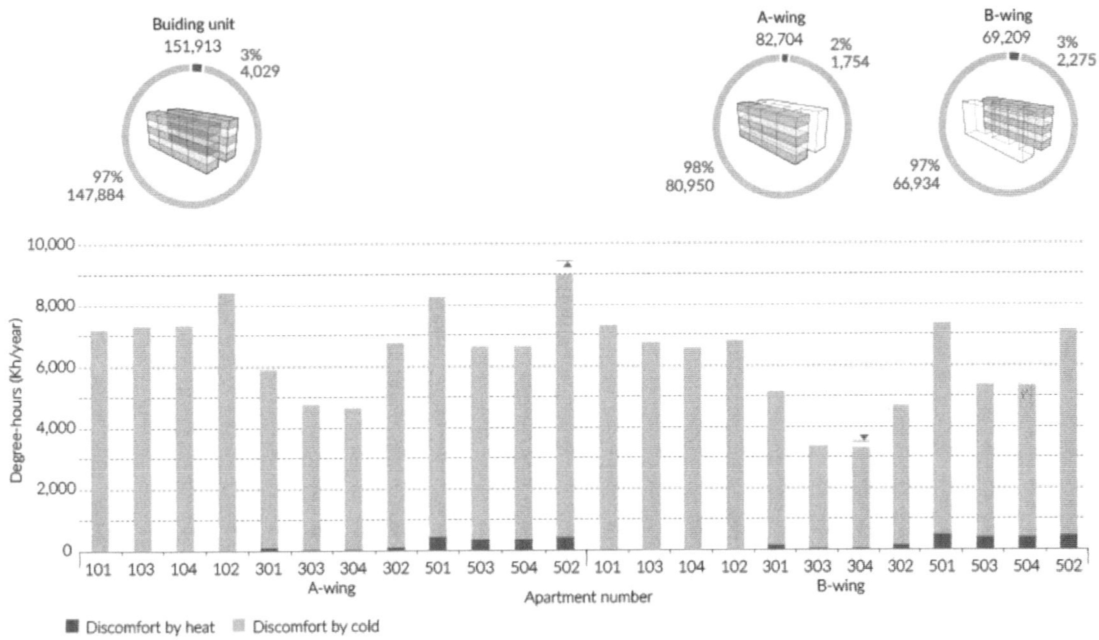

Figure 6. Total degree-hours of discomfort in the building per apartment in representative model.

3.3. Wood-based model

All construction systems with roof type R2 resulted in less DH of discomfort by heat and cold than the alternatives with roof type R1 (see figure 7). This represents the positive effect of the addition of an insulation layer to the roof system. The better thermal behavior of R2 was predicted by the TAD calculation of EN ISO 13786.

Compared to the representative model, the wood-based alternatives showed a more even proportion of discomfort by cold and heat. The proportion of discomfort by heat ranges from 28% to 40% in the wood-based models. Due to the climate conditions of São Carlos, it is expected to see discomfort by cold remaining with a bigger portion than the discomfort by heat in all simulated models. All the tested models presented a significantly higher DH of discomfort by heat than the reference outdoor temperature of the weather file EPW, indicating an overheating effect in the apartment units. On the other side, all alternatives have a reduction of more than 50% of DH of discomfort by cold from the outdoor temperature assumption. The strategy now is to reduce discomfort by heat, since it represents higher discomfort and therefore the need of air conditioning.

Among the three insulation options, mineralized wood fiber W2C showed less total discomfort level, followed by cellulose W2B. The model R1W1 resulted in a very high degree of discomfort by cold and heat, signifying the need of replacing the air gap with another material in external walls and roofs.

After the alternatives with wall system W2C (with mineralized wood fiber), W3 with three air gaps showed a better performance than the others, even better than the systems with cellulose W2B and EPS insulation W2A.

From all the tested wood-based models, R2W2C showed the best performance. This model resulted in 11% less DH of discomfort by cold than the representative model. On the other hand, the discomfort by heat is around 15 times more than the model of the representative project. The combination of these construction systems resulted in 27% more DH of the total discomfort than the representative model. These wood-based constructions behave well in cold periods (better U-values), but the tendency to overheat in hot periods increases. Following NBR 15575, W2C has the highest thermal capacity value (88.45 kJ/m^{2}K) but below the recommendation (130 kJ/m^2 K) (see table 2). According to EN ISO 13786, the dynamical thermal behavior of the roof construction R2 is good while the wall construction W2C is nearly in the recommended range (13.65, recommended is >=15), thus both constructions behave better than the others and the transfer of heat through the construction is minimized. This shows the potential for improving this model accomplished with the reduction of the DH of discomfort by heat through optimizing the model increasing the thermal storage mass and against the overheating effect.

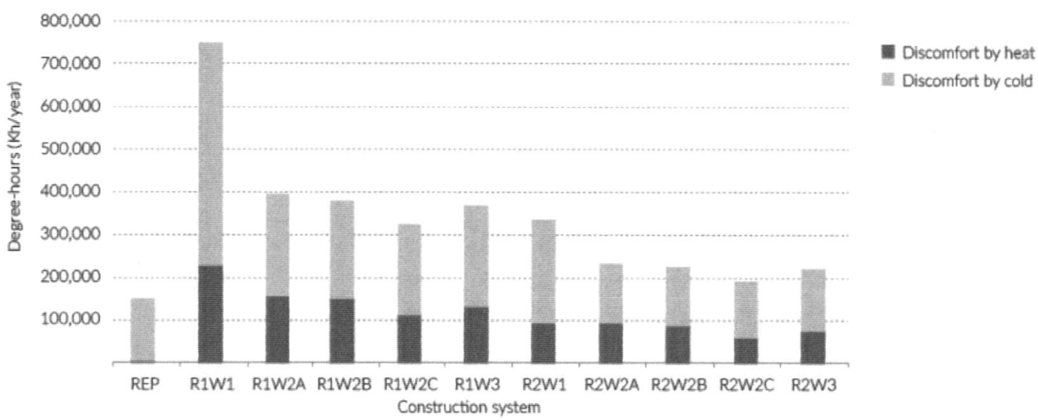

Figure 7. Total number of DH of discomfort for the whole building in representative model and with wood-based constructions.

3.3.1. Optimization of wood-based model. Five other alternatives were simulated based on the fact that R2W2C was the construction system that showed the best performance. All alternatives include the wall system W2C. W2C-1 considers the roof system R2 but replaces the layer of EPS with mineralized wood fiber. W2C-2 presents the same layers of roof system R2 but increases the thickness of the insulation layer of mineralized wood fiber from 10 cm to 15 cm. W2C-3 represents a hybrid construction with a roof system similar to the one used in the representative project. W2C-4 includes a green roof of 10 cm on top of a sandwich wooden panel with mineralized wood fiber as insulation. W2C-5 proposes a wood-based roof system similar to the one in W2C-1 (R2 system with mineralized wood fiber as insulation). In this alternative, internal floors are made of concrete slabs of 10 cm similar to the one in the representative project. This represents a second hybrid construction.

All alternatives resulted in a lower level of discomfort by cold than the representative building, with the exception of W2C-4 (see figure 8). On the other hand, the discomfort by heat remains significantly high for most of the construction systems, the hybrid model W2C-5 being the one with the lowest. The construction system with the green roof W2C-4 was the only model with a higher DH of discomfort by heat than if considering the outdoor temperature, indicating a high overheating effect. The increase of 5 cm in the thickness of the insulation layer with mineralized wood fiber (W2C-2) lead to a reduction of 6% of DH of discomfort compared to the roof system with 10 cm of the same material (W2C-1).

When summing the annual DH of discomfort by cold and heat, W2C-5 resulted in a reduction of nearly 50% of the total DH of discomfort of the representative building on a yearly basis. The addition of a concrete slab in the interior floors in W2C-5 represents an increase of the internal thermal mass, hence improving the thermal performance of the building. The model W2C-5 resulted in a higher DH of discomfort by cold (62,539 DH, 88%) than by heat (8,943, 12%) in the sum of the DH in long-stay areas of the apartment units in the ground, intermediate and top story in a year.

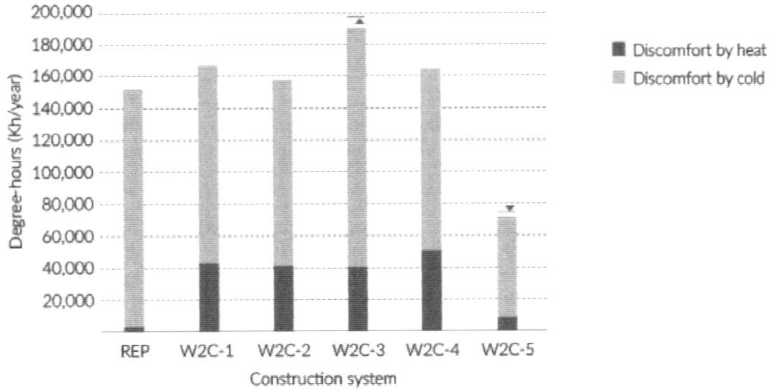

Figure 8. Total number of DH for the whole building in optimized models with wood-based alternatives.

4. Final discussion

The arrangement and the characteristics of the building units in the representative project seem to respond to the maximization of the number of units accommodated in the complex, disregarding the thermal comfort performance in the apartment units. The orientation and the proximity of the units causes high solar exposure in summer and low exposure in winter for many units. As expected, the total discomfort is mainly by cold, with the apartments in lower stories showing a higher discomfort level.

Thermal comfort standards are required to establish limits for the building thermal performance, improve comfort conditions and reduce the energy consumption for heating and air conditioning systems. The present mandatory Brazilian standard NBR 15575 could predict with simplified calculation methods the positive/negative thermal behavior of the building. However, it is important to consider not only external elements but also internal components for the overall thermal performance (e.g. internal floor and internal walls), especially in multi-story buildings. The calculation of TAD and decrement delay in EN ISO 13786 is an alternative for the Brazilian standards to have a better and more detailed prediction of the dynamic behavior of building components. The dynamic behavior of construction has great importance in many climates in Brazil where thermal mass plays an important role as a passive design strategy.

Regarding the thermal performance of wood-based models, insulation of mineralized wood fiber in wood-based constructions performed better than cellulose and EPS. This insulation material showed the highest value of thermal wave shift, as other studies have also pointed out [18]. In addition, this material is produced from wood waste and has low embodied energy and climate change potential compared to other insulation materials on the market such as EPS [19]. Insulation materials with higher density (e.g. 500 kg/m^3 for mineralized wood fiber) show a better overall thermal performance due to the thermal storage effect, even when their conductivity is lower than other insulation materials.

The computer simulation showed the existing risk of overheating when considering wood-based components in a building, especially with air cavity. However, with the optimization of these systems through the careful selection of the insulation material and the substitution of some components by "heavier" materials (such as concrete), the risk is significantly minimized. This study demonstrates, through a computer simulation, that better thermal comfort conditions are reachable with a hybrid wood-based model (wood panels with mineralized wood fiber and concrete floors) compared to the representative type of construction in social housing in Brazil. Hybrid models represent a way to increase the use of wood in buildings, without negatively affecting the overall thermal performance in certain Brazilian climates, as it is the case of São Carlos.

References

[1] Zerbini F 2014 *Cenário da madeira FSC no Brasil 2012-2013* São Paulo, Brazil

[2] Magalhães I 2014 *Nuevas Herramientas para enfrentar el déficil de vivienda en Brasil* Lima, Peru

[3] Ministério das Cidades Brazil 2009 *Plano Nacional de Habitação 2009* Brasilia, Brazil

[4] Amore C S, Shimbo L Z and Rufino M B 2015 *Minha Casa... e a Cidade?* Rio de Janeiro, Brazil

[5] Triana M A, Lamberts R and Sassi P 2015 Characterisation of representative building typologies for social housing projects in Brazil and its energy performance *Energy Policy* **87** p 524-541

[6] Peel M C, Finlayson B L and Mcmahon T A 2007 Updated world map of the Köppen-Geiger climate classification *Hydrology and Earth System Sciences Discussions* **5** 1633-1644

[7] University of California 2016 *Climate Consultant 6.0* Version 6.0.10 (Los Angeles: University of California)

[8] Givoni B 1976 *Man, Climate and Architecture*

[9] ASHRAE 2013 *ANSI/ASHRAE Standard 55-Thermal Environmental Conditions for Human Occupancy* American Society of Heating, Refrigerating and Air-Conditioning Engineers Atlanta, USA

[10] Roriz M 2012 *Arquivos Climáticos de Municípios Brasileiros* (São Carlos: Associação Nacional de Tecnologia do Ambiente Construído ANTAC)

[11] ABNT 2005 *NBR 15.220: Desempenho térmico de edificações* (Rio de Janeiro: Brazilian Association of Technical Standards)

[12] Rupp R F, Vásquez N G and Lamberts R 2015 A review of human thermal comfort in the built environment *Energy and Buildings* **105** p 178-205

[13] Brazil INMETRO 2012 *Regulamento Técnico da Qualidade para o Nível de Eficiência Energética de Edi cações Residenciais RTQ-R* (Rio de Janeiro: National Institute of Metrology, Quality and Technology)

[14] ABNT 2013 *NBR 15.575: Edifícios Habitacionais – Desempenho* (Rio de Janeiro: Brazilian Association of Technical Standards)

[15] Dornelles K 2008 *Absortancia solar de superficies opacas: metodos de determminação e base de dados para tintas latex acrilica e PVA* (Campinas: Universidade Estadual de Campinas)

[16] CEN 2007 *EN 15251 Indoor Environmental Input Parameters for Design and Assessment of Energy Performance of Buildings: Addressing Indoor Air Quality, Thermal Environment, Lighting and Acoustics* (Brussels: European committee for Standardisation CEN)

[17] DOE US 2015 *EnergyPlus Version 8.3.0* USA: US Department of Energy

[18] Schiavoni S, D'Alessandro F and Asdrubali F 2016 Insulation materials for the building sector: A review and comparative analysis *Renewable and Sustainable Energy Reviews* **62** p 988-1011

[19] Hill C, Norton A and Dibdiakova J 2018 A comparison of the environmental impacts of different categories of insulation materials *Energy and Buildings* **162** p 12-20

IOP Conf. Series: Earth and Environmental Science **297** (2019) 012020 doi:10.1088/1755-1315/297/1/012020

Primary energy and CO_2 emissions implications of different insulation, cladding and frame materials for residential buildings

Uniben Yao Ayikoe Tettey and Leif Gustavsson

Sustainable Built Environment Research Group, Linnaeus University, Växjö, SE-35195, Sweden

uniben.tettey@lnu.se

Abstract. In this study, we analyse and compare the primary energy use and carbon dioxide (CO_2) emissions associated with different insulation, cladding and frame materials for a constructed concrete frame multi-storey residential building in Sweden. Our approach consists of identifying individual materials giving the lowest primary energy use and CO_2 emissions for each building envelope part and based on that, modelling different material combinations to achieve improved alternatives of the concrete frame building with the same operation energy use based on the Swedish building code or passive house criteria. We analyse the complete materials and energy chains, including material losses as well as conversion and fuel cycle losses. The analysis covers the primary energy use to extract, process, transport, and assemble the materials and the resulting CO_2 emissions to the atmosphere. The results show wide variations in primary energy and CO_2 emissions depending on the choice of building envelope materials. The materials for external walls contribute most to the primary energy and CO_2 emissions, followed by foundation, roof and external cladding materials. The improved building alternatives with wood construction frames, wood external cladding, expanded polystyrene as foundation insulation and cellulose insulation in the external walls and roof result in about 36 - 40% lower production primary energy use and 42 – 49% lower CO_2 emissions than the improved concrete alternative when achieving the same thermal performance. This study suggests that strategies for low-energy buildings should be combined with resource-efficient and low carbon materials in the production phase to mitigate climate change and achieve a sustainable built environment.

1. Introduction

The building sector contributes largely to energy and material use through the construction, operation and end-of-life activities related to buildings. The building and building construction sectors together account for 36% of global final energy use and about 40% of total direct and indirect CO_2 emissions globally [1]. Around 24% of global [2] and 50% the European Union (EU) material extractions are used in the built environment [3]. The operation phase dominates the life cycle energy use, especially in conventional buildings. Currently, about 40% of global final energy demand is for heating and cooling in the building and industrial sectors [4]. Space heating alone constitutes more than 60% of the EU's final energy demand in these sectors [5]. Building envelopes influence the space heating and cooling demands as well as indoor climates in buildings through their thermal performance. Existing energy policies have therefore mainly focused on energy efficiency and renewable energy use within the building sector. The directive on the energy performance of buildings mandates all new buildings in the EU to be nearly zero-energy buildings from 2021 [6]. However, several studies [7-9] show that the significance of the other life cycle phases increase as more energy efficient buildings are constructed. Stephan et al. [10] noted that the production energy of a Belgian passive house was up to 77% of the production and operational energy use over a 100 year life span. Moran et al. [11] analysed different versions of a semi-detached residential building in Ireland and reported that the production energy and CO_2 emissions of the nearly zero energy building versions were up to 44% and 100% of the total life cycle energy use and CO_2 emissions, respectively. Dimoudi and Tompa [12] studied energy and environmental impacts of different construction materials for multi-storey office buildings in Greece. They showed that the production energy of the buildings' frame materials of concrete and reinforcement steel gives the largest part of the total production energy, while the external walls give the biggest contribution for the building envelope elements. The implication of various material choices for the structural frame, surface components and inner components on the life cycle energy of a hypothetical Finnish building model was studied by Takano et al. [13]. They found that the choice of materials for the structural frame results in the largest impact. Hence, as buildings are constructed to be more energy efficient, strategies for resource efficiency and material use over their life cycle should be addressed.

In this study, we analyse and compare the primary energy use and CO_2 emissions associated with different insulation, cladding and frame materials for a constructed concrete frame multi-storey residential building in Sweden redesigned to meet the Swedish building code of 2015 (BBR) or the passive house criteria (passive). Our approach consists of identifying individual materials giving the lowest primary energy use and CO_2 emissions for each building envelope part and based on that, modelling different material combinations to achieve improved alternatives of the concrete frame building with the same operation energy use.

2. Studied building description

The constructed building used in this study is a recently completed multi-storey residential building in Växjö, Southern Sweden. It is 6 storeys high with prefabricated concrete frame and comprises 24 apartments with a total heated floor area of 1686 m^2. The foundation is made up of 100 mm concrete slab, 300 mm expanded polystyrene (EPS) insulation and 200 mm crushed stone. The external walls consist of 100 mm EPS insulation sandwiched between 100 mm and 230 mm of concrete panels on the outside and inside, respectively. The intermediate floors are 250 mm concrete slabs while the roof consists of 250 mm concrete slab and 500 mm mineral wool insulation with wooden trusses and a roof covering over layers of asphalt-impregnated felt and plywood. The windows and external doors have clear glass double-glazed panels with wood frames, covered with aluminium profiles on the outside. The windows and doors have u- and g-values of 1.2 W/m^{2}K and 0.6, respectively. The ventilation is based on a balanced mechanical system with a heat recovery (VHR) unit of 76% efficiency. A photograph and floor plan of the reference building is shown in Figure 1.

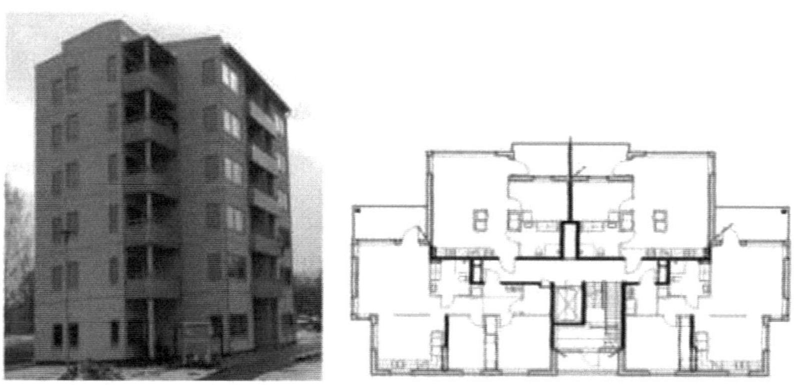

Figure 1. Photograph (left) and floor plan (right) of the reference building.

3. Methodology

Generally, the study approach includes modelling changes to the envelope characteristics of the studied building to achieve building alternatives meeting BBR standard or passive house criteria; considering different insulation, cladding and frame materials in the different parts of the two building alternatives meeting the BBR standard or passive house criteria; analysing and selecting the materials resulting in the lowest production primary energy use in each building envelope part; modelling improved alternatives of the two building alternatives based on a combination of the selected materials; and analysing and comparing the production primary energy use and CO_2 emissions of the alternatives.

Table 1. Thermal characteristics of the different envelope elements for the building alternatives to meet the BBR standard and passive house criteria.

Description	U-values (W/m^{2}K)	
	BBR	Passive
Ground floor	0.11	0.11
External walls	0.22	0.11
Windows	1.2	0.8
Doors	1.2	0.8
Roof	0.08	0.05

3.1. BBR and passive house building alternatives

We modelled changes to the envelope characteristics of the studied building, including the external walls, roof, door and windows to achieve building alternatives with thermal performances of the Swedish building code [14] and passive house criteria [15]. The BBR standard and passive criteria specify maximum limits for specific final energy use for operation of buildings, including space heating, tap water heating and electricity for ventilation but excluding electricity for household appliances and lighting. The specific final energy use depends on climate zones and the heating supply option for the building. The analysed building alternatives are assumed to have the same district heat supply as the constructed building. The constructed building is in climate zone 3 in southern Sweden, where the maximum specific annual final energy use for operation is set to 80 kWh/m^2 and 50 kWh/m^2 for the BBR standard and passive house criteria, respectively. The airtightness of the building alternatives is assumed to be 0.6 l/s m^2 and 0.3 l/s m^2 at 50 Pa for the BBR standard and passive house criteria, respectively. Table 1 shows the thermal characteristics of the building alternatives. The insulation, façade and frame materials in the different building parts of the constructed building and the considered alternatives are shown in Table 2.

The final energy use for operation of the building alternatives, including space and tap water heating as well as electricity for ventilation is calculated with the VIP-Energy software [16]. The VIP-Energy software is validated by the International Energy Agency Building Energy Simulation Test and diagnostic method (IEA BESTEST) and performs dynamic and hour by hour analysis of building operation energy use based on building orientation, geometry, heating and ventilation systems, indoor and outdoor temperature, operation schedule and thermal properties of the building envelope elements. The energy balance calculations are based on the 2013 climate data, which is close to the current average climate, for the city of Växjö and indoor temperature set points of 21°C for living and 18°C for common areas.

Table 2. Insulation, façade and frame materials in constructed concrete building and the considered alternatives.

Building part	Materials in the constructed building	Considered alternatives
Roof	Rock wool	Glass wool, cellulose fiber
External wall	EPS	Glass wool, cellulose fiber, EPS
Ground floor	EPS	Foam glass
Building frame	Concrete	Cross laminated timber (CLT), prefabricated timber modules (Modular)
Façade	-	Stucco, wood cladding, brick tiles, aluminium cladding

3.2. Primary energy and CO_2 emissions analysis

Based on building alternatives with different material choices, we estimated the quantities of the various material inputs required to construct the buildings to meet the BBR standard and passive house criteria. The analysis covers the complete materials and energy chains, including material losses and also, conversion and fuel cycle losses of the energy systems along the entire material production process. We considered the primary energy use for extraction, processing, transport, and assembly of the building envelope materials required for the building alternatives as well as the resulting CO_2 emissions based on methodology presented by Gustavsson et al. [17]. Data for specific end-use energy for material production is from Björklund and Tillman [18]. The specific end-use energy for production of selected materials are given in Table 3.

Table 3. Specific end-use energy (kWh/kg) to produce selected building materials.

Material	Coal	Oil	Fossil gas	Bioenergy	Electricity
Concrete	0.09	0.10	–	–	0.02
Plasterboard	–	0.79	–	–	0.16
Rock wool	2.00	0.36	0.02	–	0.39
Glasswool	2.87	0.52	0.03	–	2.00
EPS	0.28	3.9	3.72	–	0.63
Lumber	–	0.15	–	0.70	0.14
Particleboard	–	0.39	–	1.39	0.42
Stucco (plaster)	0.05	0.05	–	–	0.04
Brick	–	0.003	0.46	–	0.15
Steel	1.99	0.47	0.89	–	0.74
Aluminium	–	11.8	–	–	17.8

We assumed fossil fuel cycle energy inputs of 10, 5 and 5% as well as specific fuel-cycle carbon emissions of 0.11, 0.08 and 0.06 kg C/kWh for fossil coal, oil and gas, respectively based on Gustavsson and Sathre [19]. End-use electricity for building material production is taken to be from a coal based power plant with conversion efficiency of 40% and distribution losses of 2% based on Gustavsson et al. [17].

4. Results and discussion

The required quantities of insulation materials in the various building envelope parts when the building alternatives are constructed with different frame materials to meet the BBR standard and passive house criteria are given in Table 4. Glass wool and EPS result in the lowest thickness for external wall insulation, whiles cellulose fiber results in the highest for all the building alternatives. For the ground floor insulation, EPS gives a lower thickness compared to foam glass insulation. The passive building alternatives require about 56-60% and 113-158% more insulation materials for the roof and external walls, respectively compared to the BBR alternatives.

Table 4. Quantities of insulation materials required in the different building envelope parts to meet the BBR standard and passive house criteria.

		BBR			Passive		
Building parts	Insulation materials	Concrete	CLT	Modular	Concrete	CLT	Modular
Roof	Rock wool	500	465	465	780	745	745
	Glass wool	500	465	465	780	745	745
	Cellulose fiber	500	465	465	780	745	745
External walls	Rock wool	158	124	154	337	320	332
	Glass wool	150	117	146	320	286	315
	Cellulose fiber	166	130	162	355	320	350
	EPS	150	117	146	320	286	315
Ground floor	EPS	300	300	300	300	300	300
	Foam glass	315	315	315	315	315	315

Table 5 shows the mass balances while Figure 2 shows the primary energy use for production of the various insulation materials for the different building envelope parts to meet the BBR standard and passive house criteria. EPS gives a lower production primary energy use than foam glass for the foundation insulation of the building alternatives. For the roof and external walls, cellulose fiber insulation consistently results in the lowest primary energy use for production compared to the other insulation alternatives both for the BBR standard and passive house criteria, while glass wool gives the highest. The differences in production primary energy use of the insulation materials are linked to variations in manufacturing processes as well as fuel resources. Tettey et al. [20] considered different insulation materials for a Swedish residential building and found that their production process is typically fossil fuel intensive.

Table 5. Mass (kg) of insulation materials required in different building parts to meet the BBR standard and passive house criteria.

		BBR			Passive		
Building parts	Insulation materials	Concrete	CLT	Modular	Concrete	CLT	Modular
Roof	Rock wool	7140	6640	6640	11140	10640	10640
	Glass wool	3570	3320	3320	5570	5320	5320
	Cellulose fiber	7140	6640	6640	11140	10640	10640
External walls	Rock wool	5520	4330	5380	11770	11170	11590
	Glass wool	6550	5110	6370	13960	12480	13750
	Cellulose fiber	11770	9220	11490	25170	22690	24820
	EPS	4090	3190	3980	8730	7800	8590
Ground floor	EPS	4900	4900	4900	4900	4900	4900
	Foam glass	19700	19700	19700	19700	19700	19700

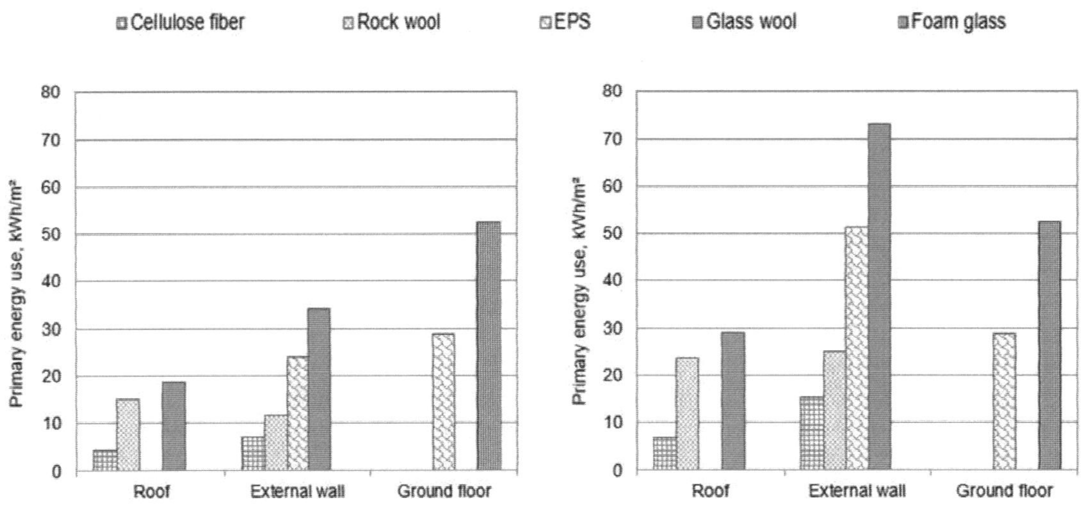

Figures 2. Primary energy use for production of the various insulation materials in the different building envelope parts to the BBR standard (left) and passive criteria (right).

The primary energy use for production of the different building frame materials are presented in Figure 3 for the building alternatives to the BBR standard and passive house criteria. The CLT building frame requires the least while the concrete frame alternative requires the most primary energy use for production. The modular and CLT building frames to the passive house criteria have about 6 and 8% more primary energy use for production than the BBR alternatives. Figure 4 shows the primary energy use for production of the different building façade alternatives. Wood cladding results in the lowest production primary energy use, followed by stucco rendering, brick and aluminium cladding. The choice of façade materials is not directly linked to the thermal performance of buildings but influences the production energy use. Façade materials such as brick and aluminium may have longer service life spans than wooden alternatives leading to variations in maintenance and repair needs [21]. The primary energy for maintenance and repairs can constitute a large share of the life cycle primary energy use, depending on the choice of façade material and the exposure conditions [22]. Hence the choice of façade materials should be combined with appropriate design strategies to minimise primary energy use for maintenance and repairs.

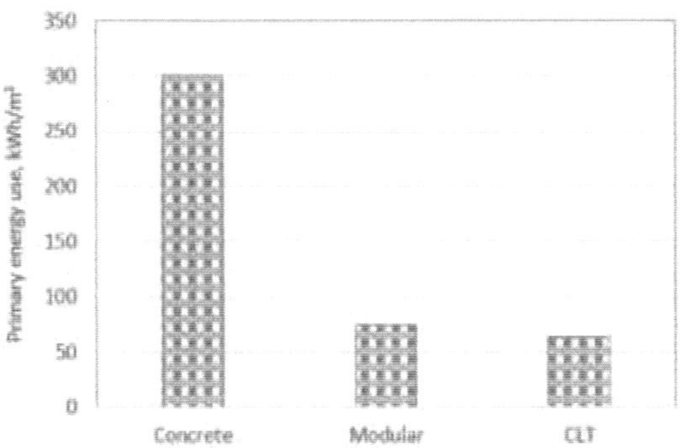

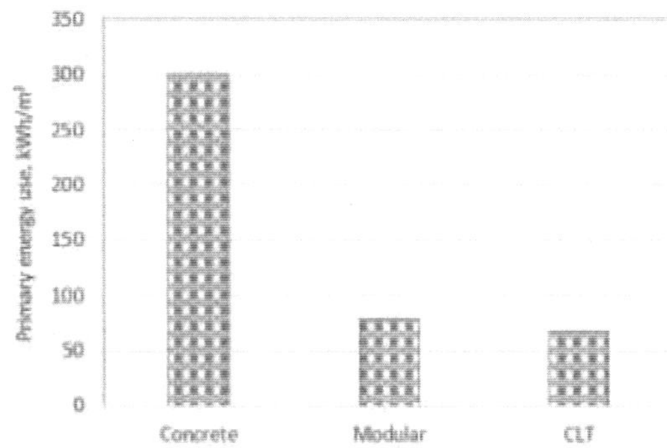

Figures 3. Primary energy use for production of the different building frame alternatives to the BBR standard (left) and passive criteria (right).

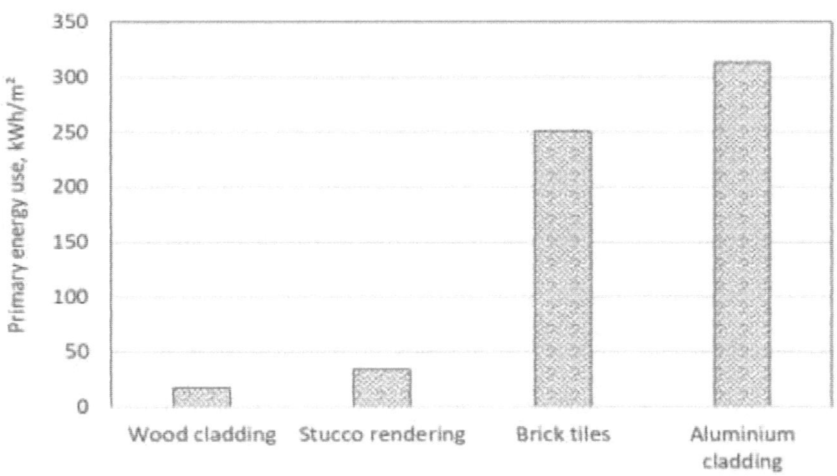

Figures 4. Primary energy use for production of the various façade alternatives.

The insulation, cladding and frame materials resulting in the lowest production primary energy and CO_2 emissions for each building part are given in figures 5 and 6 compared to the material choices in the constructed building for the BBR standard and passive house criteria. The lowest production primary energy use and CO_2 emissions for insulation is when EPS is used in the foundation and cellulose insulation in the external walls and roof. In addition, the best wall cladding choices is wood for the modular and CLT building alternatives while the concrete building alternatives do not need any external cladding, similar to the constructed building. The primary energy use and CO_2 emissions for production of the external wall materials are large compared to those of the other building envelope parts. The production primary energy use for the external wall elements constitutes 56 – 65% and 59 – 66% of the total primary energy use for the building alternatives to meet the BBR standard and passive house criteria, respectively. The best CLT and modular building alternatives give significantly lower production primary energy use and CO_2 emissions for the building envelope parts compared to the best concrete alternatives. The total primary energy use for production of the building envelope materials to meet either the BBR standard or passive house criteria are about 4%, 41% and 42% lower for the best concrete, modular and CLT building alternatives, respectively, compared to the material choices in the constructed building. Similarly, the CO_2 emissions are reduced by 4%, 47% and 51% for the best building alternatives meeting the BBR standard and 7%, 46% and 50% when meeting the passive house criteria, compared to the material choices in the constructed building.

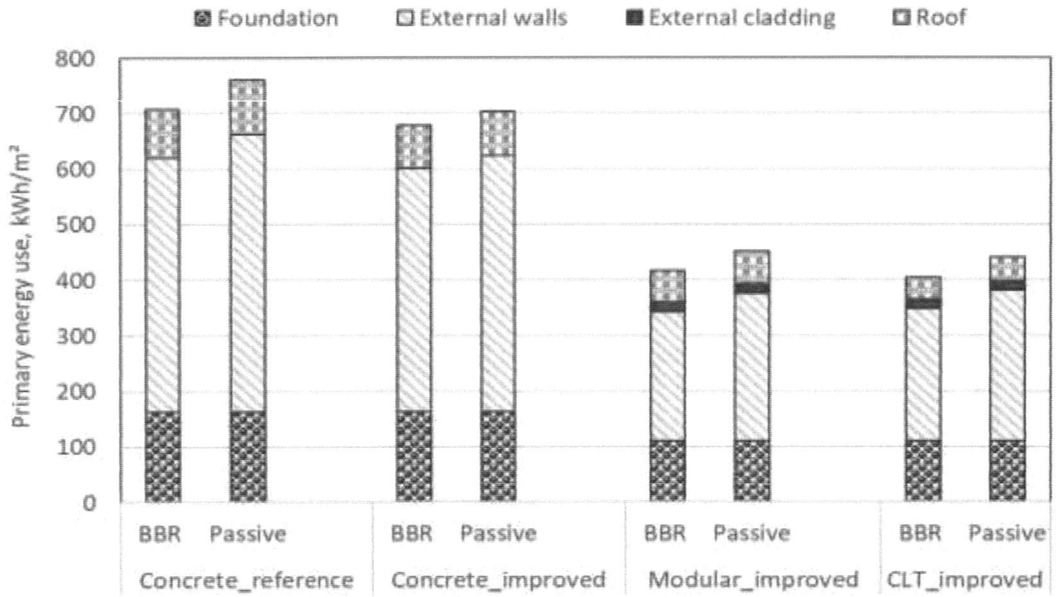

Figures 5. Production primary energy use of the different building envelope parts for material choices in the constructed building (reference) and for the combination resulting in the lowest production primary energy (improved) for the building alternatives.

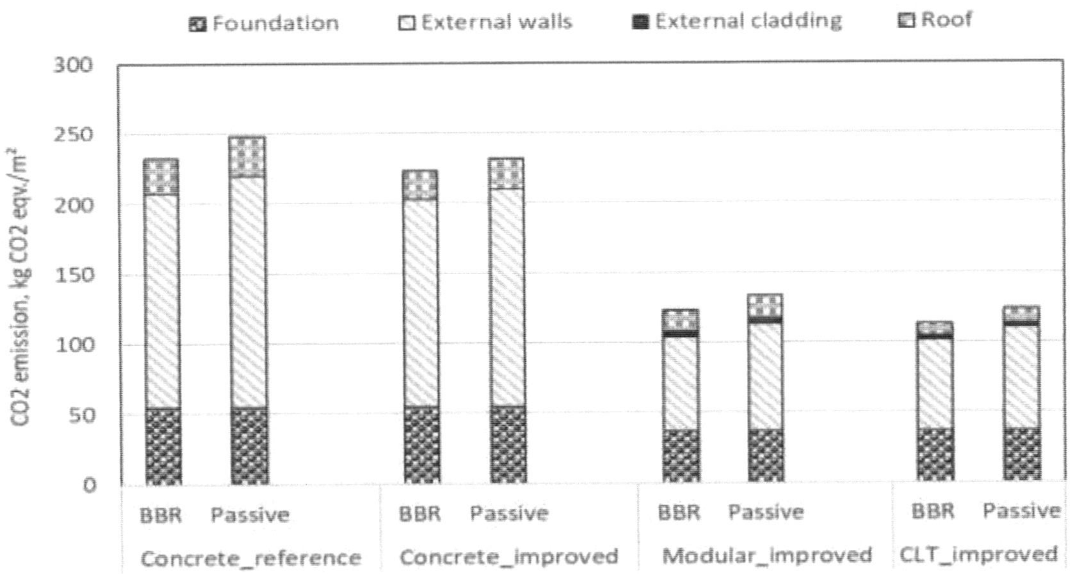

Figures 6. CO_2 emission of the different building envelope parts for material choices in the constructed building (reference) and for the combination resulting in the lowest production primary energy (improved) for the building alternatives.

5. Conclusion

We have analysed the impacts of different building envelope materials on the primary energy use and CO_2 emissions for a multi-storey residential building meeting the BBR standard or passive house criteria and explored how combinations of the building envelope materials could minimise production primary energy use and CO_2 emissions. Our results show that the choice of building envelope materials significantly influences the primary energy use and CO_2 emissions of the analysed alternatives. The materials for the external walls contribute most to the primary energy use and CO_2 emissions, followed by foundation, roof and external cladding materials. The building alternatives with wood construction frames, wood external cladding, EPS as foundation insulation and cellulose insulation in the external wall and roof were found to give significantly lower production primary energy use and CO_2 emissions than the material choices in the constructed building both for the BBR standard and passive house criteria building alternatives. The choice of building envelope materials such as insulation and frame construction systems may result in varying energy- and climate-related impacts over the full life cycle of buildings. In this study, we have focused on the primary energy use and CO_2 emissions implications of different building envelope materials during the production phase. However, the choice of building envelope materials may also influence the maintenance needs and end-of-life related impacts. Moreover, several studies [23-25] report climate and environmental benefits for timber frame buildings mainly due to significant wood material recovery over the production and end-of-life phases for energy purposes. Further studies should consider these aspects in a life cycle perspective.

Overall, the improved CLT and modular building alternatives result in about 36 - 40% and 42 – 49% lower primary energy use and CO_2 emissions, respectively, compared to the improved concrete alternative. This study suggests that strategies for low-energy buildings should be combined with resource-efficient and low-carbon materials in the production phase to mitigate climate change and for a sustainable built environment.

References

1. IEA, *International Energy Agency, Energy Technology Perspectives - Catalysing Energy Technology Transformations*. 2017.
2. Zabalza Bribián, I., A. Valero Capilla, and A. Aranda Usón, *Life cycle assessment of building materials: Comparative analysis of energy and environmental impacts and evaluation of the eco-efficiency improvement potential.* Building and Environment, 2011. **46**(5): p. 1133-1140.
3. Pomponi, F. and A. Moncaster, *Embodied carbon mitigation and reduction in the built environment – What does the evidence say?* Journal of Environmental Management, 2016. **181**: p. 687-700.
4. IEA, *International Energy Agency, Energy Technology Perspectives - Catalysing Energy Technology Transformations. Executive Summary. Available at https://webstore.iea.org/download/summary/237?fileName=Chinese-ETP-2017-ES.pdf*. 2017.
5. Saheb, Y., et al., *Energy Renovation: The Trump Card for the New Start for Europe*. 2015: JRC Science and policy reports. Web accessed at http://iet.jrc.ec.europa.eu/energyefficiency/system/tdf/eur26888_buildingreport_online.pdf?file=1&type=node&id=9069 on October 3, 2015.
6. Directive 2010/31/EU, *of the European Parliament and of the Council of 19 May 2010 on the energy performance of buildings (Recast)*. Official Journal of the European Union.
7. Chastas, P., T. Theodosiou, and D. Bikas, *Embodied energy in residential buildings-towards the nearly zero energy building: A literature review.* Building and Environment, 2016. **105**(Supplement C): p. 267-282.
8. Tettey, U.Y.A., A. Dodoo, and L. Gustavsson, *Primary energy implications of different design strategies for an apartment building.* Energy, 2016. **104**: p. 132-148.
9. Giordano, R., et al., *Embodied Energy Versus Operational Energy in a Nearly Zero Energy Building Case Study.* Energy Procedia, 2017. **111**: p. 367-376.
10. Stephan, A., R.H. Crawford, and K. de Myttenaere, *A comprehensive assessment of the life cycle energy demand of passive houses.* Applied Energy, 2013. **112**: p. 23-34.
11. Moran, P., J. Goggins, and M. Hajdukiewicz, *Super-insulate or use renewable technology? Life cycle cost, energy and global warming potential analysis of nearly zero energy buildings (NZEB) in a temperate oceanic climate.* Energy and Buildings, 2017. **139**: p. 590-607.
12. Dimoudi, A. and C. Tompa, *Energy and environmental indicators related to construction of office buildings.* Resources, Conservation and Recycling, 2008. **53**(1): p. 86-95.
13. Takano, A., et al., *The effect of material selection on life cycle energy balance: A case study on a hypothetical building model in Finland.* Building and Environment, 2015. **89**: p. 192-202.
14. *Boverkets Byggregler, Boverkets Författningssamling, The national Board of Housing Building and planning, Karlskrona. Available at http://www.boverket.se (In Swedish).*
15. FEBY 12, *Kravspecifikation för nollenergihus passivhus och minienergihus Bostäder. Available at www.passivhuscentrum.se/sites/default/files/kravspecifikation (In Swedish).* 2012.
16. StruSoft. *VIP-Energy, StruSoft AB.*
17. Gustavsson, L., K. Pingoud, and R. Sathre, *Carbon Dioxide Balance of Wood Substitution: Comparing Concrete- and Wood-Framed Buildings.* Mitigation and Adaptation Strategies for Global Change, 2006. **11**(3): p. 667-691.
18. Björklund, T. and A.-M. Tillman, *LCA of building frame structures: environmental impact over the life cycle of wooden and concrete frames in: Technical Environmental Planning Report 2*. 1997, Chalmers University of Technology: Gothenburg, Sweden.
19. Gustavsson, L. and R. Sathre, *Variability in energy and carbon dioxide balances of wood and concrete building materials.* Building and Environment, 2006. **41**(7): p. 940-951.
20. Tettey, U.Y.A., A. Dodoo, and L. Gustavsson, *Effects of different insulation materials on primary energy and CO2 emission of a multi-storey residential building.* Energy and Buildings, 2014. **82**(0): p. 369-377.

21. Erlandsson, M. and D. Holm, *Livslängdsdata samt återvinningsscenarion för mer transparenta och jämförbara livscykelberäkningar för byggnader*. 2015, IVL Svenska Miljöinstitutet 2015.
22. Piccardo, C., et al. *Comparative Life-Cycle Analysis of Building Materials for the Thermal Upgrade of an Existing Building. SBE19 BRUSSELS - BAMB-CIRCPATH "BUILDINGS AS MATERIAL BANKS – A PATHWAY FOR A CIRCULAR FUTURE", 6-7 February 2019, Brussels, Belgium*. 2019.
23. Gustavsson, L., A. Joelsson, and R. Sathre, *Life cycle primary energy use and carbon emission of an eight-storey wood-framed apartment building*. Energy and Buildings, 2010. **42**(2): p. 230-242.
24. Dodoo, A., L. Gustavsson, and R. Sathre, *Lifecycle carbon implications of conventional and low-energy multi-storey timber building systems*. Energy and Buildings, 2014. **82**: p. 194-210.
25. Lu, H.R., A. El Hanandeh, and B.P. Gilbert, *A comparative life cycle study of alternative materials for Australian multi-storey apartment building frame constructions: Environmental and economic perspective*. Journal of Cleaner Production, 2017. **166**(Supplement C): p. 458-473.

Acknowledgments
We gratefully acknowledge financial support from the Swedish Energy Agency, Växjö municipality, Växjöbostäder, Växjö Energi AB, Södra Skogsägarnas ekonomiska förening and Ronneby municipality (CEFUR). Lindbäcks and Martinsons are also gratefully acknowledged for design and data support for the timber buildings.

OFF-GRID SYSTEMS

E Pollock[1], H Kaur[2], R Shinde[2], S Batineih[2], A Dandapath[2]

[1] Lecturer, Metropolia UAS, Helsinki, Finland
[2] Master students, ConREM, Metropolia UAS, Helsinki, Finland

E-mail: eric.pollock@metropolia.fi

Abstract. The term sustainability is frequently associated with our future cities. However, in the most remote and unprivileged corners of the world off-grid systems have already established a new dimension for sustainability. With no infrastructure nor grid connections, off-grid systems offer a way to be self-sustainable and thrive with the least resources. If dealt with proper planning and caution, off-grid systems can be beneficial on a large scale and can also contribute to local integrated networks.

Off-grid systems exist in various forms such as electricity generation, bioenergy, off-grid housing, etc. As sustainability is of global interest the issue of off-grid is reviewed globally in remote locations of India and Jordan. The latest advancements in technology are considered.

Social, economic, and environmental sustainability are central themes of consideration. This paper displays how off-grid systems are advantageous in terms of the basic needs of life e.g. shelter, energy, and food. Included are a wide array of crucial case studies in different continents. The main case studies include biogas plants, and solar electricity in India and Zaatari refugee camp in Jordan.

Keywords: Sustainability, off-grid, bio-energy, water supply, electricity, waste management, refugee camps, solar.

1. Introduction

The term sustainability can be misunderstood as to how far we are prepared to sacrifice our present needs and lifestyle in order to meet the demands of the future. However, sustainability can be achieved by meeting our needs today and keeping the promises of the future alive simultaneously. One of the most significant testimonies to this fact is off-grid systems. The idea has been implemented and verified. It has also been widely accepted by government organizations, local communities and as well as NGOs (Louie, Dauenhauer, Wilson, Zomers, & Mutale, 2014).

The popularity of off-grid systems has emerged from the imbalance between supply and demand. It is a noble approach towards development being independent of any initiative from higher or external authorities. In places where there is negligence from the government or development has not reached yet with all its fruits, off-grid is critical for self-help development. Off-grid provides scope for sustainability in all three aspects e.g.-social, economic, and environmental. Furthermore, it also helps to maintain a balance between them.

One of the world's biggest concern at present is sustainable energy production. The off-grid system offers possibilities in and not limited to electricity from solar power, hydropower and wind power systems, self-supply of water and sanitation, energy generation from biogas, food production, etc. Apart from these, there is a considerable amount of innovative systems which often go unnoticed due to the small scale or remoteness of the location.

In the last thirty years, remarkable acceleration has been seen in the pace of off-grid improvements in the sunny southern hemisphere. Globalization has helped technology and innovation to spread across borders. The remote areas of India comprise one of the lower economical sections of the country. Therefore, the need to promote off-grid projects is essential to meet the demands of the people. The case studies mentioned here clearly state that the small scale off-grid projects are creating a difference in the lives of the people. If government support is given on a large scale then definite progress can be made on a vast scale.

In remote areas of India off-grid systems are used mostly in electricity generation and cooking. The previous uses of wood caused air pollution and health hazards from the cooking smoke. In many present off-grid systems utilization of charcoal improves the situation environmentally and health-wise. The biogas is used widely in domestic and community-scale in rural areas. The small-scale implementation off-grid systems in the villages in India is a recent but widely accepted phenomenon. It generates the opportunity for cheap production and social employability. Being a religious country India also witnesses waste to energy and solar systems in some major temples of the south.

Another important case is the discussion on refugee camps in Jordan. It deals with the whole life-cycle of the product of 10 years or more. The construction was started in 2011 and finished within 7 months. It has been able to provide all the facilities such as energy, water, etc. to the refugees. With no infrastructure and network in the desert, it has offered a home to 80,000 people with the help of a limited budget from the UN. The project developed through many steps like supplying water through reservoirs, installing solar panels, etc.

The need for energy in our society is ever increasing. We are having more awareness and usage of renewable energy resources. However, 85% of our energy consumption (Harvard, 2012) comes from burning fossil fuel (oil, coal, and gas). Off-grid is one of the newest and most innovative approaches in this area. The term off-the-grid (OTG) is referred to the system which is independent of the grid infrastructure such as electricity, water et al. The concept is becoming popular mainly for two reasons. Sustainability and independent energy generation. Research and application have included single houses to large communities, ranged from energy generation to distribution. Off-grid can be considered as an effective solution to meet the demand of our climate change goals.

The practical infeasibility and financial availability of grid extension into remote areas assign a remarkable contribution to the off-grid energy generation *(Sen & Bhattacharya, 2014).* Roughly, the cost of electricity generation depends upon the efficiency of the off-grid model. A third of increased demand in world energy comes from India *(International Energy Agency, 2017*). In India, HOMER or Hybrid Optimization Model for Electric Renewables is a limited and popular field of research. In many parts of the country, the Integrated Energy Renewal System (IERS) is suggested (Kanase-Patil et al., 2010) to meet the demand for electricity and cooking. India is large as a continent consisting of remote areas where till date energy from the traditional grid has not been made available.

In other developing countries the usage of the off-grid model is popular especially in electricity generation and cooking. Electricity is one-third part of energy consumption. Others are transportation and cooking etc. In southern Asia observations on off-grid has shown a mixed trend (Palit & Chaurey, 2011) in various countries. The bridging of the financial gap between affordability and electricity cost (Mainali & Silveira, 2011) can be addressed by off-grid.

Solar Energy is the most popular in all off-grid systems. It includes various applications such as solar lighting system, heating system, lanterns, cooking, etc. In climates like India where there is an abundance of solar power, it is not difficult to understand the wide implementation of PV based off-grid systems. In the numerous remote locations in India solar off-grid lighting systems (SOLS) i.e. Solar Lanterns (SLs) and Solar home lighting systems (SHS) (Choragudi, 2013) are now possible. These systems provide the opportunity to have lighting systems without the cost of electricity production in remote areas of the Indian subcontinent.

However, the scope of off-grid in developing regions is often limited (Bhattacharya, 2012). Major issues being a lack of resources, infrastructure, and poor technology. From the electricity generation to storage the off-grid model demands efficiency. There is a considerable amount of scope in improvement in off-grid systems.

2. Off-Grid Housing - Electricity generation a major factor Introduction

Around 1.6 billion people in the world will be forced to live without electricity, not because of the non- availability but due to the dispersed population living in the off-grid area. Therefore the concern of producing electricity in these areas become a relevant area of discussion (www.world statistics.com).

The major source of energy production in remote areas is from fossil fuels. The energy produced by this is not efficient enough to cater to the needs of the population and also this produces a lot of carbon dioxide ($CO2$) and Carbon Monoxide (CO) resulting in an increase in the greenhouse effect. This definitely provides a motivation to look for alternatives that do not emit green-houses gases. The challenge is to find a solution for the off-grid areas without disturbing the ecosystem.

Remote Areas of India

There are approximately 649,481 villages in India (List of villages in India, 2018). Although there is a universal electrification policy adopted by the Indian government still, one-third of the populated areas are without electrification. The census states that 93 % of urban households and 55% of the rural household currently receive electricity (India Village Directory). Therefore India needs an alternative solution to cater to this problem. Looking towards the possible options to provide electrification to such a vast country with almost 1.4 billion people, off-grid housing definitely stands out to be the best possible solution.

The ministry of new and Renewable energy (MNRE) is supporting off-grid technology and is trying to electrify India completely. The new scheme launched by MNRE is "Off-Grid and Decentralized Concentrated Solar Thermal (CST) Technologies for Community Cooking, Process Heat and Space Heating & Cooling Applications in Industrial, Institutional and Commercial Establishment" (Gupta, 2018). The main objective of this scheme is to promote off-grid technology and to meet/supplement heating /cooling systems and to generate electricity through solar and thermal energy (Off-Grid Power, 2018).

Off-Grid Projects

Bioenergy Gasifier

"Biomass is an industry term for getting energy by burning wood, and other organic matter. Burning biomass releases carbon emissions but has been classed as renewable energy in the European Union and United Nations legal frameworks, because plant stocks can be replaced with new growth" (Page, 2016).

Biomass has a high potential energy source in India, it enjoys great support from the state and the Centre. Currently, India is a leading country in Bioenergy. *Gosaba Island* gasifier is an epitome of how biomass energy can be utilized to change the lives of people in remote areas in India.

Case: Gosaba Island Gasifier

Gosaba Island is an island in West Bengal's Sundarban region and was deprived of electricity as it was not economically feasible to extend power from the grid to these widespread islands. West Bengal Renewable Energy Development Agency (WBREDA) took the initiative to provide electricity to the island. The Vadodara based Ankur Scientific energy technology in collaboration with WBREDA opened a biomass plant in the village. The island that was once lit with kerosene lamps gave way to the electric bulbs and soon the island became a small town.

Case: Turning Destructive Pine to Productive gas; Kumaon Valley

AVANI- a voluntary organization helped the people of Kumaon valley in Uttarakhand to produce electricity by the use of pine needles. The organization aims to produce about 14.65MW of electricity from the biomass (Avani, 2014). The state has the opportunity to electrify the area through the gasifier which is more accessible and cheap.

The gasifier volatize the pine needles into a producer gas which is a combination of the combustion gas. The gas is passed through the filters consisting of the sawdust and fine cloth to remove the impurities from it, the resultant gas is used to run the diesel engine which generates electricity.

Figure 1: Pine needle gasifier, Kumaon Valley (Avani, 2014)

Benefits from the Gasifier

The pine based gasifier guarantees to benefit the common household kitchen by allowing them to use charcoal effectively as it is one of the byproducts (10% of the residue) of the gasifier, this allows for a smoke-free household eliminating pollution and lung diseases. Charcoal produced by the 120 KW gasifier is sufficient to meet the household demands of at least 100 homes in the area. The villagers can pay for the charcoal or can collect the pine needles and get it in exchange. This project aims at increasing the employment growth in the area and as well as providing them with electricity and also a smoke-free alternative to cooking food. The efficient charcoal is also considered as a substitute for the LPG gas which is very expensive and limited in India. Therefore the project actually aims towards the off-grid solution leading towards sustainability. According to the estimates, the pine gasifier has the potential to generate electricity in the whole Himalayan region thus meeting the need of 1.4 families (Avani 2014).

Case: Hybrid Vermicompost Biodigesters

Karnataka is a developed state of India. Bengaluru, the capital of Karnataka is known as the silicon valley of India. However, the rural areas of this state are still deprived of modern technology and survive on open-fire cooking, which is a threat to the environment. The women generally collect wood from the forests to support their traditional cooking.

SKG Sangha (SKGS), a non-profit organization of Kolar district of Karnataka has pledged to provide safe and clean cooking gas to the people of this district and also generate a source of income by selling the fertilizers made from the biogas as a byproduct. The biogas is prepared by cow dung, it consists of an underground brick-built digester, an inlet at the ground level to feed the digester with new feedstock and two separate outlets to collect biogas and to remove the residue" (Sangha, S 2013). The galvanized steel and HDPE pipes are used to transport the gas from the plant to the kitchen stoves.

The biogas has the capacity of producing up to 4 ^{3} gas/day from an input of 50-100 kg cow dung. The liquid residue produced from the plants accounts for 36 to 72 tons and can be used directly in the fields but to transform the liquid fertilizer to the organic matter SKGS use the Vermiculture technique. This technique involves the storing of the liquid residue into a compost and then athe ddition of fibrous materials to decompose it for at least 3 weeks. After the decomposition is completed, the earthworms are added to the mixture. The top layer of the worm casts are used as vermicompost (Sangha, S 2013). Benefits of the Project:

- Smoke-free household generated decreasing the respiratory problems for women.
- Saves a lot of time for women as they search for wood is a tiresome activity.
- Generates employment for the people of rural areas.
- The vermicompost replaces the chemical fertilizers
- Increases the soil fertility and the water retention of the soil.

Case: Jatropha plant, an emerging option for rural electrification

'Ranidhera' village in Chhattisgarh is one of the electricity-deprived villages. The electricity was generated after the 'Winrock International India' started an innovative project of producing electricity by the extraction of biofuel from the seeds of the Jatropha plant. The project was collaboratively supported by the British High Commission, the Swiss Agency for Development and Cooperation and the Ministry of New and Renewable Energy. The aim of the project was to electrify around 6000 remote areas of the country.

The power plant initially (April 2007) had the capacity of producing 11 KW of electricity thus providing 3 hours of electricity for households and around 3.5 hours for the street lighting. The off-grid projects helped in producing the cheap electricity for the remote areas (Gmünder et al, 2010).

The most abundant form of energy available in India. "India is located in the equatorial sunbelt of the earth, thereby receiving abundant radiant energy from the sun" (Garud and Purohit, 2007). According to the recent report, India has a capacity of 20 GW of electricity generation (Priya S 2018). India aims to achieve 100 GW target by 2022 (Reuters, 2015). India's largest solar power plant known as 'Shakti Sthala' is installed in the State of Karnataka, India. It generated 2000 MW of electricity and is laid upon on 13000 acres covering almost 5 villages (NDTV, 2018).

Figure 2: Solar panel installed in Karnataka (Verma, P)

Case: Tirupathi, The Green Temple

'Tirumala Tirupathi Devasthanam' is the richest temple in the world. It has the largest number of devotees visiting on a single day. The temple is located in the southern part of India in Andhra Pradesh and has been trying to adopt renewable sources of energy for meeting its everyday requirements. The temple has a wastewater treatment plant that recycles and purifies the water. The canteen of the temple provides free mineral water to discourage plastic bottles and they also use solar energy for cooking. The solar cooker was installed in 2002 and has the capacity to prepare 50,000 kg of rice along with curry for 15000 persons in one cycle. The solar technology is installed by the Gujrat based company 'Gadhia Solar' and the HTT Gmbh. of Germany (Gadhia Solar).

The installation cost of the system is around 11 million rupees. The expenses were covered by the temple as well as the Union Ministry of Non-Conventional Energy sources. The solar cooker needs no modification and has a lifespan of 25 years.

IOP Conf. Series: Earth and Environmental Science **297** (2019) 012021 doi:10.1088/1755-1315/297/1/012021

Figure 3: Solar Steam cooking at Tirumala Tirupathi Temple (Terienvis, 13/10/2015)

Future Plans are for the temple to utilize the steam produced in the solar plant to meet the requirements of the Kalyanakatta, where the pilgrims do the rituals of head tonsuring. The Gadhia is also planning to install a 500 KW solar thermal system in the temple (Gadhia Solar).

3. Zaatari Refugee Camp, Jordan

Introduction

Politics play a major role in today's communities, and lately, the world has been facing critical situations and wars, unfortunately, all of that creates instability in the existing communities where conflicts occur, resulting in large populations forced to immigrate to safer areas and build their new communities. One of the latest conflicts is in Syria in 2011, which so far has forced millions to evacuate their homes and become homeless or refugees. Many refugee's camps have been established in remote areas to host those people and try to secure basic life needs for them, but with lack of resources, infrastructure, and communications it was a huge challenge; water supply, food production, energy, waste treatment, medication, education and others in the built-up camps (Fisher, 2016).

Zaatari Camp in Jordan is one of the largest camps for Syrian refugees around the world which was established in 2012 covering an area of 5.3 sq.km in northern Jordan, near to Al Mafraq which is a desert area with lack of water resources, food supply, and energy. The challenge was to secure these needs for more than 80,000 refugees. The population is growing, at least for the next 15 years, with some refugees in the camp for a lifetime, no temporary solutions would be practical or even acceptable. Off-grid housing is a convincing approach in such situations, a need to develop a self-standing sustainable community to host all these people for years. Tents were pitched as a place for them to live in temporarily which will be replaced by better housing options later on, such as shelter programs supplying donated caravans, cabins & built-up units (Fisher, 2016).

Syrian refugees numbering 461,701 have passed through the camp, and on 26 March 2015, the camp population was estimated at 83,000 refugees by United Nations High Commissioner for

Refugees (UNHCR), with an average of 80 births per week, while statistics show that 19.9 % of the population are children below 5 years' old who need intensive care (Fisher, 2016).

Water Supply

Water is essential for any living creature and it is one of the main needs for humans, without water or even lack of it no life can be assured, so organizations along with the Jordanian government had to implement a plan to secure water supply for the camp.

Figure 4: Housing in Zaatari Refugee Camp

(UN Photo/Sahem Rababah, 27 March 2017, Zaatari Camp, Jordan, Photo # 718441)

As an initial temporary plan water was supplied by trucks to the camp, although It was expensive and time-consuming and can be delayed or interrupted but it was the only solution at that time till the drilling of the two new internal boreholes near the camp are completed provides 3.2 million liters of drinking water per day (UNHCR, reliefweb, 2017). Also work is ongoing to improve the water supply in villages surrounding Zaatari, which will benefit both Syrian refugees and host communities. In addition to some expansions in MAFRAQ pipeline which feeds the nearest city to the camp will benefit 25,000 Syrian refugees and local residents (Bhai 2018).

The results and goals of these projects are that "People living here will get double the amount of water, better pressure and more reliable. The work we're doing here will not only help the Syrian refugees in the camp but also serve the people in the surrounding communities," says Mr. Hameed. Water distribution presently takes place via a fleet of approximate 82 trucks delivering water to the communal public and private water tanks (Bhai, 2018).

Energy

Electricity as a power supply is one more requirement to establish a new residential area, provides an essential lifeline for camp residents, from lighting shelters to preserving food and maintaining hygiene, "Light is absolutely essential," said Anne-Marie Grey, the executive director and CEO of USA for UNHCR, in an interview. "When you go into a camp, you realize how it's a safety issue as much as a right to light or a right to energy issue" (UNHCR 2017). As stated before, in the case of Zaatari camp the challenge was to provide enough power to supply all population in that remoted area, using generators and some connections from the local network was defiantly a temporary solution; as it was costly and insufficient; especially with a monthly bill of 800,000 $ (Pyper 2015), so UN started considering the idea of constructing a renewable power supply project to serve the area's needs.

The UNHCR's three-year Energy Strategy 2015-2018 plan was to build a solar plant which will be the largest ever built in a refugee camp, providing clean and much-needed power to the Syrian refugees in the camp. According to UNHCR "The plant will reduce annual carbon dioxide emissions from the camp by 13,000 metric tons per year, equivalent to 30,000 barrels of oil. It will also deliver annual savings of around US$5.5 million, which UNHCR – the UN Refugee Agency – will be able to reinvest in vital humanitarian assistance" (UNHCR 2017).

Figure 5: World's biggest solar power plant for a refugee camp opens in Zaatari (Luck, 2017)

Briefly, The 40,000 photovoltaic panel solar plant was constructed on the outskirts of the camp with an approximate area of 0.8 square km, arranged in rows hundreds of meters. According to UNHCR "the 12.9-megawatt peak solar photovoltaic plant was funded by the Government of Germany with a cost of 15 million euros (US$ 17.5 million)" and will be operational by the end of 2017 (Luck 2017).

Community and Social Services

The Zaatari camp hospital services include primary health care, maternity care, dental care, mental health, and nutritional care. Community centers provide psychosocial support & recreational activities (UNHCR, reliefweb, Feb 2017).

The Zaatari Camp schools have over 20,000 school-aged children enrolled. Textbooks and school supplies are distributed to all children in the camp. (UNHCR 2014).

Waste Management

Waste management considers two main categories; wastewater and solid waste, in order to maintain a healthy hygienic life in the camp it is necessary to develop a plan to control, dispose and even recycle general waste. According to UNHR approximately 2,100m3 of sludge is collected by desludging trucks every day; regularly 28–50 tons of unsegregated solid waste is collected every day. A wastewater treatment plant was established to treat approximately 80% of the wastewater generated in the camp.

Figure 6: Waste Management in Zaatari Refugee Camp (OXFAM, 2017)

Currently, all houses are connected to septic tanks, which means that desludging trucks make the rounds of the camp every day to empty septic tanks. In order to increase the efficiency of the camp's sanitation system, the next step, as from next year, will be to connect all the septic tanks to the main pipeline that will take all the waste to the treatment plant. (Aljaradin 2016). For solid waste, some efforts have been done so far, it is collected by trucks and transferred to external solid waste facilities, in order to consider recycling an approach to start separating the solid waste; plastic, paper, cardboard, glass & metal has been taking place (Saidan M, 2016).

Turn Waste into Energy

An approach to turn food and animal wastes into clean and safe fuel and fertilizer, although eliminating pathogens from the biowaste is a challenge due to it's potential to cause illness and attract disease-carrying animals, it has been possible by involving educated biogas technicians and closed-loop farmers. This will provide energy and even a chance to residents for learning, sharing and implementing clean and renewable energy solutions within Zaatari Refugee Camp. Studies proved that this approach will reduce fossil fuel usage and indoor air pollution by properly managing food waste (Initiative, 2016).

4. Conclusion

Unfortunately, the off-grid systems are sluggish and unable to meet the demand of the rapidly growing energy need (Louie, Dauenhauer, Wilson, Zomers, & Mutale, 2014). The scale of these is also tiny compared to Cities. We need more ideas, innovation, attention, knowledge, and attention to make these systems beneficial for our future 'Sustainable cities'.

The small-scale implementation of off-grid systems in rural areas of India brings much hope but is not sufficient to cater to the massive demand for energy of growing India. The government needs to come up with innovative measures to accommodate the benefits of these systems into our future cities. The economies of scale of the systems have to grow critically to be comparable to cities. Technological advancement should be adopted to integrate the on-going systems to the main grid for achieving the optimum sustainable outcomes.

The refugee camp is a unique example of off-grid systems. However, this model is crucial to study as a remedy for crisis (Natural or political) situation of the cities. It shows a quick response in developing a sustainable community which can be an effective lesson dealing with natural disasters in cities. However, one has to think about larger case solutions more apt for a sustainable city. The situation is highly undesired as a permanent or even a temporary state in any city.

References

1. Adani Group 2016 Financial express. Retrieved February 18, 2018, http://www.financialexpress.com/photos/business- gallery/536047/world-largest-solar-power-plant-adani-group-tamil-nadu-india-features-and-highlights/

2. Acted 2016. Retrieved from acted.org: http://www.acted.org/en/photostory-providing-water-refugees-za-atari-camp AFP 2012 Jordan opens the first refugee camp for Syrians. 7/29/2012, http://english.ahram.org.eg/News/48962.aspx Aljaradin, M (2016). Waste management in Zaatari refugee camp- Mafraq-Jordan.

3. Avani 2014 Sanskrit. Retrieved March 12, 2018, from www.esamskriti.com: https://www.esamskriti.com/photograph/2_2352.jpg

4. Avani 2014, http://avani-kumaon.org/programmes/clean-energy/

5. Bhai, A, (2018) UNHCR WASH Officer Water at Za'atari Camp, https://www.unhcr.org/7steps/en/water/

6. Choragudi, S. 2013 Off-grid solar lighting systems: India's sustainable and inclusive development goals. Renewable and sustainable energy reviews, 890-899.

7. Das, K N 2015 India's Modi raises solar investment target to $100bln by 2022

8. Dayalu, A, Harvard 2012, "Why we need sustainable Energy" and the United States Energy Information Administration. (2012). "Annual Energy Review 2011" http://www.eia.gov/totalenergy/data/annual/archive/038411.pdf

9. Gadhia Solar 2012 https://mnre.gov.in/file-manager/UserFiles/Ghadia_solar_cookers.pdf, www.gadhia-solar.com, ended 2012

10. Garud S and Purohit I, 2007 Making Solar Thermal Power Generation in India a Reality Research associates within the Energy and Resources Institute (TERI), New Delhi, India

11. Gmünder, S,Zah, R, Bhattacharjee,S, Classen, M, Mukherjee, P, Widmer, R, 2010 Life cycle assessment of village electrification based on straight jatropha oil in Chhattisgarh, India, Technology and Science lab, Swiss Federal Laboratories for materials and testing research EMPA, Biomass and Energy 34, 347-355

12. Gupta, A. 2018. EQ international. Retrieved February 18, 2018 http://www.eqmagpro.com/off-grid-and-decentralized-concentrated-solar-thermal- cst-technologies-for-community-cooking-process-heat-and-space-heating-cooling-applications-in-industrial-institutional-and-commercial-esta/

13. Initiative 2016 www.clintonfoundation.org/clinton-global-initiative/commitments/turning-waste-energy-zaatari-refugee-camp-jordan

14. Fisher, M, 2016 NYTimes, Straightforward Answers to Basic Questions About Syria's War.

https://www.nytimes.com/2016/09/19/world/middleeast/syria-civil-war-bashar-al-assad-refugees-islamic-state.html

15. Off-Grid Power. Retrieved March 26, 2018, from mnre.gov.in: http://www.mnre.gov.in/grid-power
16. Kanase-Patil, A & Saini, R & Sharma, M 2019. Integrated renewable energy systems for off-grid rural electrification of remote area. Renewable Energy. 1342-1349
17. List of villages in India 2018 Retrieved March 26, 2018, https://en.wikipedia.org/wiki/List_of_villages_in_India
18. Louie, H., Dauenhauer, P., Wilson, M., Zomers, A., & Mutale, J. 2014 Eternal Light: Ingredients for sustainable off-grid energy development. IEEE Power and Energy Magazine, pg 70-78.
19. Luck, T. (2017, November 13). The National. https://www.thenational.ae/world/mena/world-s-biggest-solar-power-plant-for-a-refugee-camp-opens-in-Zaatari-1.675484
20. Motasem N. Saidan A, 2016. Solid waste composition analysis and recycling evaluation: Zaatari. Science Direct.
21. Mainali, B & Silveira, S 2011. "Financing off-grid rural electrification: Country case Nepal," Energy, Elsevier, vol. 36(4) 2194-2201.
22. NDTV. 2018. Retrieved March 26, 2018, from NDTV: https://www.ndtv.com/india-news/shakti-sthala-launched-in-karnataka-is- worlds-largest-solar-park-1818803
23. OXFAM 2017 Recycling project in Zaatari Camp, Jordan [Motion Picture].
24. Page, ML 2016, September 26. The Renewable energy scam. Retrieved March 26, 2018, from New Scientist: https://www.newscientist.com/article/mg23130922-600-revealed-the-renewable-energy-scam-making-global-warming-worse/
25. Priya S 2018 Markets & Policy, Solar, Top solar and other renewable Energy News Headlines for India in Nov 2017
26. Reuters 2015 / India's Modi raises solar investment target to $100 bln by 2022
27. Shaw A 2012 Poverty facts and stats. Retrieved March 18, 2018, http://www.globalissues.org/article/26/poverty-facts-and-stats Saidan M 2016
28. http://www.academia.edu/35713876/Solid_waste_composition_analysis_and_recycling_evaluation_Zaatari_Syrian_Refugees_Camp_Jordan
29. Sangha, S. (2013). Biogas Plant Construction manual. foundationskgsangha. ttp://foundationskgsangha.org/biogastechnology4.html
30. Sen, R & Bhattacharyya, S 2014. Renewable Energy-Based Mini-Grid for Rural Electrification: Case Study of an Indian Village. 10.1007/978-3-319-04816-1_9
31. Solar 2010. https://solarenergydemystified.wordpress.com/2010/03/08/solar-steam-cooking-system-at-tirumala/

32. Terienvis 2015 Retrieved March 18, 2018, http://terienvis.nic.in/index4.aspx?ssslid=4243&subsubsublinkid=365&langid=1&mid=1

33. UNICEF2012 At Za'atari refugee camp in Jordan, as a million liters of clean water, arrive each day, focus turns to conservation and future supply of water 2012, https://www.unicef.org/wash/jordan_66227.html

34. UNHCR 2014 https://www.unhcr.org/news/updates/2014/11/5465c15d9/jordan-inter-agency-update-september-october-2014.html?query=Zaatari

35. UNHCR 2017 Jordan's Za'atari camp goes green with new solar plant Marwa Hashem

36. Verma, P, https://www.flickr.com/photos/rayspowerinfra/27580564457/

Greenhouse gas emissions from built environment development in Iceland

Jukka Heinonen[1,2], **Áróra Árnadóttir**[1], **Nargessadat Emami**[1], **Björn Marteinsson**[1]

[1] Faculty of Civil and Environmental Engineering, University of Iceland, 101 Reykjavik, Iceland
[2] Department of Built Environment, Aalto University, 00076 Aalto, Finland

heinonen@hi.is

Abstract. Without rapid and radical greenhouse gas (GHG) mitigation, irreversible damage threatening life on the globe might occur already during the next decades. One of the key sectors in finding solutions to climate change is the built environment, which currently directly or indirectly causes the majority of anthropogenic GHG emissions. The transition towards more sustainable settlements requires massive use of materials and energy, but it is not well known at all how much GHG emissions are "invested" into the development of the future low-carbon built environment. In this study we use input-output analysis to calculate an estimate of the GHGs embodied in the built environment development in Iceland. The input data consists of annual economic turnover data of different construction sectors for the years 2013-2017. The GHG estimates are derived using the EIO-LCA input-output model. We find that the built environment development emissions of Iceland are significant even though the actual emissions largely take place outside the country, being thus outsourced emissions. Surprisingly the development of the capital region did not stand out as the engine of these emissions, but the spread appeared to be relatively equal between the capital region and the rest of the country.

1. Introduction

We are facing an ever-growing need to rapidly reduce the environmental pressure we are currently causing [1]. Whereas several other important environmental problems exist and new are already in the horizon, climate change mitigation is among the most urgent. Without rapid and radical mitigation, irreversible damage threatening life on the globe might occur already during the next decades [2].

One of the key sectors in finding solutions to climate change is the built environment, which currently directly or indirectly causes the majority of anthropogenic greenhouse gas (GHG) emissions leading to climate change. However, the transition towards more sustainable settlements requires massive use of materials and energy, be it new energy efficient buildings or supporting infrastructures. Several studies have found built environment development, buildings and infrastructure, the most important source of GHG emissions in rapidly urbanizing regions (e.g. [3-6]). However, these are broad scale studies which include also other capital investments than construction and offer little details about the distribution of these emissions.

In bottom-up studies of the built environment structures, buildings have received the most interest. However, the focus has been on the use phase emissions and traditionally the emissions embodied in construction materials have not been considered of high importance [7], but they are actually becoming the key issue due to the rapidness of the need to cut the emissions and to their increasing relative importance along with improving building energy efficiency [8-12]. Furthermore, few embodied building GHG studies have included even the building specific infrastructures, such as parking lots and pedestrian pathways. It is also typical to divide the emissions equally over the life cycle of the assessment object, be it a building, street, bridge or other structure, for annual loads. This omits the important issue that the early life cycle emissions are the most crucial. Thus it is not well known at all how much GHG emissions are "invested" into the development of the built environment if looking at the bottom-up perspective.

Estimating the environmental loads embodied in construction materials has also appeared challenging [13] and despite the significant research on the environmental impacts of construction materials around the world, the reliability of the published impact estimates is still highly questionable [7,14]. Overall we don't have good understanding about the embodied impacts caused by the development of the built environment, while at the same time these impacts can hinder us from achieving the set mitigation targets [8] or postpone the GHG gains far into the future [15]. Thus, more reliable information from the construction sector is needed for advised decision-making.

In this study we use input-output analysis to calculate an estimate of the greenhouse gas (GHG) emissions embodied in the built environment development in Iceland. The input data includes annual development costs and comes from Statistics Iceland. We use annual economic turnover data of different construction sectors for the years 2013-2017 from Statistics Iceland to estimate the GHG loads caused by the built environment development in Iceland. The GHG estimates are derived using the EIO-LCA input-output model, which offers the highest resolution construction data of the available IO models, and, as inherent for the IO method, allows tracing of the full life cycle emissions through the full production and supply chains [16]. What is shown is that the built environment development emissions of Iceland are significant even though the actual emissions largely take place outside the country, being thus outsourced emissions. Surprisingly the development of the capital region did not stand out as the engine of these emissions, but the spread appeared to be relatively equal between the capital region and the rest of the country.

2. Materials and methods

2.1. *Study materials*

Two datasets were utilized to estimate emissions from built environment development in Iceland, one for transport related infrastructure and another for buildings. Data for total annual expenditure from transport infrastructure was collected from annual reports published by The Icelandic Road and Coastal Administration (Vegagerðin). The reports included expenditure data from several different categories, although the total expenditure was reported for the country as a whole and data from the Capital Region was not available separately. Table 1 shows a summary of annual expenditure (in m.kr, nominal) from the main categories reported for roads, bridges and tunnels for 2013-2017.

Table 1. Expenditure (in m.kr, nominal) by category and year.

Main category	Sub-categ.	2013	2014	2015	2016	2017
Operation & management		676	750	709	715	727
Service		4,500	4,417	4,607	5,677	4,440
Maintenance	Roads	3,900	3,890	4,550	4,600	6,500
	Bridges	355	400	500	450	600

	Tunnels	140	130	130	130	140
Initial costs	Roads	5,636	4,666	5,344	6,099	6,589
	Bridges	283	38	140	140	250
	Tunnels	1,410	3,390	3,010	3,180	3,810
Sea related infrastructure		153	356	106	1002	3029

Data for buildings was provided by Registers Iceland (Þjóðskrá). The dataset contained both annual data on area (m^2) of housing built and total expenditure (in million kr., nominal). Buildings in the Reykjavík Capital Region were reported separately from the rest of the country and the dataset included three categories with several sub categories. Residential buildings we split into detached houses, semi-detached houses and apartment buildings. Commercial housing was split into industrial housing, specialized, shops and offices and warehousing. The "Other" category included the subcategories garages and sheds, summer houses and outbuildings. Table 2 presents the annual expenditure (in million kr.) by main category and year built, Reykjavík Capital Region is separated from the rest of the country. Table 3 shows reported building infrastructure area (m^2) by main category, year built and location.

Table 2. Total expenditure (in m.kr., nominal) of new buildings by location, type and year built.

Location	Type	2012	2013	2014	2015	2016	2017
	Commercial	3,505	6,101	2,577	8,091	13,171	7,118
Capital	Residential	7,028	14,590	22,265	19,456	28,017	30,275
	Other	665	1,302	1,498	2,509	2,474	2,683
	Commercial	3,903	9,297	7,919	8,483	15,256	15,332
Rural	Residential	3,542	2,749	4,271	4,654	7,936	11,687
	Other	3,335	3,949	3,970	3,349	7,239	6,840

Table 3. Total area (m^2) of new buildings by location, type and year built.

Location	Type	2012	2013	2014	2015	2016	2017
	Commercial	23,280	35,398	12,900	36,239	62,506	33,230
Capital	Residential	37,053	74,377	114,942	102,611	137,699	127,094
	Other	4,772	3,898	5,485	3,624	9,963	5,909
	Commercial	26,992	41,190	40,778	39,012	91,512	71,630
Rural	Residential	19,630	14,940	22,108	22,626	39,225	50,780
	Other	20,976	28,166	22,375	20,773	37,20	40,043

2.2. Method

The emissions were estimated using a life cycle analysis (LCA) approach, which enables capturing the emissions from the production and delivery chains [17]. The assessment includes only the GHG emissions, and the result can thus be called carbon footprint [18].

The LCA was conducted using an input-output (IO) approach, which utilizes monetary transaction values and IO tables to estimate the emissions [19]. IO LCA greatly simplifies the assessment since the information the practitioner would have to gather from all the processes of the supply chain is now included in the input-output tables. IO LCA is also a comprehensive method including the full production and supply chains without cutoffs. Säynäjoki et al. [20] have shown that the cutoffs can be tens of percentages in the building sector. At the same time, IO approach is average based and suffers from several other problems as well, discussed further in the discussion section. In general, the method is efficient for first screening assessment, such as conducted in this study, to understand the rough magnitudes, but cannot be used for detailed analyses.

2.3. Assessment model

The U.S. economy-based IO model of the Carnegie Mellon University Green Design Institute, called the 2007 U.S. Benchmark Producer Price Economic Input-Output Life-Cycle Assessment model (EIO LCA) [21], adjusted with Icelandic data for some key local activities. The EIO LCA is among the most disaggregated models available with 388 economic sectors, which provides better sectoral fit between the model and the data than the majority of other existing models. Being a purchaser price model, the emission outputs are adjusted to the final end-user market prices. The classification follows the North American Industry Classification System (NAICS), but the fit is relatively good with the Classification of individual consumption by purpose (COICOP) system as well. The adjustments to the model are explained below and the uncertainties discussed in the discussion section.

2.4. The assessment

Table 4 shows the EIO LCA sectors utilized in the assessment and the matching with the input data. The second column uses the terminology of the input data. To enhance the match between the US economy based model and Icelandic data, the GHG intensities of the utilized sectors were adjusted with inflation between the model year 2007 and the data years 2013-2017 utilizing US sectoral inflation rates for different construction sectors [22]. The adjusted intensities were then converted to Icelandic kronor according to the annual average currency exchange rates for 2013-2017 [23].

Table 4. The EIO LCA sectors utilized in the assessment and the matching with the input data building types.

EIO LCA sector	Input data
Highways, Streets, And Bridges	Infrastructure
Manufacturing Buildings	Industrial buildings
Commercial Structures, Including Farm Structures	Shops and offices, specialized buildings
Single-Family Homes	Detached houses, summer houses
Multifamily Homes	Semi-detached houses, apartment buildings
Other Nonresidential Structures	Warehouses

3. Results

Over the five-year time span of the study the GHGs from built environment development in Iceland sum up to 1.2 million tons of carbon dioxide equivalents (tCO2e), meaning 240,000 t per annum and close to 0.8 t per capita (Figure 1). Significant growth was also detected, likely relating to Iceland's recovery after a major economic recession in the beginning of the decade. In 2013 the annual load was 180,000 t, whereas in 2017 it had surpassed 300,000 t and 1.0 t per capita. Buildings occupy a share of approximately 70% with some annual fluctuation, and infrastructure around 30%. The share of buildings includes the immediate infrastructure to the buildings, such as parking lots and yards and playgrounds, which could also be allocated to infrastructure, but could not be separated with the data in use.

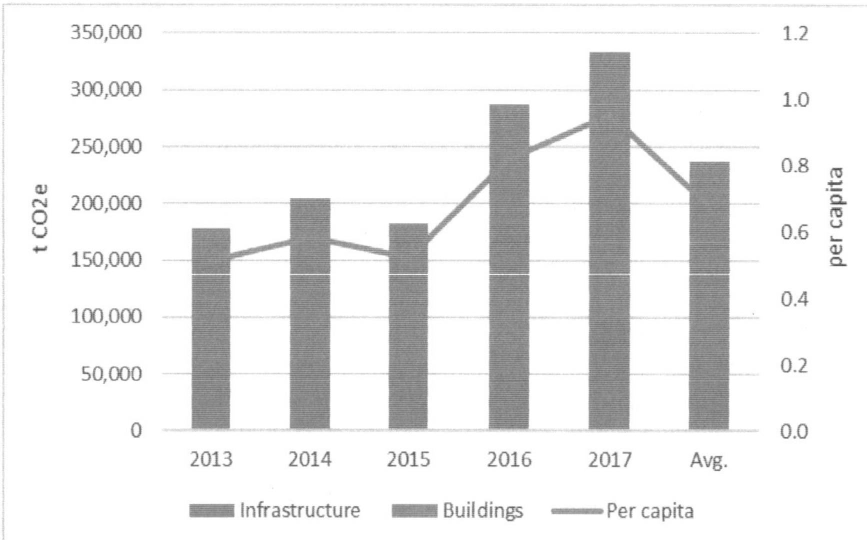

Figure 1. The GHGs from built environment development in Iceland from buildings and infrastructure in 2013-2017.

When looking at the emissions from building construction more closely, it appears rather surprisingly that the capital region does not drive the emissions as strongly as was expected (Figure 2). Due to the

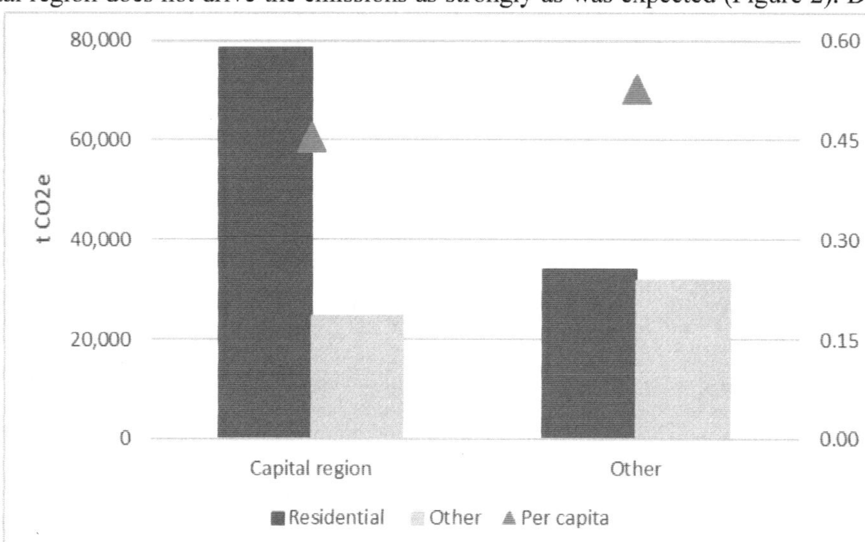

Figure 2. Average annual GHGs from building construction in the capital region and other parts of the country.

strong concentration of the population of Iceland to the capital region (about two thirds of the population living in the capital region), in absolute terms the capital region drives the emissions, but on per capita terms the emissions are even somewhat higher outside the capital region (the orange triangles in Figure 2). In the capital region it is the emissions from residential building construction which dominate, driven by the ongoing migration from other parts of the country, and abroad, to the capital region. 15 to 20% of the residential building construction in the other parts of the country are also due to summer house construction, which is likely to a large extent driven by the residents of the capital region.

4. Discussion

The study was set to assess a rough estimate for the GHG emissions from built environment development in Iceland. Typically building and infrastructure system assessments are done over the lifetime of the assessment object and to one object at a time, which gives little information about the overall annual GHG load from all building and infrastructure construction activities. This study thus provides one case example, which can in the future be used as a benchmark and complemented with other studies.

What was found is that the GHGs from built environment development should be taken into account when designing GHG mitigation strategies in the context of the built environment, such as building energy efficiency regulations and infrastructure development projects to facilitate low-carbon transport. Otherwise it may happen that the "carbon investment" in the development phase is never paid back or the payback is longer than would be acceptable. E.g. Säynäjoki et al. [8] and Chester et al. [15] have showed case examples of long payback times related to building construction and public transport infrastructure development. In addition, Kyrö et al. [24] show how building energy efficiency improvement might not be enough to payback the construction phase emissions on a city-level.

As noticed above, this study has several limitations and should only be taken as indication of what the GHGs from built environment development might be. First, the data for infrastructure is restricted to the part in the responsibility of the main road and other civil infrastructure authority Vegagerdin, thus lacking the streets and roads under the responsibility of municipalities. Second, there is no applicable IO model for Iceland, and thus the US economy based EIO LCA model was selected. To the knowledge of the authors, Iceland is only included in one multi-region IO model, Eora [25], and this model only includes 26 sectors which is too few for a one-sector study such as this [26]. The selected US model includes 388 sectors and several for different construction sector sub-sectors. The match was thus found as the best possible with the inflation and currency exchange rate corrections presented in the method section. However, Iceland is in many ways a special case. With very low population density, difficult conditions and relatively long distances it might be that particularly the infrastructure component of our assessment appears different from that in many other locations.

The IO LCA method itself has also several limitations. Because the emissions factor for each IO sector is based on the weighted average of the emissions of several actual economic sectors, with different emissions per unit of output, there is an aggregation error in the total estimated emissions [27,28]. In addition, there are concerns regarding the validity of the homogeneity and linearity assumptions. While emissions from the production of different goods in a factory can vary considerably, based on the homogeneity assumption, an IO LCA considers that the IO table's averages signify all the products manufactured under a certain sector. Moreover, according to the linearity assumption, there is a linear correlation between the market price and the environmental impacts [27].

Acknowledgments
This study was supported by Landsvirkjun (The National Power Company in Iceland).

References

[1] Steffen W, Richardson K, Rockström J, Cornell S E, Fetzer I, Bennett E M, Biggs R, Carpenter S R, de Vries W, de Wit C, Folke C, Gerten D, Heinke J, Mace G M, Persson L M, Ramanathan V, Reyers B and Sörlin S *2015* Planetary Boundaries Guiding human development on a changing planet *Science* **347** 6223 736-746

[2] IPCC *2018* Summary for Policymakers In Global warming of 15°C An IPCC Special Report on the impacts of global warming of 15°C above pre-industrial levels and related global greenhouse gas emission pathways in the context of strengthening the global response to the threat of climate change sustainable development and efforts to eradicate poverty [V Masson-Delmotte P Zhai H O Pörtner D Roberts J Skea PR Shukla A Pirani W Moufouma-Okia C Péan R Pidcock S Connors J B R Matthews Y Chen X Zhou M I Gomis E Lonnoy T Maycock M Tignor T Waterfield eds] World Meteorological Organization Geneva Switzerland 32 pp

[3] Södersten C-J, Wood R and Hertwich E *2018* Endogenizing capital in MRIO models the implications for consumption-based accounting *Environmental Science & Technology* **52** 22 13250-13259

[4] Mi Z, Zhang Y, Guan D, Shan Y, Liu Z, Cong R, Yuan X and Wei Y *2016* Consumption-based emission accounting for Chinese cities *Applied Energy* **184** 1073-1081

[5] Zhang Y, Wang H, Liang S, Xu M, Liu W, Li S, Zhang R, Nielsen CP and Bi J *2014* Temporal and spatial variations in consumption-based carbon dioxide emissions in China *Renewable and Sustainable Energy Reviews* **40** 60-68

[6] Feng K, Hubacek K, Sun L and Liu Z *2014* Consumption-based CO2 accounting of China's megacities The case of Beijing Tianjin Shanghai and Chongqing *Ecological Indicators* **47** 26-31

[7] Säynäjoki A, Heinonen J, Junnila S and Horvath A *2017* Can life-cycle assessment produce reliable policy guidelines in the building sector? *Environmental Research Letters* **12** 13001

[8] Säynäjoki A, Heinonen J and Junnila S *2012* A scenario analysis of the life cycle greenhouse gas emissions of a new residential area *Environmental Research Letters* **7** 34037

[9] Blengini AB and Di Carlo T *2010* Energy-saving policies and low-energy residential buildings an LCA case study to support decision makers in Piedmont Italy *International Journal of Life Cycle Assessment* **15** 652-665

[10] Chastas P, Theodosiou T and Bikas D *2016* Embodied energy in residential buildings - towards the nearly zero energy building A literature review *Building and Environment* **105** 267-282

[11] Crawford R H, Bartak E L, Stephan A and Jensen C A *2016* Evaluating the life cycle energy benefits of energy efficiency regulations for buildings *Renewable and Sustainable Energy Reviews* **63** 435-451

[12] Stephan A, Crawford R and de Myttenaere K *2013* A comprehensive assessment of the life cycle energy demand of passive houses *Applied Energy* **112** 23-34

[13] Anand C and Amor B *2017* Recent developments future challenges and new research directions in LCA of buildings A critical review *Renewable and Sustainable Energy Reviews* **67** 408-416

[14] Emami N, Heinonen J, Marteinsson B, Säynäjoki A, Junnonen J-M, Laine J and Junnila S *2019* A Life Cycle Assessment of Two Residential Buildings Using Two Different LCA Database-Software Combinations Recognizing Uniformities and Inconsistencies *Buildings* **9** 20

[15] Chester M, Pincetl S, Elizabeth Z, Eisenstein W and Matute J *2013* Infrastructure and automobile shifts positioning transit to reduce life-cycle environmental impacts for urban sustainability goals *Environmental Research Letters* **8** 015041

[16] Hendrickson C T, Lave L B and Matthews H S *2006 Environmental Life Cycle Assessment of Goods and Services An Input-Output Approach* Resources for the Future Press

[17] Vigon B W, Tolle D A, Cornaby B W, Latham H C, Harrison C L, Boguski T L, Hunt R G and Sellers J D *1993 Life Cycle Assessment Inventory Guidelines and Principles* US Environmental Protection Agency Washington DC EPA 600/R-92/245

[18] Wiedmann T and Minx J *2008 A Definition of 'Carbon Footprint'* In C C Pertsova Ecological Economics Research Trends Chapter 1 1-11 Nova Science Publishers Hauppauge NY USA

[19] Leontief W *1970* Environmental repercussions and the economic structure an input-output approach *Rev Econ Stat* 262-271

[20] Säynäjoki A, Heinonen J, Junnonen J-M and Junnila S *2017* Input–output and process LCAs in the building sector are the results compatible with each other? *Carbon Management* **8** 2 155-

166

[21] Carnegie Mellon University Green Design Institute *2011* Economic input–output life cycle assessment EIO-LCA US 2007 Industry Benchmark model [Internet] Available at http//wwweiolcanet/ Accessed 20122019

[22] Sedlabanki Islands *2018* Official Exchange Rates available online at https//wwwcbis/?PageId=6909b7bd-5189-45dd-bf5b-c76ea33496ef Accessed 22122018

[23] Zarenski E *2017* Construction Inflation Index Tables available online at https//edzarenskicom/2016/10/24/construction-inflation-index-tables-2017 Accessed 22122018

[24] Kyrö R, Heinonen J and Junnila S *2012* Assessing the Potential of Climate Change Mitigation Actions in Three Different City Types in Finland *Sustainability* **47** 1510-1524

[25] Lenzen M, Moran D, Kanemoto K and Geschke A *2013* Building Eora a global multi-region input–output database at high country and sector resolution *Economic Systems Research* **25** 1 20-49

[26] Su B, Huang B C, Ang BW and Zhou P *2010* Input–output analysis of CO2 emissions embodied in trade The effects of sector aggregation *Energy Economics* **32** 1 166–175

[27] Crawford R 2011 Life Cycle Assessment in the Built Environment London Routledge

[28] Suh S, Lenzen M, Treloar GJ, Hondo H, Horvath A, Huppes G, Jolliet O, Klann U, Krewitt W, Moriguchi Y, Munksgaard J and Norris G *2004* System boundary selection in life-cycle inventories using hybrid approaches *Environ Sci Technol* **38** 657–664

IOP Conf. Series: Earth and Environmental Science **297** (2019) 012023 doi:10.1088/1755-1315/297/1/012023

Opportunities for financing sustainable development using complementary local currencies

Ljubomir Jankovic

Zero Carbon Lab, School of Creative Arts, University of Hertfordshire, Hatfield, UK

L.Jankovic@herts.ac.uk

Abstract. Financing building retrofit projects that contribute to climate change mitigation has always represented a significant barrier. With 28% of global emissions coming from existing buildings, it is of paramount importance to carry out retrofit measures that lead to significant reduction of these emissions. Whilst this is perfectly possible to achieve with current methods and current technology, there is no sufficient conventional finance to carry out zero carbon retrofit at scale required for climate change mitigation. The article introduces an alternative and sustainable business model that creates new opportunities for financing zero carbon retrofit of buildings. It demonstrates that the value of solar energy falling on roofs of buildings can become a driver for new local economic systems, and discusses the requirements for practical application.

1. Introduction

Every radical intervention on an existing building that aims to achieve a retrofit to zero carbon emissions is faced with the same problem: how to raise the finance? But can we create a world in which finance is not a problem?

A recently published report on global warming by the Intergovernmental Panel on Climate Change [1] has strengthened our awareness for global responsibility and the urgent need for controlling and minimising carbon emissions within the next decade. With 28% of global emissions coming from existing buildings [2], it is of paramount importance to carry out retrofit measures that lead to significant reduction or complete elimination of these emissions. Whilst this is perfectly possible to achieve with current methods and technology [3], conventional financing methods represent a significant barrier for zero carbon retrofit at scale and for climate change mitigation. Simply, there is not enough of conventional money to carry out zero carbon retrofit of buildings at sufficient scale, and this epitomizes strategic sustainability interventions. Some researchers suggest that the barrier to sustainable development is the capitalist mode of production, and that an alternative social organization that transcends capitalism is a social requirement for sustainable development [4]. However, a much simpler and quicker solution to this problem may be achieved through regional complementary currencies, which were predominant in Western Europe for a thousand years to 1800 AD and beyond [5].

The problem of not having enough conventional finance to carry out sustainability interventions is not new. In 1813, after the Napoleon wars, the island of Guernsey was at a brink of a natural disaster and under a heavy debt. Due to the defective state of its sea banks the great extent of the island was threatened to be overflown by the sea [6]. 'The States' of Guernsey had a debt of almost twice the amount required for the repairs of the sea defences, and were burdened by paying 80% of its annual revenue of £3,000 as interest. There was no trade, no income, no attraction for visitors, and a market building was needed as market traders had to stand outside in rain and wind. Under the pressure to redevelop its infrastructure, including the sea banks, roads and the market, and with practically no conventional money left from the annual revenue, the States came up with an innovative solution in 1816 and issued £6,000 of new local currency, accompanied by the following commentary: "*In this manner, without increasing the debt of the States, we can easily succeed the works undertaken, leaving moreover in the coffers sufficient money for the other needs of the States*" [6]. The money had a withdrawal date in 1817 and 1818, and could be exchanged at that point for conventional money. The States were able to honour that exchange, having earned in the meantime enough income from the rents in conventional money on the new market building.

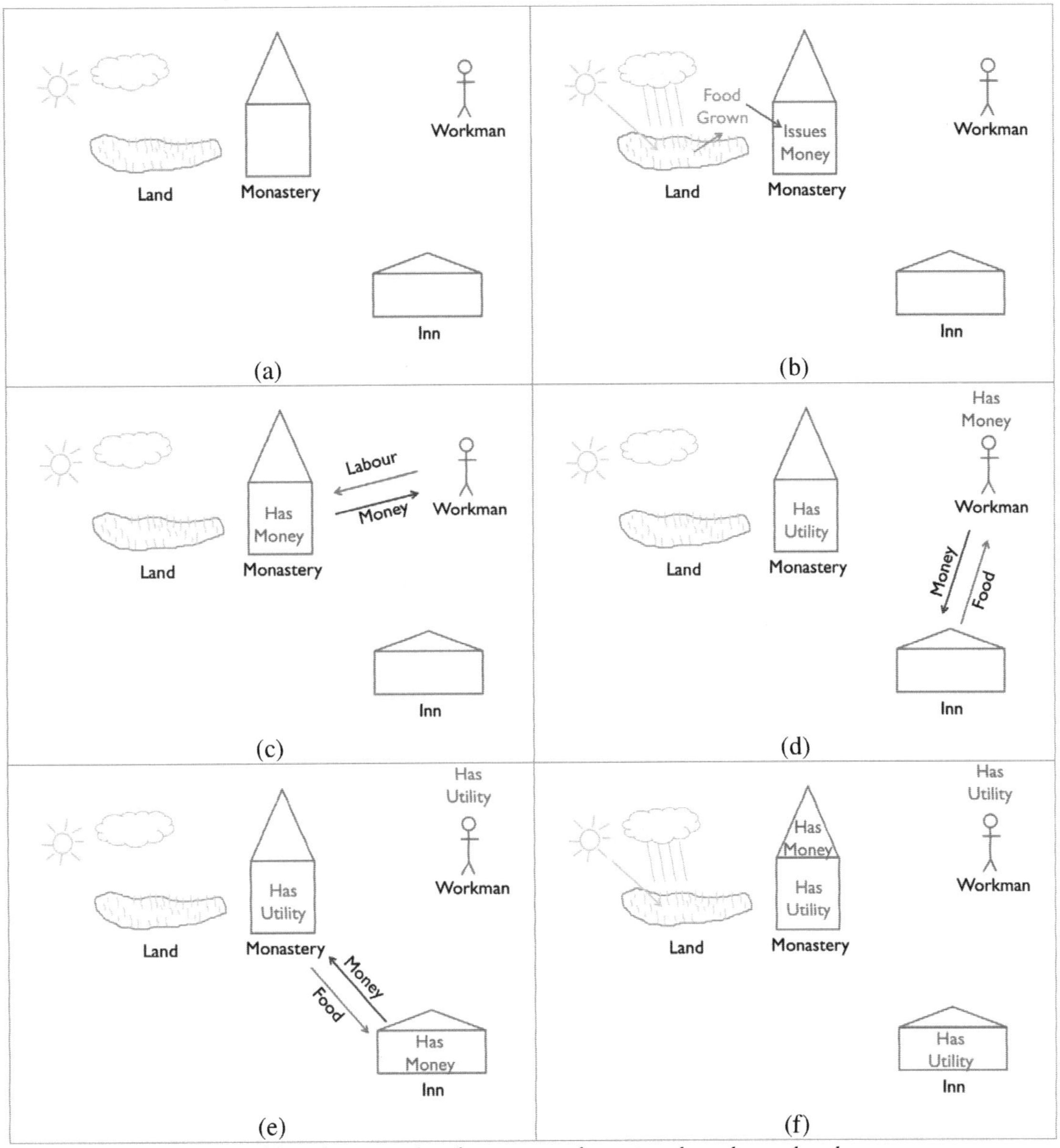

Figure 1. A historic example of an economic system based on a local currency.

Looking into this issue deeper, Thomas Greco in his seminal book 'The End of Money and the Future of Civilization' [7] introduces three prerequisites for the creation of complementary currency:

1) Money can be issued by anyone who offers goods and services for sale in the market
2) The basis for issuing money are the goods on the market or on their way to the market
3) Each issuer is entitled to create as much money as they are able to redeem by selling.

However, there is more to it than just these three prerequisites. The goods and services for sale in the market need to be generated somehow, suggesting natural resources and human labour inputs into the system.

This can be illustrated in an example of a medieval trading system in the French city of Saint-Omer [5](Figure 1), where a monastery required building work to be done and used the value of food and wine production from its land to create its own currency. Thus, the structure of this financial system consisted of the land, the monastery, workmen, and local inns (Figure 1a). Solar radiation and rainfall enabled food production and gave the monastery the opportunity to create their own currency in the form of lead coins (Figure 1b). The monastery then used the lead coins to pay the workmen (Figure 1c), who in turn used the money to buy food and wine in local inns (Figure 1d). The local inns then used the same money to pay the monastery and replenish the food and wine supplies (Figure 1e).

Thus, every actor in this system had some utility: the monastery had the building work done; the workmen were able to sustain themselves; and the local inns had paying customers (Figure 1f). Ultimately, the money ended up with the monastery as the original issuer. The external inputs into the system and the ultimate drivers were solar radiation and rainfall, which enabled food and wine production on monastery's land with labour supplied by the monks. The local currency and the utility for each of the three actors (monastery, workmen and inns) moved in opposite directions, as shown in Figure 1, and created a local economic system.

Translating this approach into the current time, this research investigates the opportunities for alternative financing zero carbon retrofit of buildings. It looks into how this approach could be made operational, and how diffusion into a wider economy could be achieved. This leads to two research questions:

1) How can a trading system be set up that enables retrofit of buildings to be carried out using a local currency equivalent to the value of energy received from solar radiation?
2) What would be required to connect such system into a wider economy?

In order to answer these questions, a model of a simple system will be set up in the Method section and analysed in the Results section.

If the method and results prove to be positive and conclusive, this work could lead to the discovery of new financing opportunities for zero carbon retrofit of buildings, and it could facilitate the creation of new economy of sustainable business models that help to mitigate the climate change.

2. Method

A model of a trading system consisting of a housing association that owns buildings, a retrofit provider, a photovoltaic (PV) system manufacturer, and a householder is created, as shown in Figure 2a.

The trading process starts when the housing association creates its own 'electric money' or 'eMoney' currency equivalent to the value of solar radiation falling on its roofs (Figure 2b) and uses it to pay the retrofit provider (Figure 2c). In return, the retrofit provider installs a PV system on the roofs of the housing association buildings (Figure 2c). This starts a circular movement between the actors, where the currency and the utility flow in opposite directions. Having run out of the PV systems, the retrofit provider purchases new PV system from the PV manufacturer, paying for it with eMoney, whilst retaining 'X' amount as its income (Figure 2d). Having sold the PV to the retrofit provider, the PV manufacturer needs energy for its production process, and purchases energy from the housing association using eMoney, whilst retaining 'Y' amount as its income (Figure 2e). The housing association also sells surplus energy to its tenants, and gets paid in conventional currency (Figure 2e). In the end of the cycle, all actors have had income and utility (Figure 2f).

At this stage, the housing association may choose to expand its PV installations, issue more eMoney and start the next 'spin' in the financial cycle, or may merely settle for income from energy sales.

In order for this system to work, an integration into a wider economic system is required. Thus, the housing association, the retrofit provider, and the PV manufacturer, need to be able to exchange their income in eMoney into conventional currency (Figure 3).

This approach will now be tested using sample calculations and the results will be presented in the next section.

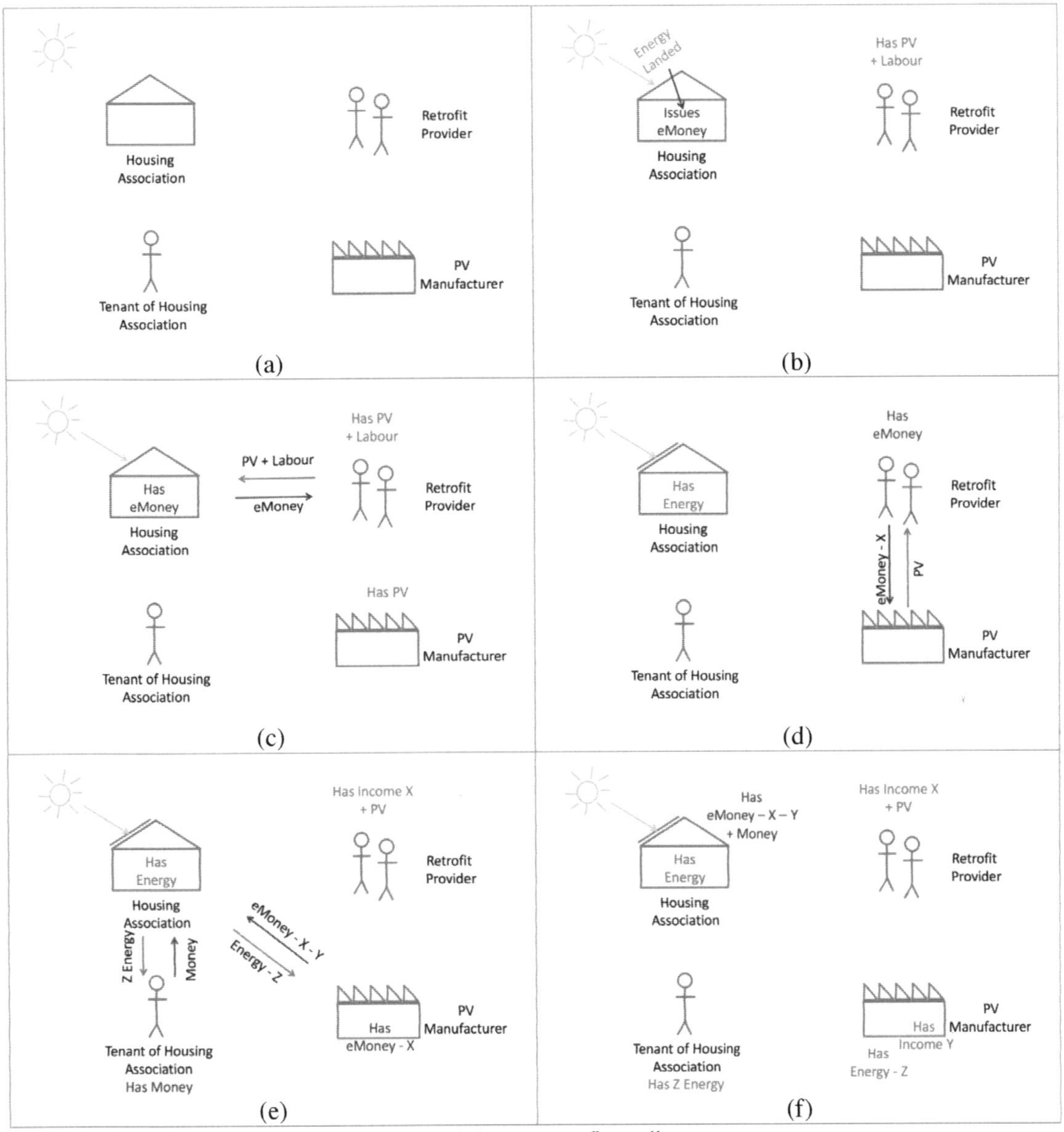

Figure 2. An eMoney retrofit trading system.

3. Results

The trading system from Figure 2 has now been run through a series of calculations and the results are shown in Table 1. The results are presented in three separate categories: credit balances, utility balances and the total system balance.

In Step 1, the starting position has been set so that the housing association creates a 1000 credit balance of eMoney units, resulting from the solar radiation falling onto its roofs; and the tenant of the housing association has a 400-credit balance of the conventional money, equivalent to the same number of eMoney units. The retrofit provider and the PV manufacturer have utility balances of 1000 units each, equivalent to the same number of eMoney units. The total system balance in the first step is 3400 eMoney units.

In Step 2, the retrofit provider fits PV for the housing association and gets paid 1000 eMoney units. Thus, the housing association credit balance becomes zero, and utility balance becomes 1000. The retrofit provider balances become exactly opposite; the credit balance becomes 1000 and utility balance becomes zero. The total system balance remains unchanged and stands at 3400 eMoney units.

In Step 3, the retrofit provider purchases PV for 800 eMoney units from the PV manufacturer and retains 20%, namely 200 eMoney units, as income. Thus, retrofit provider's credit balance is reduced to 200 and PV manufacturer's credit balance is increased from zero to 800. Again, exactly opposite situation is with the utility balances for these two actors: the retrofit provider's utility balance has increased by 800, and the PV manufacturer's utility balance has decreased by .800. The total system balance remains constant at 3400 eMoney units.

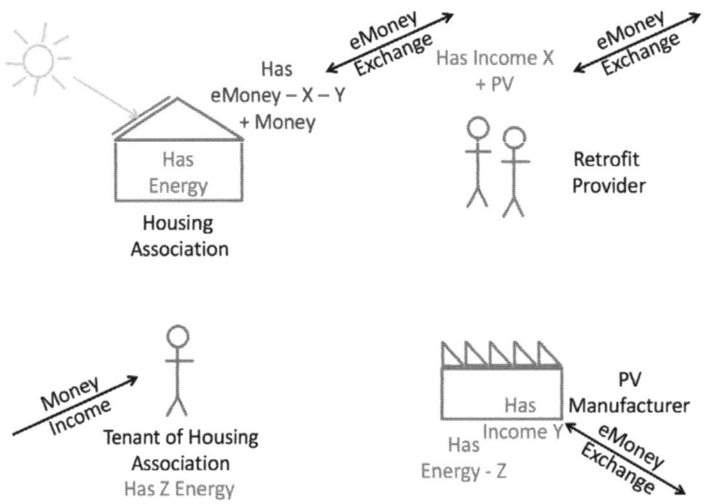

Figure 3. Links to a wider economic system through eMoney exchange into conventional currency.

In Step 4, the PV manufacturer pays the housing association 600 eMoney units for energy and retains 20% of eMoney units as income. Thus, the housing association's credit balance increases by 600 and the PV manufacturer's credit balance decreases by 600 eMoney units. The change in the corresponding utility balances between these two actors is exactly opposite: the housing association's utility balance has decreased by 600 and the PV manufacturer's utility balance has increased by 600. The total system balance remains constant at 3400 eMoney units.

In Step 5, the housing association sells surplus energy to the tenant and gets paid 400 in conventional money units, equivalent to eMoney units. Its credit balance increases by 400, and the tenant's credit balance reduces by 400. The movement in the corresponding utility balances between these two actors is exactly opposite: the housing association's utility balance has reduced by 400 and the tenant's utility balance has increased by 400. The total system balance remains constant at 3400 eMoney units.

At this stage (Step 6, Table 1), the housing association may choose to 'spin' the financial cycle again, by issuing new eMoney on the basis of additional roofs that it wishes to retrofit with the PV. This would be equivalent to adding an impulse to this trading system's 'flywheel' in Figure 2, and the process could continue until all roof surfaces belonging to the housing association are fitted with PV systems. Alternatively, the hosing association could merely continue to benefit from its renewable energy production capacity, and earn income from energy sales.

All actors with eMoney credit balance would need to convert to conventional currency at this stage, in order to connect with the conventional financial system and the rest of the world. This could be achieved by the housing association exchanging eMoney for conventional money, having earned enough conventional money from renting its accommodation. Alternatively, if or when more than one eMoney issuer emerges, a bank could be set up to facilitate the integration into the wider economy by providing exchange facilities between eMoney and conventional money.

Table 1. Transactions in the eMoney retrofit trading system.

	Credit balances				Utility balances				Total system balance
	Housing association	Retrofit provider	PV manufacturer	Tenant of hous. assoc.	Housing association	Retrofit provider	PV manufacturer	Tenant of hous. assoc.	
Step 1 - Housing association issues 1000 eMoney units	1000	0	0	400	0	1000	1000	0	3400
Step 2 - Retrofit provider fits PV for Housing association and gets paid 1000 eMoney units	0	1000	0	400	1000	0	1000	0	3400
Step 3 - Retrofit provider purchases PV for 800 eMoney units from the PV manufacturer and retains 200 eMoney units as income	0	200	800	400	1000	800	200	0	3400
Step 4 - PV manufacturer pays Housing association 600 eMoney units for energy and retains 200 eMoney units as income	600	200	200	400	400	800	800	0	3400
Step 5 - Housing association sells surplus energy to tenant and gets paid 400 conventional money	1000	200	200	0	0	800	800	400	3400
Step 6 - the process repeats…	…								

4. Discussion

What is new in this approach? The notion of complementary currency is certainly not new. It has been researched and reported extensively [7], and it has been implemented in numerous locations around the world (Figure 4). However, the novel aspect of this approach is in the application of complementary currencies to building retrofit projects, based on the production of renewable energy from the sun.

The presented trading system is not perfect and is not complete: the PV manufacturer, in addition to purchasing energy, needs to purchase materials end equipment in order to produce PV systems. There is an entire supply chain involved in this type of operation, however, for simplicity, this supply chain has been omitted from the analysis. Similar supply chain issues apply to the retrofit provider and the housing association.

Despite of these limitations, the point about complementary currency based on renewable energy has been made: the numbers add up and make sense in the simple calculation presented here and this can become the basis of an actual and more complex system.

What are the benefits of such system? First, the system provides finance opportunity where such opportunity is unavailable through conventional finance. Second, in this type of system, there is no debt, and no inflation. After the development process is completed, the complementary currency can be redeemed and the system users can revert to trading in the conventional currency.

Are there any negative taxation implications of this system – can it lead to tax avoidance? If there is no conventional finance for a retrofit project, and the project does not go ahead, there will be no opportunities for a subsequent taxation. If, however, the complementary currency creates opportunities for trading with the conventional currency after the complementary currency part of the project has been completed and the complementary currency is redeemed, that will create opportunities for taxation and full compliance with national financial systems.

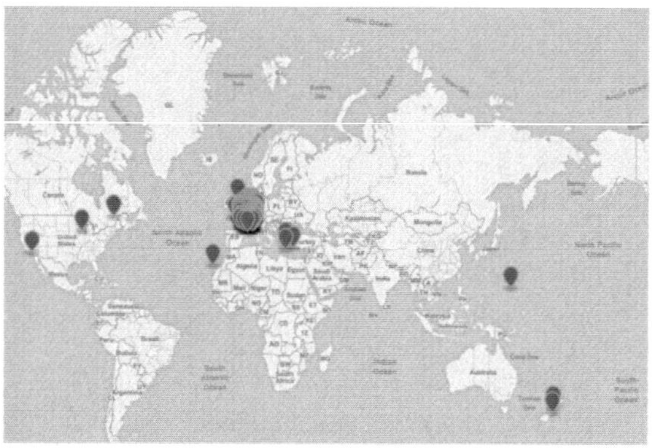

Figure 4. Complementary currency world map (source [8]).

Why is this approach not used more extensively? The reason is that it takes time to set it up and it requires the involvement of a much higher number of actors than the example in this article. In order for such system to gain critical mass and 'catch fire', at least 100 diverse actors will be needed. If there is a sufficient number of businesses involved but all of them are for instance hair dressers, they will not need much from each other - there is no business diversity to make the system operational. The experience of the process of setting up the Brixton Pound in London shows that it took three attempts, until the required number and diversity of actors was reached to keep the system self-sustaining [9]. It is encouraging that complementary currencies are finding their use as policy instruments for environmental purposes in Europe, most notably in the Belgian Science Policy INESPO Project [10]. However, the coupling of the complementary currency to smart meters and the expected behaviour change in that project represents a somewhat missed opportunity and a limited scope of what could have been a much more radical approach. Although that system was based on rewards and loyalty points only, nevertheless a useful taxonomy for a complementary currency system was devised [10]. However, based on the experience of the research introduced in this article, for a complementary economy to 'catch fire' it is essential to have a driving resource, such as solar radiation in our example, diverse actors who need goods or services from each other, and a complementary currency that circulates through the system as the means of exchange.

5. Conclusions

What can we learn from this article? Despite the urgent need to mitigate climate change, to which energy use in buildings is a significant contributor, conventional financing of projects that aim to retrofit existing buildings to buildings with zero carbon dioxide emissions presents a significant barrier.

Inspired by historic precedents of complementary currencies created on the basis of a natural resource, such as land, solar radiation and rainfall, this research investigated the use of solar radiation falling on roofs of buildings as a value generator in a theoretical model.

Thus, solar radiation falling on the roofs of a housing association was the basis for creating a model of a complementary currency. The currency was subsequently used to create a circular movement of credit and utility between the main actors in the model: the housing association, the retrofit provider, the PV manufacturer and the tenant of the housing association. The analysis demonstrated that after the credit circulated through the entire system, the complementary currency ended up with the original issuer, and each actor in the system had some utility as result: income, energy, PV or a combination thereof.

At the end of the first cycle, the housing association, being the issuer of complementary currency, has an opportunity to either issue more complementary currency, thus giving an impulse to the system that drives it through the next cycle, or to continue benefiting from energy sales income based on its installed PV capacity. It is essential for this system to connect to the conventional financial system in the rest of the world through currency conversion, and provisions for such conversion are essential for making the system work.

Although, for simplicity, supply chains have been omitted from this analysis, the case for complementary currency based on renewable energy has been made.

The system diagram and the calculation of credit and utility flows can become the basis for a practical application of such system, providing that sufficient number and diversity of actors can be achieved, and that exchange facilities are made available for converting complementary currency credits into conventional currency.

This work therefore answers both research questions and paves the way towards a practical implementation of financing sustainable development using complementary local currencies. It creates new financing opportunities for zero carbon retrofit of buildings, and it could facilitate the creation of a new economy of sustainable business models to help mitigate the climate change.

References

[1] IPCC 2018 *Global Warming of 1.5 °C* - an IPCC special report on the impacts of global warming of 1.5 °C above pre-industrial levels and related global greenhouse gas emission pathways, in the context of strengthening the global response to the threat of climate change, sustainable development, and efforts to eradicate poverty. (World Meteorological Organization & United Nations Environment Programme)

[2] Abergel T, Dean B and Dulac J 2017 *Global Status Report 2017* (Global Alliance for Buildings and Construction)

[3] Jankovic L 2017 *Designing Zero Carbon Buildings Using Dynamic Simulation Methods* (London and New York: Routledge)

[4] Liodakis G 2010 Political Economy, Capitalism and Sustainable Development *Sustainability* **2** 2601–16

[5] Kennedy M I, Lietaer B A and Rogers J 2012 *People money: the promise of regional currencies* (Axminster: Triarchy)

[6] Harris J T 2018 *An Example of Communal Currency: The Facts about the Guernsey Market House* (Creative Media Partners, LLC)

[7] Greco T H 2010 *The end of money and the future of civilization* (Edinburgh: Floris)

[8] Complementary Currency Resource Center 2017 CC World Map Complement. Curr. Resour. Cent.

[9] Thompson M 2016 Setting up the Brixton Pound - personal communication with L. Jankovic

[10] Joachain H and Klopfert F 2012 Emerging Trend of Complementary Currencies Systems as Policy Instruments for Environmental Purposes: Changes Ahead? *Int. J. Community Curr. Res.* **16 (D)** 156–68

IOP Conf. Series: Earth and Environmental Science **297** (2019) 012024 doi:10.1088/1755-1315/297/1/012024

Circular Economy in the Built Environment in Finland - A case example of collaboration

J Karhu[1] and L Linkola[2]

[1]M.Sc. (tech.), Senior sustainability specialist, Green Building Council Finland, Finland

[2]M.Sc. (tech), Circular economy expert, Ethica Oy

jessica.karhu@figbc.fi, lilli.linkola@ethica.fi

Abstract. Circular Economy in the building and construction sector is challenging the conventional habits and models of work. It demands comprehensive collaboration between stakeholders and collaborations even with unconventional partners. Nevertheless, circular economy model can give remarkable opportunities to tackle the climate change with the help of building and construction sector and helps to find sustainable solutions to the growing urbanisation and to expansion of the cities.

Very crucial obstacle in order to implement circular economy into building and construction sector´s processes is the sector´s difficulties to adopt new methods. This includes difficulties to use new recycled materials and building components, legislation obstacles, contract models do not support circular economy solutions etc. Often these obstacles are result of absence of life cycle thinking and coordination. Key to solve these kinds of situation where no stakeholder alone can take the responsibility is to gather all the stakeholders together and boost them to solve these cross sectoral obstacles.

To support the adaptation of circular economy into Finnish building and construction sector, The Finnish Innovation Fund Sitra launched a project "Circular Economy in the Built Environment" in the beginning of year 2018. The coordinator of this project is the Green Building Council Finland. Target of this project is:

1. To increase the knowledge on circular economy on the construction and building sector
2. Define how we can support circular economy in building and construction sector and what indicators we can use to authenticate this development
3. Create new collaboration between different stakeholders
4. Develop the procurement and designing processes

The project started by identifying key persons that can advance circular economy in Finland in the building and construction sector. These key persons were asked to point out more circular economy professionals. The project succeeded to collect a network with hundred professionals. With this network and workshops, we defined what circular economy means in build environment and what is our common targets in the implementation of it.

As a result, in the project we defined seven imperative targets that must happen in the building and construction sector. These targets are:

1. Construction and building sector have a common goal to advance circular economy
2. Norm and regulation steer strongly to circular economy
3. Renewing the habits
4. Life cycle thinking is the starting point of all planning and design
5. Areal planning advances circular economy
6. Procurement support circular economy
7. All spaces and locations are in maximal use

In the project we also defined action proposals how to achieve these targets.

As a conclusion we argue that achieving circular economy targets and actions that the whole sector is ready to implement willingly demands common work and the will to work together must come from the stakeholders themselves. Different stakeholder has different motives, but this does not prevent collaboration if there is a neutral platform where to co-work with colleagues.

1. Introduction

The definition of circular economy is multidimensional and comprehensive. Several studies describe it from different angles and its economical and societal concepts are not clear yet as we can see from the very recommended article by Kirchherr et al. [1]. Also, how it is implementing into built environment is under discussion and opinions and understanding are developing through experiences. The expanded definition of circular economy by Ellen MacArthur Foundation [2] underlines the re-thinking of the economical process. After Ellen MacArthur Foundation the circular economy is: looking beyond the current take-make-dispose extractive industrial model, a circular economy aims to redefine growth, focusing on positive society-wide benefits. It entails gradually decoupling economic activity from the consumption of finite resources and designing waste out of the system. Underpinned by a transition to renewable energy sources, the circular model builds economic, natural, and social capital. After Ellen MacArthur Foundation the circular economy is based on three principles:

- Design out waste and pollution
- Keep products and materials in use
- Regenerate natural systems

These principals go well with the objectives of good quality urban planning and construction. Thus, this Ellen MacArthur Foundation´s definition was a stable basis to build a definition what circular economy means in built environment.

We also find that the definition by van Buren et al. [3], is written with greatest clarity and we thus particularly recommend his definition. After van Buren unlike the current economy, which is largely based on the principle "take-make-waste" (linear economy), the focus point in a circular economy is to not unnecessarily destroy resources. She states that this implies far more than the reduction of waste through recycling, stresses the following focal points: reducing the consumption of raw materials, designing products in such a manner that they can easily be taken apart and reused after use (eco-design), prolonging the lifespan of products through maintenance and repair, and the use of recyclables in products and recovering raw materials from waste flows. After van Buren a circular economy aims for the creation of economic value (the economic value of materials or products increases), the creation of social value (minimization of social value destruction throughout the entire system, such as the prevention of unhealthy working conditions in the extraction of raw materials and reuse) as well as value creation in terms of the environment (resilience of natural resources).

This definition work already done was a good starting point to develop a definition for the built environment sector in Finland.

What is clear is that the circular economy is a powerful tool for climate mitigation. Materials production alone risks exceeding the total remaining carbon budget for a 2°C scenario. It is calculated that we can consume for the total material production 300 gigatons CO2 by year 2100 in a 2°C scenario but the calculation shows that materials emissions even with energy efficiency production will reach 918 gigatons CO2 with existing business models and concepts. [4] So, we have an enormous gap that demands zero-carbon energy production but also definitely a prolongation of material's life cycles and a shift towards circularity. There are different estimates of the climate impact of the resources usage in the building and construction sector. Material economics estimates that by 2050, just cement, steel, aluminium and plastic used for construction would result in emissions of 230 Mt CO2 if they were made with today's production processes ref. It is imperative for the sector to move to more circular resources usage and to adapt new business models that would support them. Yet we recognize that big barriers for change are lack of common understanding and joint vision of how the sector should develop to be climate neutral also from the materials perspective, lack knowledge and skills within the sector professionals on circular material choices and life cycle design, and finally lack of incentives, both economic and regulatory ones, that would drive changes in the market.

To tackle these barriers in Finland the Green Building Council proposed a project "Circular Economy in the Built Environment". The project aims to support the adaptation of circular economy into Finnish building and construction sector by gathering all the stakeholders together and facilitating them to solve cross sectoral challenges together and steering the sector towards more circular practices. The primary objectives for the project were fourfold.

First objective was to increase the knowledge over circular economy on the construction and building sector within the stakeholders in general.

Secondly, we aimed to build a joint vision over the implications of circular economy in construction sector. Together with the stakeholders we clarified the definition of circular economy in the building and construction sector, and defined means to support the change in the sector and indicators, that can be used to authenticate this development.

Thirdly, we facilitated the networking process in order to create new collaboration between different stakeholders. Finally, we pushed the stakeholders to draw together concrete plans and develop the procurement and designing processes for the whole industry.

2. Data collecting and methods to build a network

The project started by identifying key persons that can advance circular economy in Finland in the building and construction sector. In order to recognise the experts in Finland that were both interested in circular economy and were experts on field of sustainable built environment we used a snowball technique as it used to recognize members of a rare population described by Tompson [5] and again by Handcock [6]. We approached ten known experts and asked them to join the network and also to point other experts. To collect their answers, we used an internet form that could be answered anonymously.

Our target in the beginning was to form a network of a hundred experts. That number of persons was more or less reached during the spring 2018. The yammer group that was establish for the network includes today exactly 100 members. The network consists from designers, engineers, city officers and from experts from different backgrounds.

The Green Building Council Finland organized two full day stakeholder workshops for the identified expert network. The goal of the first workshop was to introduce the concept of circular economy and build a joint vision and goals for circular economy in the building and construction sector together with the network. The goal of the second workshop was to translate the previously set goals into an action plan for the whole sector to take forward. Working groups were appointed in the expert network to develop the action plans further and lead the change in the industry.

The workshops were held a moth apart from one other in order to ensure that the participants will have enough time to develop their knowledge over the topic and finalize as a network the joint definition of circular economy in the building and construction sector and satisfy with the set goals. The workshop results were uploaded to the network's yammer group for a review by all the network members not only the workshop participants.

3. Results

After two workshops in spring 2018 was, the experts agreed and came to conclusion, that a circular economy cannot be achieved in the built environment without a circular economy of materials. The goal is for Finland to build its competitiveness by means of sustainable material use. The demand for virgin materials will be minimised, the length of material and product life cycles maximised, and products designed so that they can be maintained and reused at the end of their first useful life.

After the experts, the real estate and construction industry has a downright obligation to move towards a circular economy to ensure that the built environment – as a major consumer of energy and resources – will not exceed the carrying capacity of nature. A circular economy provides the industry with great potential for development and change. The principles of a circular economy should be embedded in the entire ecosystem, and the industry will create system-level innovations that challenge current practices and business models.

The main findings of the two workshops were encapsulated in seven essential objectives that we have to reach in Finland in order to make happen the change to circular economy in the built environment:

Table 1. *The seven essential objectives to reach circularity in built environment*

Essential objective	Definition of the objective
1. Objective: Pursuing a common goal to promote a circular economy in the real estate and construction industry	- The real estate and construction industry have defined a common goal and targets for a circular economy. All operators are aware of the goal and committed to reaching it. - The industry has defined generally accepted circular economy criteria to support planning and decision-making. Criteria have been set, among others, for the following sectors: o Planning o Construction products

	o Land use planning o Architecture o Procurement o Infrastructure o Use of space (affects construction demand and methods). - Based on these criteria, a list has been drawn up of measures required to promote the circular economy. The implementation of these measures has started, and progress will be monitored.
2. Objective: Policy instruments and legislation efficiently promote the circular economy	- The state and cities have recognised their role in promoting the circular economy in the property and construction industry. This is reflected in concrete actions. - The Ministry of Economic Affairs and Employment, the Ministry of the Environment, the Ministry of Agriculture and Forestry, the Ministry of Education and Culture, the Ministry of Finance and other administrative organisations have set common objectives for promoting the circular economy in the property and construction industry and agreed on measures to achieve the objectives. These objectives should be maintained across electoral terms. - Political parties have recognised their role as drivers of a circular economy, and their own government programme objectives include the promotion of a circular economy in the real estate and construction industry. - A circular economy is strongly featured in the Land Use and Building Act that is currently under preparation.
3. Objective: New practices in the industry promote the circular economy	- Synergies between projects and the overall optimisation of life cycles will be ensured through efficient exchange of information within and between projects. - New circular economy companies are being created in the building and construction industry. - Best practices are promoted through international co-operation. - Each organisation will be appointing a circular economy expert, and these experts collaborate via a shared forum for the industry. - Activities to promote the circular economy will be rewarded (e.g. industry awards). - An operating model for increasing social capital has been created and taken in use

	(more open and transparent urban planning and development).
4. Objective: Life-cycle thinking guides planning in infrastructure and building construction	- We have common calculation rules that cover the whole life cycle, a common, open and up-to-date database as well as scenarios for realising end-of-life recycling and reuse. - Key opportunities for promoting the circular economy are being identified at each stage of planning. - Life-cycle costing directs procurement and decision-making, and operators are moving away from partial optimisation. - Each structural design includes a plan for recovery and recycling.
5. Objective: Land use planning and zoning promote the circular economy	- Urban circular economy is supported with a versatile urban structure. - Processes to change the purpose of use of a property are smooth. - Land mass logistics in earthworks are locally coordinated and appropriate. - There is efficient information exchange and co-operation between land use planning, infrastructure construction and building construction: synergies are identified and utilised.
6. Objective: Procurement processes and purchasing expertise are developed to support the circular economy	- Municipalities are using the upcoming Ministry of the Environment's guidance on public construction projects. - Purchasing organisations are aware of their own opportunities to promote the circular economy and include the circular economy in their procurement criteria. - The property and construction industry have developed common circular economy criteria for purchasing organisations and procurement in different sectors: o Infrastructure construction o Regional planning o Building construction
7. Objective: Use of space is efficient	- Operators managing spaces (private and public buildings and other spaces) are providing a wide range of usage opportunities through different interfaces. - Design of spaces is taking into account and enabling multiple and shared use.

	- Use of spaces is as efficient as possible and target levels have been set. New ways of using spaces will be developed, alongside with means of continuous improvement (measuring, monitoring and setting new targets). - Restrictive legislation and operating models have been identified and abolished.

These findings were published by the Green Building Council Finland as general recommendations and guidelines for the industry and other stakeholders to adopt circular economy in build environment.

4. Discussion

In the built environment, the circular economy refers to an operating model adopted by the industry and policies supporting the model. Together, these two aspects drive a transition towards a sustainable built environment. The built environment plays a key role in the circular economy in Finland, particularly in optimising material and energy cycles, returning resources into these cycles and producing new resources.

The concept of a circular economy will steer us to design our environment in a way that the changing needs of users can be met, and new forms of use adopted. Thus, we will be able to support a sharing economy. When designing environments in accordance with the principles of a circular economy we create more sustainable and more low-carbon neighbourhoods.

The collaborative process proved to be successful in disseminating the ideas of circular economy in the building and construction sector. By letting the stakeholders to define together the importance of circular economy for the sector and formulate the most important objectives and actions for themselves and their peers, we think to witness a much faster pace of adoption of the ideas within the sector than we would have achieve with a more traditional top-down awareness raising campaign. The workshops gathered a lot of enthusiasm from the participants, and that resulted to a quite good participation rate to the subsequent seven working groups. However, after a while the working groups suffered some loss in participation as the first sparks of enthusiasm for the subject at hand vanished.

We also note that it requires a lot of effort to keep up the discussion among the network members between the face to face meetings. The discussion forum Yammer was rather quiet, yet functions as an important clue for the network, something tangible that concretizes the network.

5. Conclusion

While writing this, this project is still going on. The current work is focusing to concretise the actions needed in order to reach the essential seven targets. The goal is also to get the right stakeholders to commit to the change towards the circularity.

For further research it would be useful to compare the similar approaches in other Nordic countries. Norway and Denmark have similar circular economy targets but the methods to implement these targets to the national building and construction sectors are different than what has been done in this project. By comparing different implementing methods, we could point out the successful concepts to commit the construction and building sector to the circular economy and the business models that support the change to circular economy in built environment.

Acknowledgments

We appreciate the support of the Finnish Innovation Fund Sitra for funding this "Circular Economy in the Built Environment" project.

References

[1] Kirchherr J, Reike D and Hekkert M 2017 Conceptualizing the circular economy: An analysis of 114 definitions *Resources, Conservation and Recycling* Volume 127 pp 221-232

[2] Ellen MacArthur Foundation 2018 What is a circular economy? A framework for an economy that is restorative and regenerative by design available online at https://www.ellenmacarthurfoundation.org/circular-economy/overview/concept

[3] Van Buren N *et al.* 2016 Towards a Circular Economy: The Role of Dutch Logistics Industries and Governments *Sustainability* 8(7), 647 available online at https://www.mdpi.com/2071-1050/8/7/647/htm#B2-sustainability-08-00647

[4] Enkvist P-A Klevnäs P 2018 *The Circular Economy a Powerful Force for Climate Mitigation* 2018 Transformative innovation for prosperous and low-carbon industry, Materia Economics Stockholm available online at https://media.sitra.fi/2018/05/04145239/material-economics-circular-economy.pdf

[5] Thompson S K 2002 Sampling. *Wiley, Second edition*,

[6] Handcock M S and Gile K J 2011 On the Concept of Snowball Sampling available online at https://arxiv.org/pdf/1108.0301.pdf

Appendix 1

Organisation that has participated to the workshops.

As the participants have not signed the permission to publish their names, the writers are not presenting them here. There were often several persons participating from the same organisation, working in different thematical groups. Thus, here the organisation is mentioned as many times as there were persons joining the workshops and the thematical work after the workshops.

1) aalto yliopisto, EKO SAFA	42) INFRA ry
2) ACRE (Aalto)	43) INFRA ry
3) asoypalvelut.com	44) INFRA ry
4) asoypalvelut.com	45) INFRA ry
5) Delete Oy	46) INFRA ry
6) Delete Oy	47) Kliffa
7) Delete Oy	48) Kuntaliitto
8) Destaclean Oy	49) Logicor Oy
9) Destaclean Oy	50) Lujatalo
10) Destaclean Oy	51) LUKE
11) Espoon kaupunki	52) LUKE
12) Espoon kaupunki Tilakeskus	53) NCC Finland Oy
13) Espoon kaupunki Tilakeskus	54) Nomaji maisema-arkkitehdit Oy
14) FCG Finnish Consulting group Oy	55) Nomaji maisema-arkkitehdit Oy
15) Fira Oy	56) Porin kaupunki / CIRCWASTE
16) Fira Oy	57) Puhas Oy
17) Gaia Consulting	58) Pääkaupunkiseudun Kierrätyskeskus Oy
18) Gaia Consulting	59) Pääkaupunkiseudun Kierrätyskeskus Oy
19) Gaia Consulting	60) Rakennustuoteteollisuus
20) Gaia Consulting	61) Ramboll
21) Gaia Consulting	62) Saint-Gobain Finland Oy
22) Green Building Council Finland	63) Saint-Gobain Finland Oy
23) Green Building Council Finland	64) Senaatti-kiinteistöt
24) Green Building Council Finland	65) Sitowise/ Maapörssi Oy
25) Green Building Council Finland	66) Sitowise/ Maapörssi Oy
26) Green Net Finland	67) Skanska Oy
27) Helsingin kaupunki	68) Skanska Oy
28) Helsingin kaupunki	69) Skanska Oy
29) Helsingin kaupunki	70) Skanska Oy
30) Helsingin kaupunki	71) Suomen Erityisjäte oy
31) Helsingin kaupunki	72) Suomen Erityisjäte oy
32) Helsingin kaupunki	73) Syke

33) Helsingin kaupunki	74) Sykli
34) Helsingin kaupunki	75) Sykli
35) Helsingin seurakuntayhtymä	76) Tulevaisuuden tutkimuskeskus / Turun yli-opisto
36) Helsingin kaupunki	77) Tulevaisuuden tutkimuskeskus / Turun yli-opisto
37) HSY	78) Valonia
38) HSY	79) Valonia
39) HSY	80) Vantaan kaupunki
40) IAH - arkkitehtuuritoimisto	81) VTT
41) IAH - arkkitehtuuritoimisto	82)

Sustainable Transportation in Metropolitan Cities; Berlin, Helsinki, New Delhi and Pune

S Rajesh[1], P Shashank[1], D Abhirup[1], A Tolu[1], D Zorro[1], A Zakarya[1], M Albo[1]

[1] Master students, ConREM, Metropolia UAS, Helsinki, Finland

Abstract. If one comprehends a city as a human body, then transportation would be its arteries and veins. It makes mobilization possible. In the age of rapid globalization transportation does not only concern itself with a single city but transportation deals with a global scale today. This paper focuses on a comparative analysis of transportation methods for four cities in terms of how new sustainable transportation concepts can be applied to make our future better. The discussion includes a vast range of possibilities of our future sustainable cities in transportation. Environmental and technological aspects as well as politics and policies are inspected. Cities flourish, develop and revolve around transportation. Our future sustainable cities would be no different. We need to approach transportation from all the three aspects of sustainability e.g. social, ecological, and environmental. We have chosen our cities for case studies on a broad range from low-density in Helsinki to very high population in Delhi. The rapid growth in transportation has remarkably increased the carbon footprints of our cities. While planning for our future cities making transportation sustainable is of paramount importance. To understand the issue of transportation for our sustainable cities we discuss the present state of transportation globally and analyse transportation in major cities such as Berlin, Delhi, Helsinki and Pune.

Keywords: Berlin, Helsinki, Delhi, Pune, Transportation Policy and Sustainable Transportation.

1. Introduction

A majority of the world's population is living in urban areas today. We already have a world urban population more than ever before. It is inevitable that transportation is going to grow more critical in our future sustainable cities. Looking at history we would find the major cities on the face of earth have developed around mobility routes. Urban transportation is the single most important component instrumental in shaping urban development and urban living. While urban areas may be viewed as engines of growth, urban transport is, figuratively and literally, the wheel of that engine. The test of urban governance depends upon the quality of life the city or town offers. Since transport is one of the prime determinants of quality of life, it is for the government to articulate the need for mobility and facilitate it through an appropriate mechanism. In fact, the efficiency of cities greatly depends on the development of transport systems; as urban transport is a catalyst for overall development. All these have very heavy economic, social and environmental costs. The need of the hour is therefore a sound urban transport policy. The major thrust of such an urban transport policy should include integrated planning, an optimum share between public and private modes, the choice of relevant technology for public transport systems, optimal use and management of available resources, restructuring of monetary and fiscal policy to encourage and promote public transport and establishment of institutional arrangements, at all levels of governance, particularly at the city level, for planning, development, operation, management and coordination of urban transport systems. Much of the confusion in these matters is due to a lack of professional expertise. [1]

Although policy measures that involve restraining the use of private cars and two-wheelers are likely to be unpopular, a gradualist approach of progressively introducing restraints on road use, while at the same time improving public transport, is more likely to lead to greater acceptance. It is believed that improved public transport and more efficient management of demand would help to combat the trend away from public transport vehicles towards greater use of personalized modes. The application of Transport System Management (TSM) strategy such as one-way systems, improvement of signals, traffic engineering improvement measures for road network, intersections, bus priority lanes and

IOP Conf. Series: Earth and Environmental Science **297** (2019) 012025 doi:10.1088/1755-1315/297/1/012025

suitable policies and development of intermediate passenger transport as a short-term measure should be introduced in all cities especially in metropolitan cities so that the existing road capacity and road user safety is increased. [1]

2. Berlin

Berlin is divided into three Fare zones A, B and C. Area A is based on a variety of aspects. They represent the area in the middle of the city. Area B represents the city and Area C boundaries representing the airport and the surrounding areas of Berlin. The main reason for this division is to know the ticket prices. There are also trains. For example, trains run continuously in the city center. Berlin is divided into three tariff zones: AB, BC und ABC. Tariff zone AB includes the urban area to the city boundary. Zone ABC additionally includes Berlin's surrounding area and Potsdam main railway station. [2]

Figure 1: Travel Zone Berlin, (Visit Berlin, 2018)

The Berlin S-Bahn is faster and easier to use, especially through the Ringbahn route, which is a great way to travel around the city in a circular and very fast pace. It consists of 15 lines and 166 stations. On weekdays, the S-Bahn starts running around 4.30 a.m. and stops service at 1.30 a.m. Depending on the time of day the trains run in five-, ten- or 20-minute intervals. On weekends, S-Bahn trains run 24 hours, at night in 30-minute intervals. The eastern line of the city is the central axis of the transport system which connects the stations Westkreuz and Ostkreuz. The most important are peak periods where the S-5, S-7 and S-75 lines operate in standard time periods of three minutes, making it easier to navigate throughout the city. This is where the most beautiful features such as stations Zoo and Alexanderplatz are especially worth seeing. There, the S-Bahn passes landmarks like the Victory Column, Central Station, Museum Island and HackescherMarkt. [2]

The link line between the north and south of the city is represented by the S2 and S1 which connects Oranienburg and Wannsee Bernau and Blankenfelde. The so-called Ringbahn runs around Berlin's city centre on a route of about 37 kilometres. It stops at 27 stations and needs approximately 60 minutes to circle around the city. Many stations of the Ringbahn have links to S-and U-Bahn lines running to the city centre or outskirts of Berlin. Important junctions of the Ringbahn are the stations Westkreuz, GesundbrunnenOstkreuz and Südkreuz. The Ringbahn consists of the S-Bahn lines S41 and S42: S41 moves clockwise, the S42 anticlockwise. The trains run throughout the day, during the rush hour in five-minute intervals, in the evening in ten-minute intervals.

The Underground (U-Bahn) and S-Bahn operate on weekends and holidays throughout the night, ensuring that late-night travellers are safely home. On regular days of the week, trains operate daily from 4 am to 1 am. In addition, buses continue to link stations throughout the night. During the day the train arrives at each station for 5 minutes. At night, the train arrives at any station every ten minutes. At weekends, trains operate nonstop throughout the day. The Berlin trams are very much available in East

Berlin. Consisting of 22 lines 9 of them operate 24 hours and the remainder only works at regular times. Tram and Metro lines run on different schedules. Furthermore, Metro tram lines operate during the night even on weekdays. The BVG Metro has nine lines (1 to 17) which are supplemented by 13 tram lines (12 to 68). The lines of the Metro are marked with an "M". During the day these lines run at least every ten minutes, and at night (from 0.30 a.m.) they run in 30-minute intervals. [2]

The very sophisticated system of buses in Berlin serve roads that are not covered by other means of transport. It also acts as a link between train stations. In Berlin, buses do not operate from one area to another within the city. For buses, it is the slowest public transport but it is a sufficient way to take you to almost any area in Berlin. The double-decker busses stop in each lane, while the express buses stop at the main stations only. You can visit many of Berlin's tourist attractions through a few bus routes, something every tourist should do. Most buses in Berlin run throughout the day nonstop. The distinguishing feature of these buses is the letter N located at the front of the bus. These buses operate at night where they connect the metro stations when the trains are not operating.

3. Delhi

Delhi is an ancient city situated on the banks of river Yamuna, founding its first mention in mythologies of Mahabharata, as the capital city of Pandavas then known as Indraprastha dating 3000 BC. It has an area of 1483 km^2 and has a radius of approximately 50 km. Delhi has one of the robust transport systems among Indian cities. Derived from the town plan of Architect Edward Lutyens, Delhi has radial road networks. Being an ancient Mughal capital, it has a rich cultural heritage to bestow. [3]

In terms of the transport system, there are several high points which the city feels proud of, but still it lacks on certain grounds which give scope for ample improvement in the future. Delhi transport includes the Delhi Metro (DMRC), the Delhi EMU ring rail network (Indian Railways), the buses, auto-rickshaws, taxis and cycle-rickshaws and battery-operated rickshaws. [3]

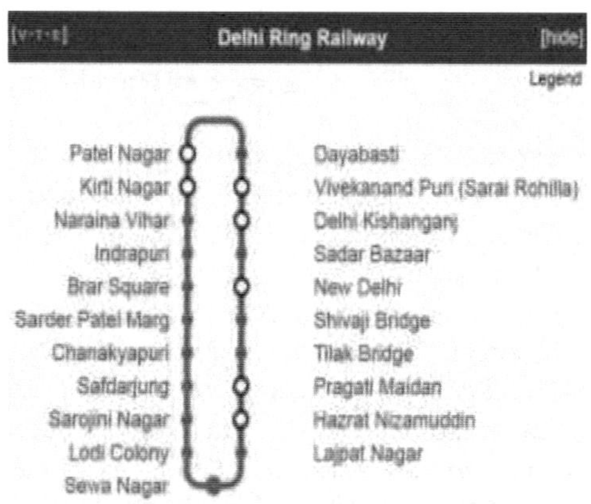

Figure 2: Delhi Ring Railway, (Delhi United, 2018)

It is said that "If you have a car, you are treated like a king. If you come by bus or metro, then you are a second-grade citizen." Despite Delhi's evolving transport system, the financially effluent classes are reluctant to use public transport like in other parts of the world. Although Delhi metro only accounts for 2.76 million riders per day, but this basically consists of people earning less than 1200 Euro per month (98% of the total ridership). Travelling by personal car or taxis gives people last mile connectivity which is often found missing with a metro ride. Harsh extreme climate, pollution worries and safety concerns also play a role in people opting for car rides. But lately, government as well as private institutions have been working hard to spread awareness regarding need for acceptance of public mode of transport system. Car free week was largely advertised in recent year as part of this campaign.

Real estate around metro stations has primarily two models of metro and real estate relationships. For instance, in Mumbai, metro stations were built first and extensive markets started to build up on both sides of the stations. The station became the focal point of any sub-region in the city. The influence is so large that for example areas got named as Malad (east) and Malad (west) owing to its direction on the either side of the rail station. And this phenomenon is seen in all stations of Mumbai city. On the other hand, city of Delhi got Metro rail installed way after the establishment of core areas be it retail or residential in nature. The last 15 years of metro engagements has greatly influenced the real estate market of the nodal areas. There has been a sharp increase in property prices in vicinity as well as notable distances of the metro stations. It has not been restricted to commercial or retail sector but also impacted the residential commodities. [3]

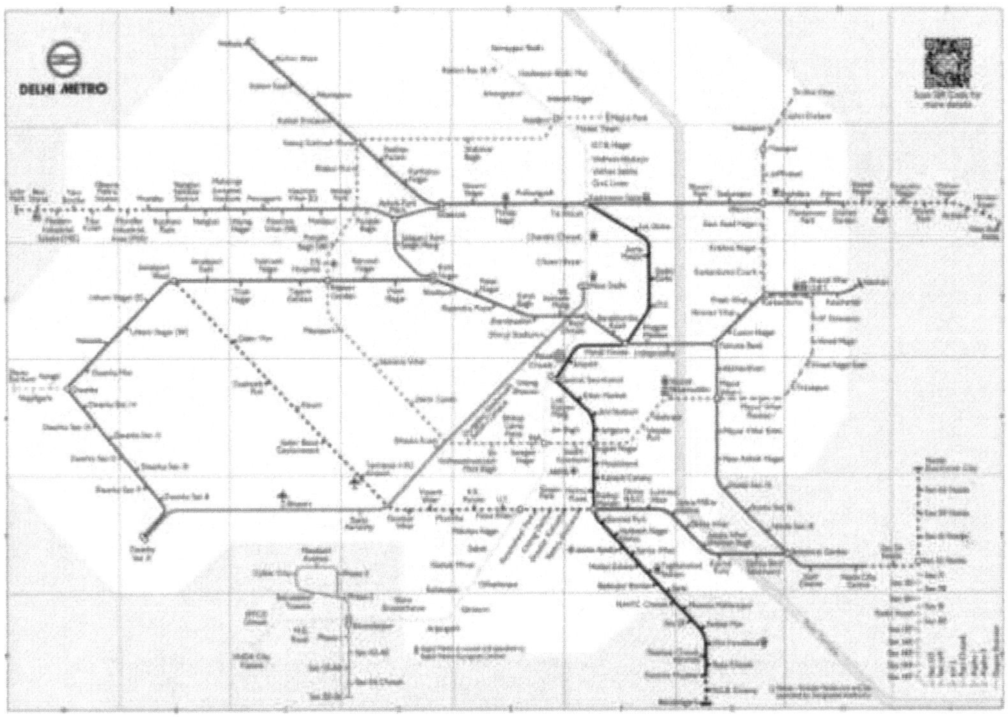

Figure 3: Delhi Ring Railway, (Delhi United, 2011)

Due to better connectivity, employment in all areas has thrived giving way to a demand for housing. This has seen positive developments in the residential sector. FAR (floor area ratio) also has been increased in transit corridors. This paved the way to new developments over the existing structures as well as vacant lands. Property prices in western areas of Dwarka almost doubled from 2009 to 2017. This kind of rise was seen in all areas of Delhi as well as its satellite towns. Due to better connectivity, employment in all areas has thrived giving way to a demand for housing. This has seen positive developments in the residential sector. FAR (floor area ratio) also has been increased in transit corridors. This paved the way to new developments over the existing structures as well as vacant lands. Property prices in western areas of Dwarka almost doubled from 2009 to 2017. This kind of rise was seen in all areas of Delhi as well as its satellite towns.

Ownership of land in Delhi is a mix of private and central government. For the development of infrastructure and transport, the government acquires land using the "Right to Fair Compensation and Transparency in Land Acquisition, Rehabilitation and Resettlement Act, 2013 (LARR)" which has recently replaced redundant “Land Acquisition Act of 1894.” In Delhi all auto rickshaws have been converted into CNG fuel-operated mode. This has helped in the reduction of PM10 emissions. Recently battery-powered rickshaws serve as feeder for last mile transport from metro stations. They contribute to low emissions operating with 250W battery which runs for 50 kms on charging it for 4-5 hours. Despite various public transport options, most of Delhi’s population still prefers to travel in their personal vehicles such as a car, taxi, auto-rickshaw and two wheelers. These roads are also instrumental in carrying freight on a daily basis.

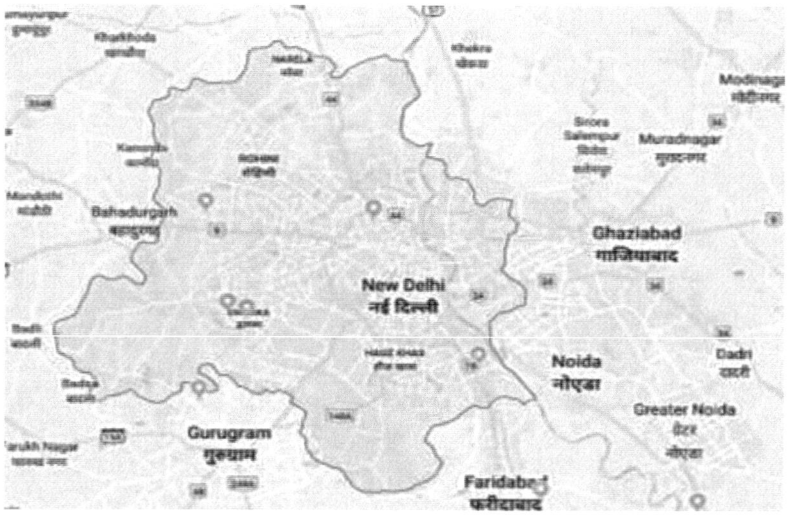

Figure 4: Delhi Major National Highways, (Google Maps, 2018)

The class system in railways are inherited from the British colonial mind-set. There have been several voices suggesting the unification of varied classes in the railways, but this will mean adopting a common rail fare that will be average of existing fare systems. This might not be necessarily acceptable to the masses especially the weaker sections.

S. No.	Class	Reservation	AC/Non - AC	Type
01	General class	Non-reserved	Non-AC	Sitting
02	Sleeper class	Reserved	Non-AC	Sleeper
03	3 tier AC	Reserved	AC	Sleeper
04	2 tier AC	Reserved	AC	Sleeper
05	1 tier AC	Reserved	AC	Sleeper
06	Chair car 01	Reserved	Non-AC	Sitting
07	Chair car 02	Reserved	AC	Sitting

Figure 5: Class System in Delhi

The Inter-State bus depots are in Maharana Pratap ISBT in Kashmiri Gate (North Delhi), Swami Vivekananada ISBT in Anand Vihar (East Delhi), Vir Hakikat Rai ISBT in Sarai Kale Khan (South/East Delhi) and Dwarka ISBT (West Delhi).

The Delhi airports are Indira Gandhi Airport, Indira Gandhi International Airport, Safdarjung Airport for government use and Palam Airport, used by the Indian Air Force.

IOP Conf. Series: Earth and Environmental Science **297** (2019) 012025 doi:10.1088/1755-1315/297/1/012025

Delhi's transport system has evolved over the years trying to meet up with its ever-increasing demographic structure. The Influx of population from other states burdens the transport system. Although there have been fewer success stories like Delhi Metro, still a lot needs to be done in Delhi Metro and especially so in other sectors of transportation. The roads of Delhi are primarily good only for use of cars. It has almost no consideration for pedestrians, bicycle users and even motor bike users. In most places there are no separate lanes for each and non-car users struggle to find their voices heard. Bicycles and motor cycles also run in highly vulnerable circumstances and often meet fatal accidents. The Delhi bus rapid transport system (BRT) inaugurated in 2008 ahead of 2010 commonwealth games failed to live up to its expectations. It was scrapped and dismantled in 2016 by the state government. The BRT failed due to poor implementation, people violated the lane system by entering into bus lanes thereby defeating the purpose of the system.

The flyovers built in last three decades have started to wear down and major faults are being observed in a few of the flyovers due to soil settlement and temperature variations. Pressure on Airport infrastructure has increasing ridership. Air ridership has seen a sharp increase due to cheap flight tickets and rising middle classes. However, the airport infrastructure has not been able to keep up with the same pace. Major airports and the state government are facing this challenge to upgrade the infrastructure as per recent demands of the consumers. Transport all over India and especially in Delhi finds the challenge to provide adequate safety for women. Although in Delhi metro condition has seen improvement, still teasing and groping are major challenges in buses and pedestrian areas. The Delhi government promised to install cameras in all buses and provide marshals but the concept is still in the pipeline even after 3 years. CNG vehicles and the Delhi metro helped in curbing city's pollution to a certain extent but due to ever increasing number of private vehicles pollution due to transport is increasing.

4. Helsinki

Helsinki is a compact, well-planned city and it grows due to internal and external immigration which cause its metropolitan area to extend rapidly; currently it covers 14 municipalities, spread over a total area of 3700 km2. With 1.8 million inhabitants, the main problem in transportation system is the low density of people. [6]

Helsinki Region Transport Authority (HSL) was found on 2010 and since that it takes responsibility of managing the transportation system and this include developing transport plans (short and long term). HSL service covers cities of: Helsinki, Espoo, and Vantaa, Kerava and Kauniainen and the municipalities Kirkkonummi, Sipoo, Tuusula and Siuntio. [6]

Figure 6: Helsinki Metropolitan Area, (HSL, 2016)

Integrity and good public transport connections are the main goals of HSL policy, so the various transport means operates to insure seamless journey. HSL transport management system based on hiring contractors to operate the various public transport means.

There are three types of tickets: single ticket, day ticket and travel card, you can issue or recharge your ticket via ticket machine, sms, mobile ticket or direct payment. No gates in ferry, metro or commuter train stations; that means the rider does not need to show ticket unless required by spot checkers, which may result in a fine. In buses the ticket must be shown to the driver. HSL provides a comprehensive IT solution that works on electronic devices to facilitate the journey. The trip planner tool enables the user to determine nearest point to reach public transport and guide him/her to the next means of transport, until the rider reaches the destination. [6]

Helsinki Metro consists of 25 stations (six stations lie in Espoo), six of these stations runs below ground, while others are at street level. All the stations aligned in two lines: M1 (Vuosaari – Matinkylä) and M2 (Mellunmäki – Tapiola). Tracks - of total length equal to 35 km) - run in two directions - east to west and vice versa. Construction of the metro started on 1969 and the first official service operation was in August 2, 1982 and metro trains ran from Itäkeskus to Hakaniemi. From that time until now the line extended to the east and west. The last extension was to the west towards Espoo and opened to the public in 2017, final route as shown the figure below. The figure shows the metro stations in Helsinki before the last extension in the orange colour and last extension in green colour and the future planed stations in light green colour. Importance of metro line comes as it is the main public transportation mean that crosses Helsinki from east to west. The fact that metro is the fastest means to transport inside Helsinki [3] metro has many other features that attracted more than 127 million passengers in 2015[4]. The most important feature is the connectivity of metro station to other transportation means, for instant there is bus station or at least bus stop at each metro station which enable passengers to continue their trips smoothly. Other features are that metro allows bicycles and pets in its trains and it offers continuous free Wi-Fi internet in the stations and in the trains. Metro starts first journey at 05:00 am and last one at 23:20 pm with headways varying from 2.5 min. to 10 min. [6]

Figure 7: Helsinki Metropolitan Route Map, (Yle, 2018)

Most of Helsinki streets have a bike lane, and there are several bike stations attached to metro stations and many other places, bikes also allowed in metro and commuter trains. HSL has its city bikes, which are available for common share during summer period and can be borrowed by residents or visitors via special ticketing system. These bikes are in 225 bike stations, 150 stations in Helsinki and 105 in Espoo. [4] HSL policy aims to increase the use of bikes and for a lifestyle for those who care about their health and travel time, also to benefit from bikes rides with neither carbon emissions nor noise. [8]

5. Pune

The city of Pune is the second most growing metropolis in the state of Maharashtra after Mumbai. The urban sprawl of the core city is situated on the Deccan plateau on the banks of river Mula and Mutha. It has an area of almost 243.84 square kilometres. This city has been the land of administration since its establishment in 18th century by the great Maratha warrior "Chhatrapati Shivaji Maharaj". The land has been the political centre and was ruled back then by the "Peshwas". So, the city has a strong influence of planning and administration over the last four centuries. However, with the increasing footprint is not able to cater the norms. Mumbai is the economic and financial capital of India situated 150 Km apart, this makes the city of Pune more commutable and accessible. The city is a major contributor to economy, which is western part of the state. It is one of the most populated cities of India, with over 8 million residents. The rapid modernization, setting up of international IT companies, migration of skilled labour from other states, world-class educational institutions and a good standard of living are among few factors, which make Pune a preferred city to live. The city is transcending from last decade to achieve this kind of world-class amenities and status. [1]

There is a proportionate growth of sectors in infrastructure, business and transportation. Hence, there is an increasing demand for public transport for the mobility of people, which is termed as inter-city and intra-city connection. However, the research areas for this report are only focused on intercity transport and would be focusing more on latest updates and changes in transportation policy in the city of Pune. This research focuses only on three public transportation sectors. They are as follows:

"Streets of Aundh, Pune" – An Initiative to promote cycle and pathways.
"Feasibility study of Hyperloop One" – Case Pune and Mumbai in 20 minutes"
"Pune Metro Rail – "Maha Metro"

We are already in the age of revolution of the transportation industry. One such example is Hyperloop One. Although the concept has reached only testing of wind tunnels the start-up which was founded by visionary "Mr. Elon Musk" has taken a sudden take off in the last two years. The company has gone from 3 people to 100 people working on the design and feasibility study. The company has made so far eight MOU (Memorandum of understanding with the specific country's Governments) for understanding the feasibility of the project. One of the routes on which it focuses is Pune – Mumbai in 20 minutes. This is a revolutionary idea in the field of transportation and if implemented for these two cities then it changes the concept of commute time for every transport. The feasibility study is being carried out by Indian railway ministry and the company.

The much-awaited Pune metro has begun its construction (excavation and foundation phase) from last two years, since its' beginning in 2000. The metro suits the city since the average travel time in city is more than 100 min per person. The proposal has 3 metro lines two in PMC area and one in PMRDA region. Total numbers of stations proposed are 53. The metro is being elevated at 23 stations and will go underground at 5 stations. Since land acquisition is a critical part for setup the metro got escalated for the city of Pune. The estimated date of completion for line 1 and line 2 is around the year 2021. The total area covered would be 54.58km. The PMC region is 31.25 km long while the PMRDA is 23.33 km long. The total estimate for the project is 845 million Euros. Table 3 explain the demand of metro for the city.[7]

The fare collection would be done on an automated system which would have a smart application. The estimated service of day is around 19 hours/day. The estimated travel speed would be 31-33 Km/h. The maps developed for the routes relate to the main stops and a comprehensive mobility plan has been developed for the same. According to surveys and urban experts the TOD (Transit oriented development area) would boost the real-estate sector since there are high chances of FSI (Floor space index) to be raised to 4.

CRITERIA	EXISTING SCENARIO	WITH METRO SCENARIO
1. Traffic Congestion	Bumper to Bumper Traffic In peak hours	No congestion of roads in peak hours
2. Pollution	Air quality is going down	Will improve as vehicles will start getting low in number
3. Road Accidents	High rate of accidents	Will drastically reduce due to less commute time
4. Convenience	Traffic jams	Highly comfortable and reliable
5.Weather Extremities	Affecting Drivers Health	Distraction free environment
6. Travel Time Cost	Excessive more than100 min average	Travel time reduction by 50 %
7. Energy Savings	High due to Fuel consumption	Use only 1/5th of the energy according to DPR

Table 1: Transportation Scenario in Pune, (Pune Metro Rail, 2018)

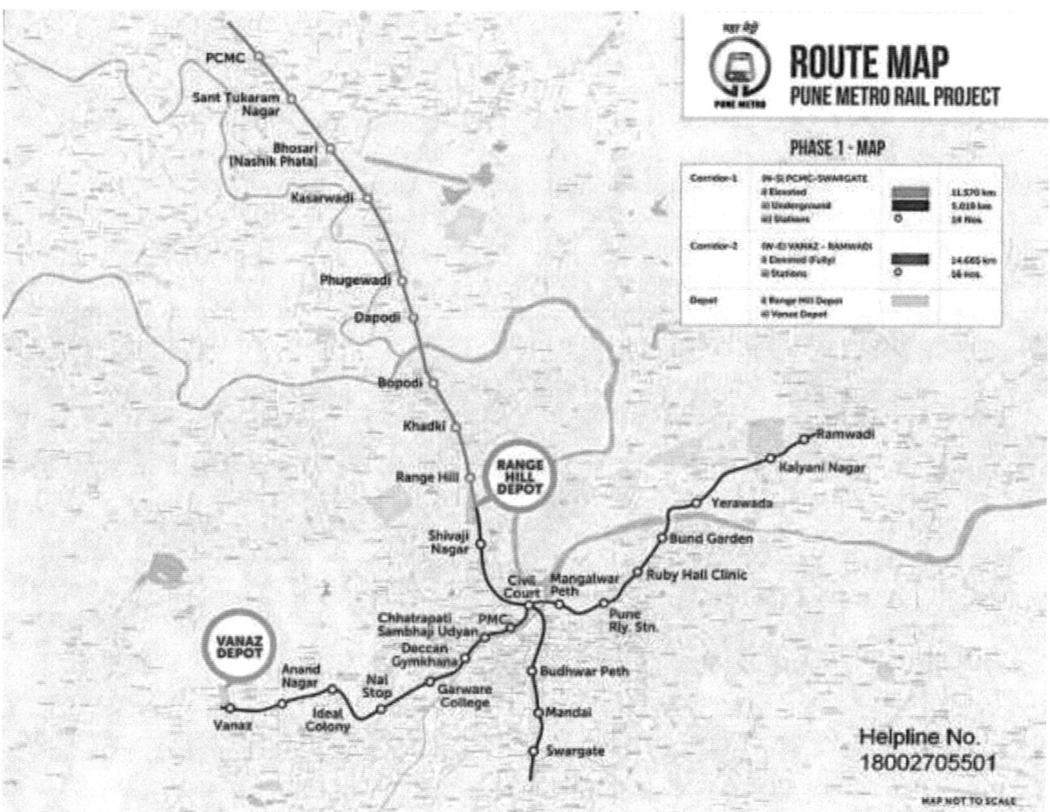

Figure 8: Pune Metro Map (Pune Metro Rail, 2018)

This figure is lucrative and a lot of debate is happening around the urban policy makers on what could be the solution. The metro is to be built on BOT (Build operate and transfer) basis, with a maintenance contract of 35 years. The success of the metro will be determined based on the users experience in terms of commute time and fair prices. A comprehensive report for mobility has been developed by the regional municipal authorities for efficient implementation of systems of transport. [7]

Another project which recently got appreciated related to street development in an urban setting is the streets of Aundh; Pune, Maharashtra, India. It was developed with the help of public participation and was designed by urban designer "Architect Prof. Prasanna Desai". The project aimed at developing a breather space on the heavily occupied traffic region for the residents. The pedestrian walkway stretches to 3M in length having cycle tracks of 1.5M. The street has an end to end connection, hence is used to full potential by residents. "The public transport plan is supplemented with an extensive bicycle network that is both interconnected and continuous that reinforces and feeds into the public transport corridors. To improve the mobility of the corridors and pave way for the public transport corridors an effective parking management plan that complements the corridor plan is identified. The parking plan in addition to freeing the right of way for the public and non-motorized transport would also act as a demand management tool. A strong and reformed institutional framework is identified to help achieve the mobility plan and the vision".

These individual plan elements are detailed in the subsequent sections". (Pune Municipal Corporation, 2018). Additionally, there has been initiative of strongly promoting bicycling in city. An association with a car rental company "Zoom car" and PEDL have made it possible to rent bicycles on hourly basis. There are cycling hot spots set around the areas of city. It's just 15 cents per hour thus by promoting the cycling experience in city. [8]

Figure 9: Cycle and Pedestrian Pathways (ITDP, 2016)

The increase of the population in the 20th century past to 220 million to 2.8 billion. According to the UN statistic, by 2050 6.9 billion people will be living in the urban areas, almost the 70% of the global population. By the hand of this increase of population, the number travels inside the cities also increase. The transportation systems in all urban cities must grow in a sustainable and responsible way. But, for the transportation system work efficiently, other actors must be involved. The planning of the cities, the policies, and the new technology must work together.

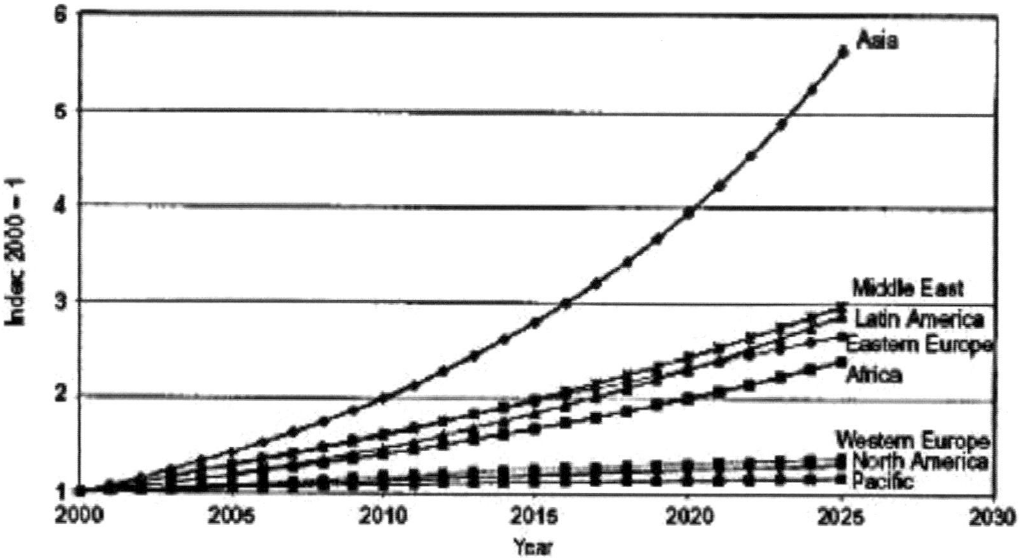

Figure 10: Estimated changes in road traffic levels relative to year 2000, (Dargay, 2002)

Asia will be the continent with the major increase in the road traffic levels and it is also the one who has 1/3 of the global population. Nevertheless, the increase will be considered around the world and its one of the problems that every citizens suffer every day. Six cases or stages were presented to a city to be considered friendly to the environment based on the United Nations Reports of Sustainable Development:

1. Reducing greenhouse gas emissions
2. Improving air quality / reducing noise
3. Improving transport safety
4. improving Access / Reducing exclusion
5. Reducing Congestion
6. Crafting wealth / supporting the economy

6. Comparison Delhi vs. Berlin and Pune vs. Helsinki

The population of Delhi has 4.3 times the population and 2.75 more density than that of Berlin. In the airport Cargo Tonnage, Delhi has18 times more than Berlin. In trains, both cities have almost the same number of stations; Berlin has more lines (15 to 6) and the system length is longer (327 km to 213). Both have the same number of metro stations, almost same lines, but in length, Delhi has 80 km more than Berlin. Only Berlin has a tram. Delhi has more trips per day than Berlin.

The population of Pune has 3 times the population and 2.4 more density than that of Helsinki. At the airports, the passenger movement in Helsinki is 2.65 times faster than in Pune. In Cargo Tonnage it is a big contrast, Helsinki has 5.44 times more than Pune and 3.52 more aircraft movements. In trains, Helsinki has 15 lines and Pune only 2; Helsinki has 70 stations and Pune only 17. Daily ridership in Helsinki move 2 times faster than in Pune; Helsinki has Metro and Tram systems. The Pune Bus rapid transit has 6 lines, but unfortunately, the project has been a disaster.

	Delhi - India	Berlin - Germany
Population	26.454.000	6.000.000
Area (km2)	1.483	892
Density (hab/km2)	11.312	4.100
Airport		
Passenger Movements	35 millions	29.5 millions
Aircraft Movements	397.799	254.952
Cargo Tonnage	857.419	48.000
Train Sub-Urban		
Number Lines	6	15
Number Station	160	166
System Length (km)	213	327
Daily ridership	3.000.000	1.142.465
Metro		
Lines	7	10
Length (km)	231	146
Number of station	173	173
Daily Ridership	2.760.000	1.389.041
Tram		
Lines	-	22
Length	-	189
Number of station	-	404
Daily Ridership	-	495.890
Bus Service		
Bus Rapid Transit	No	No
Public and Private Bus	Urban and InterCitys	Urban and InterCitys
Taxi Services		
Taxi	Yes	Yes
Three Wheeler Motorcycle	Most Popular transport	No
Uber and Apps	Yes	Yes

	Pune - India	Helsinki - Finland
Population	4.600.000	1.500.000
Area (km2)	1.109	3.697
Density (hab/km2)	7.214	3.003
Airport		
Passenger Movements	6.787.391	18.000.000
Aircraft Movements	46.932	165.330
Cargo Tonnage	35.312	192.204
Train Sub-Urban		
Number Lines	2	15
Number Station	17	70
System Length (km)	63	99
Daily ridership	100.000	194.500
Metro		
Lines	-	2
Length (km)	-	35
Number of station	-	25
Daily Ridership	-	201.400
Tram		
Lines	-	11
Length	-	39
Number of station	-	293
Daily Ridership	-	156.164
Bus Service		
Bus Rapid Transit	6 Lines	No
Public and Private Bus	Urban and InterCitys	Urban and InterCitys
Taxi Services		
Taxi	Yes	Yes
Three Wheeler Motorcycle	Yes	Yes
Uber and Apps	Yes	No

Table 2: Data Delhi vs. Berlin and Data Pune vs. Helsinki

7. Conclusions

A modern city runs on its mobility. In today's world, railways are essential in the daily life of citizens. Metro is one of those basic infrastructures on which common people heavily depend. The Delhi Metro is the lifeline of the citizens. The basic purpose of public infrastructure like Metro is to serve the public a safe and convenient transport. The policy makers must not forget that Delhi metro is also keeping the pollution and the traffic of the city in control. The management of transportation systems is of high importance in cities like Delhi and Berlin. In smaller cities like Helsinki, a nice modular mix of transportation without dedicated lanes for pedestrians and bicycles works. In Delhi, auto-rickshaws are widely used. The accident rates are also high among people travelling in these transportation modes where the traffic density is not accommodated at all by the road lanes. There is a need for remarkable improvement and proper integration of mixed transportation in Delhi. The Delhi Metro is one of the few good examples the city provides for future sustainable cities.

The Pedestrian safety should be given priority in the sustainable development of transportation. In older cities pedestrian lanes can be narrow. On the other hand, for the making of future sustainable cities the transportation planning should have emphasis on pedestrian mobility. No matter how much technology develops in transportation systems pedestrians will always be prime in sustainability of future cities.

Berlin is an economic and transportation centre in Europe. Berlin transportation has witnessed change from old slow cars to modern fashioned high-speed commuter trains. Compared to the increasing population Berlin is coping well in its transportation services. The development of technology and sensible implementation is critical for sustainable transportation. The aspects of renewable energy and tech- savvy transportation systems such as hyperloop should be considered carefully for our future sustainable cities.

Although policy measures that involve restraining the use of private cars and two-wheelers are likely to be unpopular, a gradualist approach of progressively introducing restraints on road use, while at the same time improving public transport, is more likely to lead to greater acceptance. It is believed that improved public transport and more efficient management of demand would help to combat the trend away from public transport vehicles towards greater use of personalized modes. The central government should assist local governments for effective implementation of such measures. In fact, there is pressing need to strengthen institutions in the transport sector. Central government should provide training and technical assistance to local governments to prepare and implement sound policies and programs. The implementation methods, proper monitoring is equally important as the policy making.

Transportation development largely depends on policy and budgeting. There should be proper structure of transportation budget which is essential for structural development of sustainable cities. In a country like India, where policy making depends and changes on electoral polling, construction and implementation of transportation policy suffer greatly from the shift of power among political parties. Also, the industry, housing, and jobs will increase. To reduce the pollution and the CO2 in these cities, the government is planning to invest in green technology for the public transportation. For example, Helsinki is planning to change the diesel of the buses for new biodiesel, gas and electric. A more radical solution is Berlin that is planning to get free the public transportation for the population. According to the United Nations, 66% of the global population will live in cities by 2050 (UN, 2014). It is important to create urban plans that are "green cities". The new ideas, new technologies must be implemented and grow "shoulder to shoulder" with the cities.

The infrastructure of the cities must change to improve the public transportation and break the walls and the belief that the public transportation is for "Poor" people. The private car will never disappear, but with political laws, the used of them can be reduced by investing in public transportation.

Helsinki is a well-planned city and it has strong transportation infrastructure represented by the streets, highways, foot paths, bike lanes, metro, and commuter trains. All these means form together a solid integrated mobility network that gives the residents a wide range of travel options and this lead to more reliability on public transport system. Helsinki has developed short-term and long-term strategic planning for the present and future aims mainly to increase the sustainable journeys. Some goals are to maximize the cycling, offer car parking close to metro and train stations, utilize IT solution to facilitate journeys via trip-planner tool, use e-tickets and mobile apps.

Urban transport policies should encourage the need for developing 'green' modes like bicycling and walking with pedestrian paths and cycle tracks especially in new development areas of larger cities and small and medium towns which should be integrated with the transport network. The application of Transport System Management (TSM) strategy such as one-way systems, improvement of signals, traffic engineering improvement for road network, intersections, bus priority lanes and suitable policies and development of intermediate passenger transport as a short-term measure should be introduced in all cities especially in metropolitan cities so that the existing road capacity and road user safety is increased.

Acknowledgments
Authors wishing to acknowledge assistance or encouragement from the Professor Eric Pollock, and our classmates for their cooperation in the development of this paper which is based on the book *Sustainable Cities* developed for the Master´s Programme in Construction and Real Estate Management (ConREM).

References

[1] Urbanisation and Urban transport in India: the sketch for a policy, S Padam and S Singh, www.deas.harvard.edu/TransportAsia/workshop
[2] Travel Zone Berlin, https://www.berlin-welcomecard.de/en
[3] https://www.quora.com/What are some lesser known interesting facts about Delhi Metro
[4] Touru, T. (2016). Helsinki region transport system. Urban node concluding workshop (p. 26).Helsinki: hsl.
[5] HSL (2018, N.D. N.D.). City Bikes. Retrieved from HSL: https://kaupunkipyorat.hsl.fi/en
[6] HSL (2018, N.D. N.D.). Commuter Train Services. Retrieved from HSL: https://www.hsl.fi/sites/default/files/uploads/hsl_lahiliikennelinjat_pysty_700x1000_2017_v04_esikatselu_0.pdf
[7] Pune Municipal Corporation. (2018, 03 27). PMC. Retrieved from www.pmc.gov.in: https://pmc.gov.in/en/comprehensive-mobility-plan
[8] The Times of India. (2016, December 8). Pune metro rail project gets go-ahead from Centre. Retrieved from Pune metro rail project gets go-ahead from Centre: https://timesofindia.indiatimes.com/city/pune/Pune-metro-rail-project-gets-go-ahead-from-Centre/articleshow/55865879.cms
[9] Shivatare, C. D. (2017, October N.D.). Pune Traffic Problems & Control Measures. Editor in chief Dr. G.R. Kulkarni: http://www.ejournal.aessangli.in/ASEEJournals/CIVIL86.pdf

IOP Conf. Series: Earth and Environmental Science **297** (2019) 012026 doi:10.1088/1755-1315/297/1/012026

Optimum tank size for a rainwater harvesting system: Case study for Northern Cyprus

Mustafa Ruso(1), Bertuğ Akıntuğ(2), Elçin Kentel(3)

(1) Sustainable Environment and Energy Systems Graduate Program, Middle East Technical University, Northern Cyprus Campus, Kalkanli, Guzelyurt, Mersin 10, Turkey.
(2) Center for Sustainability, Civil Engineering Program, Middle East Technical University, Northern Cyprus Campus, Kalkanli, Guzelyurt, Mersin 10, Turkey.
(3) Civil Engineering Department, Middle East Technical University, Ankara, Turkey.

bertug@metu.edu.tr

Abstract. The available freshwater is limited on earth. On the other hand, available water resources on earth have been depleting and being polluted due to climate change and population growth. In order to reduce the risk of water scarcity and water resources contamination, Integrated water resources management (IWRM) is required. IWRM is a concept to manage water resources that aims to balance economic efficiency, social equity, and environmental sustainability. When rainwater harvesting systems (RWHS), one of the techniques of IWRM, are implemented, the stress on water resources is reduced. Since the installation cost of rainwater harvesting systems significantly depends on the size of the rainwater storage tanks, in the implementation of rainwater harvesting, the selection of tank size is one of the main concerns for the feasibility of the system. This study aims to investigate the feasibility of domestic rainwater harvesting systems for a single house. In order to find the optimum storage tank size of the rainwater harvesting system, a linear programming (LP) optimization model is employed. As a case study, the LP model is applied to six regions from semi-arid Eastern Mediterranean island Northern Cyprus, where water resources are limited. The model considers thirty-seven years monthly rainfall data, the roof area of the building, the water consumption per capita, the discount rate, the cost of the rainwater storage tank, and the number of residents. The results of the selected study areas show that the implementation of the RWHS for a single house is infeasible due to the substantial installation costs and maintenance expenses. The financial losses caused by the implementation of the RWHS are found higher than the installation costs and maintenance expenses for all regions. In addition to economic analyses, environmental benefits of the RWHS should be included into the feasibility analysis.

1. Introduction

Population growth, rapid urbanization and industrial developments around the world have resulted in increased water demand in municipal, industrial, and irrigational use [1]. Climate change further increases uncertainty and adds stress on existing water resources. As a result, in some regions, the quantity decreases and the quality of water deteriorates [1]. More frequent extreme events and temperature and precipitation variations in seasons are being faced and they boost the tension on limited water resources. At this point, integrated water resources management (IWRM), which will improve and accelerate the adaptation of water resources to climate change, has become an attractive concept to develop, manage, and distribute water resources equitably without disturbing ecosystems [2]. Furthermore, IWRM is a framework that aims to promote countries concerning economy and social wellbeing, which helps countries to achieve sustainable development goals defined by the United Nations.

Rainwater Harvesting Systems (RWHS), one of the actions of implementation of IWRM and sustainable water resources management, collect rainwater in storage tanks for future consumption such as non-potable household use (washing clothes and dishes, toilet flushing) [3], agricultural irrigation [4], and potable use [5,6]. The fact that the quality of harvested rainwater is expected to be higher than the quality of the surface water and groundwater results in the low treatment cost of harvested rainwater to be used for the potable purpose [7]. Harvested rainwater may decrease storm water runoff and prevent watershed contamination in addition to its multipurpose use and relatively better quality [4]. Rainwater may be stored during high precipitation seasons and supplied to the regions during dry periods.

The traditional rooftop RWHS consist of a roof catchment to receive rainwater, a filter to prevent sediments in harvested rainwater from getting into the system, pipelines to carry rainwater from the roof catchment to the storage tank, and a storage tank called cistern to store harvested rainwater. Depending on RWHS design, a water pump can be necessary to transport harvested water from storage tanks to the desired place. Among these components, the storage tank size is the critical variable that affects the net financial benefit gained from RWHS. Previous studies have analyzed the optimum size of the storage tank for rainwater harvesting based on RWHS design, feasibility analyses, environmental considerations and water conservation [3,6,8,9,10,11,12]. [6] investigated the optimal tank size using a computer program named Neptune to assess potable water savings for nine cities in Brazil. [8] evaluated the cost-effectiveness of RWHS for a single house and apartment building using the Plugrisost software. In a recent study, [9] developed an analytical model to determine the optimal tank size by minimizing the total cost without explicitly considering RWHS' maintenance and electricity costs. In other study, [10] developed a generalized curve to select the optimal tank size for the Greater Melbourne region, Australia. [11] examined four RWHS located in humid North Carolina, USA and underlined that the implementation of RWHS in humid regions is not reasonable. Therefore, rainwater harvesting studies have been conducted for different regions of the world. In line with these studies, [3] proposed an optimization model for Girne, Northern Cyprus that uses a linear programming approach to select the optimal storage tank size of a single house for domestic rainwater harvesting. The model was built in Excel Solver, and the optimum storage tank size was calculated in OpenSolver Software by minimizing freshwater purchase costs [3]. [12] explored the whole-life cost performance of rooftop domestic RWHS at the single-building scale for 16 different storage tank sizes and noted that the rooftop RWHS generally ended up with financial losses because of their high capital cost and maintenance expenses. Most of the time, studies focused on the design, water savings, and environmental aspects of RWHS. However, some variables like electricity and maintenance costs that affects the cost-effectiveness of RWHS were ignored.

In this paper, our objective is to determine the optimum storage tank size for rainwater harvesting to satisfy the domestic water need of a single house and carry out the feasibility analysis of RWHS. As a case study, six different regions in Northern Cyprus, a water-poor country due to the low average annual precipitation regime, are selected. Optimization model developed by [3] is modified according to the new water tariff scheme in the country, new costs, and water consumption values. Our study differs from [3,9,10] such that the maintenance cost of RWHS and electricity cost of pumping, which affect the feasibility of RWHS, are included into the optimization model in addition to the capital cost of RWHS.

In Section 2, the linear programming model is provided. Then, the study area and input data are explained in Section 3. In Section 4, results are demonstrated, and finally, conclusions are provided in Section 5.

2. Linear programming model

The mathematical formulation of the optimum storage tank size determination for the rainwater harvesting problem is developed in this section. It is assumed that harvested rainwater is used to meet the domestic need of a single house. The problem is modeled as a linear programming (LP) problem. Whenever the amount of harvested rainwater is not enough to satisfy the domestic water need, necessary additional water is provided from the utility network. Thus, our goal is to minimize the cost of water supplied from the utility network and the costs of building a RWHS.

The costs of building a RWHS consist of fixed costs and a variable cost. Fixed costs comprise the installation cost and the maintenance cost. The installation cost includes the mounting of a storage tank, two filters, pipelines, and valves. The maintenance cost contains the overhauling cost of the whole system during the useful lifetime of the rainwater storage tank. Since the size of the rainwater storage tank is determined after the LP is solved, the cost of the rainwater storage tank is named as the variable cost. An additional freshwater pump is used in many residential units because freshwater from the utility network has low pressure in Northern Cyprus. Therefore, it is assumed that existing freshwater pumps in residential units can be used to transfer both harvested rainwater and water from the network to the single house. Harvested rainwater is transferred from the rainwater storage tank to the main water tank by the gravitational force. The cost of electricity to transfer harvested rainwater from the rainwater storage tank to the residential unit is neglected because the same cost occurs when water is supplied from the utility network. Figure 1 illustrates the plan of a RWHS for Northern Cyprus in details.

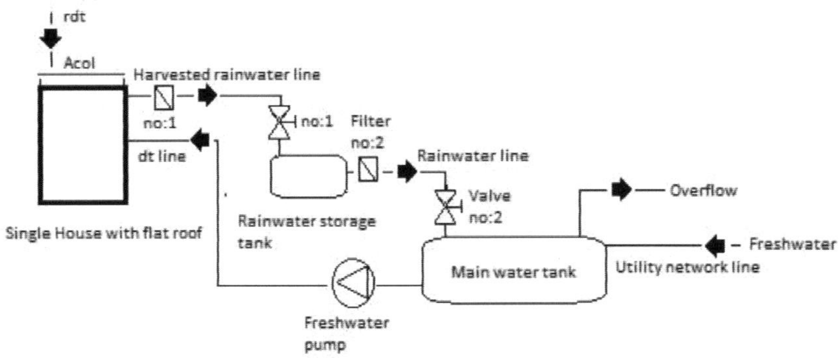

Figure 1. Diagram of a rooftop RWHS for Northern Cyprus.

An increasing block rate (IBR) tariff scheme is used for the water supplied by the utility network to maximize water saving in Northern Cyprus. When the volume of water purchased from the utility network increases, the cost of water per m^3 also increases in the IBR tariff scheme. The cost of water supplied from the utility network can be calculated as follows [3]:

$$CPt = \sum_{j=1}^{k-1} b_{tj}\left(Vj - V_{j-1}\right) + b_{tk}\left(V' - V_{k-1}\right) \tag{1}$$

where k is the highest price level, b_{tj} is the unit cost of water supplied from the utility network in period t at the price level j, J refers to the number of price levels, V_j is the maximum cumulative volume of freshwater that can be supplied at the jth price level, $V = 0$, and V' is water purchased from the utility network at any period t with k. Since there is a direct proportion between the volume of water purchased from the utility network and the unit price of water, $b_{t1} < b_{t2} < \cdots < b_{tj}$ for any period t. For more details see [3]. In addition to the IBR tariff scheme, for each m^3 of water supplied by the utility network a wastewater cost (w_c) is charged by municipalities in Northern Cyprus to support wastewater collection system. The volume of rainwater that can be harvested in period t can be calculated as follows [13]:

$$r_t = c_f \cdot A_{col} \cdot rd_t \cdot 10^{-3} \tag{2}$$

where c_f is the dimensionless runoff coefficient, A_{col} is the area of the rooftop in m^2, and rd_t is the measured rainfall depth in mm in period t (see Figure 1). The monthly domestic household water demand d_t can be calculated as follows [3]:

$$d_t = W_d \cdot n \cdot Nt \tag{3}$$

where W_d is the water usage in $m^3/day/capita, n$ is the number of residents, and N_t is the number of days in month t.

By minimizing the objective function, the LP model is designed to find the optimum rainwater storage tank size to be build (T_{cap}), and decision variables are the amount of water purchased from the utility network at the jth price level in period t (P_{tj}), the amount of rainfall stored by the rainwater tank in period t (R_t), the amount of water used from the rainwater tank to satisfy demand in period t (U_t), and the inventory level in the rainwater storage tank at the end of the period t (I_t). LP model is designed for monthly time steps, ranging from $t = 1$ to the end of the useful lifetime of the rainwater storage tank (τ). The mathematical model used in this study is a modified version of [3]:

$$Z = a \cdot T_{cap} + \sum_{t=1}^{\tau} \sum_{j=1}^{J} (b_{tj} \cdot P_{tj} + (Ptj - U_t) \cdot w_c)(1+i)^{-t} \tag{4}$$

Subject to

$$I_t = I_{t-1} + R_t - U_t \qquad \forall 1 \le t \le \tau \tag{5}$$

$$R_t \le r_t \quad \forall 1 \le t \le \tau \tag{6}$$

$$I_t \le T_{cap} \quad \forall 1 \le t \le \tau \tag{7}$$

$$R_t \le T_{cap} \quad \forall 1 \le t \le \tau \tag{8}$$

$$T_{cap} \le s_{max} \tag{9}$$

$$U_t + \sum_{j=1}^{J} P_{tj} = d_t \quad \forall 1 \le t \le \tau \tag{10}$$

$$P_{tj} \le V_j - V_{j-1} \qquad \forall 1 \le t \le \tau,\ 1 \le j \le J \tag{11}$$

$$I_O = 0 \tag{12}$$

$$T_{cap}, R_t, U_t, P_{tj}, I_t \ge 0 \qquad \forall 1 \le t \le \tau \tag{13}$$

where Z is the objective function in Turkish Lira (TL), a is the unit cost of the rainwater storage tank in TL/m^3, T_{cap} is the optimum size of the rainwater water storage tank in m^3, τ is the useful lifetime of the rainwater storage tank in months, J is the total number of price levels, b_{tj} is the cost of water per m^3 purchased from the utility network in TL in period t in jth price level, P_{tj} is the volume of water purchased from the utility network in TL in period t in jth price level, U_t is the volume of water that is used from the rainwater storage tank in m^3 in period t, w_c is the wastewater cost in TL/m^3, and i is the monthly discount rate, I_t is the inventory level in the rainwater storage tank in m^3 at the end of period t, R_t is the collected water in the rainwater storage tank in m^3 in period t, r_t is the maximum harvested rainwater in m^3 for period t, s_{max} is the maximum allowable rainwater tank size to build in m^3, d_t is the domestic water demand in m^3 in period t, V_j is the maximum volume of water that can be provided from the utility network in jth price level in m^3.

The monthly discount rate (i) is used in the objective function (Equation (4)) to compute the present value of costs because all costs that will occur during the estimated lifetime of the rainwater storage tank should be in the same monetary value. The objective function (4) is the sum of the cost of the rainwater storage tank built, the total cost of water purchased from the utility network, and the total wastewater cost during the useful lifetime of the rainwater storage tank. Starting from period $t = 1$, the water level in the rainwater storage tank, also referred to the inventory level at the end of period t (I_t) is determined using the continuity equation as given in Equation (5). Constraints (6) guarantee that the amount of harvested rainwater cannot be higher than the amount of rainfall for the same period. Constraints (7) ensure that the amount of harvested rainwater saved in the rainwater storage tank cannot be higher than the volume of the rainwater storage tank in any period. Constraints (8) state that the amount of harvested rainwater in the storage tank cannot be higher than the volume of the rainwater storage tank in any period. Constraint (9) guarantees that the volume of the rainwater storage tank cannot be higher than the allowable volume of water tanks for non-commercial use due to the local space restrictions in Northern Cyprus. Constraint (10) emphasize that water purchased from the utility network or rainwater supplied from the rainwater storage tank in a period should satisfy the domestic water demand in that period. Constraint (11) ensure that in any period, water purchased from the utility network at the jth price level cannot exceed the permissible volume of water that can be purchased at that price level. Constraint (12) states that the inventory balance level is zero at the beginning because the rainwater storage tank is assumed to be empty at the simulation period. Finally, constraints (13) express that all decision variables are positive.

We used the open source optimization software for Microsoft Excel named OpenSolver Ver. 2.9.4. to solve the LP problem. OpenSolver is a user-friendly software that can quickly solve large linear and integer problems. It solves instantenously to optimize one instance of the model which has up to 2220 variables.

After obtaining the optimum rainwater storage tank size (T_{cap}) and the objective function (Z), fixed costs, which are the installation costs (f_{ini}) and the maintenance costs (m_c), are added to the optimum objective function value to calculate the total discounted cost of the RWHS. If $T_{cap} > 0$, the net financial benefit of building the RWHS is calculated by subtracting the total discounted cost of the RWHS from the total discounted utility network cost without the RWHS (C_{UN}). The equation of C_{UN} is calculated using:

$$C_{UN} = \sum_{t=1}^{\tau}\sum_{j=1}^{J}(b_{tj} \cdot P_{tj} + d_t \cdot w_c)(1+i)^{-t} \tag{14}$$

If the net financial benefit ($NFB = C_{UN} - Z - f_{ini} - m_c > 0$) is greater than zero ($NFB > 0$), building the RWHS is considered to be feasible for the selected area; otherwise, it is not.

3. Study areas and input data

LP model has been applied to a number of locations across semi-arid Northern Cyprus located in the Eastern Mediterranean. The average rainfall per month in Northern Cyprus between 1978 and 2009 is calculated as 30 mm using data collected from 19 precipitation stations. Such low precipitation causes extreme water scarcity in the country. Until 2016, the country met its water demand from its limited groundwater resources and insufficient surface water. Currently, water is transferred from the maritime neighbour Anamur, Turkey to the country via pipelines constructed 250 m below the surface of the Mediterranean Sea through a total distance of 80 km [14]. The pipeline system was put into service in 2016, as a result of that, the price of water per m^3 provided from the utility network has slightly increased in Northern Cyprus. Water transferred from Turkey to Northern Cyprus initially enters to a pump station. Then, the water is pumped to Geçitköy Dam in Northern Cyprus. From this reservoir, amount necessary to supply the demand is transported to the treatment plant and then to the network. Electricity is consumed at each step of this procedure. Thus, the unit price of water in Northern Cyprus is directly proportional to the energy price.

In this study, Northern Cyprus has been divided into six regions namely Alevkayası, Dipkarpaz, Girne, Güzelyurt, Lekoşa, and Mağusa (see Figure 2). These areas represent various hydraulic characteristics of Northern Cyprus. Figure 3 illustrates the monthly rainfall data measured in these six regions between 1978 and 2015. Thirty-seven years monthly rainfall data is obtained from precipitation stations located in the selected regions, and used as input for the linear programming model. Box-plots of precipitation for these six regions is given in Figure 3.

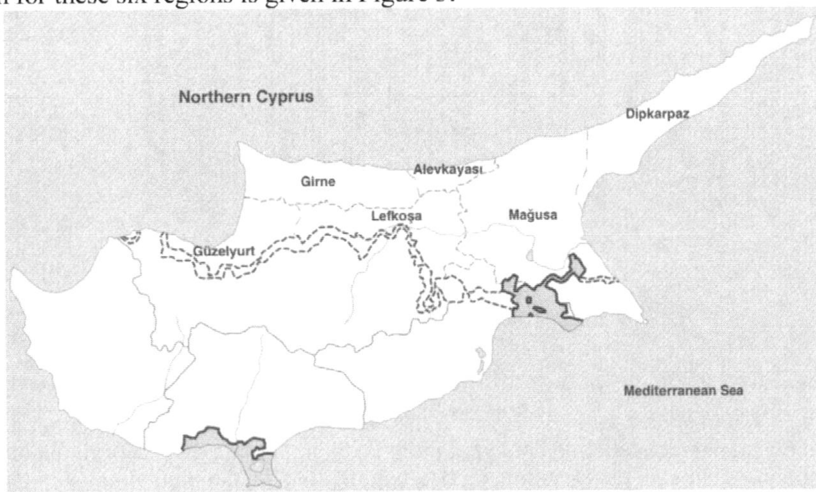

Figure 2. Study areas in Northern Cyprus.

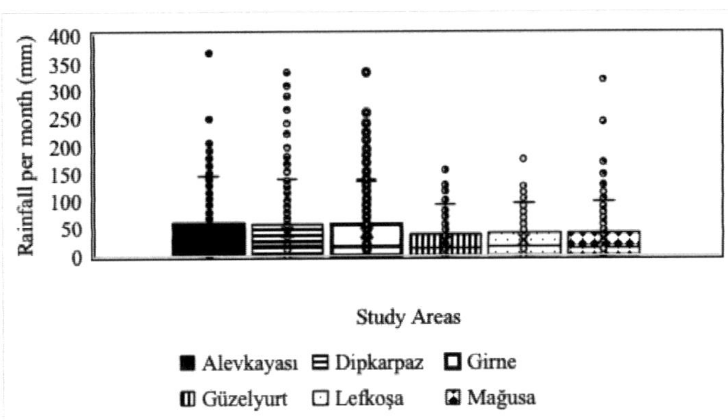

Figure 3. Monthly rainfall data in the study areas measured between 1978 and 2015 (Data provided from the Meteorological Office of Northern Cyprus).

Municipalities in Northern Cyprus adopted water volume sensitive utility network tariff scheme to protect the country's water resources from overuse. The utility network tariff scheme of the study areas for 2017 is illustrated in Figure 4 (Data is retrieved from the local newspaper, Havadis, 2017). Moreover, for each m^3 of water purchased from the utility network a wastewater charge of 1 TL/m^3 is applied. Therefore, implementing RWHS will decrease wastewater cost as well, which will provide an additional financial benefit for the consumers.

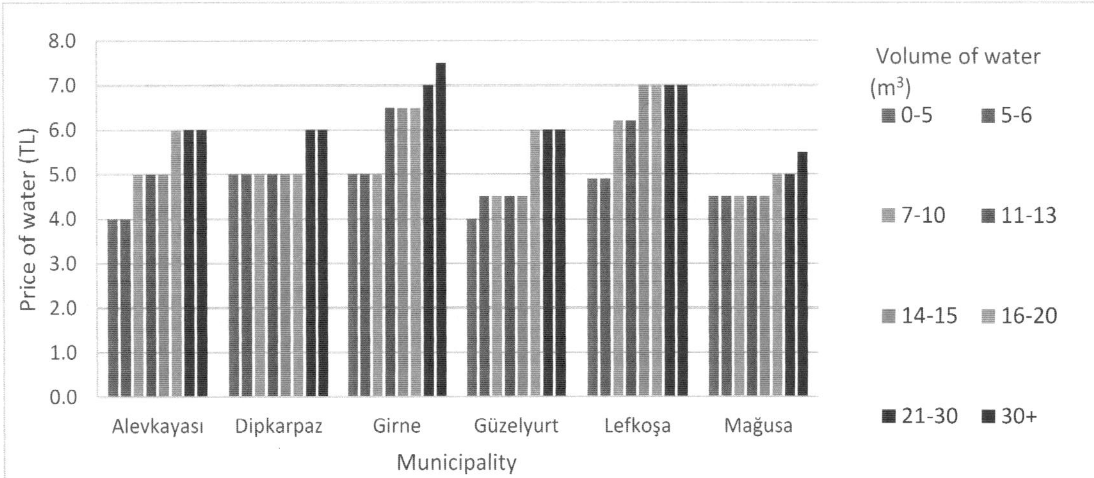

Figure 4. Utility network tariff scheme in the study areas.

In this study, we assume the number of residents (n) living in a single house is four. The flat roof area (A_{col}) of single houses is taken as 200 m^2. The equivalent runoff coefficient (c_f) for a flat roof is selected as 0.90 [15]. The installation cost (f_{ini}), the fixed monthly maintenance cost (m_t), the cost of the storage tank per m^3 (a), and the lifetime of the rainwater storage tank (τ) are taken as 1500 TL, 8.5 TL, 600 TL, and 444 months, respectively, suggested by a local supplier in Northern Cyprus. The fixed monthly maintenance cost occurs throughout the simulation period which is taken as 37 years for this study. Therefore, the present value of 37 years maintenance cost is estimated as 1514 TL using a monthly discount rate (i) of 0.05 % [16]. The maximum storage tank size (s_{max}) is limited up to 20 m^3 because of the local restrictions in the study areas. The average daily water consumption per capita (W_d) in Northern Cyprus is taken as 0.25 m^3 [17].

4. Results

The optimum rainwater storage tank sizes for each region are shown in table 1. As expected, the optimum rainwater tank sizes for different regions vary. According to the input data given in Section 3, the optimum rainwater storage tank sizes in six regions range between 1 and 5 m^3. Since some of these optimum rainwater storage tank sizes are not available in the market, they are revised according to the available water tank sizes sold in the market (see the third column of table 1).

Table 1. Financial assessments of RWHS and optimum rainwater storage tank sizes in the study areas.

Area	Rainwater storage tank size (m^3)	Available tank size in the market (m^3)	NFB in 37 years without fixed costs (TL)	Financial loss in 37 years (TL)
Alevkayası	4.60	5	692	2322
Dipkarpaz	3.83	4	477	2537
Girne	4.77	5	871	2143
Güzelyurt	2.50	2.5	320	2694
Lefkoşa	3.87	4	756	2258
Mağusa	1.36	1.5	297	2716

After obtaining the optimum domestic rainwater storage tank sizes for a single house with 4 residents, we evaluated the feasibility of installing domestic RWHS. Without the maintenance and installation costs, installing the RWHS is found feasible for all cases. However, for all regions, when fixed costs, which are the installation and maintenance costs, are added to the optimum objective function values, installation of the RWHS becomes infeasible ($NFB < 0$) as shown in the last column of table 1. High maintenance expenses and installation costs is the main reason of infeasibility. For all regions, the economic losses are greater than 2000 TL, which is more than the cost of the installation, 1500 TL, and the maintenance expenses, 1514 TL. This situation supports the conclusions of [12] that without noteworthy incentives, installation of RWHS for domestic uses in a single housing unit results in financial losses.

5. Conclusion

In this study, we have calculated the optimum rainwater storage tank sizes for a single house to meet the domestic water demand of 4 residents. We developed an LP model to minimize the total discounted cost. Then, we analysed the feasibility of installing the RWHS for each case study site located in semi-arid Northern Cyprus. The results show that without the installation and maintenance costs, installation of RWHS are feasible for the regions but when installation and maintenance costs are considered, installation of the RWHS turns out to be not cost-effective. Although currently, RWHS are found out to be infeasible for Northern Cyprus, their feasibility may change with changing electricity prices. As mentioned earlier, since water comes from Turkey to Northern Cyprus, a number of pumping stations are used in the network. Therefore, future changes in electricity price are going to affect the unit cost of water price in Northern Cyprus. Identification of optimum RWHS will become more critical and important as the electricity price increases.

Although the results of this showed that domestic RWHS are infeasible to implement due to economic analysis, we believe that the environmental benefits of RWHS should be considered in this analysis as well. RWHS is a noteworthy tool in integrated water resources management to protect the decreasing water resources around the world as a result of the rising population and climate change. Therefore, as a future study, we plan to integrate the environmental benefits of RWHS into our analysis to provide a complete evaluation.

To conclude, as displayed in the case studies, RWHS may become an attractive option to cope with water scarcity in semi-arid regions if the governments compensates certain portion of the maintenance and capital costs.

Acknowledgement
The authors thank the Meteorological Office of Northern Cyprus for providing the required rainfall data.

References

[1] The Intergovernmental Panel on Climate Change 2019, www.ipcc.ch
[2] United Nations 2019 Water Resources Management (IWRM), www.un.org
[3] Okoye C O, Solyalı O and Akıntuğ B 2015 Optimal sizing of storage tanks in domestic rainwater harvesting systems: A linear programming approach *Resources, Conservation and Recycling* **104** 131–40
[4] Ghimire S R and Johnston J M 2013 Impacts of domestic and agricultural rainwater harvesting systems on watershed hydrology: A case study in the Albemarle-Pamlico river basins (USA) *Ecohydrology & Hydrobiology* **13** 159–71
[5] Khandelwal P, Tiwari K and Goyal R 2014 *International Journal of Engineering Research & Technology*
[6] Ghisi E, Bressan D L and Martini M 2007 Rainwater tank capacity and potential for potable water savings by using rainwater in the residential sector of southeastern Brazil *Building and Environment* **42** 1654–66
[7] Che-Ani A I, Shaari N, Sairi A, Zain M F M and Tahir M M 2009 Rainwater Harvesting as an Alternative Water Supply in the Future European *Journal of Scientific Research* **34** 132–40
[8] Pinzon T M, Rieradevall J, Gasol C M and Gabarrell X 2015 Modelling for economic cost and

environmental analysis of rainwater harvesting systems *Journal of Cleaner Production* **87** 613–26
[9] Pelak N and Porporato A 2016 Sizing a rainwater harvesting cistern by minimizing costs *Journal of Hydrology* **541** 1340–7
[10] Khastagir A and Jayasuriya N 2010 Optimal sizing of rain water tanks for domestic water conservation *Journal of Hydrology* **381** 181–8
[11] Debusk K, Hunt W and Wright J 2013 Characterizing Rainwater Harvesting Performance and Demonstrating Stormwater Management Benefits in the Humid Southeast USA *JAWRA Journal of the American Water Resources Association* **49** 1398–411
[12] Roebuck R M, Oltean-Dumbrava C and Tait S 2010 Whole life cost performance of domestic rainwater harvesting systems in the United Kingdom *Water and Environment Journal* **25** 355–65
[13] Aladenola O O and Adeboye O B 2009 Assessing the Potential for Rainwater Harvesting *Water Resources Management* **24** 2129–37
[14] The General Directorate of State Hydraulic Works 2019, www.dsi.gov.tr
[15] California Water Boards 2019, www.waterboards.ca.gov
[16] Okoye C O and Atikol U 2014 A parametric study on the feasibility of solar chimney power plants in North Cyprus conditions *Energy Conversion and Management* **80** 178–87
[17] Gozen E and Turkman A 2008 Water Scarcity Impacts On Northern Cyprus And Alternative Mitigation Strategies Environmental Problems of Central Asia and their Economic, Social and Security Impacts NATO Science for Peace and Security Series C: Environmental Security 241–50

IOP Conf. Series: Earth and Environmental Science **297** (2019) 012027 doi:10.1088/1755-1315/297/1/012027

Sustainable Development approaches in Egypt

Ahmed Adel Abdelrahman Hussein[1], Eric Pollock[2]
[1] Corresponding Author: Master student, Metropolia University of Applied Science, Helsinki, Finland. HTW Berlin University, Berlin, Germany.
[2] M.Arch. Lecturer, School of Real Estate and Construction, Metropolia University of Applied Science, Helsinki, Finland
Email: Eng.ahmedadeel1990@gmail.com

Abstract. In conjunction with the current Egyptian sustainable and economic reform plan to enrich the country resources and face future challenges, the megaprojects have been pursued to improve the infrastructure and encourage investment. Starting with renewable energy planets and passing by 14 new sustainable cities all over Egypt to absorb the future expected population and improve the living standards of the Egyptian citizens. One of those cities is the New Administrative Capital which is being built as a smart city on an uninhabited strategic area between the Suez Canal and the Nile River and it will become the new administrative and financial capital of Egypt, housing the main government departments and ministries, as well as foreign embassies. This paper attempts to review the United Nations Sustainable Development Goals in the Egyptian megaprojects with a special focus for the Egyptian New Administrative Capital city.

Keywords: Egypt, New Administrative Capital, United Nations Sustainable Development Goals, UNSDG

1. Introduction:

Before 1952, Egypt was depending on agriculture for a massive part of its economy. That drifted to an ambitious industrial planned program in the1960s. Most of the reform plans were going as planned until the war of 1967, or the second Arab Israel war, when Egypt lost an important strategic part of its land (Suez Sinai) and the whole country economy went for military reform effort to retrieve the occupied land. At the 1980s, the re-orientation of the investing strategy took place but at that time most of the reform effort wasted until the famous revolution in 2011.(Washington 2012)

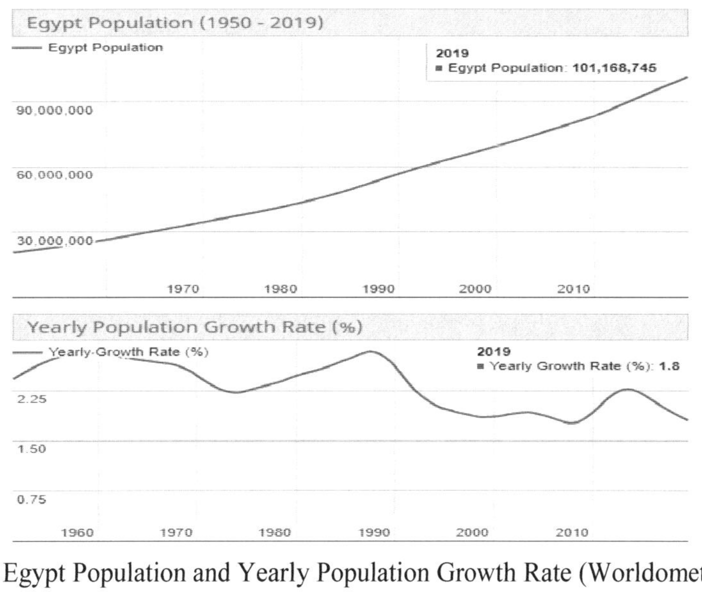

Figure 1. Egypt Population and Yearly Population Growth Rate (Worldometers 2019)

IOP Conf. Series: Earth and Environmental Science **297** (2019) 012027 doi:10.1088/1755-1315/297/1/012027

As shown in figure (1), the total population of Egypt in 2019 exceeds 100 million people and with a growth rate of 1.8% annually in the population, it's predicted to range from 121.8 to 151 million people by 2050 and could exceed 200 million in 2100 (Worldometers 2019). All age groups in Egypt's demographic pyramid are set to increase in the next decades. (Nations 2015) However, the increase is relatively higher in younger strata than the older one (Zaky 2009), which means that policies should be aimed towards constructing new cities in accordance with the future needs of new housing units.

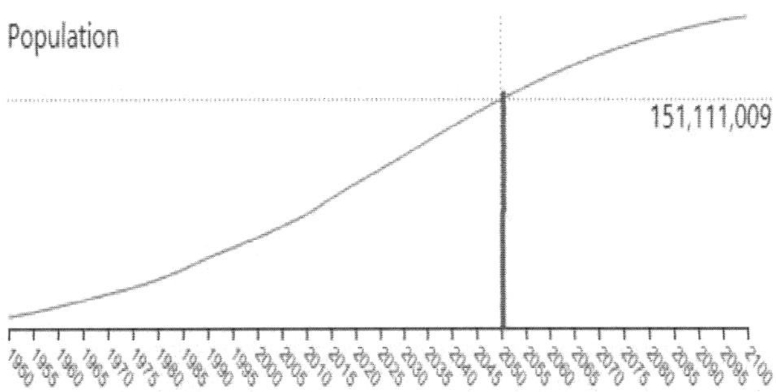

Figure 2. Egypt Population Graph (1950-2100) (Nations 2015)

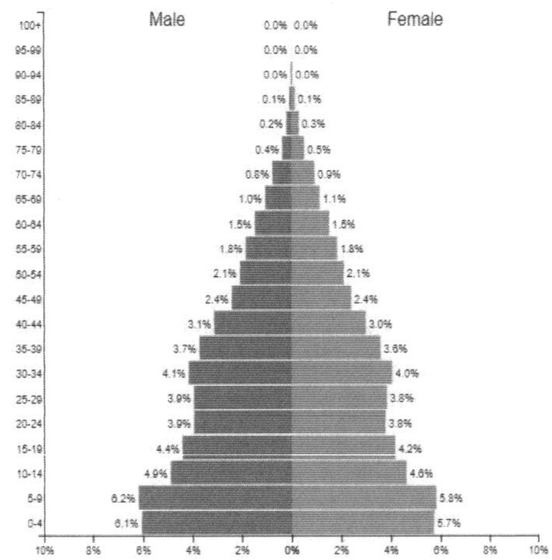

Figure 3. Egyptian demographic pyramid (Nations 2015)

Moving towards achieving the United Nations Sustainable Development Goals (UNSDG) (NATIONS 2018)The current Egyptian administration doctrine is to capitalize on Egypt's big market. The aim is to make Egypt a safe competitor country that attracts investors from all over the world. This is only possible by constructing mega infrastructure projects, adopting an open market policy and deregulating the state's currency (devaluation) while easing regulations and hurdles for foreign investors.

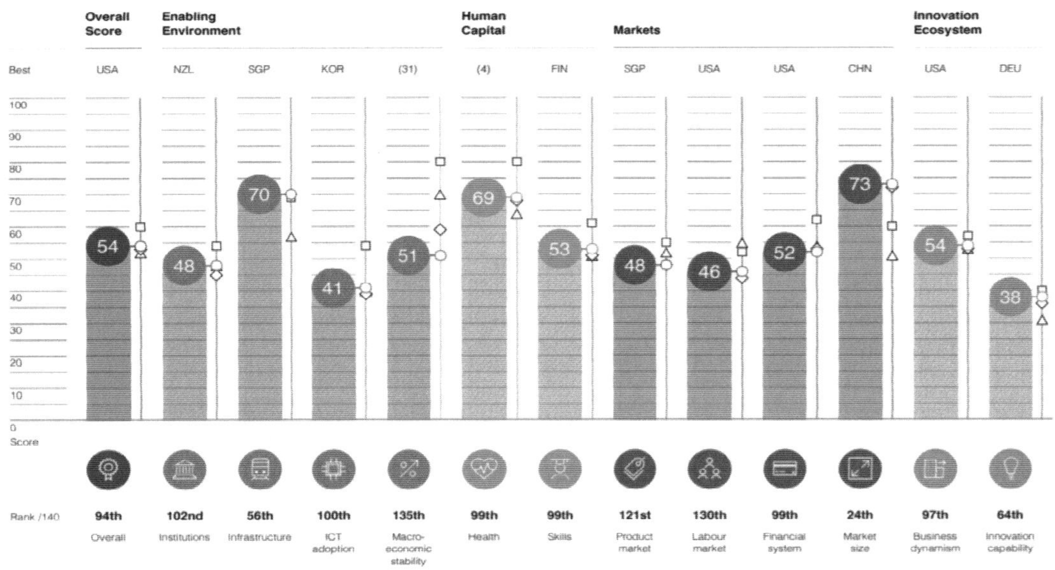

Figure 4. Egyptian Economic performance in 2017
(Global competitiveness Index, World Economic Forum 2018)

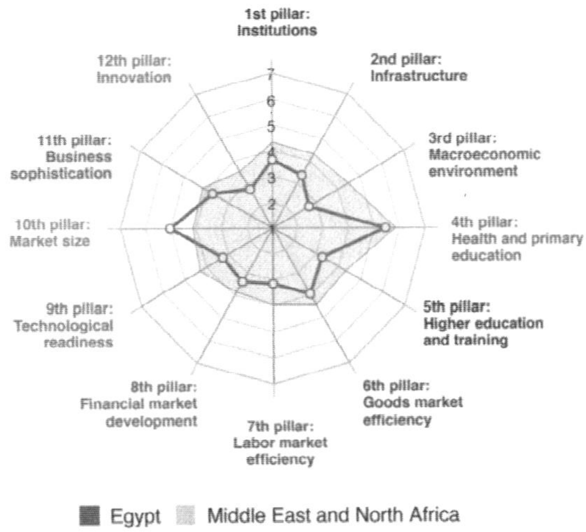

Figure 5. Egyptian and MENA region economic performance in 2016
(World Economic Forum 2016)

Egypt main potential for a strong economy is its large market size (figure 4, market size=73), proximity to the rich European market and it is a relatively developed infrastructure (figure 4, infrastructure=70). The change of macroeconomic environment to a free market economy will put Egypt on the path to catch up with its neighbours in MENA region. (figure 5).

Egypt has climbed steadily in the ease of doing foreign business index in the past 8 years which shows a good indicator of the potential growth in all sectors. The world economic forum survey in 2016 showed Policy and Government instability and Access to financing as the main hurdles of doing business in Egypt. As shown in Figure (6), It's clear what should the current Egyptian Government work on to improve the overall investment atmosphere. (World Economic Forum 2011, 2016, 2018)

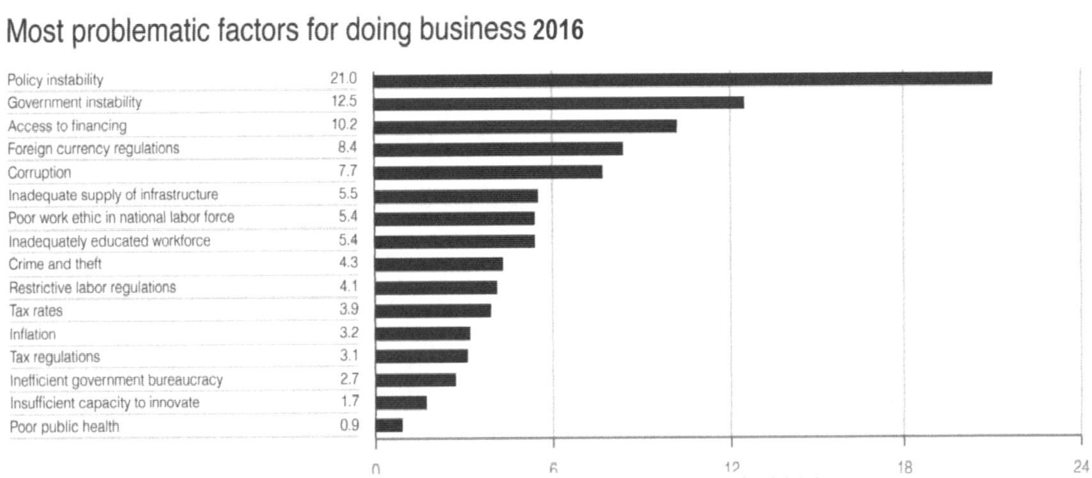

Figure 6. Problems facing business in Egypt in 2016
(World Economic Forum 2016)

2. Renewable Energy projects:

During 2014, Egypt was facing a crisis that threatened all key sectors as there was a continuous never-ending electricity blackout which made daily life for most citizens becomes more and more complicated. The political leadership categorized the electricity and power demand as a national security demand and an immediate plan has been developed to increase the capacity of the national electricity grid and provide power to all sectors in the country and as soon as possible. (LIVE 2017)

An overall budget of 515 billion Egyptian pounds (EGP) was allocated for improving electricity sector of which 433.5 billion EGP was allocated for the generation, 54.5 billion EGP to support the transmission network and 27 billion EGP to upgrade the distribution network. And in summer of 2015, the dream became true and the government was able to provide reliable power to all sectors without any further blackouts which had a great impact for providing the investing environment and political stability. The current Egyptian administration is eager to maximize the use of renewable energy as a source of clean and affordable energy in the long term. (LIVE 2017)

Sustainable renewable power sources (like wind and sun-based power) will decrease the contamination caused by the regular thermal power producing stations which may produce vast volumes of oxides of nitrogen and carbon (CO2). (Shumkov, 2015)

2.1 Wind Energy

The Gulf of Suez (Northwestern shores of the Red Sea) is viewed as a prime worldwide area for on-shore wind energy because of its unique air currents marked by its high speed. The Egyptian government, in coordination with the Japan International Cooperation Agency (JICA), established the biggest wind farm in the Middle East and Africa located in Jabal el Zeit on an area of 700 sq.km. Featuring the latest turbine's technology in the world with a capacity of over 200 megawatts.(Egypt Energy 2018)

This installation will be the first wind farm executed under a 'BOO' scheme ('Build,' Own 'and' Operate '). Kfw Bank in Germany and the European Investment Bank (EIB), as well as the European Commission, financed the project with more than 270 million USD setting a project schedule over a 30-month period. The plan contemplates private sector participation that will reach 60%. (Shumkov, 2015)
Jabal el Zeit Wind Park will provide up to 800 GWh of electricity for more than 500,000 citizen annually. (Shumkov, 2015)

2.2 Solar Energy

To reach the 2022 Goal of having 20% of the total electricity grid from renewable resources, the largest solar complex in the world is being constructed in Benban village located in Aswan governorate. The Benban solar park Located over 37.2 km^2 (14.4 square meters) and consists of 41 plants with a total capacity of 2000 megawatt. The private sector investors have signed agreements to implement solar projects with a capacity of 1465 megawatt. (Raven, 2017)

The task got financing from Bayerische Landesbank (BLB) for 85% with the other 15% from Arab African International Bank [ar] (AAIB). The German government has given Euler Hermes trade credit ensure (ECG), covering the BLB advance. As a major aspect of the German outside exchange advancement program, it bolsters Egypt's objective of meeting more than 33% of its vitality necessities by 2035 through sustainable power sources. (Prabhu, 2017)

The Benban solar power plant will support the implementation of the Egyptian ambitious plan for achieving self-sufficiency for generating electricity in the next decades. From another perspective, that would also support a more sustainable environment by limiting the polluting traditional source of power ''fossil fuels''. It would also provide 4000 job vacancies once it is fully operated (Mbogo 2018)

Figure 7. Jabal el Zeit Wind Farm (MacFarlane 2018)

Figure 8. The Benban solar park (Mbogo 2018)

3. Roads and Transportation:

Despite its large population, and compared to similarly developed cities, Cairo has a very small metro network and the highest number of passengers per kilometer compared to cities of similar size (as shown in figure 10). The population density led to the necessity of the overuse of the metro transportation system. Therefore, there is an ongoing metro line extension project to meet the needs of the growing population. .(Bank 2018)

Many bridges and tunnels have been constructed over the past few years all over Egypt to connect the main Egyptian cities supporting internal trade lines and encouraging foreign investment opportunities. The new Rod el Farag Bridge is the new life artery for Egyptian prosperity as it connects greater Cairo metropolitan area from its far east (Nasr city) to its west (October city) and continues to reach Cairo-Alexandria desert road.

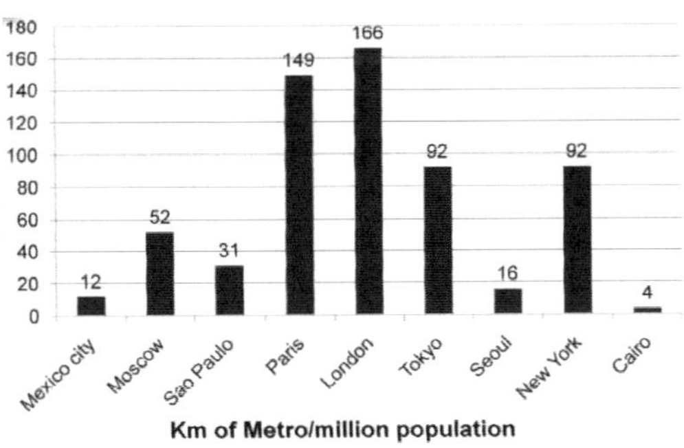

Figure 9. Comparison of Km of Metro-line/population in different cities (Bank 2013)

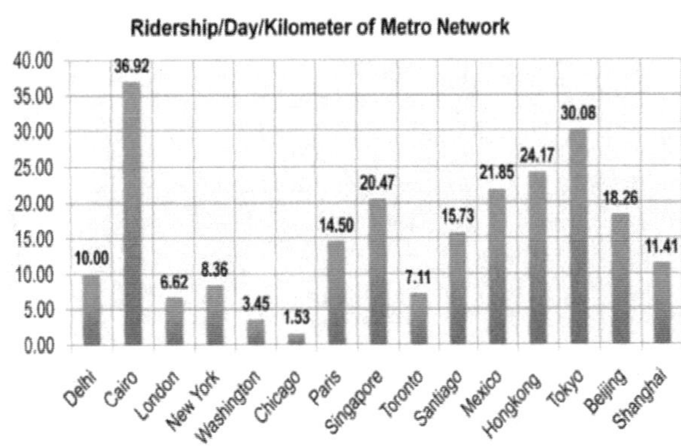

Figure 10. Comparison of Ridership/day/kilometer in different cities (Bank 2013)

4. The Construction of the New 4th Generation Cities:

4.1 Introduction

Considering the fact that most of the Egyptian population are living in just 7% of the total Area of Egypt land (1.01 million Km2). Currently, the housing capacity does not meet the future requirement and considering the growth rate will remain the same, the current Egyptian administration should act fast to expand the urban areas. An integral part of the investment plans should be directed to the housing purpose and building new cities away from the overcrowded old cities. Therefore, today is the day for constructing megaprojects that will improve the country infrastructure and achieve its future ambitious goals as well as supporting a stable environment for investing in Egypt. (LIVE 2017)

With the Egyptian Ministry of Housing, Utilities and Urban Communities (MoHUUC) remarkable efforts to develop the living standard for the Egyptian citizens, 14 new cities are being constructed all over Egypt in the next decades. On an approximate total area of 580,000 acres to accommodate 30 million people, the new cities have the title of the 4th Generation cities as there were previous waves of construction new cities from the mid-20th century until now. It is predicted that a significant decline of housing unit's prices will occur after the construction and delivery of these cities because of the rise of supply in the Egyptian real estate market. Taking into consideration the United Nations Sustainable Development Goals, every city will have a sustainable design that provides high living standards for its residents.(LIVE 2017)

Some of the 4th generation Egyptian new cities are:(Agency 2018)

- The new administrative capital city, (the biggest new city) with an approximate area of 170 000 acres, the construction has been started in its first phase over an area of 40 000 acres.
- New Alamein City, with an area of 49000 acres. Its first phase spans over 14000 acres.
- New east Port Said City, with an area of 16000 acres.
- New 6 October city, with an area of 78 000 acres and the first phase spans over 9000 acres.
- New Luxor city, with an area of 10000 acres.

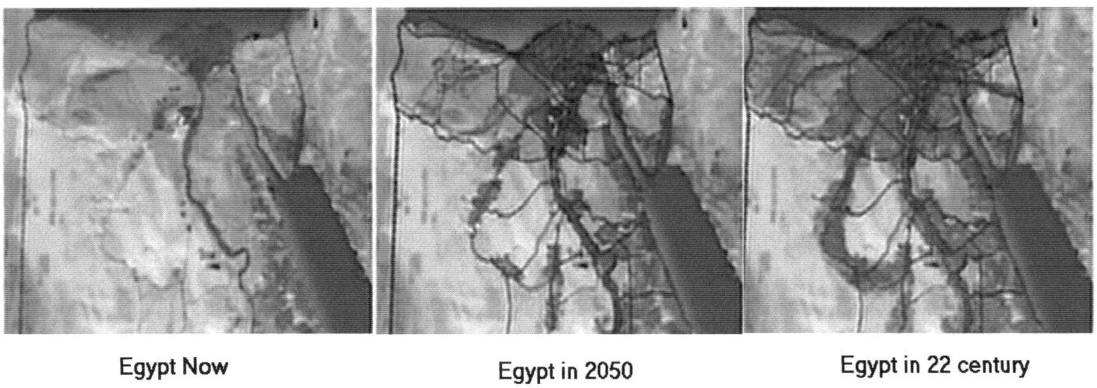

Figure 11. Government's prediction of the urbanized area in Egypt in the future (LIVE 2017)

5. The New Administrative Capital of Egypt (NAC):

5.1 Introduction

Cairo is one of the most crowded cities all over the world. Relieving the congestion in Cairo was the main reason for constructing the New Administrative Capital (NAC). Cairo has a population over 20 m people that expected to be double in the next few decades.(Anonymous 2019b). The population density reached 98 people per km2, which is the second highest in Africa. In 2017, the Egyptian president declared that overpopulation one of the national crisis.(Al-Wahaidy 2018)

With half of Egyptians living in cities and the other half living in rural areas, the urbanization rate is estimated to be 1.86, one of the highest in the region. Together with the population growth, Cairo's population is expected to reach 24 million by 2027 while with a current population of 19.5 million people in 2019 its one of the most crowded cities around the world (the largest in the middle east, second largest in Africa, and one of the top ten most crowded cities around whole world). The already overused transportation system and infrastructure will not be able to contain this changes in the city's demographics. (Agency 2018)

With problems such as traffic jam, air pollution from carbon dioxide emission and noise pollution in Cairo metropolitan area (Divided by the Nile river, Greater Cairo consists of Giza City on the left bank of the river and Cairo City on the right bank), the standard amenities and services cannot be reached. A study by the World Bank estimates the monetary loss to exceed 2,442 million USD because of traffic jams alone. (Bank 2013)

Figure 12. Crowded Greater Cairo Metropolitan Area (STREETS 2013)

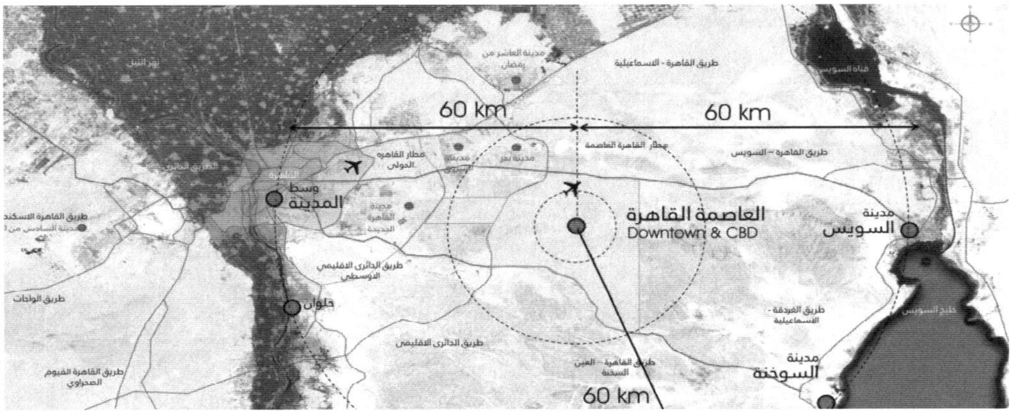

Figure 13. The Location of The New Administrative Capital (ZAINEGYPT 2018)

On an uninhabited strategic desert located 60 kilometers east of Great Cairo center and 60 Kilometers west of the Suez Canal, a new administrative capital for Egypt is being build. It would eventually have a population of seven million people on a total area exceeds 1800 acre (while the great Cairo metropolitan area didn't exceed 1000 acre). March 2015, the president of Egypt announced that the New Administrative Capital will be built by the national Egyptian firms. 21 April 2016, the New Administrative Capital for Urban Development (ACUD) has been established and assigned for constructing the national megaproject. (ACUD interview 2018)

"The city is planned to have skyscrapers and home to Africa's tallest tower and would become the new administrative and financial capital of Egypt, housing the main government departments and ministries as well as foreign embassies. The city will also have a green river with artificial lakes and 90 km2 of solar energy farms. It will be built as a smart city that relies heavily on automated services instead of manpower to supply its citizens with services such as, traffic and transportation systems. It is planned that the transfer of parliament, government ministries and foreign embassies will be completed between 2020 and 2022 at a cost of US $45 billion for the first phase.'' quoted by Egyptian Prime Minister Mostafa Madbouly.

Figure 14. The New Administrative Capital (ACUD interview 2018)

5.2 Inspiration

Inspired by some other countries that have changed their historical capital for different reasons, Like: Brasilia City in Brazil and Astana City in Kazakhstan. Egypt was not the first country to construct a new administrative capital. In fact, the idea was first discussed in 1970 but the implementation of the project suffered a failure in reaching its main goal due to lack of providing housing units for employees that should operate the governmental buildings. (ACUD interview 2018)

5.3 Components of the new capital

The new administration capital has an overall area of more than 170,000 Fadden. It would be equals to the size of Singapore, 7 times over Paris area, 4.5 times over Washington D.C. The city would become the new financial and administrative capital of Egypt, housing the main government and ministries buildings, as well as foreign embassies. It also contains about 2,000 educational institutions, 663 hospitals and clinics, over 1000 mosque and churches, 40,000 hotel rooms, a major park four times the size of Disneyland, 90 km^2 farms of solar energy. (UNIVERSAL 2019) It has been divided into three construction phases. The first phase would be over an area of 40.000 Fadden at a cost of US $45 billion. A full cost and timescale for the overall project have not been disclosed yet. (ACUD interview 2018)

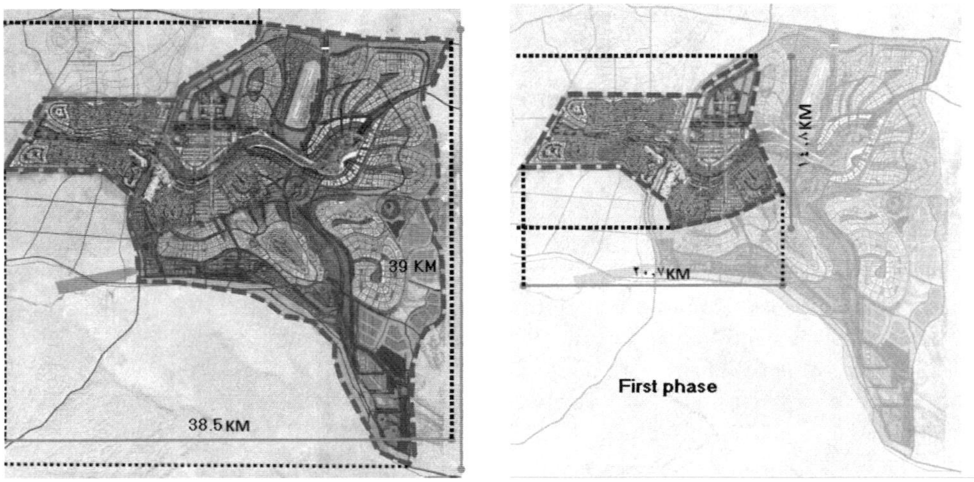

Figure 15. phases of the new capital (ACUD interview 2018)

5.3.1 The Governmental District.

The governmental district will be constructed over an area of 566 acres and includes the new Cabinet's headquarters, the parliament building, and 34 other buildings designated for each ministry. The transition of the new Parliament Building, presidential palaces and government ministries will fold out over two stages; each stage will transfer 57,000 employees. (ACUD interview 2018)

Figure 16. The Governmental District. (ACUD interview 2018)

Figure 17. The new Egyptian Parliament Building (ACUD interview 2018)

Figure 18. The new Cabinet's headquarters (ACUD interview 2018)

5.3.2 The Central Business District.

The first phase of the central business District would contain more than 20 residential, administrative, commercial and services' sky scanners with a total Build-up area of 1.7 million m2 and it will be home to Africa's tallest tower with 385 m height. Five towers for residential purpose with a total area of 318,000 m2. Five towers for Hospitality purpose with a total area of 81,000 m2 (38,000 m2 four-star hotels and 43,000 m2 for five-star luxurious Hotels) and other 207,000 m2 Building area for Retail and commercial purpose. (ACUD interview 2018)

Figure 19. Central business district building's heights (ACUD interview 2018)

6. Conclusion:

Although the Egyptian administration has a very ambitious reform economical plan to enhance the country infrastructure by constructing megaprojects, there is many challenges facing them like energy. the energy demand is rapidly increased as a vital catalyst to wider social and economic development enabling health, education and creating jobs. The era of heavy reliance on fossil fuels has come to its end. The renewable energy solutions in mitigating climate change threat have been widely proven. Transitioning to renewable energies resources supports significant progress on the Sustainable Development Goals (SDGs) not only for the seventh goal, affordable and clean energy, but also for other goals related to global temperatures, air quality and human health. Sooner or later all nations will adopt the strategy of clean and renewable energy. Projects like the Benban solar power plant and Jabal el Zeit wind power plant will support the implementation of the Egyptian ambitious plan for achieving self-sufficiency for generating electricity in the next decades.

There are many challenges facing the new capital megaproject. The first challenge is operating a highly modern smart city technologies for the first time in Egypt and the entire Middle East while Egypt has a humble experience in operating smart cities. The Egyptian Administration should consider hiring an international company with unique experiences for operating smart cities to manage the new capital and transform their experiences to Egyptian management over a 5- or 10-years period. Another challenge is the high cost of the smart infrastructure in the city and the ways of funding of these constructions. While the New Administrative Capital will provide thousands of job opportunities, Housing for these employees should be close from their workplace to minimize the cost of daily travelling for thousands between old Cairo and New Administrative Capital.

While some opinions are against the concept of constructing a new Administrative Capital because, as they conclude, its massive cost made most of the Egyptian citizens suffer from painful austerity measures, others see it as a necessary solution for developing the country's infrastructure and attract investors. Those who agree for the concept of an expensive new capital debates whether to demolish the old ministries and plant green gardens in its places to revive old Cairo or leave these building as a historic heritage. From another perspective, there is a demand from the Egyptian government to allocate more budget for the development of education and healthcare systems besides constructing megaprojects as there is a huge shortage in these two essential sectors in Egypt.

7. Acknowledgments:

I would like to thank every individual contributed to fulfilling this research paper. I would like to express my gratitude to my supervisor M. Arch. Eric Pollock for his supervision and guidance. Special thanks for ACUD Egyptian company for their detailed information about the planning and construction phases of the new administrative capital of Egypt. Finally, I would like to dedicate my work to my beloved father who passed away on 21 August 2018. And to my beloved mother for her moral support, encouragement and patience.

8. References:

1. Agency, Central Intelligence (2018), 'The world fact book', https://www.cia.gov/library/publications/the-world-factbook/fields/2212.html
2. Al-Wahaidy, Fatima (2018), 'What's being done about overpopulation in Egypt?', http://www.egypttoday.com/Article/2/40295/What%E2%80%99s-being-done-about-overpopulation-in-Egypt.
3. Bank, World (2013), 'Cairo Traffic Congestion Study : Final Report.', https://openknowledge.worldbank.org/handle/10986/18735.
4. EGYPT, ENERGY (2018), 'Wind Farm | Energy Egypt', https://energyegypt.net/tag/wind-farm/.

5. LIVE, ON (2017), 'Egyptian prime minster mostafa madbouly speech while opening the new adminstrative capital project", https://www.youtube.com/watch?v=mYjzvHEFFr4&t=1690s.
6. MacFarlane, Drew (2018), 'Wind and Solar Farms Can Produce Their Own Weather, Like More Rain Over the Sahara', https://weather.com/news/news/2018-09-07-wind-solar-farms-precipitation-sahara.
7. Mbogo, Moreen (2018), 'Egypt constructs world's largest solar park in Benban', https://constructionreviewonline.com/2018/02/egypt-constructs-worlds-largest-solar-park-benban/.
8. Nations, United (2015), 'United Nations, Department of Economic and Social Affairs, Population Division. World Population Prospects', https://www.populationpyramid.net/egypt/2050/.
9. Nations, United (2018), 'About the Sustainable Development Goals', https://www.un.org/sustainabledevelopment/sustainable-development-goals/.
10. Shumkov, Ivan (2015), 'Egypt inaugurates 200-MW wind farm', https://renewablesnow.com/news/egypt-inaugurates-200-mw-wind-farm-503636/.
11. STREETS, EGYPTIAN (2013), 'Egypt: Living Under a Grey Sky', https://egyptianstreets.com/2013/01/17/egypt-living-under-a-grey-sky/.
12. UNIVERSAL, NBC (2019), 'Egypt builds a new capital city to replace Cairo', https://www.nbcnews.com/news/world/egypt-builds-new-capital-city-replace-cairo-n893606.
13. Washington, D.C.: The World Bank (2012), '"World Development Indicators."', http://data.worldbank.org/data-catalog/world-development-indicators.
14. WORLD ECONOMIC FORUM, & HARVARD UNIVERSITY (2011), 'The global competitiveness report', World Economic Forum, (2016), 'The global competitiveness report',
15. (2018), 'The global competitiveness report',
16. Worldometers (2019), 'Egypt Population', http://www.worldometers.info/world-population/egypt-population/.
17. ZAINEGYPT (2018), 'the capital way', http://www.zainegypt.com/property/capital-way/.
18. Zaky, HHM (2009), 'The timing of the demographic dividend in Egypt', Proceedings of the 2009 Hawaii International Conference on Social Sciences, Honolulu, HI, 175-88.
19. Raven, A. (2017). A New Solar Park Shines a Light on Egypt's Energy Potential. [online] Ifc.org. Available at: https://www.ifc.org/wps/wcm/connect/news_ext_content/ifc_external_corporate_site/news+and+events/news/cm-stories/benban-solar-park-egypt [Accessed 30 Mar. 2019].
20. Prabhu, W. (2017). IFC-Led Consortium Invests $653 Million to Develop 13 Solar Projects in Egypt - Mercom India. [online] Mercom India. Available at: https://mercomindia.com/ifc-led-consortium-invests-653-million-to-develop-13-solar-projects-in-egypt/
21. ACUD Interview, Most of the Information written in this paper was gathered by conducting an Interview (December 2018) with persons from different department working in the new administrative capital. For example, but not limited,
Dr Fadel Digham, R&D director at National Telecommunication Regulatory Authority (NTRA), Egypt. https://www.linkedin.com/in/fadel-digham-0b54456/
Dr Mohsen Badawy, Consultant for the ACUD Company, Egypt.
Colonel Khaled El Hoseny Soliman, official speaker of the ACUD Company, Egypt. https://www.linkedin.com/in/mohammed-khalil-02a4482b/

IOP Conf. Series: Earth and Environmental Science **297** (2019) 012028 doi:10.1088/1755-1315/297/1/012028

Towards more sustainable patterns of urban development

F P R Marino, F Lembo, V Fanuele

School of Engineering, University of Basilicata, 10 Viale dell'Ateneo Lucano, 85100 Potenza (Italy)

francesco.marino@unibas.it

Abstract. DETR, 1998, wrote in *Planning for Sustainable Development. Towards Better Practice*: "*New settlements will enjoy a high quality of urban and landscape design. As well as integrated open space, there should be habitat areas, and environmental gains such as energy efficiency measures introduced in layouts and individual buildings*"; but this claim is remained largely dissatisfied. There is a great interest (and large investments in research) for the *smart city*, which seeks to optimize the existing cities, while people go on to design urban developments with building typologies set ninety years ago by Modern Movement or little more. Urban Planning compares existing cities' models and asseverates that densification has positive role for sustainability, but do not turns enough to account passive typologies for heating and cooling, which can turn down near to zero power-consumption, and at the same time can raise urban environment quality. Famous settlements too, as Malmö or Vauban in Freiburg, has been built with building typologies conventional, after all; German *passive Haus* have generally two, three or four building fronts and in reality, are simply hyper-insulated traditional buildings, furnished with mechanical controlled ventilation. For an urban extension of Potenza, we have used for the urban planning the sustainability assessment categories of ITACA's Protocol, derivative from GB*Tool* of GBC, people normally use to assess *ex-post* sustainability of projects and of realized buildings. The result is a settlement in which pedestrian, cycle and public transport's network is fully integrated with adjacent urban areas; effective landscaping connects public and private green and kitchen-gardens/orchards are everywhere; buildings are made with new semi-underground typologies, nZEB and made with local, re-cyclable materials; rain water is collected, *in-loco* fito-depurated and reused; *in-loco* renewable energies (sun, earth, wind) satisfies remaining necessities.

1. The state-of-the-art of the sustainable city planning and the objectives of the research

"*At least in Europe … a clear institutional position – even if not shared by everybody – has been taken by the European Community, which openly declared for the compact city and for the strengthening of polycentrism. That's what sustainable urban development and sustainable urban planning became over the Nineties key points of urban planning and integral part, sometimes a bit redundant and repetitive, of his vocabulary*" [1]. "*A high quality of urban design not only represents a general objective of planning, but it is a specific requirement for sustainable development. The success of urban regeneration to higher densities, for example, will depend on the skill of creating a more appealing environment*" [2].

To tell the truth, beyond declarations of intentions, planning and urban design continue to plan urban expansions with building typologies developed ninety years ago from Modern Movement, in his search of a "democratic" balance between east and west exposure ("equisolar axis") (both negative in terms of passive solar energy, as they cannot heat enough in winter, and are a source of overheating in summer)

[3]. Many recent urban development show little or nothing more [4], although already in 1991 a publication of widespread use showed that buildings, specifically designed to have a significant energy contribution from the sun, actually limit this contribution to 25-40% if they have a double-face type, with rooms facing either south, that to the north or other directions [5]. Urban planning specialist compare the models of existing urban cities and theorize about the greater sustainability of the denser ones [1], but do not enhances typologies that use passive systems for heating and cooling [6] [7] [8], which can lead to wipe out energy requirements and consumptions and, in the same time, to improve the quality of urban environment. Famous settlements, like that of Malmö or Freiburg [9] resort to building typologies all thing considered conventional, and the German *Passive Haus*, beyond calls to the necessity of a correct exposition, have as a general rule typology with exposition toward two, three of four fronts and, in reality, they are nothing else than conventional buildings, but hyper-insulated and provided with systems of mechanical ventilation with heat recovery [10, see especially Chapter "Urban Design", pp. 68-75]. Solitary stands, as the paradigm of a new way of thinking the typology of multi-storey building too, the Ove Arup GmbH realization in Wittenberger Straße in Berlin [11], which has already more than twenty years.

And the *smart city* culture is still playing with informatic devices that would make intelligent cities that continue to be designed in a stupid manner. It is a typical case of what scholars of process optimization methods call "the inversion between the spoon and the bulldozer": to open a way, it is first necessary to work with bulldozer and graeder; only at the end it is possible to work with the spoon to finish the gutter. That is, it is in no way useful to rationalize details of a non-rationalized process [12].

From the sustainability assessment point of view, international panorama offers methods both rating based (BREEAM, LEED, GB*Tool*, CASBEE, HQE), and analytical ones (Norms EN ISO 14000 series, in particular 14025 on EPD and 14040 on LCA), which allows of assess in an objective manner sustainability of a building both at the design level, and as built. In Italy, the norm adopted is a rating based one, a GB*Tool* of Green Building Challenge version, named ITACA Protocol, which expresses the judgment on the basis of a series of criterions, from the site quality to the materials and constructive choices. The challenge of the research has been that of adopt ITACA Protocol not as assessment criterion, but as design method, and of structure up with his guide the planning of a whole neighbourhood in Potenza, in the centre-south of Italy, from the town-planning level to the building and to the constructive one, with the target of checking possibility, in that manner, of attaining the result of a settlement not only sustainable, but also appealing and delightful, according to Louis Kahn's lesson: "*What is measurable is in the service of what you can't measure*"; and to Kevin Lynch's one: "*It is necessary to learn what is desirable as much as to study what is possible: to act without target can be useless as much the idealism without power. Even the range of possible things can be extended by greater knowledge of what is desirable*" [13].

2. The site: environmental and town planning commitments

The site is a great area, covering 128.057 m^2, named *Zona "C2"* in the Urban Regulations of Potenza Borough, at the northern border of the city, at an altitude between 815 and 880 m o.s.l, with an average slope of 24%, lat. N 40,65°, long. E 15,79°, and a good exposure to the sun toward South (figures 1, 2, 3). Scheduled territorial density is of 0,50 m^3/m^2, which allows building 64.028 m^3, primarily dwellings, but also offices and shops. It is scheduled that buildings can have a maximum height of 10,50 m and a minimum distance from the vehicular roads of 7,50 m. It is planned the realization of neighborhood services, made up of primary school in an area of 3.170 m^2, common interest services (administrative and cultural) in an area of 2.000 m^2, public equipped green on 23.164 m^2 and public parking on 2.785 m^2. The soil is made of clay and silt grey-blue; it is stable, but it needs of foundation on piles. It is seismic zone I, with high dangerousness (acceleration with probability of exceeding of 10% in 50 years: $a_g \leq 0,25$ g). Potenza has a Mediterranean montane climate: cold and snowy in winter time, tepid in summer time. The coldest month is January, with an average temperature of +3,5 °C, hottest July and August, with average temperatures of 20 °C. Average rainy condition of the past decade has been of 753 mm/year, raising 100 mm with respect to the previous decade. Important is the relative average annual

IOP Conf. Series: Earth and Environmental Science **297** (2019) 012028 doi:10.1088/1755-1315/297/1/012028

moistness, equal to 71%: for the most of months (but July and August), she seizes between 60% and 88%. Therefore, it is necessary to forecast systems of mechanical controlled ventilation equipped not only with heat recovery device but also with a dehumidifier. During summer months the winds come from 240° (West-South-West) and in the rest of the year mainly from 220° (South-West), with speed between 2 and 5 m/s. Solar radiation is high, thank to air clearness. It is climatic zone E, with 2.472 degrees-day (heating from October 15 to April 15, 14 hours daily).

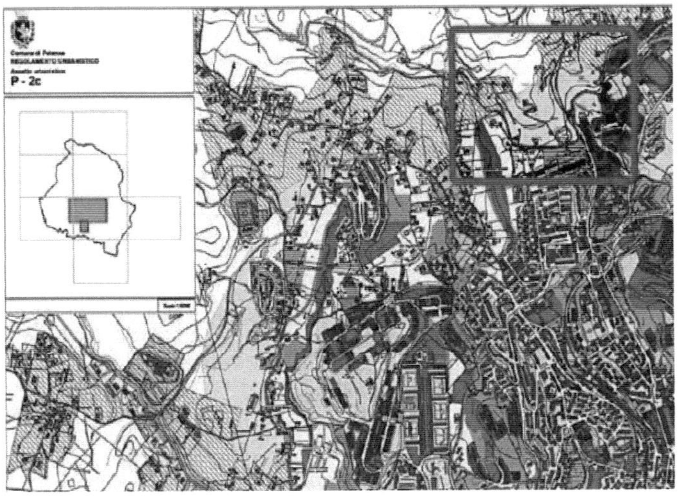

Figure 1. Excerpt from the urban regulation of the Municipality of Potenza.

Figure 2. Orthophoto of the area under study

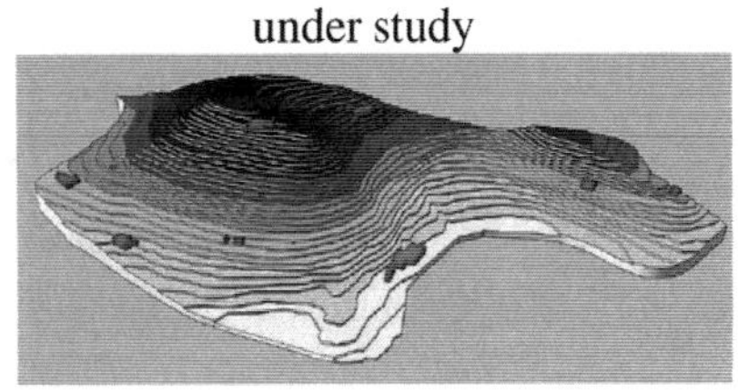

Figure 3. 3d reconstruction of the intervention area

Borough Administration has prearranged an Operative Plan [14] as guidance to the development of the area (figure 4), which presents all the defects typical of urban contemporary extensions, meaningless and lacking of soul, but also of culture of sustainable planning, of which we wrote in the premise: it lacks an overall conception and the hierarchic organization of the areas; there is not a pre-arrangement of a public transport network; the development of the vehicular roads is excessive (the majority of the parcels is served by two roads, both from downstream than from upstream); green areas, despite great surface on which they stand, are not connected in a unitary logic and are not easily enjoyable; school (grey zone in the figure) is set in the worse place as slope, exposure to the sun and accessibility; the Plan lacks an adequate mix of activities and an overall working logic; even though the layout of the building follows roughly contour lines, most of the buildings are not exposed to South and scheduled building typologies are all "*not-passive-solar*" (little towers with two or four flats every floor, oriented to the four cardinal points).

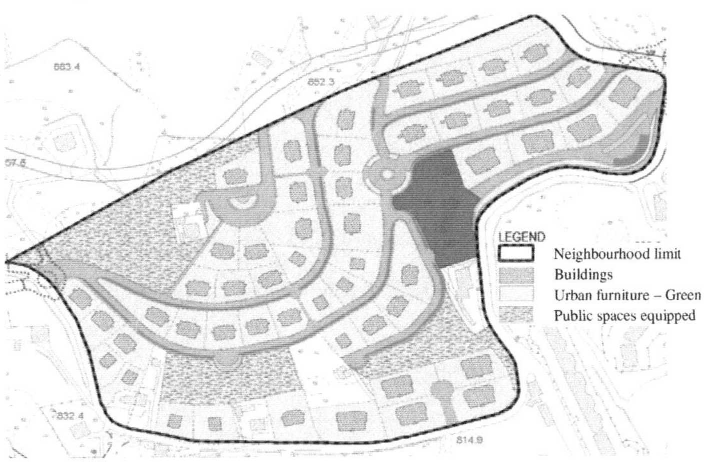

Figure 4. Operative Plan of Zone "C2".

Demonstration of the fact that also a site with good exposure can be badly utilized and can be the Public Administration to stimulate disorder and fall out of love with the city, promoting the urban *sprawl*. As a matter of fact, it is impossible to understand why citizen would must undergo the greater cost of being involved in a "planned" urban extension, if he can get a product similar (or probably worse) than that he would obtain building by himself a house in the country.

3. The use in urban planning of the assessment criterions of ITACA Protocol

ITACA Protocol and its Standard Procedures [15] [16] [17] deal with five "areas" or significant macro-themes: A) site quality; B) resources consumption; C) environmental loads; D) indoor environmental quality and E) service quality (table 1).

Table 1. ITACA Protocol criteria applied and related impact levels

Criterion code	Criterion name **(Area A)**	Impact level P_k
A.1.5	Land re-use	18
A.1.6	Public transport accessibility	12
A.1.8	Functional mix of the area	4
A.1.10	Adjoining to infrastructures	6
A.1.12	Settlement sprawl	6
A.3.3	External equipped common use areas	4
A.3.4	Support to bike use	8
A.3.7	Use of local arboreal species	4
A.3.10	Incidence on the urbanized context	12
Criterion code	Criterion name **(Area B)**	Impact level P_k
B.1.2	Non-renewable primary energy	27
B.1.3	Total primary energy	27
B.3.2	Renewable energy for thermal uses	18
B.3.3	Energy in situ produced for electrical uses	18
B.4.1	Re-use of existing structures	27
B.4.6	Materials recycled / reused	18
B.4.7	Materials from renewable sources	12
B.4.8	Local materials	12
B.4.10	Materials recyclable or dismountable	18
B.4.11	Certified materials	12
B.5.1	Drinkable water for irrigation	18
B.5.2	Drinkable water for indoor uses	18
B.6.1	Thermal energy useful for heating	27
B.6.2	Thermal energy useful for cooling	27
B.6.3	Average global coefficient of thermal exchange	18
B.6.4	Control of solar radiation	18
Criterion code	Criterion name **(Area C)**	Impact level P_k
C.1.2	Expected emissions in exercise	27
C.3.2	Produced solid waste in exercise	12
C.3.3	Reuse of the soils	8
C.4.1	Greywater piped in the sewage system	8
C.4.3	Permeability of the soil	8
C.6.8	Heat island effect	12
Criterion code	Criterion name **(Area D)**	Impact level P_k
D.2.5	Ventilation and air quality	9
D.2.6	Radon	9
D.3.1	Summer thermal comfort in air-conditioned rooms	6
D.3.2	Operative temperature in summer	6
D.3.3	Winter thermal comfort in air-conditioned rooms	6
D.4.1	Natural lighting	6
D.5.6	Acoustic quality of the building	6
D.6.1	Electromagnetic fields at an industrial frequency (50 Hertz)	6
Criterion code	Criterion name **(Area E)**	Impact level P_k
E.2.1	Services equipment	6

E.3.5	B.A.C.S.	18
E.3.6	Domotic plants	6
E.6.5	Availability of technical documentation of buildings	6
E.7.1	Design for all	6

Following methodological order set by ITACA Protocol, in the settlement centre a mechanized public route (cable railway with automatic control) has been arranged (figures 5 and 6); it allows to link fast, across intermediate and final stations, pedestrianised areas and bikeways, set beside the contour lines, to the existing city, also across a pedestrian bridge which, at the bottom of the neighbourhood, crosses the busy way lying at South border. The automotive traffic is confined to the border of the settlement and is linked to underground parking's, in strategic manner set to the West side and to the East side, connected to the pedestrianised areas and bikeways and covered by public urban vegetable garden and orchards (n. 78 parcels of 40 m^2 each) and private ones (n. 160 parcels of 40 m^2 each) in two great areas (9.520 m^2).

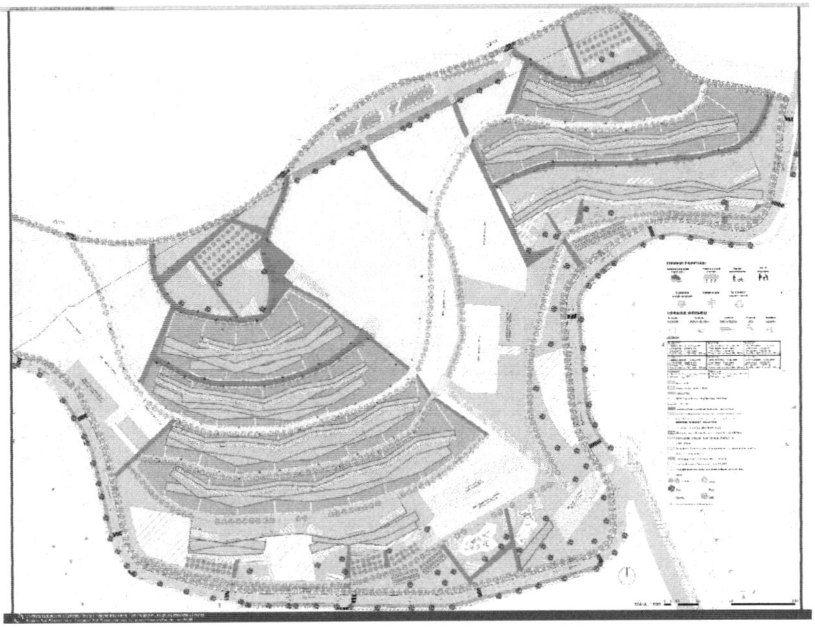

Figure 5. General Plan of the District.

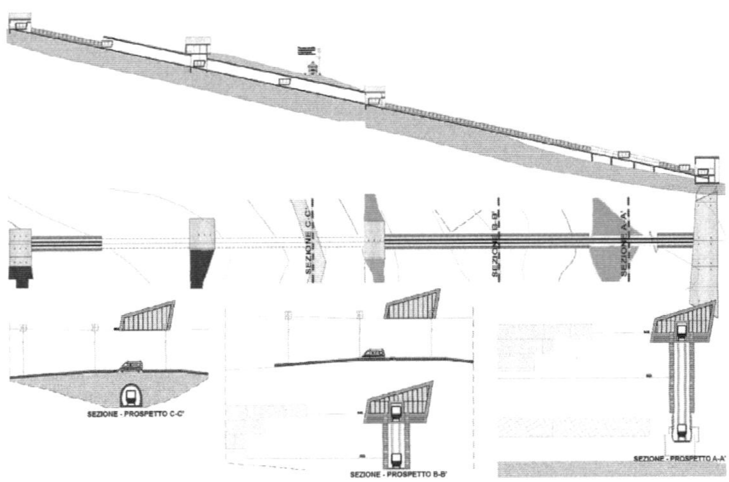

Figure 6. Plan and cross-sections of the cable railway with automatic control.

Parking's are intended for the most part to those residents in the neighbourhood and hence are equipped with closed stalls, each for two cars, in addition to those open, intended to the visitors. Only one road, however, paved with grating and grass, goes lengthwise, from West to East, at the centre of the settlement: it is intended to the security service (ambulance, Fire Brigade) and to the separate refuse collection. Green fabric (figure 7), structured on a wide range of local essences, both of first height (*Fagus Sylvatica*, *Quercus Cerris*), and medium / small ones, with permanent leaf or deciduous and flowering, makes a pervasive belt along the cable railway and spreads on branches in horizontal matter, with a playground on the top and a stronger presence at the bottom, where is also a little artificial lake, connecting to the gardens and private vegetable gardens, set in front and upon the houses, becoming a zone for free physical activity and "*life course*" for fitness in the zone with the outmost slope, on the roof of a great parking. In the central bottom part are set out offices and shops, with the school and common interest services, you must cross for connecting with the existing city.

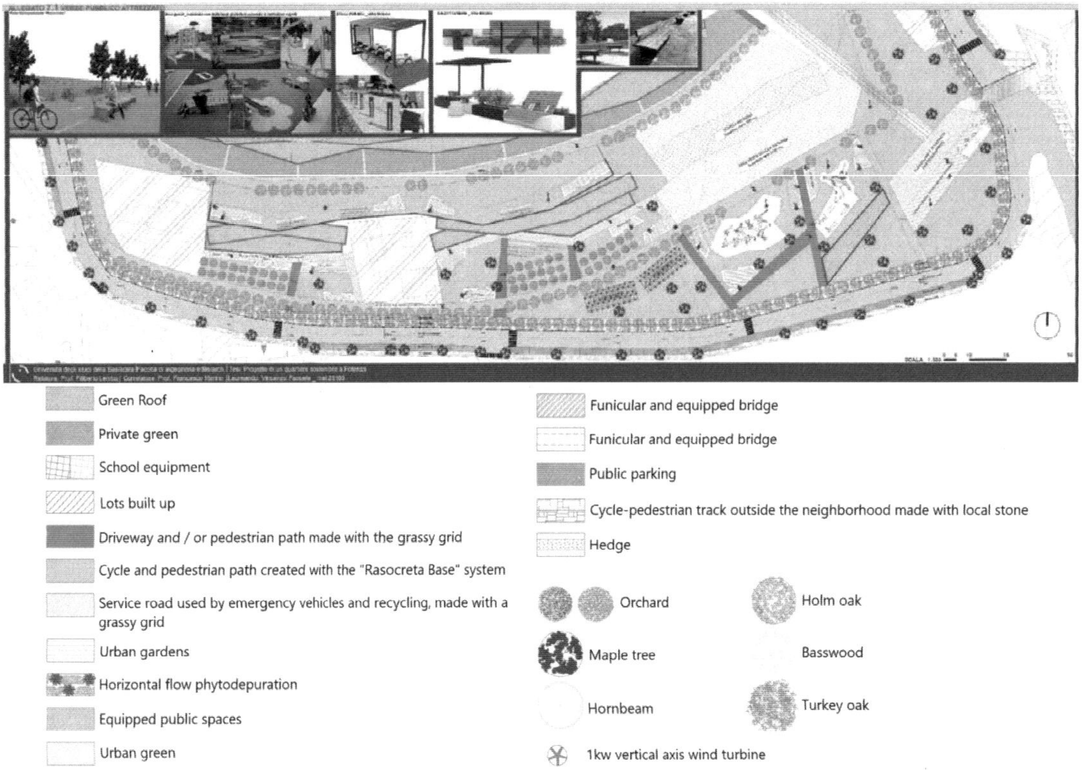

Figure 7. Planning of public green areas

Building's typologies forecast many sizes of dwellings, from 45 m^2 to 95 m^2, and are designed based on established criterions of solar passive architecture (figure 8), so that they warm themselves "at direct gain", and they get cold themselves with radiative passive cooling from the ground. The dwellings (figure 9) have only ground floor, partially under-ground, set diagonally following the slope of the hill so that the roof of a dwelling is the garden of the dwelling upon.

The top floor has *maisonettes* in which roof are set out thermal solar panels (producing hot water for sanitary purposes and supplement to heat pump heating) and PV panels. The interface with the ground and the retaining walls are in reinforced concrete. Buildings have been designed with structures pillar-beam in local glued laminated timber made of hardwood *Quercus Cerris*, thermo-hydrometrically modified following procedure set by the Authors, with very high structural performances (it is a GL 42), with external walls highly insulated, 20 cm mineral wool, and ventilated rainscreen for maximum

IOP Conf. Series: Earth and Environmental Science **297** (2019) 012028 doi:10.1088/1755-1315/297/1/012028

protection from the dampness and the sun.

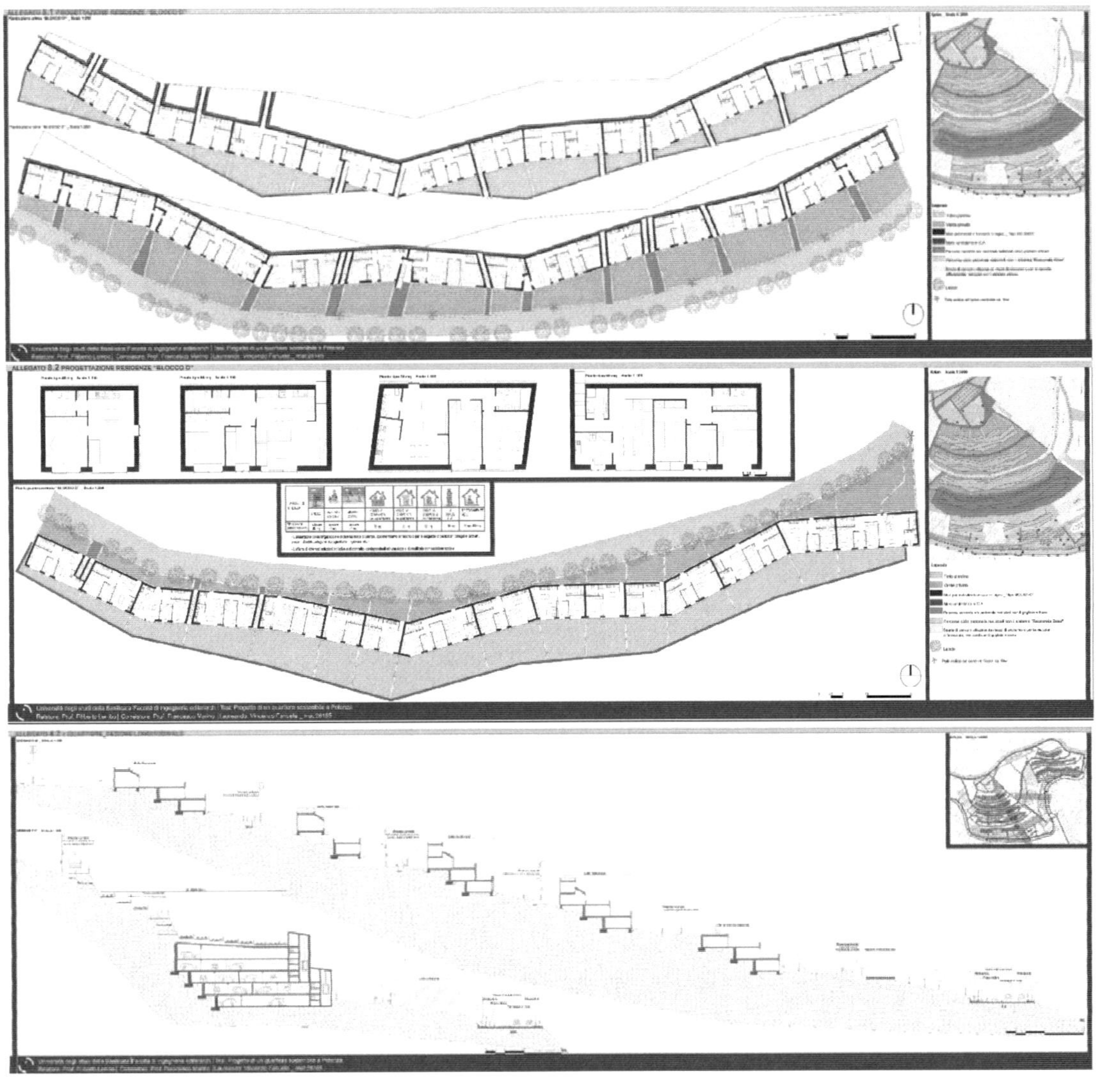

Figure 8. Building's typologies: plan view of the first and ground floors (top), second floor and distribution schemes (in the middle), dwellings' and parking longitudinal sections (below).

Dwellings offer all the main rooms to the light, the air, the sun, even if they are covered by earth for reducing the ΔT. Insulating shielding with opening "to the knee" protect doors and windows from excessive solar radiation, isolate by night and protect against breaking and entering, given the layout at ground floor of the dwellings. Only bathrooms and corridors, set toward the earth, are not naturally illuminated. Gypsum boards containing Phase Changing Material in microcapsules, boards for false ceiling with paraffin PMC, insulation 30 cm thick in wood fibers at roof, panels in local hardwood with dry joists, without glues, 180 mm thick at the perimeter and 120 mm thick for dividing internal spaces and wooden floors, 260 mm thick, maximize thermal inertia of the passive accumulation system.

They are all nZEB, with U ranging from 0,151 W/m^{2}K of the external walls and roof garden, to 0,128 W/m^{2}K of the walls against earth. Windows and doors are in wood/PU/aluminium with triple glass, with Uw= 0,62 W/m^{2}K. The insulation against earth is only perimetrical, extended 3,00 m outside the built contour (figure 10). Sailing sunshades for living external zones, Controlled Mechanical Ventilation with heat recovery device and dehumidifier, heat pump air-water with very high efficiency. Local materials, you can recycle in full.

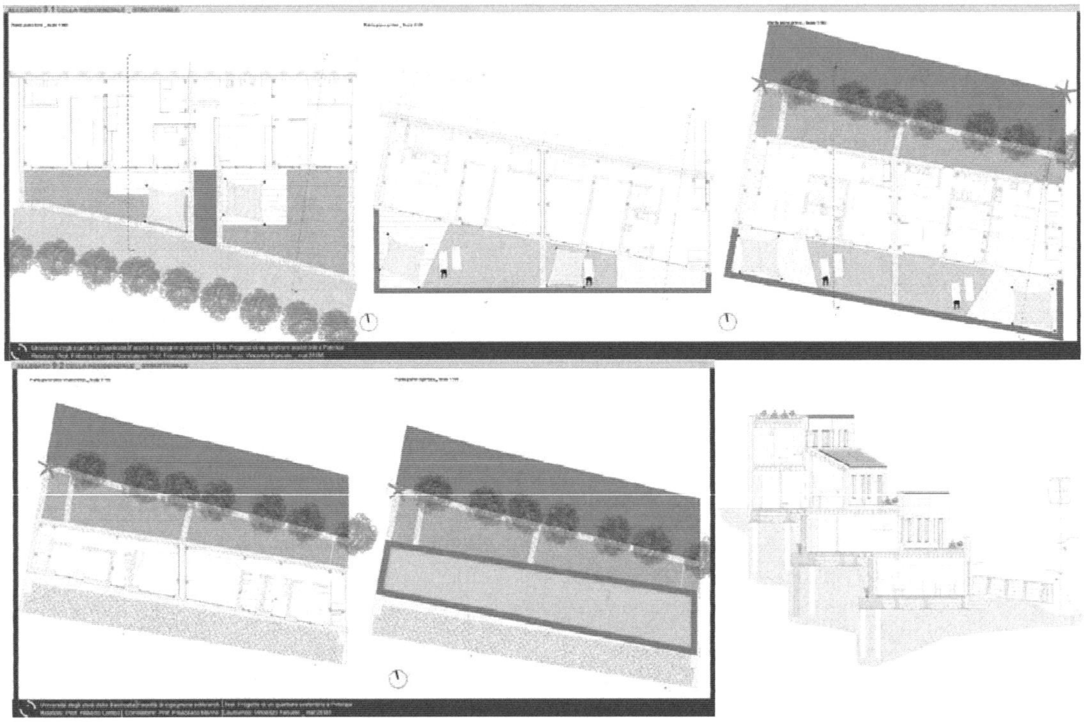

Figure 9. Residential cell type, distributed on three levels (top view) and a *maisonette* (bottom left) for a total gross area of 600 m^2; roof plan (bottom center) and cross-section (bottom right)

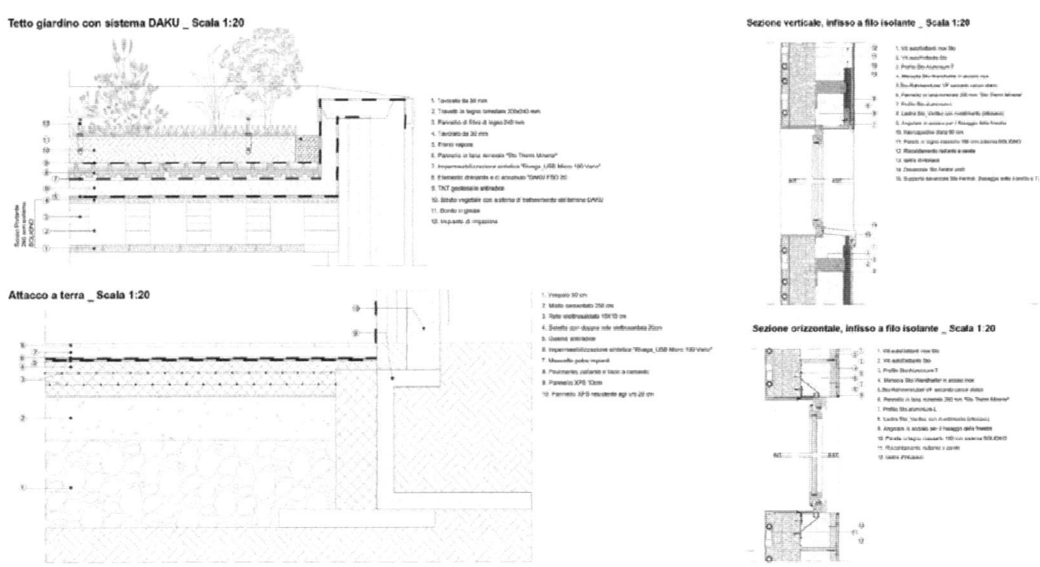

Figure 10. Details: "roof garden" coverage (right top); ground attack (right, bottom); external vertical wall in the vertical section (upper left) and horizontal (bottom left) on the frame.

It is scheduled the recovery of meteoric water, collected, to be fito-purify with a vertical submerged flow system and piped in accumulation reservoirs in the service of different buildings, re-used for feeding flushing cisterns of wc, washing machines and irrigation of gardens and vegetable gardens. Grey waters originated from houses and from other activities (average 600 people in all) it is expected to collect and pipe to be fito-purify in two plants of 160 m^2 each, able to handle 42 m^3 / day of grey water, re-utilized not for vegetable gardens, but for decorative ones. Public lighting installation rest on energy acquired by mini-Eolic plants with vertical axis, integrated into the street lamps; their electrical production, more than demand by led light, it is scheduled to supplement cable railway supply.

4. Conclusions

Assessed with ITACA Protocol criterions, the settlement gains rating of 4,5/5. It does not rank at full score only because it is placed on a virgin area, not before urbanized or polluted. The research shows the usefulness and effectiveness of the use of the ITACA Protocol as a support methodology for the design of sustainable urban developments.

It is a neighbourhood where it is possible to move walking or by bike, or by a mechanized public transport service; public and private green, orchards and vegetable gardens are everywhere, on the roof of the parking's and of the homes too; the "heat island" effect is banned; buildings are nZEB, built with local materials, re-usable, re-cyclable; meteoric waters and grey ones are collected, fito-purified and re-used, and renewable energy available *in loco* (sun, earth, wind) satisfy limited requirements which remain after having used passive solar typologies for heating and for cooling. In fact, the climatic study and the privileged solar exposure of the site have directed the design towards unified elongated volumes developed along the East-West axis, with the South front totally free and the North front buried to optimize the thermal contributions and reduce the need for artificial lighting and heat loss, so as to minimize the use of heating and cooling systems. The buildings are terrace type that provide a variety of solutions from the typological point of view: they are planimetrically articulated according to a geometry with broken lines, redrawn on the level curves that shape the existing ground and on the basis of generating lines that are incident to them, and thus allow to take advantage of the high slope inclination between 20 and 30 degrees.

Research indicates the need, in order to achieve the highest energy efficiency, as a first rule, to overcome the building types of nineteenth-century city and of Rationalism and Modern Movement.

But, most of all, it remains established that, if just you want, it is possible to plan a more sustainable city, more joyful and healthier too, also remaining in the actual city planning Regulations, and that the true *smart city* is not that stupid existing, just cabled and informatized, but that which use at the best resources of a specific culture and of intelligence.

References

[1] Galanti A 2010 *Forma urbana, sostenibilità, pianificazione* (Roma: Aracne Editrice) p 29

[2] DETR – Department of the Environment Transport and the Regions 1998 *Planning for Sustainable Development: Towards Better Practice* (London)

[3] Samonà G 1973 *La casa popolare degli anni '30*, ed. Manieri Elia M (Venezia: Marsilio Editori)

[4] Berrini M and Colonnetti A 2010 *GREEN LIFE. Costruire città sostenibili* (Bologna: Editrice Compositori)

[5] Steemers T and The ECD Partnership 1991, *Solar Architecture in Europe. Design, Performance and Evaluation*, Commission of the European Communities, Publication n. EUR 12738 EN, Bridport, Dorset, Prism Press

[6] Lembo F and Marino FPR 2012 *Raffrescare naturalmente un edificio con le tecniche costruttive storice dell'architettura pugliese*, PROTECTA, vol. 4, pp. 70-73, ISSN: 1121-3124

[7] Lembo F and Marino FPR 2013 *Semi-underground buildings: design models for the contemporary ities*, in Zhou, Cai & Sterling (eds), "Advances in Underground Space Development", Proc. ACUUS 2012 Congress (Singapore: Research Publishing, The Society for Rock Mechanics & Engineering Geology) pp 139-149, ISBN: 978-981-07-3757-3, DOI:

10.3850/978-981-07-3757-3_RP-111-P154
[8] Lembo F and Marino FPR 2014 *Typological choices for underground architecture* Proc. ACUUS 2014 - 14th World Conf. of Associated Research Centers for the Urban Underground Space (Seoul: Associated research Centers for the Urban Underground Space) pp 404-410
[9] Spanu S 2017 *Città del futuro. Il modello di Friburgo in Breisgau* (Milano: Franco Angeli Editore)
[10] Gonzalo R and Vallentin R 2014 *Passive House Design. Planning and design of energy-efficient buildings* (Munich: DETAIL Green Books)
[11] Cody B 1998 Low energy apartment building in Berlin *The Arup Journal* Vol. 3 (London) pp. 14-17
[12] ILO – International Labour Office 1970 *Manuale di studio del lavoro, dei metodi e dei tempi*, 5^ ed. (Milano: Franco Angeli Editore)
[13] Galanti A 2010 *op cit* p 30
[14] Operative Plan of the "C2" Zone (Contrada Botte) of the General Regulatory Plan 2007 *Technical Implementation Rules* (Municipality of Potenza: approved on 13/04/2007 with City Council resolution no. 24)
[15] Reference practice UNI/PdR13.0 2015 *Environmental sustainability in construction- Operational tools for the assessment of sustainability - General framework and methodological principles* (published on 30 January 2015 and corrected on 22 June 2016) ITACA/UNI, ICS 91.040.01
[16] Reference practice UNI/PdR13.1 2015 *Environmental sustainability in construction- Operational tools for the assessment of sustainability - Residential buildings* (published on 30 January 2015 and corrected on 22 June 2016) ITACA/UNI, ICS 91.040.01
[17] ITACA Protocol 2015 *Non-residential buildings* (published on 12 November 2015, it fits into the context of a general framework and methodological and procedural principles described in UNI/PdR13:2015) ITACA/UNI, ICS 91.040.01

Contributions

Filiberto Lembo coordinated the research and provided the objectives. Francesco Paolo R. Marino developed the research project, methodological and operational tools, analysed and verified the results. Vincenzo Fanuele made specific analyses in his degree thesis. The authors' contribution in drafting and writing the text of the article was the same.

IOP Conf. Series: Earth and Environmental Science **297** (2019) 012029 doi:10.1088/1755-1315/297/1/012029

RTS EPD's – Different Methods of Producing Reliable Environmental Information of Building Products in Finland

Laura Sariola
The Building Information Foundation RTS sr, Po.Box 1004, 00100 Helsinki, Finland, Europe, laura.sariola@rakennustieto.fi

Ari Ilomäki, Confederation of Finnish Construction Products Industries RTT, Eteläranta 10 (Po.Box 381), 00130 Helsinki, Finland, ari.ilomaki@rakennusteollisuus.fi

Abstract. The Building Information Foundation RTS sr launched a national EPD program in the year 2016 and after that there have been produced several RTS EPD's published by the RTS. The Information gathered from the Environmental Product Declaration (EPD) is verified by a third party and therefore neutral and reliable. Environmental performance of construction products is compiled according to the standard EN 15804+A1 and A2 and with a guidance's and methodology processed in the committee PT18 RT EPD's. The name of the principal publication is "Methodology (PCR) for compiling environmental declarations for building products: EPD's published by The Building Information Foundation RTS (2019)" and in addition there are guidance for the application process and verification processes.

The Building Information Foundation got a full membership status of EcoPlatform in September 2018 and therefore the guidance has been revised. PT18 noticed the future demand of easy way of producing EPD's and therefore it compiled a guidance of verification process of customized and product specific EPD-tools used in connection to production processes. PT18 considered also amendment of the EN15804 standard (A2) in advance as far as possible and updated the RTS methodology accordingly. The EPD's are created in Finland by the Principal Committee PT18 of The Building Information Foundation RTS. A work group of the Committee PT18 will audit and approve/reject the applications. All RTS EPDs have been verified, there is no difference if the EPDs have been complied with the EPD tool or without it.

The eco-profiles and other environmental aspects provide input data for the assessment of environmental impacts of the buildings. The purpose of the declarations is to provide environmental information of construction products in a standardized and transparent way. The EN-standardized method developed by CEN/TC350 is utilized in other European countries as well by different EPD-programs. The purpose of different EPD-tools is to provide easy way to produce EPD information of similar products and to use gathered information in multiple product information services and data modelling tools.

1. Introduction

RTS EPDs help building contractors, constructors and planners to better understand environmental issues related to construction products and materials. The environmental impacts of a product during its life cycle and making better use of natural resources have become relevant information in order to assess the environmental performance of a building or other construction works. EPDs according to the European standard EN 15804:2012+A1:2013 are voluntary, transparent and well- established way to present fundamental and verified environmental information based on life cycle assessment about the environmental impacts of a product or product group.

An EPD for construction materials can be drawn up for many different applications. The manufacturer of a construction product can make a declaration:

- For the company's own use, in order to provide information about the environmental impacts of products.
- An EPD can also be made for other companies, in which case the declaration provides the company with essential information from the point of view of the information needs of other companies.
- for EPD tools and basic information (concerning calculation data) in it.

RTS EPDs are made primarily for "business-to-business" use. The declarations are Type III EPDs, i.e. they provide quantified environmental information about the life cycle of a product in order to enable comparison at the building level between products that fulfil the same functional and technical requirements. The life cycle assessment is carried out in accordance with EN ISO 14044 and the construction sector-specific amendments contained in EN 15804:2012 + A1:2013 (e.g., system boundaries, modularity, etc.). An EPD can also be made for consumers. In this case, the most technical information can be left out to focus exclusively on what is essential for the consumer (e.g., carbon footprint and energy consumption).

Figure 1
a RTS EPD-logo

2. Content and process of the RTS EPD

EPDs shall contain all the information that, in accordance with EN 15804:2012 + A1:2013, must be presented in EPDs in the context of "business-to-business" communications. They can be produced by one or more organisations for one product or a product group containing similar products. Product information and environmental impact calculations shall be presented in EPDs as defined in EN 15804:2012 + A1:2013. All RTS EPDs shall be verified by a third party. The information contained in them has been checked and proved to be a reliable source of environmental information.

The indicators referred to in the standard EN 15804:2012+A1:2013 are used to describe the environmental impacts caused during the product life cycle, including:

- climate change impact, which takes account of the so-called carbon footprint
- ozone-depleting substances, which make the ozone layer thinner
- acidifying emissions to soil and water bodies, which damage ecosystems and the built environment
- depletion of non-renewable energy resources and mineral flows resulting from the exploitation of these resources.

There shall be presented all the modules shown in a table and marked with x.

Product stage			Construction process stage		Use stage							End of life stage				Supplementary information beyond the life cycle		
A1	A2	A3	A4	A5	B1	B2	B3	B4	B5	B6	B7	C1	C2	C3	C4	D	D	D
☒	☒	☒	☒	☐	☐	☐	☐	☐	☐	☐	☐	☒	☒	☒	☒	☒	☒	☒
Raw material supply	Transport	Manufacturing	Transport	Construction-instsallation process	Use	Maintenance	Repair	Replacement	Refurbishment	Operational energy use	Operational water use	De-construction demolition	Transport	Waste processing	Disposal	Reuse	Recovery	Recycling

Figure 2 Building process stages according to CEN/TC 350 standards

Approval shall be carried out by a workgroup approved by the PT18 RTS EPD Committee. The workgroup will need application form fulfilled and with a signature, RTS EPD declaration document compiled according to EN 15804, product information, verification checklist filled up, dialog filled up. The verification data and the EPD drawn up by the company shall be checked during the approval process. The approved EPDs are published on the website of the Building Information Foundation RTS sr at http://epd.rts.fi.

Approval process flow

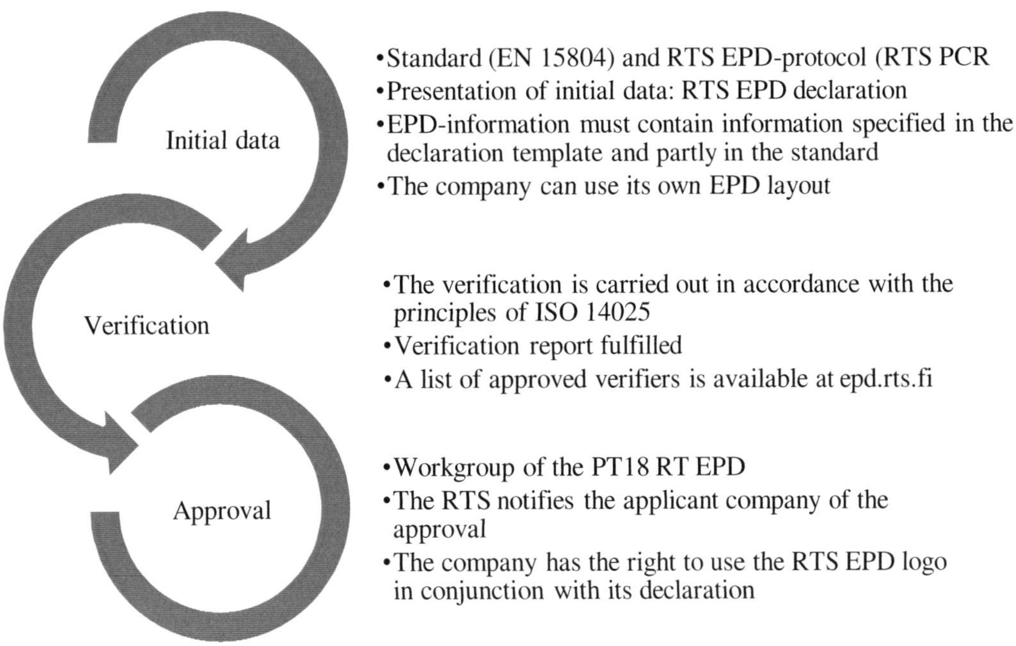

Figure 3: RTS EPD approval process

After few years publishing process of RTS EPD's the PT18 work group has accepted 25 compiled RTS EPDs covering approximately hundred products. It is obvious that without great amount of published EPDs, e.g. carbon footprint calculations cannot be provided appropriately. There are a lot of other EPD organizers in the Europe and according to latest calculations, amount of published EPDs have increased to 6000 pieces.

3. The EPD-tool and verification process

Environmental impacts of a product during its life cycle have become relevant information to assess environmental performance of a building or other construction works. Therefore, access to the data should be as simple as possible. EPD-tools will provide manufacturers to demonstrate environmental impacts and flows and to create environmentally better products to meet increasing requirements placed for the products.

The RTS EPDs comprised with an EPD-tool are verified. The PT18 RT EPD will accept EPD-tools and with this option the Building Information Foundation RTS sr will establish process to reduce the amount of work in collecting data, performing LCAs and creating RTS EPDs for similar product types. When creating a new EPD-tool, the purpose and target product groups of the tool shall be determined in detail.

The verification of EPD-tools differs from the verification of independent EPD's. The LCA-report/PRE-EPD shall contain all relevant product and production information in suitable form (e.g. data sheet programme, machine readable) to be easily added to the EPD-database and later use e.g in building information modelling.

The EPD-tool cannot be created for wide product groups, the products shall be classified in the same group with similar ingredients. If the tool is suitable for different product groups, the details concerning specific products shall be separated. The EPD-tool shall be created in such a way that product information in different product groups cannot be mixed together.

The EPD-tool can be developed by a single company or by several companies like industry organizations. EPD-tools that will be used by several companies will lead to comparable EPDs that reduce the possible systematic error in EPD-tools used only by one company. It is important for both the company comprising RTS EPDs and for the Building Information Foundation RTS to emphasize that RTS EPDs created using EPD-tools shall have the same quality as EPDs created without EPD-tools.

The quality of the EPD-tools and process will be checked continuously by appointed data collector and verified every three years by approved verifier. SAMPLE-EPDs verification will be carried out every three year in connection with verification of the EPD-tool. The quality of the EPDs will be ensured by verifying every EPD compiled in the EPD-tool. The verification process of single EPDs has been simplified and does not cover all parameters verified at the first time. The EPD-tool can use background information e.g. ready-made and approved LCA databanks mentioned in the RTS PCR like GaBi (http://www.gabi-software.com/databases/gabi-databases/) and ecoinvent database (www.ecoinvent.ch). The background LCA-tool shall be verified with the foreground LCA-tool and the combination (Background LCA and foreground LCA databanks) is valid for three years. Reverification is after three years. When verifying the EPD-tool for the first time, both the tool and SAMPLE-EPDs' (three different kind, extremities included) will be verified.

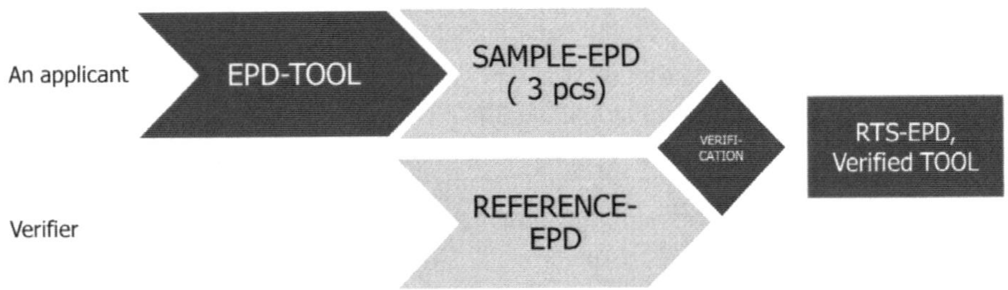

Figure 4 Verification process of an EPD-tool and SAMPLE-EPDs

PRE-EPD verification process, in other words, simplified verification process can start after verifying the tool and SAMPLE-EPDs for the first time. The verifier shall use the checklist for the verification process of PRE-EPDs. The checklist differs from the SAMPLE-EPD and tool verification checklist and concentrate on the modified details. The verifier should check basic information and modified information using the log book as a map of all changes.

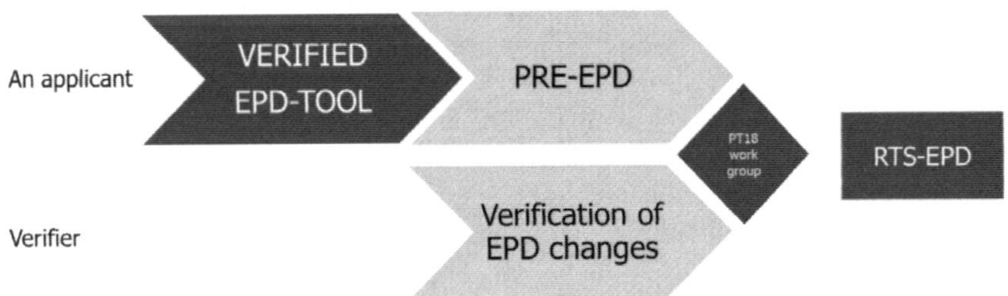

Figure 5 Simplified verification process of an EPD printed from the EPD-tool

4. Connections to other systems

The building sector is digitalizing the information needed for the building process and for other processes during the building life cycle with accelerating speed in Finland.

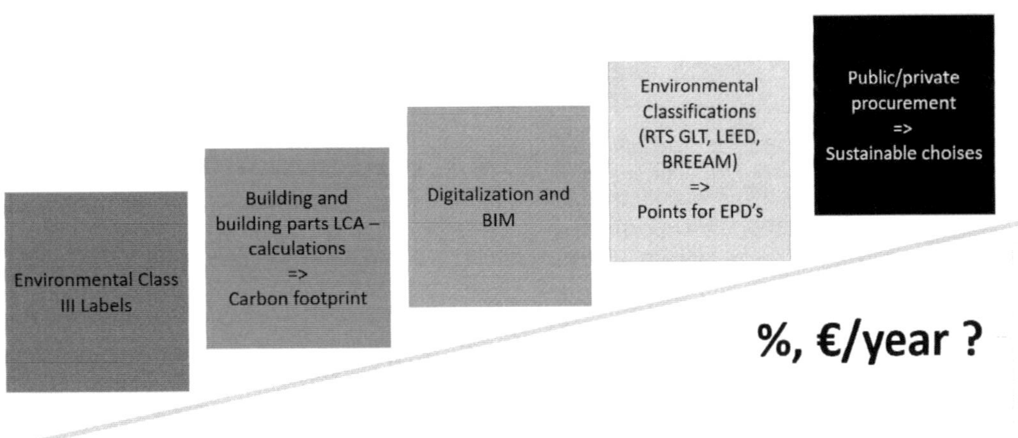

Figure 6 Information gathered to the RTS EPD's can be used in different levels

The need for specific information in certain phase of the designing process can be achieved because of digitalized designing tools and databases linked to them. The tools help designers, contractors, building owners, real estate management and other participants to decide in an early phase e.g. with the following parameters: used materials, used amounts, targets for carbon foot print, targets for energy efficiency in construction and in use phases, targets for resource efficiency. The outcome will be more precise target values and more precise price for the project. Altogether the major power in decision making process is in public and private procurements, in the second place are environmental classifications like RTS Environmental Classification.

4.1. The Finnish RTS Environmental Classification

The purpose of the RTS Environmental Classification is to manage, monitor and lead the construction project efficiency and achieve more sustainable buildings and to avoid problems in the future. The expected outcome of the tool is an enhanced environmental performance of the final "as-built" building. The Finnish legislation has been considered in the tool, i.e. every target is above the minimum regulatory requirements. In addition, there are few key classifications and certifications utilized to create the criteria for the Green Leadership Tool. These key factors are the Building Performance Indicators of GBC Finland, the Emission Classification of Building Materials and Cleanliness classification of HVAC devices (M1-label), Classification of Indoor Environment, Energy Audit (E-Value), Green Factor Tool (GF) and Drychain10. The real outcome and constructed premise will be certified. The certification after construction phase is mandatory if the owner of the project like to use certification label.

RTS EPD's are presented in the part Y1.2 Resource efficiency, where are considered material efficiency related to used materials in the construction project. There is possible to gain points also of used EPD's in the projects to encourage material producers to conduct information related to environmental product declarations.

4.2. The Building Performance Indicators

The Building Performance Indicators were developed to react to the demand of key indicators that assesses environmental efficiency of properties and to ensure principles of sustainable development. The project was implemented with support of SITRA and published by the Green Building Council Finland in the year 2013. The eight indicators are mainly based on the European CEN/TC 350 standards including Life-Cycle Carbon Footprint, Life-Cycle Cost, energy performance indicator (E-value), Indoor Air Class, Measured Energy Consumption, Operational Carbon Footprint, Baseload Demand, User Satisfaction. RTS EPD information is basic data provided by the manufacturers.

4.3. Carbon footprint calculations and calculators

In Finland there are different LCA-calculation tools to assess carbon footprint of buildings. In principle carbon footprint calculations can be modelled in the calculation programs, but the simplest way to calculate is a special program where the basic data information is converged to the tool and checked all the time.

4.4. Eco-Platform system

The Building Information Foundation RTS sr has been approved into the ECO Platform system, which contains hundreds of approved EPDs. The holder of an RTS EPD declaration has the possibility to access the list of approved declarations published in the ECO Platform system and add the ECO Platform logo to the declaration. It guarantees the approvability of the declaration within the EPD programmes operating in all European countries as these programmes are ECO Platform members. ECO Platform membership is also mentioned on the Building Information Foundation's website as well as on the list of approved declarations.

5. Conclusions

The members of the principal committee PT18 represent the highest expertise as well as all parties concerned by the system in Finland. A declaration is granted for five years depending on the time for verification document. If the composition and method of manufacturing the product have not changed, the declaration can be renewed by application without new LCA calculations. Both new and continuation applications will always be examined in accordance with the current methodology and guidance. RTS EPD's created using EPD-tool are granted for five years, the EPD-tool and SAMPLE-EPDs' are granted for three years.

The RTS EPD is a voluntary and public document providing comparable and impartial information on the environmental impacts of construction products. It is a source of information for users, designers and constructors. In the future demand for environmentally, socially and economically sustainable buildings will increase and for that purpose shall be created lot of transparent, comparable and third party verified environmental declarations to be used in digitalized designing processes.

6. Acknowledgement

The methodology for compiling RTS EPD is a document and guidance tool created by the enthusiastic and indefatigable committee PT18 RT EPD appointed by the Director General of The Building Information Foundation RTS sr.

7. References

[1] ISO 14025, ISO 14025:2011-10 Environmental labels and declarations. Type III environmental declarations. Principles and procedures
[2] EN 15804, EN15804+A1:2014 Sustainability of construction works. Environmental Product Declarations. Core rules for the product category of construction products
[3] The Building Information Foundation RTS, Methodology (PCR) for compiling environmental declarations for building products: EPD's published by The Building Information Foundation RTS (2018), http://epd.rts.fi
[4] the Construction Products Regulation (Regulation (EU) No 305/2011)
[5] The Building Information Foundation RTS sr, The Green Leadership Tool http://glt.rts.fi
[6] Eco-Platform https://www.eco-platform.org/
[7] The Building Performance Indicators http://figbc.fi/elinkaarimittarit/

IOP Conf. Series: Earth and Environmental Science **297** (2019) 012030 doi:10.1088/1755-1315/297/1/012030

Equation for zero emissions -achieving carbon-neutrality in building sector

Antti Ruuska[1] and Markus Mäkinen[2]

1 Co-creation manager, Doctor of Science, VTT Technical Research Centre of Finland Ltd, Espoo, Finland

2 Research Scientist, Master of Science, VTT Technical Research Centre of Finland Ltd, Espoo, Finland

E-mail: antti.ruuska@vtt.fi

Abstract. Energy has a central role in people's lives. However, access to affordable, clean and reliable energy is not the case yet for all. As more people have access to energy, the global energy consumption is set to rise, along with the global greenhouse gas emissions.

The building sector is one of the most significant ones in terms of energy consumption and GHG emissions (Ruuska, 2018). Buildings consume vast share of global energy and produce significant share of greenhouse gas emissions. Thus far the improvements in energy efficiency have not been able to offset the increase in energy use due to increased amount of buildings. In order to cap the global warming to 1.5 °C (IPCC), thorough changes are needed in the way we consume and produce energy. For the building sector, it is suggested that the whole of the building sector should decarbonize by 2050 (WGBC, 2018). This paper assumes that the building sector can decarbonize by 2050 and looks for ways to make it happen.

Technological means to move towards clean, low-carbon energy in buildings are many. On the supply side, moving away from fossil fuels and increasing the share of renewable energy are the big goals. On the demand side, energy use can be reduced through better energy efficiency in buildings. Furthermore, smart energy management can help to operate buildings optimally, and to shift and reduce peak loads. There are also opportunities related to, for example, energy storages. However, this paper assumes that the key towards broader implementation of the technological means may lie in user-friendly services and business models.

This paper further assumes that decarbonizing construction industry is not a technological problem as such, but more of a systemic problem. Therefore, this paper seeks to understand the interactions and feedback loops in the sector, by focusing on the level of a single building and its owner/user. This paper shows the general dynamics of the energy systems around a single building, and the connections they have to the broader energy system. Furthermore, this paper presents a simple equation for zero emission building stock and links this equation to system dynamics around single building's energy systems.

All of this is done from the viewpoint of a single building and its owner/user to illustrate how the situation looks from the user/owner perspective and to see if some factors arise that could speed up the transformation towards decarbonized building sector.

1. Background and problem formulation

The equation for zero emissions from building sector is mathematically very simple. We can either zero the operational energy consumption from the sector, or we make the consumed energy emission-free. At the same time, we need to look at the building life-cycle to make sure that the embodied carbon from material production, construction, renovations and other life-cycle activities are zero. You can write the equation in the form:

$$GHG = E_E * E_A + GHG_M = 0 \qquad \text{(Equation 1)}$$

GHG = Greenhouse gas emissions
E_E = Emissions from energy production
E_A = Amount of energy use
GHG_M = GHGs from material-related sources, or embodied carbon.

Equation 1 is very simplified version of the work by Ruuska (2018) [1], but serves the purpose of this paper. In principle, almost any building owner could zero their emissions already today. Theoretically, moving to use clean and green energy is a choice that a building owner/user could make any given day. Likewise, energy efficiency improvements are available at the markets and the owners can choose to select low-carbon materials in their new construction and renovations. The building owners and users can also compensate any occurring emissions with different means, for example by purchasing emission allowances through the EU ETS and keeping them off the market [2]. This paper understands that in practice, the options for building owners and users may be only limited and those available may not be cost-efficient.

At the same time, this paper understands that on global and societal levels the building sector is one of the most significant GHG emission producers and global energy consumers [1]. Thus far, sector's energy efficiency improvements have not been able to offset the increase of energy use due to increased amount of buildings. In order to cap the global warming to 1.5 °C [3], thorough changes are needed in the way we consume and produce energy. There's also viewpoints that the whole building sector should decarbonize by 2050 [4] to keep us on track towards 1.5 °C. At the moment, the progress on sectoral level is not fast enough as we are falling behind even from the 2.0 °C goals in this sector [5, 6].

In order to catch up and reach the challenges of 1.5 °C, we need to find new ways to rapidly take action in the built environment. Furthermore, massive investments on global scale will be needed to boost these actions. This paper assumes that the building sector can decarbonize by 2050 and looks for ways to make it happen. It hypothesizes that system-level impacts and their value could offer a way to speed up the progress in the sector and actions at the level of single buildings.

2. Research methods

2.1. System dynamics of building energy ecosystems

This paper builds on earlier research by Nieminen, Åkerman and Mäkinen, 2018 [7] and focuses on the viewpoint of a single building owner as part of the energy system. In their work, the authors developed system dynamics for 'evolving building energy ecosystems' and studied the prerequisites and dynamics of such as system in Finland. The authors' model was based on expert review of different stakeholder's strategies in the sector. The authors also conducted a total of 36 interviews of various actors in the field of energy and construction (public authorities, big and small energy companies, construction companies, solution providers and users/dwellers). The interpretations and the system dynamics were validated in two separate expert workshops, held with relevant stakeholders (Nieminen, Åkerman and Mäkinen, 2018).

In their work the authors:

- showed various interactions and feedback loops in the system
- defined the general dynamics which makes possible the emergence of new energy regime and ecosystems

- indicated how number of actions & actors affect the transition.

Figure 1 illustrates the system dynamics of the evolving building energy ecosystems (Nieminen, Åkerman and Mäkinen, 2018)

Figure 1. System dynamics of the evolving building energy ecosystems (Nieminen, Åkerman and Mäkinen, 2018).

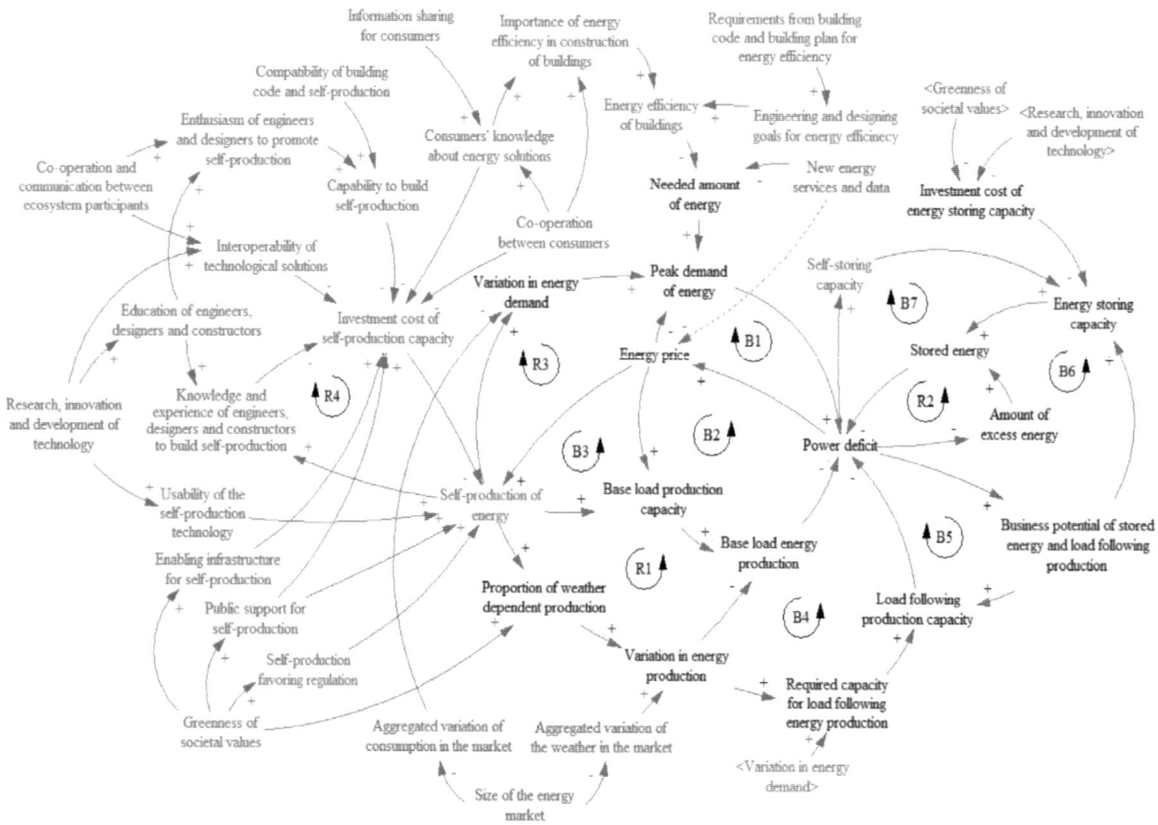

2.2. System dynamics of the energy systems of a building

This paper utilizes a system dynamic model from past research [8] and develops it to focus around the owner/user of a single building. This paper looks to:

- show interactions and feedback loops of an energy system that is centered around a single building and its owner/user
- define general dynamics that could make it possible to move towards carbon-neutral building sector by connecting the system dynamics to the different factors in *Equation for zero emissions*, shown in Equation 1
- draw conclusions about what could speed up the transition towards carbon-neutral building sector, focusing on the owner/user perspective

The modelling work in this paper was done as expert work and its interpretations have been validated through expert discussions. The model does not aim to be a complete presentation of the system dynamics in this field, but it aims to serve in focusing future efforts in the field. The model, along with most relevant interactions and the feedback loops in the system are described in more detail under the 'Results' section.

3. Results

3.1. System dynamics around the energy system of a single building

The work of this paper is largely based around the system dynamics of energy systems of a single building that is presented in Figure 2. The model also shows how different factors of Equation 1 are linked to model. The figure gives a visualization of how different feedback loops impact the emissions of a single building.

Figure 2. System dynamics of energy systems of a single building, along with linkages to Equation 1. *GHG= Greenhouse gas emissions of a building,* E_E *= Emissions from energy production,* E_A *= Amount of energy use and* GHG_M *= GHGs from material-related sources, or embodied carbon*

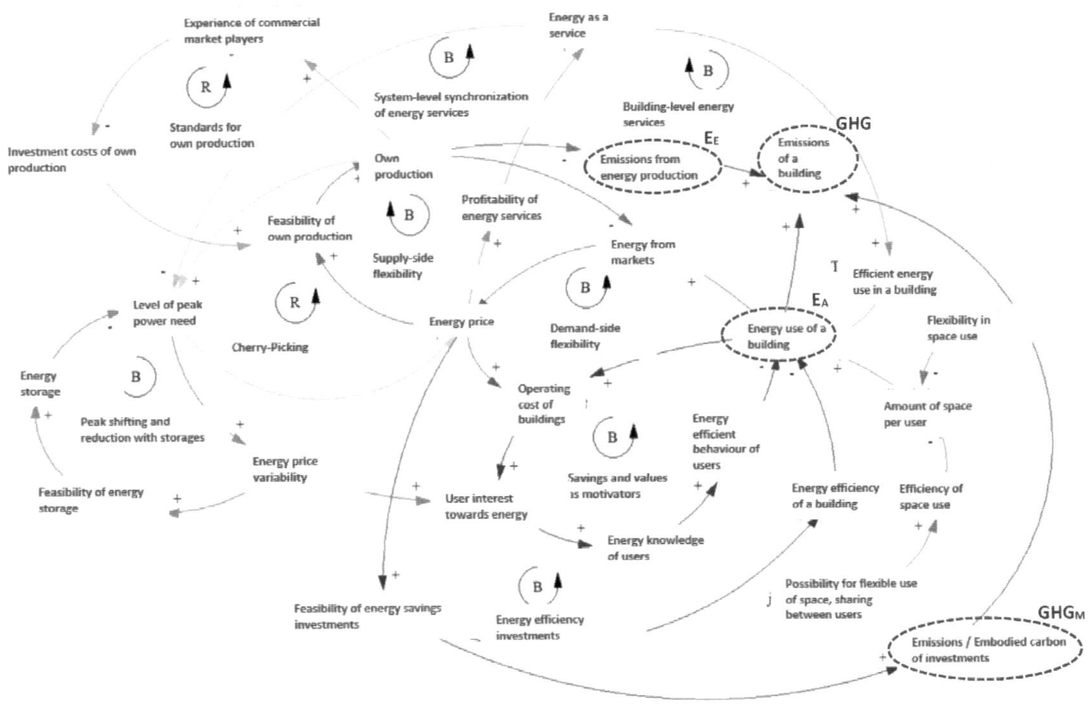

As Figure 2 illustrates, system dynamics around the energy system of a single building form a complex model with multiple interlinked feedback loops. Based on the modelling results and expert assessment, a total of nine feedback loops were identified. The following goes through the key loops, while the remainder are discussed in the *discussion*-section.

The key loop for making things happen on building level is the one centred on the owner/user of a building. This is called *Savings and values as motivators.* In short, values and savings are strong drivers for this loop, as every action on building level requires motivated actions from the owner/user. This loop can activate or deactivate any of the other loops that impact the emissions of a building (here, we mean loops impacting Equation 1 factors E_E, E_A and GHG_M).

IOP Conf. Series: Earth and Environmental Science **297** (2019) 012030 doi:10.1088/1755-1315/297/1/012030

The other two key loops in the system are: *demand side flexibility* and *energy efficiency investments.* First of these loops, *demand side flexibility,* along its related sub-loops, has an impact on the energy need (or supply) of a building (E_A), and the timing of any such need / supply. This loop also impacts emissions from energy production, E_E, although this link is not as apparent. Demand side-flexibility can help to shift energy consumption to more favourable times from the energy system perspective, for example, away from the morning peak hours. Likewise, it can supply clean energy to the grid, for example, through solar panels.

The second of these loops, *energy efficiency investments,* can lower the energy consumption of a building, but also result in material-related carbon emissions. This means that the benefits of energy savings (lower E_A) need to exceed the negative impacts (GHG_M) in order for energy efficiency improvements to be beneficial from the viewpoint of decarbonisation efforts of the sector.

These feedback loops are discussed in more detail in the discussions section. It is understood here that their naming of the loops could be different, some of the loops could be combined with each other and some of the relationships could be modelled in a different way. However, this paper sees this modelling work is sufficient for the purpose of this paper.

4. Discussion

This paper utilized a system dynamic model from past research and developed it to focus around the owner/user of a single building. This paper aimed to: 1) show interactions and feedback loops of an energy system that is centred around a single building and its owner/user, 2) define general dynamics that could make it possible to move towards carbon-neutral building sector by connecting the system dynamics to the different factors in Equation for zero emissions, shown in Equation 1, 3) draw conclusions about what could speed up the transition towards carbon-neutral building sector, focusing on the owner/user perspective.

4.1. Savings and values as motivators

The key loop for making a change happen on building level is the one that's focusing on the owner/user of a building. This is called *Savings and values as motivators.* In short, nothing will happen at the level of a single building without some actions from the owner/user.

While the single building is the focus of this paper, it is in principle possible that the whole building sector could decarbonize without any actions from the owner/user. This would require that the energy in the broader system would become clean. In other words, if E_E could be zeroed, used energy would be emissions-free.

This would require that the GHG_M should also become zero, meaning that the building material industry, construction industry, transportation sector, etc. would also need to decarbonize. Were these things to happen, the owner/user would not have to take any action, as the building sector could decarbonize.

However, this paper sees that user/owner actions, such as *demand-side flexibility* and *energy efficiency investments* can help to decarbonize the broader system through their connection to the broader energy system.

4.2. Demand-side flexibility

Demand side flexibility is a linked to the operating costs of a building. In practice this means that the buildings could offer flexibility towards the energy networks in terms of lowering their demand temporarily. In order for such a feedback loop to form, the building users/owners should be rewarded for offering flexibility. At the moment, such rewards are not commonplace for the users/owners. In order to boost this loop, the user/owner should be rewarded for such activities. Furthermore, as the technologies that are related to the demand-side flexibility are many, it is seen here that there is a space for service providers who could combine for example: *building-level energy services* (such as smart control of indoor air conditions), *supply-side flexibility* (or production of on-site renewable energy), and *peak shifting and shaving with storages* (for example batteries) into meaningful and beneficial services for the building users/owners.

It is also seen here, that whereas the user/owner may benefit from some actions, there may also be system-level benefits through *system-level synchronization of energy services*. Aggregating loads from multiple buildings into larger loads that can be connected to the energy markets is a concrete example of this[1]. Another example is bundling demand-side flexibility of heating loads from multiple buildings into larger entities that can help district heat companies decrease use of peak power plants[2]. It is understood that some things, such as *standards for own production* could boost up the uptake of all of these solutions. However, a more contradictory aspect is that some of the building-level actions may be seen as *cherry-picking*. Some solutions, like heat pumps replacing district heating, may lead to unwanted results on system-level, regardless of being more beneficial for the building owners/users.

4.3. Energy efficiency investments

Energy efficiency investments are investments into, for example, better insulation, more effective building heating, ventilation and air conditioning systems, or in smart control. However, any investments in energy efficiency also result in embodied energy and carbon, meaning that the amount of such emissions needs to be calculated with life cycle assessment to ensure that such investments actually result in payback in terms of energy and emissions.

The system dynamics of Figure 2 also show that one aspect of emissions from buildings is the amount of space per user. This is also a contradictory feedback loop, as it may actually increase the energy consumption and emissions on building level but decrease them on the system level. Fitting more people, or utilizing buildings more effectively increase their energy consumption and emissions, but bring benefits through decreased overall need of buildings.

5. Conclusions

In order to catch up and reach the challenges of 1.5C, we need to find new ways to rapidly take action in the built environment. This paper assumes that this can be done and shows how the different feedback loops on building-level can help achieve this. This paper sees that the complexity needs to be decreased towards the users/owners of the buildings with services that improve the indoor conditions and bring value for them. Technologies on building-level are many, but their benefits and/or pitfalls from greenhouse gas emissions may be apparent only on the system-level.

References

[1] A. Ruuska, Sustainability of buildings -New perspectives on material-related environmental impacts of buildings. Doctoral Dissertation.

[2] CO_2Esto Ltd, "CO2 Elimination by CO2 Esto Ltd", Web-pages: https://co2esto.com/english. Accessed 30.12.2018.

[3] Intergovernmental panel on climate change (IPCC), "Global warming of 1.5°C - An IPCC Special Report on the impacts of global warming of 1.5°C above pre-industrial levels and related global greenhouse gas emission pathways, in the context of strengthening the global response to the threat of climate change, sustainable development, and efforts to eradicate poverty - Summary for Policymakers", Switzerland, 2018. Available online at:

[1] For commercial example, see [8]

[2] For commercial example, see [9, 10]

https://www.ipcc.ch/site/assets/uploads/sites/2/2018/07/SR15_SPM_High_Res.pdf. Accessed 30.12.2018.

[4] World Green Building Council (WGBC), "Delivering the Paris Agreement – The Role of the Built Environment", Europe, 2018. Available online at: https://www.worldgbc.org/sites/default/files/2050%20Letter%20Final.pdf. Accessed 30.12.2018.

[5] International energy agency (IEA), Tracking Clean Energy Progress 2017, Energy Technology Perspectives 2017 Excerpt Informing Energy Sector Transformations, Release 6 June 2017. Available online at: https://www.iea.org/publications/freepublications/publication/TrackingCleanEnergyProgress2017.pdf. Accessed 30.12.2018.

[6] International energy agency (IEA), Tracking Clean Energy Progress 2017, -background data and online tracking of progress, www.iea.org/etp/tracking2017. Accessed 30.12.2018.

[7] Nieminen, M., Åkerman, M., & Mäkinen, M. (2018). Dynamics and structures for emergent building energy ecosystems in Finland?. Energizing Futures, Tampere, Finland.

[8] Leanheat Oy web-pages. Available online at: https://leanheat.com/company-info/. Accessed 30.12.2018.

[9] Fingrid Oy web-pages. Available online at: https://www.fingridlehti.fi/en/sello-shopping-mall-targets-the-reserve-market/. Accessed 30.12.2018.

[10] Siemens Oy web-pages. Avalable online at: http://www.siemens.fi/fi/media/uutiset/inauguration-of-the-virtual-power-plant-.htm. Accessed 30.12.2018.

IOP Conf. Series: Earth and Environmental Science **297** (2019) 012031 doi:10.1088/1755-1315/297/1/012031

Participatory service design and community involvement in designing future-ready sustainable learning landscapes

N Sandström[1] A Nevgi[1] and S Nenonen[2]

[1]Campus Learning and Development Initiatives hub (Caledonia hub), Faculty of Educational Sciences, University of Helsinki, Finland
[2]Department of Construction Management and Economics, Research Group of Responsible Construction, Tampere University

E-mail: niclas.sandstrom@helsinki.fi

Abstract. The United Nations Sustainable Development Goals (SDGs) work as a new agenda for sustainable development globally. Many if not most of the SDGs can be combined with different levels of education. This paper leans on previous work in Sustainable Education Design (SED), which looked at sustainability from its multifaceted angles with a broad global scope. The context of the study is a campus at a research-intensive Finnish university. The methodology entailed participatory service design approaches. For piloting, one classroom was chosen as a test bed. The data consist of workshops, use walks and structured interviews. The analysis started from identifying KPIs of sustainable learning environment creation, after which these were tested against Sustainable Education Design Criteria described in a manual book earlier. The key findings include nine preliminary alternative KPIs that were merged with previous SED criteria and related SDGs. The alternative KPIs were trialled in the test bed environment. These proposed alternative KPIs can be used as indicators for sustainability, innovation and learning during participatory change processes and in evaluating the outcome.

1. Introduction

The United Nations Sustainable Development Goals (SDGs) work as a new agenda for sustainable development globally. Many if not most of the SDGs can be combined with different levels of education. The relationship between learning landscapes and SDGs is interconnected and integrated both to learning processes and built digital and physical learning environment. We use Dugdale [1] (p. 52) to define what we mean by learning landscapes:

> [T]he Learning Landscape approach defines a future campus by envisioning overlapping networks of compelling places and hubs, which can offer choices to users and generate synergies through adjacencies and the clustering of facilities.

This paper leans on previous work in Sustainable Education Design (SED), which looked at sustainability from its multifaceted angles with a broad global scope. The special country focus was on Namibia [2]. The project identified various indicators (for instance, Resource Wisdom, Connectivity, Vitality, Communality and Recycling, see Table 1) for sustainable education, educational building and assessment thereof. It is postulated that both weak and strong sustainability [3, 4] require co-design [3] and community involvement as well as shared decision making [5] to become more tangible and to have an impact. Additionally, another related project, DOLO Learn, focuses on primary education in

Namibia, also applying the SED tool [2, 6]. The approach in this paper focuses on higher education and campus retrofitting as a means to operationalise the SDGs. In the present work, we pay attention to stakeholder participation and service design as tools to co-create learning landscapes that are attractive, healthy and sustainable.

Sustainability in our approach heavily entails social, rooted involvement that in turn produces user satisfaction and ownership in developing the campus learning landscapes further [7]. We test and develop further in a Finnish context the SED indicators and assessment tool identified earlier and presented in SED manual. The tool is a crystallisation of the levels of focus and participating stakeholders that are involved during different phases of designing and constructing educational facilities. The SED carousel is presented in Figure 1 below.

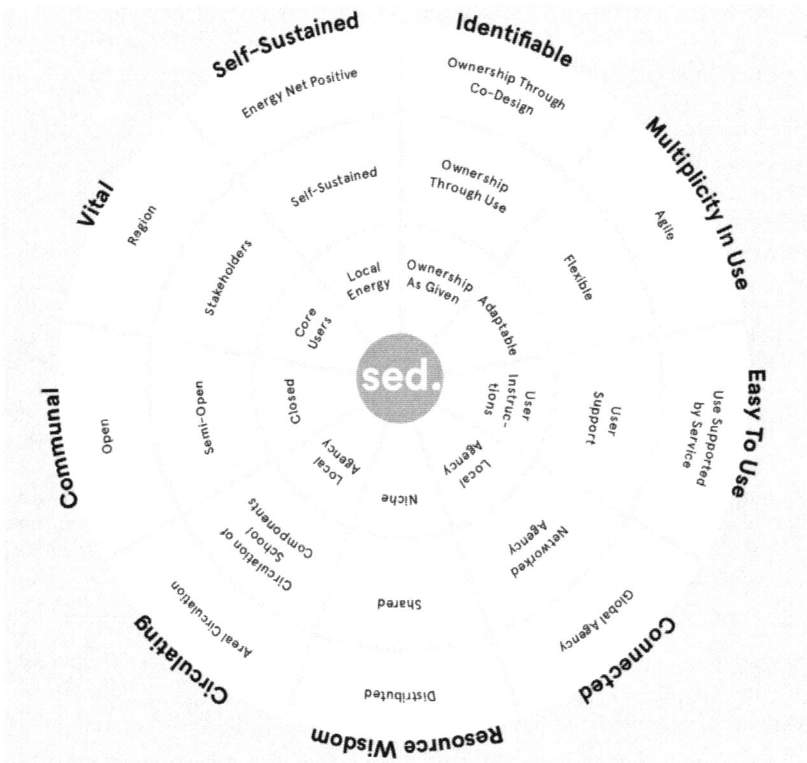

Figure 1. The SED carousel used for sustainability assessment and evaluation (Sandström & Nenonen, 2018, p. 21).

The paper concerns enhanced awareness of sustainability and how this awareness through learning in the process can be implemented in future processes in the form of culturally and socially more sustainable solutions (for instance equal possibilities to participate in the design of spaces and curricula; transparent decision-making). Universities and campuses as their concrete facilities are places for alternative and innovative ways of thinking, and thus, also places for future-proof solutions. In terms of sustainable futures, the built environment can work as a platform for increased awareness related to sustainability issues. This can take place both through the solutions that are chosen and through assessment of different aspects of the change process in the space and the impact that the changes have. The sustainability of campus retrofitting can be reinforced using participatory methods that enable community involvement and being heard of all stakeholders. The main goal in this paper was to test the framework in finding alternative Key Performance Indicators (KPIs) that can be used for sustainability considerations and learning and assessment of sustainability during campus development.

They can be applied when designing for learning landscapes and/or when evaluating the outcome, the learnings, the services produced etc. They can also promote awareness of sustainable issues in the participants - this way creating learning opportunities. As Gelfand [8, p. 22] writes regarding schools, "The qualifications necessary for planning and designing a sustainable school should include experience and skills in community outreach and community-based design." School or educational site as campus entails dimensions that differ from many other types of buildings. Despite being outside sensitive habitat [8], the institution has a link to the surrounding environment, and can use other services and places in the neighbourhood for learning and creative collaboration.

The social/cultural dimension of sustainability [7; 9] is stressed in the present paper, although the discussion regarding which components form part of cultural sustainability [9, p. 217] is still ongoing:

> Capabilities such as literacy, creativity, critical knowledge, sense of place, empathy, trust, risk, respect, recognition, to list a few, could then be understood as cultural components of sustainability.

Sustainable learning landscape is here defined being sustainable as a process (culturally, also inclusive), as a material selection, and as a learning platform: it engages users in sustainability awareness. Sustainable school architecture [8], as an example of built environment and sustainability, likewise considers communities important agents in delivering solutions and practices that promote sustainability.

As platforms for learning about and for sustainable development, schools and educational institutions are prime places because they are a major construction activity; they model long-term changes in attitudes; and they are used by entire communities. Sustainable design delivers better educational facilities. Gelfand [8, p. 18] makes a case in point in writing that

> Schools that tie sustainability and education into a way of inhabiting the campus support both the running of the facility and the educational activities. As buildings and landscapes that are found in every community, schools constitute a piece of the building sector that is uniquely suited to provide leadership in sustainability.

The educational facility can thus provide both leadership about sustainability and solutions that promote communal activities with sustainability as one of the goals.

The context of the study is a campus at a research-intensive Finnish university. At the university studied here, the teachers and students have so far not had big agency in developing the learning environments. Hence, the need for participatory design and co-design to produce future-ready learning landscapes was identified early on in the process. The overarching aim of the current approach was to build alternative KPIs using participatory service design. The research question is *what kinds of alternative KPIs can promote the design of future-ready and sustainable campus learning landscapes*.

2. Method

For piloting the participatory service design process, one classroom K113 was chosen as a test case because of its known problems and general undesirability. The aim of the participatory design was to retrofit the classroom K113. During the participatory service design the key stakeholders were involved into the process by interviews (N = 19; FM, teaching personnel and students), and by 4 participatory workshops and user journeys. The participatory workshops were recorded and transcribed verbatim. The interviews and user journeys were not recorded but the notes of the interviews were written by 2-3 interviewers and combined immediately after the interviews and user journeys. The data were mirrored against the SED indicators and the SED assessment tool (see Figure 1), testing and validating the tool further. The test case classroom is a pilot that will be scaled up in other campus contexts after analysis and conclusions about the process and feasibility of the outcomes in this study.

2.1. The Interviews

The key *stakeholders* were chosen so that they would represent all the groups involved in *using or managing the facilities. The interviews began with a question focusing on how a typical day* proceeded starting from waking up at home and leaving home for work. The next topic of the interview concerned what a key stakeholder first did when s/he arrived to the workplace and what s/he did during her/his workday. The interviewees were also asked to select from the campus the most pleasant room or other space and the most unpleasant space and to describe what in these spaces made them like or dislike them. The informants were continuously asked to reason their choices through the question "Why?". In the end, the interviewees were guided to the test case classroom K113, and when they entered the classroom, they were asked to describe what they saw and how they felt towards the space, in what ways they as a teacher/student/FM/maintenance acted in the classroom, what they did, how and why. Finally, two picture series - design characteristics and material characteristics with four to five sets of pictures - were shown to the interviewees and they were asked to reflect on the pictures, to select and argue for some that they found inspiring, pleasing and making them feel good and some that made them feel the opposite such as anxious or bored.

2.2. Participatory workshops

Participatory workshops were run by a service designer. The aim of the workshops was to map out the characteristics of well-functioning and ideal working and teaching/learning environment, i.e. what kinds of features of both the physical and digital environment supported the participants' work. Furthermore, the workshop looked at current and pipeline classroom technologies with the view of understanding the guidelines, experiences, and e.g. frequently occurring problems.

2.3. User journeys: Simulations of an experience of first-time visit to the campus

The visiting sustainability expert and interior architect organised the simulation of the experiences of a new student arriving to the campus and how s/he finds her/his way to the spaces. The simulation of an experience consisted of a walk starting outside the campus and leading inside. During the walk, the participants noticed and discussed about the need to update the information board, the lack of guides how to find one's way inside the building and other drawbacks. The methodology is described in more detail in a book in print by the interior architect [10].

2.4. Analyses

The first step of the analysis started from identifying from research literature KPIs of sustainable learning environment creation [1, 2, 6, 11] and then these were tested against Sustainable Education Design Criteria [2]. The first two authors carried out intensive co-creative research sessions where they compared their separate individual understanding regarding how they interpreted the possible alternative KPIs that could be formed from the data. After this they reflected the preliminary suggestions to the SED tool and started discussing how the alternative KPIs corresponded to the dimensions present in the SED carousel. Then they proceeded to drawing a comparison table that depicts the proposed alternative KPIs in the form of questions that eventually became the foundation for the alternative KPIs. Then the questions were tested against the participatory service design process and outcomes of the process. These were then crystallised as credits (e.g. innovation intensity or future-readiness), that were in turn coupled with corresponding SDGs. The coupling and assignment of points on a 1–5 scale (addressing how well the alternative KPI had been covered or successful in the process) was done in intensive iterative researcher meetings. The analysis process resulted in a combinatory table that is presented as Table 2.

The previous work described in [2] laid out a ninefold-table with sustainability dimensions. These are presented in Table 1 below. In the analysis process, three key themes were identified as essential for the participatory process and how it could contribute to sustainable solutions on campus. The themes were reflected upon key performance indicator (KPI) literature [10] and the SED manual [2]. In the manual, the 9-fold table (Table 1 below) takes into account the economic, environmental and social aspects of sustainability, which in turn are opened up in three layers divided into three questions.

The dimensions, leading to the proposed alternative KPIs, were: co-design & participation; socio-cultural performance; and values and post-occupancy evaluation (POE) and related guidelines for planning and assessment.

Table 1. Modified sustainability dimensions in holistic design of sustainable education.

	ECONOMIC	ENVIRONMENTAL	SOCIAL
Cultural	**vitality**	**recycling**	**identifiable**
	The solution connects to the surrounding area and brings vitality to the area as a natural part of it	The solution is ecological and sustainable on many levels	The identity of the solution fits its surroundings and culture and represents as a continuum the vision of the activities and supports the goals set for its use
Digital	**resource wisdom**	**easy to use**	**connectivity**
	The practices and physical and digital choices of the solution are based on local sourcing and sharing	The solution is intelligent and intuitively usable	The solution is connected and it connects physically, digitally and socially
Physical	**multiplicity in use**	**self-sustained**	**communality**
	The solution responds to different needs of various users and enables different uses	The solution is self-sustained and produces energy	The solution supports communal practices by its physical and digital affordances

3. Findings

The results highlight the involvement of all stakeholder groups in order to source for user perspectives and agency in learning landscape retrofitting. The original nine-fold table (Table 1) was merged with the themes identified in the present study in terms of participatory design and co-design. The merger resulted in a new set of key topics (credits in Table 2) corresponding to the original topics, and aligned with the dimensions of environmental, economic and social sustainability. These questions form the backbone for alternative KPIs that are further developed using subsets of assessment questions that can be assigned points. In addition, the alternative KPIs and credits were mirrored against the Sustainable Development Goals (SDGs) to see how the SDGs can be operationalised in participatory service design processes. The preliminary alternative KPIs presented in Table 2 below need not be in the past tense.

When formulated in the present, they can work as guidelines and a sounding board for the planned process.

Table 2. Alternative KPIs, credit and related SDGs as well as a tentative evaluation of the pilot campus environment.

Alternative KPI	Label	Credit	Related SDG	Points*
How were the SDGs considered or visible during the process?	A	sustainability goals and values	**3** – Good health and well-being	2,5
What and how was learnt during the process?	B	learnings and assessment	**4** – Quality education	5
How were the stakeholders heard and involved in the process?	C	stakeholder perspectives and participation	**8** – Decent work and economic growth	4
What kinds of services and/or infrastructure were produced during the process?	D	services and infrastructure	**9** – Industry, innovation and infrastructure	2
How did the process reduce inequalities and promote equality in participation and contribution?	E	equality and contribution	**10** – Reduced inequalities	5
How did the process & product support local innovation and sustainable community development?	F	innovation intensity	**11** – Sustainable cities and communities	4
How was future-readiness part of the process and product?	G	future-readiness	**12** – Responsible consumption and production	2,5
How did the process promote justice and resilience?	H	resilience and justice	**16** – Peace, justice and strong institutions	4
How did the process promote open communications, networking and connectedness?	J	sharing and partnerships	**17** – Partnerships for the goals	5

*Scale of the points assigned: 1 = No evidence, 2 = Some evidence, 3 = Moderate evidence, 4 = Strong evidence 5 = Very strong evidence

For instance, for alternative KPI **D** (services and infrastructure), there might be tentative new services provided by the transformed outcome, yet they might not be accessible or ready. This yields to less points for alternative KPI D. The assignment of points for the test case was done by two researchers that were involved during the whole process and studied and observed the actions and co-design procedures.

IOP Conf. Series: Earth and Environmental Science **297** (2019) 012031 doi:10.1088/1755-1315/297/1/012031

Assigning points

A	2,5

The interior architect did implement sustainability criteria in the material selection and process, but she did it secondarily, and the issues were not discussed as part of the process as such.

B	5

The campus architects repeatedly stated that they found it very useful and important to contribute and participate in the process also from a user perspective, and they felt they learnt new skills for future use.

C	4

Most stakeholder groups were involved in the beginning phases of the process, and their input was integrated in the master plan, however, they were not consulted between participatory processes and ready layouts.

D	2

The process produced e.g. the idea and eventually implementation of a new HUB, Campus Learning and Development Initiatives HUB (Caledonia). The management process was however not transparent nor consistent, and the process stopped when it was supposed to be implemented.

E	5

Gender distribution was even, and the participants thanked the team for having having invited them and felt they could really contribute.

F	4

The new HUB was started and completely new networks were created thanks to the process. Also, people started reaching out after the participatory process, making the outcome broader than the mere pilot, and the FM of the university are looking forward to the retrofitted testbed in order to learn from the case study and be able to scale the outcomes up in other currently ongoing vast construction and retrofitting processes.

G	2,5

The master plan is future-ready and flexible in furniture and other solutions, but the process has for external reasons come to a halt, and the outcome cannot be assessed yet.

H	4

The process produced several documents that are being communicated to the university management, and there will be a new process that is intended to be applied in future campus development and retrofitting projects.

J	5

There are several new developments that are directly or indirectly related to the campus change process described here, including "Supervision Corner", a novel type of a meeting point for student services and students.

An overview of the trialled alternative KPI tool and points yielded was decided to be visualised as a spider diagram. The diagram shows the points assigned on a scale 1 - 5 for each alternative KPI. The spider is presented in Figure 2

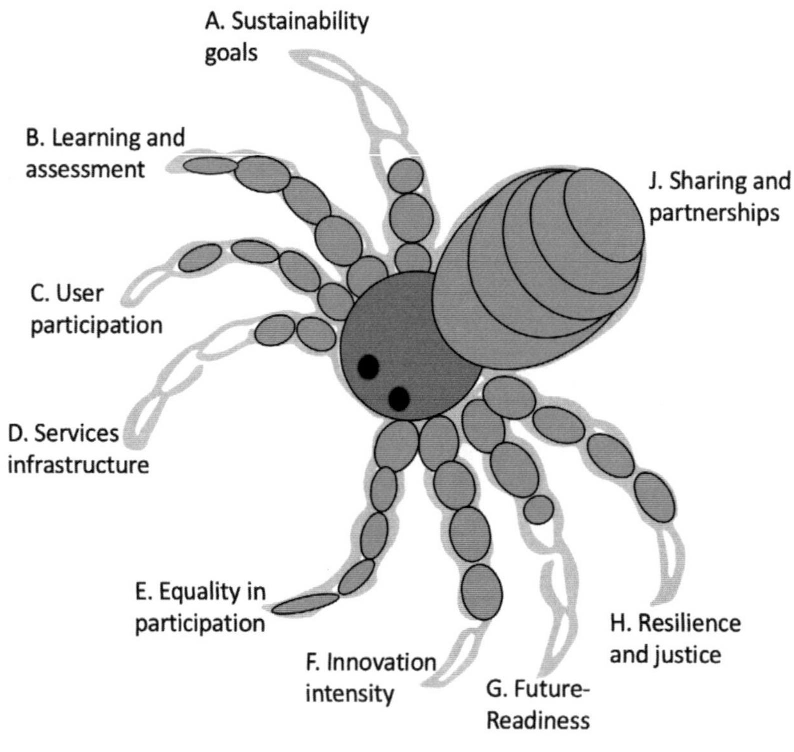

Figure 2. Spider diagram illustration of the assessment of the test case.

4. Discussion

One of the tenets of the current approach was to build alternative KPIs. One dimension to the work is, how the sustainability evaluations and related/relevant user experiences and perspectives can be extracted from participatory design, and be used for alternative KPIs, in addition to existing ones [11; 12; 13] that inform design and retrofitting processes. The present work looked at co-design and participatory service design as means to design sustainable learning landscapes. We hold that participation is the key to success in reaching the SDGs. Without participation, there is no contribution, and without contribution, no change. This surely works in contexts other than learning landscapes, as well [8].

The approach had to do with existing spaces and facilities that are retrofitted, and the measures that the participatory process aimed to develop do not include e.g. site selection, initial cost or indoor air quality, otherwise important aspects to be kept in mind while constructing sustainable, healthy buildings [8].

New kinds of co-design processes and alternative KPIs can contribute to learning about the SDGs, making people more aware of their importance for life on land and in water. The alternative KPIs can also inform management during change processes, thus giving necessary feedback and the possibility to redirect the process. In terms of the SDG number 17 (Partnerships for the goals), for instance, it was found that intensive, multiprofessional participatory processes can produce mutual understanding of different professional approaches, and it can lay the foundations for a common language and a common ground to be able to apply different dimensions of sustainability, as well.

Education plays a crucial role in attaining sustainable development goals, in fostering individuals and vital communities with change agency and a vision for a better tomorrow. The tool in progress that is presented in this paper works in informing the management before, during and after a learning landscape change process. The dimensions and the related SDGs can be considered well before the initiation of transformative organisational processes, so as to pave the path towards a better implementation of sustainable solutions in the outcome.

These proposed alternative KPIs can be used as indicators for sustainability, innovation and learning during participatory change processes and in evaluating the outcome. They can increase understanding regarding the impact that the physical-digital environment has on learning and well-being. The alternative KPIs can inform learning landscape - and other - change processes preemptively when they are being planned, as a provisionary tool for sustainability implementation through co-design and community involvement [14]. The scope of the present test case was one campus and specifically, one change process of one test bed. Thus, the assigned points as to the alternative KPIs and their credits are naturally indicative and not claimed to have been covered exhaustively.

In the process, the users learn to become more aware of their spaces and surroundings (physical and social), and to be active agents in transforming them and adjusting them to suit their needs and purposes. This is central sustainable development - agency and a sense of contribution - and agile spaces make the transformation possible on many levels (well-being, sustainability, attractiveness, suitability to needs etc.) [10; 15]. It would be beneficial also for sustainability that learning landscapes be designed in a participatory fashion. Only through participation, the learning landscapes can be custom made to meet user requirements.

The results show how the built environment can be seen as a platform in producing both sustainability and learning about sustainability through awareness. In this work, participatory service design can be an effective approach. We portray the healthy learning landscapes for social/cultural sustainability through co-design. Validating the alternative KPIs and through them, assessment methods for sustainability, remain key also in future studies building on our present findings. Co-design is a laborious and demanding process, and we need to develop tools that make the process less work-intensive and more accessible for different organisations.

The overall outcome was a new kind of operationalisation of the SDGs in the context of a campus learning landscape change process. The related SDGs cover what could be called the social/cultural dimension of sustainability. For instance, we claim that well-being [10] can be produced through participation, and participation produces a sense of agency and belonging [16]. These basic needs - see e.g. [17] - are quintessential for human prosperity and in leading a meaningful life. In the future, we will develop assessment criteria based on the credits presented for the alternative KPIs, and these criteria will be tested in more cases and different contexts globally to validate the framework further.

Acknowledgements

This study was partially supported by the Faculty of Educational Sciences, University of Helsinki, by a grant to development work regarding digital solutions and spatial retrofitting.

We would like to extend our thanks to all the participants for their valuable input during the test process and co-design. Campus architects Pirjo Ranta and Ari Nisonen are appreciated for their professional participation and expert discussions. Our team members, interior architect and sustainability expert Elina Grigoriou and service designer Jaana Tarma, are warmly thanked for an immensely valuable input, without which the foundations for this research paper would not exist.

References

[1] Dugdale S 2009 Space Strategies for the New Learning Landscape Educause Review March/April 51-63

[2] Sandström N and Nenonen S 2018 Sustainable Education Design - A Manual. (Tampere: Tampere University of Technology. Laboratory of Civil Engineering. Report 24)

[3] Ott K 2003 The Case for Strong Sustainability. In: Ott, K. & P. Thapa (eds.) Greifswald's Environmental Ethics. Steinbecker Verlag

[4] Pelenc J Ballet J and Dedeurwaerdere T 2015 Weak Sustainability versus Strong Sustainability. Available at: https://sustainabledevelopment.un.org/content/documents/6569122-Pelenc-Weak%20Sustainability%20versus%20Strong%20Sustainability.pdf, accessed on 13th April, 2019

[5] Elf M Fröst P Lindahl G and Wijk H 2015 Shared decision making in designing new healthcare environments—time to begin improving quality BMC Health Services Research 15 114 1–4

[6] Nenonen S, Niemi O and Agrawal R 2019 Co-creation of the future sustainable learning environment. A paper to be presented in this conference SBE19 Helsinki - Sustainable Built Environment Conference at Helsinki May 22nd - 24th 2019

[7] Vals A E J and P B Corcoran 2006 Sustainability as an Outcome of Transformative Learning. In J Holmberg and B E Samuelsson (ed.) Drivers and Barriers for Implementing Sustainable Development in Higher Education. Göteborg Workshop December 7-9, 2005. U N E S C O: Education for Sustainable Development in Action Technical Paper N°3, September 2006 103–108

[8] Gelfand L 2010 Sustainable School Architecture (New Jersey: John Wiley & Sons)

[9] Axelsson R Angelstam P Degerman E Teitelbaum S Andersson K Elbakidze M Drotz M K 2013 Social and Cultural Sustainability: Criteria, Indicators, Verifier Variables for Measurement and Maps for Visualization to Support Planning AMBIO 42 215–228

[10] Grigoriou E 2019 [in print, to be published in May 2019] Wellbeing in Interiors: Philosophy, design and value in practice (London: RIBA Publishers)

[11] Chan A B C and Chan A P L 2004 Key performance indicators for measuring construction success Benchmarking: An International Journal 11 No2 203-221

[12] Brown M Cevetello J Dugdale S Finkelstein A Holeto R Long P Meyers C 2017 Learning Space Rating System Version 2 EDUCAUSE Learning Initiative Available at: https://www.educause.edu/~/mediParmenter D a/files/educause/eli/initiatives/lsrsv2.pdf?la=en

[13] Parmenter D 2015 Key Performance Indicators: Developing, Implementing, and Using Winning KPIs (New Jersey: John Wiley & Sons)

[14] Murphy K 2012 The social pillar of sustainable development: A literature review and framework for policy analysis. Sustainability: Science, Practice, & Policy 8 15–29

[15] Læssøe J 2008 Participation and Sustainable Development: The Role and Challenges of Mediating Agents. In: Reid A Jensen B B Nikel J and Simovska V (eds) Participation and Learning. (Dordrecht: Springer)

[16] Sandström N and Nevgi A 2019 From needs to deeds: Where is pedagogy in a university campus change process. Manuscript submitted to be published

[17] Maslow A H 1970 Motivation and Personality (New York, NY: Harper & Row)

Resource strategies - discussions on remodelling modernistic housing and related social impacts

Inge Vestergaard, Kari Moseng and Birgitte Geert Jensen

Aarhus School of Architecture, Noerreport 20, 8000 Aarhus, Denmark

Corresponding author: vestergaardark@gmail.com

Abstract. A growing tendency to remodel social housing areas can actually be recognized in the Nordic countries. The paper analyses a case where a vulnerable modernistic housing area is remodelled based on social arguments. Through findings it is discussed how more responsible circular strategies can save the inhabitants and the environment from ′wasting′ already valuable welfare and already existing spatial structures and materials. The starting point is taken in the actual transformation of Gellerup, a New Town area built from 1968 to 1972 in Aarhus. A dramatic demolition has already taken place to avoid social disorder. This was done to provide room for rethinking, for changes in demography and for mixed ownership, and thereby solve several social problems. The process is under development, but already before this remodelling has finished and will perhaps fulfil its purpose, the government and the city council have predicted further demolishment. The tenants are increasingly reacting to this development, which erases their homes and the history of Gellerup. The loss of the areas memories and the fact that they feel they are not involved in decisions regarding their area worry them.

1. Background

The UN Sustainable Goals for 2030 serves in this paper as a basic understanding to discuss sustainable cities and communities, sustainable consumption and good wellbeing which are formulated within the seventeen main goals [1]. The paper deals especially with the coherence between the themes of a clean planet and happy people in shared prosperity [2]. The possibilities for emerging concepts for a Sustainable Built Environment are here displayed at a scale of a neighbourhood in the case of a local transformation of a modernistic city. However, the intention is to draw some general recommendations regarding future transformation projects.

1.1 Intentional architectural influence on the Danish welfare suburb

During the years after WWII the development of the welfare society accelerated a special humane interpretation that was developed in the Nordic countries. The political and economic foundations for this development were initiated by the Social Democrat' politicians in Denmark [3]. At the time there was an urgent needs for housing for the mid and low income population. This development was organized and handled by non-profit and public/private housing associations. The planning and development was carried out in cooperation with municipalities and was developed all over Denmark within a set of governmental restrictions.

A large increase in the number of homes was designed based on the ideas of New Towns [4], where inspiration from Europe such as the CIAM [5], especially the architect le Corbusier [6] was predominant. A large increase in new and modern housing blocks in relatively large settlements was seen in Denmark. This development was influenced by the Garden City Movement [7], ideas of separating different traffic elements followed the Radburn traffic system [8] - and the planning and building ideas of le Corb's 'Towards a new Architecture' [9]. In Danish Architecture the setup of modern ideas met local humble design attitudes and tradition, and the influence was moderated [10].

From 1965 until 1973 building construction developed from craftsmanship into industrial concrete panel production, which, combined with montage on site, were the most common ways of constructing a large number of housing estates in the surroundings of the existing cities.

1.2 Translation of international ideas to The Gellerup Park

West of Aarhus City, in the suburb of Brabrand an outstanding and modernistic housing estate was designed at the office of the architect Knud Blach Petersen [11] and was named The Gellerup Park [12]. The planning of this New Town [4] stipulated 7,000 inhabitants, and both the planning and the architecture were heavily influenced by the modernist's ideas. Part of the planning principles of that time was the segregation of working places and living places. The public spirit was characterised by social ideas, collective thinking and strong beliefs in development, trust and welfare.

The plan offered a huge park surrounded by housing for all kinds of families, to provide spaces and flats designed according to Danish ideas and culture and to add more functionality. A physical framework for living a whole lifespan in the estate was established: it should be possible for the elderly to live within the community, children should have easy access to kindergartens and public school, and everybody had easy access to theatres, a church and even a glass-covered swimming pool - together with modern conference facilities and library. All these programmes were built in the park. The traffic was divided according to the differentiated system to secure safe movement between the recreational areas, institutions and housing blocks without crossing car lanes [12].

The realization of the plan went well; the sizes of the flats were diverse, ranging from one room flats to five room flats. All flats had the modern transition to green recreational areas around the blocks, as large windows draw the inside to the outside and the daylight and the view to the inside. All flats were designed based on thorough functional studies: the organisation of the spaces was outstanding for the time. The housing blocks were built at a high rate of industrial speed, and the flats could easily be rented by working and middleclass people. The tenants were people from the surroundings, both people from Brabrand and from the inner, dense city structure.

1.3 The Welfare City´s development

Another new movement of the Welfare City became popular at this time: detached houses in larger settlements in the suburb. These houses were partly economically subsidised by State loans. The detached houses were affordable for the middle class and even for some working class families; at the same period the wealth of families increased, as many women joined the labour market.

In the lay-out of the plan several welfare issues are recognised: new living possibilities, growing economy, internal competition between renting a flat and owning a detached house, and not least, that the activities in the overall society were so optimistic that the Danish production facilities invited people from abroad to work in production. This conjuncture of circumstances led to a situation where the Gellerup housing blocks were not as attractive as expected. And the municipality decided to use its rights to assign flats to the new working class, which in this case were people from abroad. A Turkish migration to Gellerup followed at the end of 70s. Later came the Vietnamese refugees, and in the 80s the Palestinians [13]. Besides this migration flats were also attractive to many Danish students, who found the large five room flats affordable for collective living. In the 90s new migrants from Somalia arrived [13]. At this time the first migrants to have arrived were slowly moving out of Gellerup, as their relation to the Danish society and their financial situation improved; they also wanted to live in detached houses in the sprawl.

1.4 Gellerup today

The population in Gellerup is diverse, 80% are of other ethnic origins than Danish, and majorities are well integrated in the Danish society. However, Gellerup is a multi-ethnic settlement where many cultures live together. This mix of people have sometimes created internal problems in Gellerup, but in a united effort the housing association, people living there, the police and public associations have created an optimistic atmosphere for working together to shape a calm and welcoming humane climate in all of the area. Nielsen describes Gellerup as a stepping stone to Danish society [13]; the recognition that everybody is migrants in the Arrival City [14]. In Gellerup one can meet people who cannot use a smartphone because of illiteracy, you can meet educated people, and migrants who have lived there their whole lives and have children. The majority seems to be loyal to the area. 38% of the population is children under 18 years old; others have not worked since they came to Denmark thirty years ago and do not speak proper Danish. Others have only just arrived and have not frequented schools for education yet. 20% of the population in Gellerup is of Danish origin [15], and many of them leave during daytime for work and studies.

Gellerup is described by the government as a socially disadvantaged housing estate and is included in the governments list of hard ghettoes [16]. After many efforts to normalise these problems, it was decided to remodel the area in order to change the demography and the composition of the inhabitants. The housing organisation and Aarhus Municipality have worked together on this remodeling, and they have chosen to embrace the theories for eliminating social disorder and provide better living conditions described in the Helhedsplanen [17]. The plan comprises in two sets of plans: the housing social plan, which involves strong interaction in order to raise the level of education, employment and integration of women into the working marked. The physical plan aims to eliminate social disorder by using the instruments of tearing down and rebuilding new structures.

1.5 The state of the transformation

The actual transformation is in a midway stage: five blocks have been demolished, new buildings containing the municipality's technical department, the housing association; an entrepreneurial hub, a parking house etc. are under construction. An ambitious landscape transformation has been established in the middle of the estate, and two housing blocks are under renovation in order to meet the requirements for energy efficiency, building technical demands and architectural quality. A housing social plan is being implemented parallel to this physical change; new offices take care of the education of grownup migrants in order to promote language skills and cultural understanding, and unemployed tenants are educated to be interns for later employment, etc..

Furthermore, the municipality and the housing association are developing new plans for a further transformation and for reducing the share of rented flats to 40% of the total inhabitants [18]. Seemingly this looks like it will result in further demolitions, new housing forms and new ownerships.

1.6 The research questions

The focus of the research is: what is the outcome of tearing down and adding new structures to Gellerup in relation to the social impacts? It is to look for circularity in the actual transformation process and to be aware of the tenants' feelings and meanings. As the process is at a halfway stage, it is possible to draw attention to the ambivalent wonderings of the tenants. It will be investigated whether there are aspects of the transformation that have been forgotten and could be taken into consideration in the future.

2 Methodologies

The methodologies used intend to draw overall attention to the different developments through the Gellerup's life span, to pick up some interesting mid-term investigations and develop a humanistic discussion.

Retrospective studies of architectural history are recalled and inspiration is found in the architecture of the early modernistic movement. Material from the early modernism is examined;

Danish architectural history and thoughts on influences from this period combined with the remodelling of the actual buildings are noticed.

Literature of modern technology from the industrial period, especially relating to the case of Gellerup, is retrieved. The methodologies which are used are literature studies and interviews with people who were involved in the original design and construction of Gellerup.

An investigation of the local housing association Brabrand Boligforening's intentional material for Helhedsplanen is carried out. Material from newspapers and the internet media is followed during the planning and remodelling period.

Interviews of tenants are performed. The interaction between the housing association, the municipality, the consultants and users are referred in refereed publications; especially the view points on tenant's homes are supplemented with references to interviews done by another researcher in anthropology.

Reviews of scientific sources are visited in regard to the environmental impact of transformation sites, reuse of space and energy consumption when buildings are demolished. The material has added scientific knowledge to the subject.

3 Learnings from Gellerup

Three aspects related to the remodelling of Gellerup are highlighted and their implications on social impact and memory of space are noticed: environmental learnings from demolition, architectural inheritance learning and the importance of smooth communication.

3.1 Environmental learnings

The transformation's design tools create possibilities of erasing physical structures in the form of buildings and landscapes, and in combination with new buildings, of adding changes to the overall impression of The Gellerup Park. The learnings will illustrate circularity from a position partly through the actual ongoing process. These learnings are supplemented by some tenant's comments and feelings, which are translated to social impacts.

The physical transformation started by relocating the tenants to other flats in either Gellerup or in other social housing associations nearby.

In 2014 the five blocks were vacated and the demolishers started a cleaning process: facades, cabinets and toxic substances were removed (Fig. 1); the concrete structures were torn down and crushed into gravel (Fig. 2) to be recycled as material for use under new roads. After cleaning the site a new architecture with other programs replaced the old housing blocks (Fig. 3) and a new park was established with many thorough ideas imbedded in the design.

The two transformation strategies lead to huge investments; the environmental account is not figured out here, but it is obvious that both investments and the consumed efforts, economy and energy by demolition are enormous, not to forget environmental investments in new programs for the rebuilt compound of private owned housing, rentiers housing and office and business offices. The strategies create a mixed functionality in the city. At the same time some existing building structures are lost, well-functioning flats are erased and enormous amount of materials are lost, resulting in a great waste of energy, in order to prepare new building sites at the area. Not to mention noise and dust. In Denmark two million tons of concrete waste is produced every year on average, which corresponds to 25 % of the total waste from building sites [19]. From a circular viewpoint a great deal of this waste could have been turned into resources in new buildings, if wisely reflected.

It is a key instrument in Helhedsplanen that social dysfunctionality can be changed by physical interventions [20], which are illustrated by relocating tenants in other parts of the city. But the social problems will then naturally also be relocated at the risk of introducing the dysfunctionalities to other areas. The State Building Institute has recently reported that evidence shows that large scale demolition and forced rehousing aboard just move the problems [21]. This fact is off course not circular, and cannot be a solution to follow.

Figure 1: From demolition starting phase Photo: Ryolf

Figure 2: Concrete turned to gravel Photo: Ryolf

Figure 3: Blocks under remodelling. Photo: Vestergaard

The overall impressions from inhabitants living in Gellerup, some of them even working here, are expressed by Bach and are supplemented through interviews taken for this paper. Bach refers the tenants' feelings for the transformation [22]; these feelings are also described in recent interviews taken by authors. Some of the tenants express that they can no longer recognize their settlement; they cannot find the qualities which were connected to their earlier period of living in Gellerup. One tenant cannot find the places as he knew before, which means that he is not walking outside any more, he cannot recognize his Gellerup now. It is a common feeling that places are gone and nobody cares about cleaning around the blocks and the area as such.

A huge amount of tenants have very strong family relations. These tenants are complaining about the risk being relocated, which threatens the family pattern.

Another example of environmental loss is that Gellerup for almost 50 years had trees and thickets which had growth power. These were erased as a starting point of the transformation; this was done to provide good views and proper lines of sight in all parts of the landscape. The trees created wind barriers and spatial qualities as a path across the park. The row of trees was telling a specific story of how to cross through the landscape from one end of Gellerup to the other. Nobody thought of the importance these plants had for the narrative of Gellerup.

Also self-grown gardens between the blocks are either erased or in danger of being erased. This worries the tenants and destroys initiatives and ownerships to something of great value, although these initiatives are unplanned activities.

As a concluding learning, it is of great importance in order to create ownership of the transformation to pay attention to memory in architecture, retaining the local narratives and places to create identity and emotional feelings and a sense of belonging to the area.

3.2 The inheritance of modernistic architecture

Gellerup is originally one of the most ambitious projects from the birth of the Danish Welfare Society. The period represents the largest building epoch in Danish history. This period deserves to be treated as architectural heritage, and, at least, the most important neigbourhood deserve to be handled as such. It is of great architectural value, that one can retrieve this narrative in the city environment, and in all terms imagine how the society was enlarged after the two wars in the 20th century.

Figure 4: The repetitive façade in the park - before demolition Photo: Vestergaard

The values of Gellerup are multiple and also have a history which is expressed from 1972 until today; here is a whole city of housing with all common facilities represented in the surroundings of a great park. The parks values where great, now a huge and planned park has replaced the old park but there could still be remains from the original park, which should be retained; an example that is mentioned is the responsibility taken by the tenants to rearrange an area of urban gardening between two blocks. The places of memory referring to the tenants and are at risk of disappearing; it is important to keep this value for the future. Another case was a row of 50-year-old trees. Other areas of great value could be retained wild and untouched.

Huge potentials and values are imbedded in keeping and restoring the buildings; they all have excellent aesthetic and physical qualities, they have spatial qualities and are made of good materials (Fig. 4). Furthermore, they are based in the Danish tradition of concrete as developed in the 60s. But an argument of this kind is not intended to turn settlements such as Gellerup into museums by conservation; on the contrary, it aims to bring the settlement and housing into a future, where respect for the modern ideas is handled in an architectural treatment. This aims to bring the architectural blocks up to a modern and environmental level which corresponds to the building standard.

Add to this architectural history that the buildings and spaces have stories to narrate that are related to the tenants; they have all their stories and places, where good and bad things happened. It is of great importance for people to keep important places to connect to narratives.

3.3 Communication shortcomings

Through tenants interviews it is clear, that communication about the transformation is lacking. Several tenants point out that there are shortcomings in the communication; some of the vulnerable tenants believe that the housing association makes a profit from their rent, although the organization is a non-profit organization. Others cannot comprehend a written message; even the highly educated Danish people have difficulties understanding and to getting in touch with the service and building workers, when some installations are not functioning in their new flat. Even to understand that a transformed flat has less square meters after the transformation and that you have to pay more in rent can be difficult for a well-educated Danish tenant.

During the building period tenants are also annoyed because of noise, dust and difficult access to cars and to buildings where they have their homes. These things are to be expected, when you live at a building site, but are referred to as more annoying than expected.

The shortage of consideration given to the tenants leads to negative thoughts regarding the remodelling. The lack of information and communication leads to negative feelings: insecurity and unconsents. The missing understanding of decisions leads to apathy on several levels. It seems as if communication in the area demands a specific insight and care in order to be smooth.

One of the tenants pointed out that this lack of good communication also had the advantage, that the tenants in the newly renovated block had become more closely connected than before, because they had to communicate and guess at the staircase about what had happened and what was going to happen, now that they had received no information. Off course this was a positive point, but she could also see that such experiences did not put the housing organisation in a good light.

4 Discussion

The transformation strategies for hard ghettoes used in Gellerup are the most radical restructuring of Danish neighbourhoods from the 60s. Environmental, cultural, social and humane impacts are at risk of being forgotten when such a harsh demolition instrument is used with the intention of erase social problems by physical interventions. To qualify the research, three circular aspects are addressed through the themes of clean planet, happy people and shared prosperity.

4.1 Clean planet

To strive for a clean planet in recent initiatives regarding circularity in European Union [23] as well as in Denmark, the government has predicted new initiatives for turning waste into resources for development. Circularity is defined by a value hierarchy [24] in the form of a staircase. The upper steps relate to the city, the landscape, the infrastructure and the city's spaces or the nearby building places. All buildings relate to the higher steps. Lower steps are components and materials, and at the ground floor step we find demolished materials, just before the lowest point of the staircase: toxin fragments and leftovers for land fill. Transferred to circularity in an architectural understanding of the staircase it is apparently predicted to keep the values of an architectural object as close to the top-level as possible.

Looking at the Gellerup block we see a housing machine carefully shaped with a relatively high differentiation of flats. From an environmental viewpoint there is no reason to demolish; the concrete material serves its purpose as load bearing structure, as spatial organisation of the flats and their interrelations. The images of the blocks are all well-proportioned and functional, although the blocks are criticised for the repetitive expression they have: they fulfil their purpose for living comfort, but should be renewed to present a more attractive expression and to meet modern standards.

The Collective City of Gellerup [25] represents an exemplary way of reusing a building: this building is located close to the centre and the local square and was originally designed as a collective building with hotel, collective housing for the elderly and a library for the whole city. The architect Exner has recently drawn a prospective for the building, where all physical structures are reorganized, spatial qualities are both kept and transformed, and modernistic colours are added to the facades [25]. The content of the building is reorganized based on a collective idea that entrepreneurs and small business initiatives form a community as a collective hub [25]. Such initiatives should be respected and transformed into other projects for the new Gellerup.

When modern architecture is taken care of and being kept in relation to experience and functions, this strategy will obviously save a lot of economy, energy and primary materials, which a clean planet can benefit from.

4.2 Happy people

The key reason for demolition is argued based on in beliefs that social disorder can be solved by physical interventions. The segregation and social disorder will obviously be moved to other places [26]. Apart from this, an enormous amount of money is spent to do this exercise. It seems as if this strategy can be an ongoing machine in Denmark, if the society has money enough. This fact makes it extremely important to look closer at the subject of happy people.

This research shows that the different aspects of demolition transfer unsafety and lack of trust into the neighbourhood. It is well known that migrants form a homogeneous occupant crowd, where strong family ties are characteristic. These ties have embossed strong social bonds, and the families have a high rate of social capital [27]. Obviously, people are worried about their relations or about whether they themselves are the next to be relocated by force. In this research is learned that trust between the people who decide the physical project and the tenants living in Gellerup is suffering.

International research has not identified whether forced displacement ensures a social mix in a neighbourhood in form of interaction between different groups of populations [26]. What Bachs research shows is that the inhabitants have a sense of belongings for their valuable spaces, and are annoyed if the spaces are lost [22]. It is of great importance to secure the tenants' ownership of the area and thereby secure a positive future for the neighbourhood.

4.3 Shared prosperity

Despite the stories told in the city of Aarhus that Gellerup is a socially disordered city within the city, the tenants who have informed this research seem on the contrary to love their neighbourhood. They strongly want to secure that they have lots of memories of the settlement as it was thirty years ago. A high rate of earlier migrants feels they are Danish citizens and have adopted Danish attitudes, such as education, attended the labour marked and trying to solve problems in common, as a friendly way to manage everyday relations. One of them underlined that he wanted to literally fight for Gellerup as this place was his place since he came to Denmark from a war-filled area in the Middle East.

These inhabitants carry very strong narratives of Gellerup of both the bad times with gangs and the good times now, where they have also been involved together with the local police in pushing the bad gangs away. Their pride of these stories was obvious. For people outside Gellerup these stories are maybe not known, and one can doubt that the municipality or even the government are aware of them. At least these narratives give hopes for other and more respect full ways of interacting in such neighbourhoods. Related to these narratives and to the lost places this research describes that there is a strong recognition, that Gellerup shared memories of welfare.

For any given person, places are infused with memory. When a personal memory or place-association is shared with others, we see signs of a collective memory. This is a way of understanding architecture...
Embedded in the architecture and the physical environment are memories that can be read, understood, discussed and transformed...
Collective memory, together with physical traces of history and culture is the foundation underpinning identity and a sense of belonging... [28]

The modernistic architecture in Gellerup tells an extraordinary story, which is in danger of disappearing due to the demolition. But it is also the authors' aim to speak up about the modern period's birth in the Danish suburbs and suggest a more moderate intervention. The architecture should be recognized as a row of values and the actual interventions should be related to these values.

The memories of the period must not be forgotten, and the Collective City [25]; shows a building-scale reference to a collective memory. At the same time as the inheritance is kept, there is a visual and sensitive possibility to narrate the stories of the past. No doubt the shared memory of the Gellerup of the past is remembered in the population of today. One can only hope that the future stories have not suffered too much through this radical demolition period.

5 Conclusions

The findings show that there exist several values and remains of the original Gellerup, which are of great importance for the new version of Gellerup. These remains can be of physical nature such as a building character or city character; and they can be stories and good memories of how neighbours' handled difficult situations during the big transformation. These narratives are of great value for the collective feelings and belonging. Such experiences will lead to pride and ownership and contribute to the future living quality.

Through the discussion it is argued, that a wealthy society such as the Danish is behaving in a contra dictionary way when handling the so called ghetto neighbourhoods with the use of instruments like political force and radical demolition.

Democratic human understanding and involvement of people are preferable: fundamentally society should strive for understanding and democracy; and by thus create ownership through social activities and interactions. These should be the learnings from the welfare society, as well as common sense in handling existing physical and social structures and reusing their qualities.

It seems that the term sustainable is often used for greenwashing when dealing with local issues in isolation and is thereby in conflict with a global balance: it is urgent to think in relations, although decisions are taken at a local level.

Circular acting demands knowledge of larger systems.

Acknowledgement

Authors acknowledge a row of tenants from Gellerup, who have informed us with knowledge of how living conditions have changed since the 80s.

References

[1] UN Sustainable Goals (https://sustainabledevelopment.un.org)
[2] SBE 19 Conference *Clean Planet, Happy People & Shared Prosperity* (Helsinki: https://www.ril.fi/en/events/sbe-2019.html)
[3] Bendsen JR, Kleis B and Morgen MA 2015 *Bellahøj, fortællinger om en bebyggelse* (København: Strandberg Publishing) pp 18
[4] New Town (http://denstoredanske.dk/Geografi_og_historie/Geografi/Kulturgeografi/Bebyggelsesgeografi/new_town)
[5] CIAM(https://da.wikipedia.org/wiki/Congr%C3%A9s_Internationaux_d%E2%80%98Architecture_Moderne)
[6] Corbusier 1927 *Towards a new architecture* (London: The Architectural Press)
[7] Garden City (Encyklopædia Britannica (https://www.britannica.com/topic/garden-city-urban-planning)
[8] Radburn design (https://en.wikipedia.org/wiki/Radburn_design_housing)
[9] Corbusier 1927 *Towards a new architecture* (London: The Architectural Press)
[10] Lund NO 1993 *Nordisk Arkitektur* (København: Arkitektens Forlag) p 41
[11] Ejlers E 1994 *Knud Blach Petersen* (Weilbachs Kunstnerleksikon)
[12] Danish Arkitektur 1974 *Gellerupplanen, Brabrand* (København: Arkitektens Forlag) pp 289 -
[13] Nielsen T 2018 *Gellerup: 50 år gammel og først ved at blive voksen* (Viby J: Jyllands-Posten 5.6.2018)
[14] Saunders D 2011 *Arrival City* (London: Windmill Books)
[15] Samvirket for Gellerup og Toveshøj (https://www.samvirket.dk/fakta)
[16] Danish Governments ghetto list 2018 (København: https://www.trm.dk/da/nyheder/2018/ny-ghettoliste)
[17] Helhedsplanen 2007 (Gellerup: https://helhedsplangellerup.dk/)
[18] Gellerup Udviklingsplan 2018 (Gellerup: https://bbbo.dk/media/2230/bbbo-udviklingsplanen.pdf)
[19] VCØB 2019 *Videnscenter for Cirkulær Økonomi I Byggeriet* (Taastrup: https://vcob.dk/byggeaffald/materialefraktioner/materialer/#1684)
[20] Bjørn N 2014 *Evidens for sociale effekter af fysiske indsatser i udsatte områder* (Københavns Kommune & Arkitektforeningen)
[21] Jepsen MB and Nielsen RS 2018 *Ufrivillig fraflytning fra udsatte boligområder* (Aalborg: SBI 2018:10Statens Byggeforskningsinstitut Aalborg Universitet) p 13
[22] Bach JS 2019 Ambivalent Hopes - Residents′ Views on Architectural Transformations in a Socially Disadvantaged Estate (Nordic Journal of Architecture 2019:1)
[23] European Union. 2008. *Directive 2008/98/EC of the European Parliament and of the Council of 19 November 2008 on waste and repealing certain Directives* (http://eur-lex.europa.eu/LexUriServ/LexUriServ.do?uri=OJ:L:2008:312:0003:0030:en)
[24] VCØB 2019 *Videnscenter for Cirkulær Økonomi I Byggeriet* (Taastrup: https://vcob.dk/byggeaffald/affaldshierarkiet/)
[25] Exner AM 2018 *Kollektivbyen vision for kulturcenters genopståen - et alternativ til nedrivning* (Gellerup: https://docs.wixstatic.com/ugd/94c37a_70032af9645240d2b05c5efa5b7113b8.pdf)
[26] Jepsen MB and Nielsen RS 2018 *Ufrivillig fraflytning fra udsatte boligområder* (Aalborg: SBI 2018:10Statens Byggeforskningsinstitut Aalborg Universitet) p 17
[27] Putnam RD 1995 *Bowling Alone: America's Declining Social Capital* (Journal of Democracy, January 1995) pp 65-78
[28] Albertsen N et al. 2018 *Concepts of Transformation* (Aarhus: Boris SD, Hjortshøj R, Kragh MM and Olesen K (Ed.) Arkitektskolens Forlag, Aarhus School of Architecture) p 6-7

IOP Conf. Series: Earth and Environmental Science **297** (2019) 012033 doi:10.1088/1755-1315/297/1/012033

Testing of PEF method to assess the environmental footprint of buildings – results of PEF4Buildings project

C Spirinckx[1], M Thuring[1], L Damen[1], K Allacker[2], D Ramon[2], N Mirabella[2], M Röck[3] and A Passer[3]
[1] VITO/EnergyVille, Boeretang 200, B-2400 Mol, Belgium
[2] KU Leuven, Division of Architectural Engineering, Department of Architecture - Faculty of Engineering Science, Kasteelpark Arenberg 1 box 2431, B-3OO1 Leuven
[3] TU Graz, Working Group Sustainable Construction Graz University of Technology, Rechbauerstraße 12, 8010 Graz, Austria

Abstract. The aim of the PEF4Buildings project, commissioned by the European Commission, was to assess if the Product Environmental Footprint (PEF) method and related guidance documents developed in the framework of the Environmental Footprint pilot test phase are applicable at the building level. The project has been coordinated by VITO with TU Graz and KU Leuven as research partners. As a first step, two PEF studies on new office buildings were performed. The assessment of two office buildings allows to test the PEF method in different geographical contexts. In a second step, a possible approach to benchmark office buildings and to define classes of performance is developed. The approach is developed based on the findings of the PEF study on two new office buildings, the PEF guidance and dedicated desk research. It covers issues like how to define the reference building, how to define system boundaries, how many reference buildings should be defined and how to handle this range of different options. A third step of this study was devoted to the assessment at the building level and on how to link the assessment of the environmental performance of construction products to the assessment of a building by using the PEF method.

1. Introduction

Due to the enormous energy and resource consumption linked to the construction sector assessment and reduction of associated environmental impacts has become a top priority on multiple levels [1]. The PEF4Buildings project fits within the European Commission's Single Market for Green Products Initiative. The European Commission proposes the Product Environmental Footprint (PEF) and Organisation Environmental Footprint (OEF) methods as a common way of measuring environmental performance. The approach is tested between 2013-2017 together with more than 280 volunteering companies and organisations. The aim is to understand the real potential of the methods before proposing new policies.

The PEF4Buildings project is commissioned by the European Commission and is carried out by the Flemish Institute for Technological Research (VITO), KU Leuven and TU Graz. The aim of the project is to test the applicability of the PEF method Guide (Commission Recommendation 179/2013 on the use of common methods to measure and communicate the life cycle environmental performance of products and organisations) and the latest versions available of the related guidance documents developed in the framework of the Environmental Footprint pilot test phase [2] to a new office building, and to provide an overview of pros and cons of alternative possible approaches to the definition of the benchmark and classes of performance for the typology of buildings within the scope

of the study. The assessment and the overview will contribute to the development of a final approach to develop benchmark and classes of performance for different typologies of buildings.
As a first step, a PEF study on two new office buildings is performed. The type of buildings (offices) was specifically requested by the EC in the request for proposals. Two buildings from different geographical regions (Belgium vs Austria) are selected in order to allow to test he PEF method to buildings located in different climatic contexts, with a variation in energy performance and built according to different national building codes. The first building is the office building BelOrta of BelOrta CVBA, designed by the architectural firm ar-te and built in Belgium. The second building assessed in this study is the office BE2226 of Baumschlager & Eberle, Lustenau, Austria. The BelOrta building is a business-as-usual new office building, while the BE2226 building is a nearly-zero-energy building.
In a second step, a possible approach to benchmark office buildings and to define classes of performance for newly built office buildings is developed. The approach is developed based on the findings of the PEF study on two new office buildings, the latest version of the PEF Guidance document available at the that time [2] and a dedicated desk research which can be downloaded from the EC's website (http://ec.europa.eu/environment/eussd/smgp/pdf/Matrix_lit_sources.xlsx). It covers issues such as how to define the reference building, how to define system boundaries, how many reference buildings should be defined and how to handle this range of different options.
A third step is devoted to the assessment at the building level, and more specifically how to link the assessment of the environmental performance of construction products to the assessment of a building by using the PEF method. The assessment starts from the results of the PEF study on the two new office buildings, but is extended to other possible European building typologies.
Several workshops with stakeholders were organised during this project where the different steps were systematically presented, followed by an interactive discussion with the stakeholders.

2. Research questions

The main research questions of the PEF4Buildings projects were:

- How can the assessment of the environmental performance of construction products used in buildings be linked to the assessment of the building by using the PEF method?
- How can the results and experiences from the PEF assessment of a specific newly built office building be used to develop an approach for benchmarking office buildings and for the definition of performance classes?
- How can the results and experiences from the PEF assessment of a specific newly built office building be used to extend to other possible typologies of buildings?
- What are the lessons learnt and the recommendations based on the experiences and results of the PEF study for a new office building based on the assessment of the environmental performance of construction products?

The Product Environmental Footprint (PEF) is a Life Cycle Assessment (LCA) based method to quantify the relevant environmental impacts of products (goods and services) and organisations. It builds on existing approaches and international standards [3]. One important feature of the PEF method is that it sets the basis for comparability of the results.

3. PEF study on two new office buildings

3.1 Two building case studies

In order to test the applicability of the PEF method to a new office building (as requested by the EC), two case studies were selected, thereby ensuring the findings are not case specific nor geographical context specific. The first building is the office building BelOrta of BelOrta CVBA, designed by the architectural firm ar-te (figure 1). The BelOrta building is an office building built in 2013-2015 in Belgium. It is a compact building with an inner patio around which the workspaces are organised. The architectural firm ar-te provided the necessary data and inputs for the PEF study of this building. We have chosen this building because (a) a BIM (Building Information Modelling) model was available which enhanced the inventory phase of the PEF study and (b) it is a business-as-usual building regarding typology, energy performance and technologies used. The second new office building assessed in this study is the office BE2226 of Baumschlager & Eberle, Lustenau, Austria (figure 2). BE2226 is a nearly-zero-energy-building without active cooling and heating. Also for this building a BIM model was available. This second case study allowed to test the PEF method on a building with a very high energy performance located in a different geographical context.

Figure 1. BelOrta building of ar-te in Belgium

Figure 2. BE2226 building of Baumschlager & Eberle, Lustenau in Austria

3.2 Element assessment method

A hierarchical decomposition of the building (element method) was used in order to avoid data gaps as much as possible and in order to allow for inventory data of various sublevels to be used for the assessment of the environmental performance at the building. The model structure is based on a hierarchical subdivision of the building in smaller entities: building elements, sub-elements and materials (figure 3).

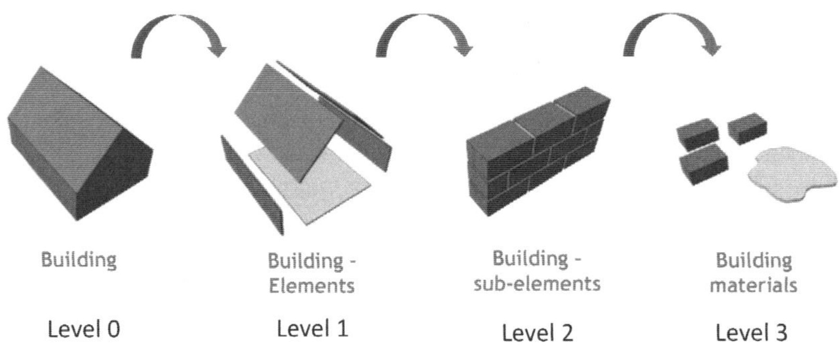

Figure 3. Hierarchical de-composition of the building [4]

3.3 Functional unit and reference flow

The functional unit is one office building with a reference service period (RSP) of 50 years, assessed from the bill of materials according to the element method for cost control [7] (Table 1).

Table 1. Key aspects used to define the functional unit

Aspect	Definition
What?	Office building excluding the surroundings
How much?	One office building

How well?	Energy performance and thermal comfort requirements met, relevant technical and functional requirements
How long?	50 years RSP

The reference flow is the amount of products needed to fulfil the defined function and shall be measured in m², m³, kg and pieces as necessary for the various building elements and components. All quantitative input and output data collected in the study are calculated in relation to this reference flow. For the purpose of this study the building as a whole is considered, without the infrastructure for accessing the building. The reference flow is inventoried at the level of building element, sub-element and materials.

3.4 Scope

The system boundaries for the two PEF studies were defined from the perspective of the designer/architect. This means that everything that the designer/architect can influence has been included in the system boundaries, but everything the designer/architect does not control directly is excluded (e.g. paper, furniture, commuter transport). It was moreover chosen to exclude the parts of the office building which are not necessary for the functioning of the office building, such as kitchen/catering. This is seen as important to make fair comparisons of the environmental footprint of office buildings (i.e. benchmarking purposes) which not all comprise a kitchen. Furthermore, the PEF studies are limited to the building as such and exclude the surroundings (e.g. parking lot).

Figure 4 presents the system boundary diagram for the assessment at the building level. The diagram indicates the three levels of situations according to the PEF Guidance document indicating the need for specific data. The diagram moreover distinguishes foreground and background processes and material and energy flows.

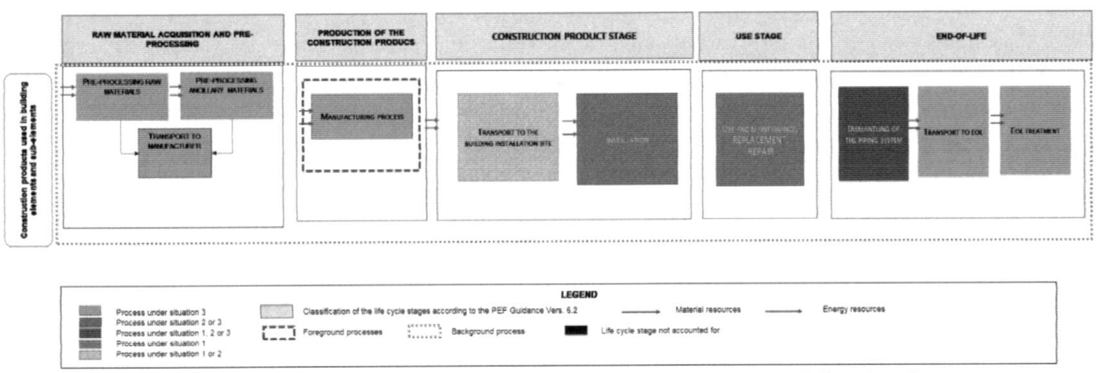

Figure 4. System boundary diagram for assessment at building level

Based on a cross-analysis of the life cycle stages defined in the draft PEFCRs for the construction products within the PEF pilot phase and the EN norms related to construction products (EN 15978/EN15804) the life cycle stages (LCS) presented in Table 2 have been defined for the PEF assessment of the two office buildings. So the assessment considers all upstream stages (production of building materials/energy equipment), including transport to the construction site as well as downstream stages (demolishing of the buildings after reference service life and end of life treatment). For the use phase energy and water consumption, as well as replacement, maintenance and refurbishment activities are included. So the assessment is performed from the cradle to the grave.

Table 2. Life cycle stages for the assessment at building level

LCS name	Description
PEF_A1	Pre-processing and acquisition of raw materials and packaging of raw materials

PEF_A2	Transport of the raw (engineering) materials to the production site
PEF_A3	Manufacturing of the construction products and the related packaging
PEF_A4	Transport to building site
PEF_A5	Construction - processes necessary for the construction of the building, including all ancillary materials, End-of-Life (EoL) of the packaging material disposed, any losses during construction)
PEF_B1	Use stage
PEF_B2	Maintenance
PEF_B3	Repair
PEF_B4	Replacement
PEF_B5	Refurbishment
PEF_B6	Operational energy use
PEF_B7	Operational water use
PEF_C1	Dismantling
PEF_C2	Transport to EoL
PEF_C3/C4	Disposal at EoL (sorting towards the EoL treatment, recycling, incineration and landfill of all materials at the EoL of the life of the building)

For each life cycle stage specific scenarios were developed for the assessment of the environmental impact at building level. In defining such scenarios, the PEF Guidance as well as the draft PEFCRs, EN norms and national norms have been used. These scenarios are mainly focusing on:

- Assumptions related to construction process stage: e.g. transportation of construction products from manufacturers to building construction site, but also the construction process;
- Assumptions related to the use stage: scenarios for the use of the installed construction products in the building, maintenance of the building, repair activities, replacement activities, refurbishment activities, operational energy use and operational water use;
- Assumptions related to End of Life stage: scenarios for the de-construction, reuse, demolition, recycling and disposal; including all transport.

For the purpose of the PEF assessment rules for allocation at building level and at product level have been identified:

- At building level, the allocation of the impacts of re-used building elements could be discussed but was not part of this study as the focus was on new office buildings. Defining allocation rules related to the impacts shared between previous system boundaries and next system boundaries will be necessary when making a PEF study of renovated buildings;
- At product level the allocation follows the approach for the handling of multi-functional processes in the PEF Guidance [2].

3.5 Data model

The complexity of an assessment at building level requires the use of several tools during the various steps of development. Therefore, a process flow connecting the relevant data, processes and tools was elaborated for this project (Figure 5). First of all, sources to define the necessary information on building elements, building materials, scenarios, use phase were identified in the data gathering process. In a second step, a building element catalogue of the building components was developed in a spreadsheet (Excel) and are the main input for the life cycle impact assessment (LCIA) in the LCA software (SimaPro).

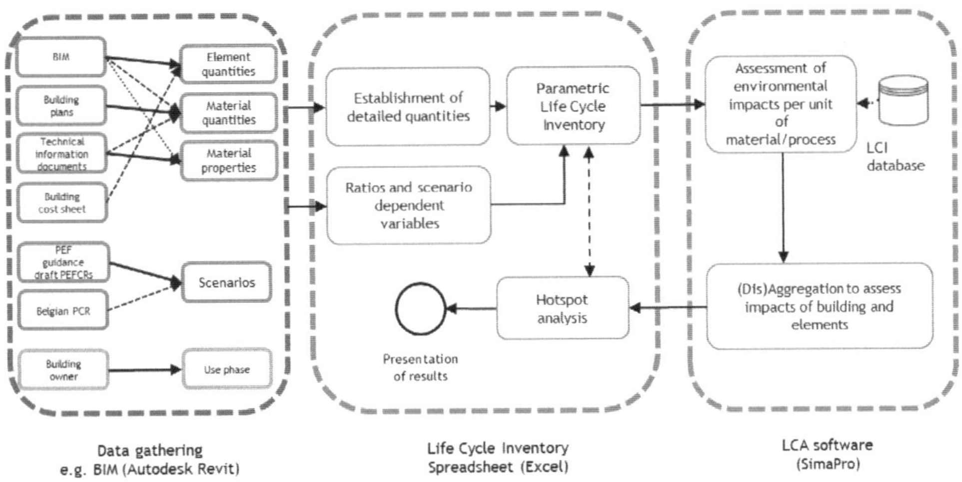

Figure 5. Data model development process

3.6 Results and conclusions

The aim of this PEF4Buildings project is to test the applicability of the PEF method to a new office building, as specifically requested by the EC. With this goal in mind the aim is not to focus on the PEF results for the two office building as such but to learn from the assessment in terms of pros and cons (e.g. methodological challenges) when applying the PEF method to office buildings and to develop some recommendations for future developments in this area. That is why we will not discuss the environmental profiles of the two buildings but focus on the methodological challenges during course of the assessment as specifically requested by the EC. So the results in this paper are presented and interpreted with this overall goal in mind.

The challenges and issues that people meet while using LCA at the building level are also valid for PEF assessments. PEF specific issues are mainly the availability of PEF compliant datasets and the quality of generic datasets (not always PEF compliant). As the PEF endorses the use of the Circular Footprint Formula (CFF) it is recommended that all generic datasets allow the application of the CFF formula. This should be a recommendation for further development. The PEF compliant datasets purchased by the EC are seen as an important step in that direction. Other learnings gathered during the life cycle inventory phase are:

- Clear guidelines are needed to model the use phase;
- National scenarios are needed (e.g. for products for materials/building elements for which no draft PEFCR are available);
- During data gathering, the BIM model proved to be an important source with added value.

The LCIA results of both case studies were calculated according to the PEF Guide. It is important to note that two different modelling approaches were applied to the two case studies in order to learn from the differences. The BelOrta Simapro model follows strictly the hierarchical structure presented in Figure 3, while the BE2226 Simapro model is made on material level in order to test implications of different modelling approaches. The difference in modelling has mainly consequences during the interpretation of the results. More specifically, the BelOrta model allows to gain insight in the contribution of a material applied in a certain element to the total building impact, while the BE2226 model provides insight in the contribution of a certain material to the total building impact. For example, the BelOrta model gives the impact of concrete in inner walls and in outer walls separately, while the BE2226 model provides the impact of concrete (in all building elements aggregated).

As this project didn't intend to focus on the absolute values and results obtained, but aimed at understanding the methodological aspects of calculating a PEF of an office building, the numerical results of the study are not reported here. Although the intention of the two assessments of the two cases is not to compare their LCIA results in detail, some important outcomes can be highlighted:

- Firstly, the overall weighted score of the two buildings differ significantly: the BE2226 building clearly show a lower impact than the BelOrta building. This was expected as the second case (BE2226) is an advanced building (NZEB) while the first case (BelOrta) is a business-as-usual case. However, it should be noted that the building control system (including sensors, control panels…) and piping is not modelled for the second case because of lacking data. Furthermore, the two buildings have the same use and a similar size in terms of floor area but represent different building typologies;
- Secondly, the most relevant impact categories identified differ between the two cases. The most relevant impact categories hence clearly depend on the building (energy performance and materials used), but also on the location (electricity mix). This is important if the aim is to limit the impact categories of PEF4Buildings to the most relevant ones. We can furthermore conclude that several of the most relevant impact categories are 'additional' impact categories which are not included in the current version of the EN 15804 and EN 15978;
- Thirdly, for both buildings the operational energy use and the pre-processing of raw materials have been identified being amongst the most relevant life cycle stages. However, for the BelOrta building, the use phase contributes significantly more to the life cycle impact than for the BE2226 building. As the use phase in the BE2226 building causes lower impacts than in the BelOrta building, additional life cycle stages become more relevant;
- Fourthly, the identification of the most relevant processes leads to a different set of most relevant processes for both buildings. This was expected as both buildings consist of different materials and elements, but are also located in a different location and hence a different electricity mix is used during the important use phase of the building.
- Finally, the inclusion/exclusion of the toxicity impact categories clearly influences the results to a significant extent, especially for the second case study.

4. Proposal for an approach for a benchmark and classes of performance for office buildings

4.1 Research method

In a second step, a possible approach to benchmark office buildings and to define classes of performance was developed. The approach that has been developed is based on the findings of the PEF study on two new office buildings, the latest version of the PEF Guidance documents available at that time [2] and a dedicated desk research on existing approaches. It covers issues like how to define the reference building, how to define system boundaries, how many reference buildings should be defined and how to handle this range of different options. The desk research focused on existing reports on building typologies, sustainability schemes, LCA guidelines (such as the EN 15978 norm and the draft PCR for buildings) and national methods or legal requirements (such as the E+C- method in France, DGNB and BNB in Germany, GWW in the Netherlands, MMG in Belgium). The results of the desk research can be downloaded from the EC's website (http://ec.europa.eu/environment/eussd/smgp/pdf/Matrix_lit_sources.xlsx). The researched methods are compared on different aspects: the meaning and approach of the benchmark that is defined, how reference buildings are defined, the considered system boundaries, functional unit and reference flow, and the approach used to define performance classes.

4.2 Results

Several important differences were identified in the meaning and related approaches used for defining the environmental benchmarks of buildings. The literature review (http://ec.europa.eu/environment/eussd/smgp/pdf/Matrix_lit_sources.xlsx) revealed that various meanings of benchmarks of buildings are used, ranging from limit values (minimum acceptable performance), over a reference value (present state of the art, e.g. average) and best practice value (values reached in experimental or demonstration projects) to target values (values that can be reached in medium- or long-term perspective). Depending on the meaning chosen, different approaches and sources are used to calculate the benchmark values, ranging from laws prescriptions, standards (for limit values), over statistical values and reference buildings (for reference values) to political target values or economic and technical optima (for target values). Building certification schemes, such as BREEAM or DGNB, typically use various values, i.e. limit, reference and target values, to define their benchmark and classes of performance. Current national benchmarks mostly represent a level-playing-field with a lower limit as a starting point. Their goal is to allow buildings with a lower value on the one hand, but also to avoid free rider behavior on the other hand. Nevertheless, benchmarks can evolve in time as technology advances and knowledge of the building stock grows.

Another differentiation noticed in approach followed in literature is the use of external versus internal benchmarks. The former is the most common and compares a certain building with a reference building differing in layout and material choice (sometimes also in energy performance). The inner benchmark is more rarely used (e.g. by LEED) and compares the impact of the design building with a reference building with only variation in materials (meaning that the propose-building is identical to the reference building in terms of layout, size and energy performance).

Some systems furthermore have defined a benchmark for each environmental impact category separately (e.g. DGNB), while others have defined a benchmark for the aggregated single score (e.g. The Netherlands). In the systems with benchmarks for each impact category separately, the final score is obtained by weighting the different scores on the different impact categories. The weighting of the different impact categories is then part of the methodology.

From the desk research we can conclude that reference buildings to determine the benchmark can either be real buildings representing the reference building, or can be virtual buildings (i.e. based on statistical analysis of a representative part of the national building stock). Such statistical analysis differentiates various building types, sizes, materials and energy performances.

"50 years" has been found as the most common reference study period for office buildings. A commonly used reference flow is "per floor area, per year". Some studies suggest to take the function of the building into account, e.g. the number of work stations in the office building or the amount of people in full time employee equivalent. No relevant assumptions on cut-off rules or scenarios were found in the desk research.

In an issue paper of the PEF Technical Helpdesk guidance is giving on how to calculate performance levels for products (based on most relevant parameters and by determining market realistic minimum and maximum values). The desk research aimed at exploring if other methods are used to define performance classes for buildings in existing approaches. We can conclude that the aim of performance classes can be different and determines the approach to define performance classes. For the definition of performance classes, a lot of information is needed on the relevant buildings to be "represented" in different classes, to be able to set the desired thresholds. For that, just as for benchmarking, one needs to have an overview of the environmental impacts of the various buildings to be covered by the performance classes, which can be either real buildings or virtual buildings. In the case of performance classes, as the amount of data needed is large, generally virtual buildings are used. To set the thresholds between different performance classes, some systems rely purely on the relative improvement compared to the benchmark (e.g. 5%, 10% or 15% better than the benchmark). Other systems are based upon the statistical spread of the analysed building stock.

4.3 Recommendations

Based on the learnings from the desk research, we recommend to define a common EU method to calculate the environmental impact of buildings and a common EU methodology on how to define environmental benchmarks. This is needed in order to reach a harmonized approach in Europe. This could be reflected in a PEFCR for buildings which is in line with the PEF method, but provides more

specific guidelines for buildings. As sustainable building is more than environmental impact solely, such method could moreover be integrated in a broader assessment framework such as the LEVEL(s) framework [5]. If this is aimed for, the Level 2 "Comparative performance assessment" could be considered as a starting basis to further develop a common EU methodology for environmental benchmarks for office buildings in order to make meaningful comparisons between functionally equivalent office buildings. The Level 2 framework lays down rules to support the comparability of results at national level (without flexibility to reflect in methods between Member States) or building portfolio level. The PEF method can be used in this framework to fix certain key parameters and input data and assumptions used for calculations. The experiences gained and the tools developed during the PEF assessment of the two office buildings can give input to the so-called "overarching assessment tool: Cradle to grave LCA" within the LEVEL(s) framework.
It is furthermore recommended to calculate the benchmark values at national level in order to take into account external factors such as climate, construction practice and culture influencing the environmental impact of a building. It is also recommended to use a stepwise conservative approach, meaning that initially benchmarks are defined representing lower limit values and which gradually become more severe in time. This allows all stakeholders to get used to the method first and to ensure that the full market is included in a transition to sustainable building.

In order to define an environmental benchmark for office buildings, insight is needed in the average environmental impact of office buildings in Europe and on the variation of it throughout the building stock. This insight can either be gained from real cases or virtual cases. The first conservative benchmark should be set to allow all office buildings that fulfill minimum legal requirements on energy, water, fire safety, etc. To define the environmental impact fulfilling the improvement or target value, this can be based on best-practice buildings (experimental or demonstration projects) or based on virtual buildings with for example 30% less impact than the reference value (i.e. representing business-as-usual). In the second approach, the percentage reduction could either be based on a statistical analysis of building practice in the specific Member State or based on political targets set.

For the definition of reference buildings, building typologies must be defined that represent the building practice in a specific Member State. To calculate the environmental impact of each of the building types, the influence of the market variations on the impact of the building needs to be identified per building type in a dedicated statistical study. Market variation in terms of size, materials used and technical solutions are to be considered.

The desk research revealed that benchmarks can be defined to cover both material and energy impact of the building, or a separate benchmark can be defined for each. The project team recommends to have one benchmark including both material and energy impact. However, taking practical implementations into account, it might be easier to separate both (in a first phase) because energy benchmarks are currently already established in the EU Member States. Although material and energy impact can be separated in a first implementation stage, it is recommended that the same environmental indicators are assessed in order to allow for a smooth aggregation in the second implementation stage.

The desk research provided an overview of insights gathered, but it is clear that additional data is needed at the national level in order to define explicit recommendations on certain aspects that are relevant to defining reference buildings. Most countries lack a large database on non-residential buildings, and therefore the gathering of more data and the creation of databases are recommended.

Providing a clear reference flow for the assessment of buildings will greatly improve comparability of future studies. From a scientific point of view, the project team would recommend to define an appropriate functional unit for benchmarking that takes the function of office buildings into account, e.g. full time equivalents per year. However, energy performance requirements of buildings already established in all member states use m^2 floor area for all benchmarks defined in current certification schemes of buildings. Therefore, taking practical implications into account and in order to allow consistent analyses of buildings, the project team recommends m² floor area per year as the most appropriate reference flow. When opting for m² floor area, it is recommended to define clear and strict rules on how to define the floor area: e.g. gross, net or heated floor area.

Further, it is recommended to assess the complete building and its elements: foundations, building envelope, inner walls and intermediate floors, including all finishes as well as technical equipment. Furniture, desks, IT equipment, kitchen and parking space should be excluded. Most importantly, clear rules should be defined in a PEF method for buildings. PEF recommendations can be followed related to cut-off rules and scenarios.

Although a benchmark should not differentiate between design (for obtaining a building permit) and post-construction phase (e.g. when 2 years in use), we recommend to define clear and different calculation rules on how to calculate the impacts in the design and post-construction phase when comparing these to the benchmark.

To avoid burden shifting or trade-offs between impacts, the project team recommends to have a benchmark for each impact category separately, as well as at the aggregated level. Once the most relevant impact categories for buildings have been identified, the benchmarks could be limited to the relevant impact categories only.

As data quality requirements are seen as crucial for benchmarking purposes in LCA, it is recommended to follow the same approach as in the PEFCRs. This ensures the necessary data quality and therefore the representativeness of defined benchmarks.

Starting from the benchmarks defined at national level, performance classes can be defined. If environmental assessment data is available on a large amount of buildings representing the office building stock, data on existing buildings should be used. If such data is not available, few representative buildings should be chosen and virtual variations of these buildings analysed. Market variations in terms of size, materials used and technical solutions are to be considered.

When the legal requirements define the lower limit of the first performance classes, and a statistical analysis defines the upper limit of the best performance classes, this range can be divided in five performance classes, as defined in the PEF Guidance.

5. Guidance for the assessment at the building level

A third step of the PEF4Buildings project was devoted to the assessment at the building level. It gives guidance on how to link the assessment of the environmental performance of construction products to the assessment of a building by using the PEF method. The assessment started from the results of the PEF study on two new office buildings, but has been extended to other possible European building typologies.

Existing methods to assess the environmental performance at the building level were screened during desk research. The results and references to the respective methods are published at the website of the EC (http://ec.europa.eu/environment/eussd/smgp/pdf/Matrix_lit_sources.xlsx)

It can be concluded that national systems are very valuable since they usually take national guidelines into account and therefore have the highest potential of complying with the local realistic situation. In case national Product Category Rules (PCRs) for construction products provide guidance on the definition of the functional unit, this results in an easy link between the Environmental Product Declarations (EPDs) of construction products and the LCA at the building level.

It is recommended to develop a PEFCR(s) at building level in the future, which forms the basis for all (new) PEFCRs for construction products and should be aligned to existing PEFCRs at the construction product level, for example by consistent definition of life cycle stages. Currently, the number of life cycle stages considered is different in the various construction PEFCRs. The project team recommends to consider the life cycle stages as presented in Table 2. A clear reference flow should be defined, as described in the previous section, and clear rules should be defined at the building level for what is included or excluded and on cut-offs. Different PEFCRs should be developed for different building typologies and include national scenarios, for example on the use phase and for the dismantling/demolishing of the building. A differentiation in rules/guidelines is needed for PEF studies for design support/building permissions and ex-post construction. Clear guidelines to model the use phase and the reference study period and related references service lives are crucial. At building level, the PEFCR(s) for building(s) shall be clear on the procedures related to the allocation of impacts of re-used building elements (mainly important during renovation projects) by setting up the principles for the allocation of impacts between previous system boundaries and next system boundaries is necessary (such as the reuse of pile foundation of previous building). At product level, the allocation should follow the approach for the handling of multi-functional processes in the latest version of the PEF Guidance document.

A more extended list of generic datasets is needed to model an entire building. It is therefore recommended to provide a common EF database for the most frequent processes, e.g. using PEF compliant datasets purchased by EC, with integration of the CFF formula in the datasets.

Another aspect that has the potential to increase the feasibility of an LCA at the building level, is creating a link with design tools (CAD systems and BIM systems), assessment tools and PEF compliant databases. This can make data gathering a lot easier and greatly reduce the time and effort spent by the LCA practitioner. As modelling a building is very complex, a more flexible parameter structure for the amounts of elements and materials has to be found to further support a parametric approach. We recommend a hierarchical decomposition of the building as shown in Figure 3.

Today only a few assessment tools can calculate all PEF-required environmental impacts. However, all PEF categories could be integrated into these tools in the future to avoid burden shifting. Most of the current assessment tools are focused on the construction phase because there is a lack of information from environmental profiles on building processes (maintenance, repair, refurbishment, etc.). All life cycle stages should be taken into account. It is recommended to use PEF/LCA software that allows to directly extract the LCIA results at the level of the building, at the level of the life cycle stages, at the level of the element in each life cycle stage, and at the level of the material/process. It would be helpful if in future the PEF/LCA software would allow to do the hotspot analysis (identification of the most relevant impact categories, life cycle stages and processes) directly in the software.

A critical review of all PEF studies of buildings is recommended to ensure that the assessment is fairly, completely and accurately reported and in line with the latest version of the PEF guidance document.

6. References

[1] Passer A *et al* 2015 Environmental product declarations entering the building sector: critical reflections based on 5 to 10 years experience in different European countries. *The International Journal of Life Cycle Assessment* 20(9), pp.1199–1212. Available at: http://link.springer.com/10.1007/s11367-015-0926-3

[2] European Commission (EC) 2017 PEFCR Guidance document - Guidance for the development of Product Environmental Footprint Category Rules (PEFCRs), version 6.2.

[3] Chomkhamsri K and Pelletier N 2011 Analysis of existing environmental footprint methodologies for products and organizations: recommendations, rationale and alignment.

Institute for Environment and Sustainability (November 2011), p.61. Available at: http://ec.europa.eu/environment/eussd/pdf/Deliverable.pdf%5Cnhttp://scholar.google.com/scholar?hl=en&btnG=Search&q=intitle:Analysis+of+Existing+Environmental+Footprint+Methodologies+for+Products+and+Organizations:+Recommendations,+Rationale,+and+Alignment

[4] Trigaux D *et al* 2017 Life cycle assessment and life cycle costing of road infrastructure in residential neighbourhoods. *The International Journal of Life Cycle Assessment* 22(6), pp.938–951. Available at: http://link.springer.com/10.1007/s11367-016-1190-x

[5] Dodd N and Garbarino E and Gama Caldas M 2017 Level(s) – A common EU framework of core sustainability indicators for office and residential buildings - Parts 1 and 2: Introduction to Level(s) and how it works (Draft Beta v1.0), 1–68. Retrieved from http://susproc.jrc.ec.europa.eu/Efficient_Buildings/docs/170816_Levels_EU_framework_of_building_indicators_Parts.pdf

Acknowledgments

The research presented in this paper was funded by the European Commission "Study on the Application of the PEF Method and related guidance documents to a newly office building (ENV.B.1/ETU/2016/0052LV)". The authors appreciate the constructive collaboration within the project team and with the representatives of the European Commission DG ENV (Dr. Michele Galatola, Product Team Leader at European Commission DG ENV and his colleagues). Furthermore, we also would like to thank the stakeholders for their valuable feedback throughout the project.

IOP Conf. Series: Earth and Environmental Science **297** (2019) 012034 doi:10.1088/1755-1315/297/1/012034

Analysis of thermal comfort in residential buildings in the city of Ha Tinh, Viet Nam

Ngo Hoang Ngoc Dung[1], Elias Jacobs[1], Emilia Motoasca[2], Alexis Versele[1], Hilde Breesch[1]

[1]Construction TC, Department of Civil Engineering, Faculty of Engineering Technology – TC Ghent, KU Leuven, Belgium
[2]Department of Electrical Engineering, Research group Energy & Automation (E&A), Faculty of Engineering Technology – TC Ghent, KU Leuven, Belgium

hoangngocdung.ngo@kuleuven.be

Abstract. The study was performed in Ha Tinh city of Vietnam and aims to investigate the current thermal condition inside houses of different urban neighborhoods by field measurement and survey. This is a starting point to develop sustainable housing and neighbourhood typologies as adaptation to climate change in the city. The survey was conducted in the summer of 2016. The analysis of the survey data showed that the satisfaction level of local inhabitants with actual thermal condition was relatively high, and the perception of those living in various neighborhoods did not differ significantly. To monitor the comfort level during summer in houses of Ha Tinh city, nine houses were selected based on their typologies and locations represented by local climate zones (LCZs). The investigation was conducted from late August to mid-October 2017. With data obtained from the field measurement, psychrometric analysis was performed and based on modified adaptive comfort model for South-East Asian people to evaluate thermal condition of considered houses. The analysis evaluated and showed explicit difference in comfort level between these housing typologies. Result of the analysis also showed a dependence of thermal comfort in different rooms of a house on their position and exposition to solar heat gained, but relation of comfort level to LCZs was not clear.

1. Introduction

The city of Ha Tinh is urban part of Ha Tinh province in the Northern region of Central Vietnam. Located at the latitude of 18 [1], Ha Tinh lies on tropical monsoon climate zone as classified by Koppen-Geiger [2] and climatic region I_D of Vietnam [3]. Its climate is prevailed by the extremely high amount of solar radiation and is typically exposed to hot-dry wind in summer [3]. In recent years, the city has been suffering from more intensive harsh weather conditions including heat stress and typhoons [4] due to impact of climate change while its resilience requires more improvement under the pressure of economical and spatial growth [5].

Various efforts in technical and academic fields have been made to provide solutions to these issues, including the research project WAMADE (Water Management and urban Development) [6] which, from the perspectives of hydrology and urban studies, researches the urban development of Ha Tinh city in relation to climate change and provides scientific support in the form of tools to develop sustainable development plans.

The fulfillment of the aforementioned objectives requires not only studies at large urban and neighborhood scale, but also those at housing level since houses are fundamental elements for the composition of urban areas. In order to improve resilience and sustainability of urban-areas, housing typologies should be developed in a way that brings their occupants thermal satisfaction while at the same time reduces power consumption. Such developments could only be properly realized when investigation of existing thermal condition of houses as a starting point is conducted. Therefore, a measurement- and survey-based research to provide insight into the current condition of houses in studied city and their effects on the occupants has been conducted. From this research, further studies can be conducted towards the improvement of thermal comfort and sustainability of buildings and neighborhoods.

The aim of this paper is to present results of measurements and survey on thermal comfort of houses in Ha Tinh city. At first, the method section explains the principle of houses selection and describes how the measurement was performed. Also in this section, basis for the evaluation of thermal comfort is introduced. The next part of the paper shows key results of surveys and measurements, and from the result, findings are drawn and used as uptake for the evaluation of thermal comfort levels in these houses.

2. Method

At first, a number of different housing typologies in the city were chosen. During almost two months of 2017 (from late-August to mid-October), field measurements were conducted at the location of the selected houses in order to obtain more insight in actual thermal condition of the houses in summer time. The evaluation of thermal comfort level was done by a psychrometric analysis. An additional analysis was made with data from survey taken in different neighborhoods in Ha Tinh city [7] in 2016, which provides information of dwellers' thermal perception during the whole year. In this section, the principle of house selection and the method to assess thermal comfort by measurements and surveys are described.

2.1. Selection of houses

The determination of houses played an important role to the relevance of the study. The first selection criterion was the deviation between neighborhoods of the city which were identified according to the local climate zone (LCZ) scheme developed by Stewart and Oke [8]. The second criterion was the housing typologies which are characterized by numerous factors including function (used for either only dwelling purpose or combined with business), arrangement (detached or terraced), number of floors (having only 1 or 2 to 4 floors), shape of roof (pitched, flat, or combination of pitched and flat roof), the presence of a shed, and finally wall materials (bricks and cement, raw earth with straw or concrete).

Table 1. Description of the seven LCZs in Ha Tinh city [9]

LCZs	Description (Built/ land cover type)
LCZ3	Compact low-rise: Dense mix of low-rise buildings (1-3 stories); Few or no trees; Land cover mostly paved; Stone, brick, tile and concrete construction materials
LCZ6	Open low-rise: Open arrangement of low-rise buildings (1-3 stories); Abundance of pervious land cover (low plants, scattered trees); Wood, brick, stone, tile and concrete construction materials
LCZ9	Sparsely built: Sparse arrangement of small or medium-sized buildings in a natural setting; Abundance of pervious land cover (low plants, scattered trees)
LCZA	Dense trees
LCZB	Scattered trees
LCZD	Low plants
LCZG	Water

In the study on LCZs of Ha Tinh city [9], seven LCZs were identified and denominated as LCZ3, LCZ6, LCZ9, LCZA, LCZB, LCZD and LCZG based on the difference in built and land cover type.

The first three zones followed by a number (LCZ3, 6 and 9) were recognized by the property of built land while the other four were classified by types of land cover. This research only focused on the zones LCZ3, 6, and 9 where the variety of housing density and typologies were significant and therefore could have feasible effect on indoor thermal comfort. Description of LCZs is found in Table 1.

For the next step, typical housing typologies should be determined based on functional and physical properties. The study followed the housing typologies classification [7], but made modification so that all selected houses were single-family houses and functioned as dwelling place without incorporating business activity (house only). Finally, nine typical typologies in three LCZs were chosen for field measurement and each typology was named in abbreviation form presenting functional and physical properties. For example, typology 1 was labeled as HSO2-4PS to explain that it was used for Housing purpose only, was in Semi-Opened arrangement, with number of stories in between 2 - 4 (3 stories), Pitched-roof structure and attached with a Shed. Table 2 provides details of selected houses for this study. Positions of measured houses were shown on the map of the city on which LCZs were illustrated (see Figure 1).

Table 2. Description of the nine selected houses

House	Picture	Detailed description of typology	Typology	LCZ
1		Used for Housing purpose only, Semi-Opened, 3 floors, Pitched roof, Shed attached	HSO2-4PS	LCZ3
2		Used for Housing purpose only, Semi-Opened, 1 floor, Flat roof, No Shed attached	HSO1FNS	LCZ3
3		Used for Housing purpose only, Terraced, 3 floors, Flat roof, No Shed attached	HT2-4FNS	LCZ6
4		Used for Housing purpose only, Detached, 1 floor, Pitched roof, Shed attached	HD1PS	LCZ6
5		Used for Housing purpose only, Detached, 2 floors, Pitched roof, Shed attached	HD2-4PS	LCZ9
6		Used for Housing purpose only, Terraced, 2 floors, Pitched roof, Shed attached	HT2-4PS	LCZ3
7		Used for Housing purpose only, Terraced, 1 floor, Pitched roof, Shed attached	HT1PS	LCZ9
8		Used for Housing purpose only, Terraced, 3 floors, Pitched roof, Shed attached	HT2-4PS	LCZ9
9		Used for Housing purpose only, Detached, 2 floors, Pitched roof, Shed attached	HD2-4PS	LCZ6

IOP Conf. Series: Earth and Environmental Science **297** (2019) 012034 doi:10.1088/1755-1315/297/1/012034

Figure 1. Satellite map of Ha Tinh (Vietnam) with location of local climate zones [9] and selected houses for measurement.

2.2. Field measurement and psychrometric analysis
The measurement was performed from late-August till mid-October 2017, and the main goal was to evaluate the summer comfort in selected houses. Psychrometric analyses are the backbone of the assessment, but their distributions are influenced by the utilized comfort model. In the study, the modified adaptive model for South-East Asian people introduced by Nguyen [10] was adopted as basis for analysis and evaluation. Nguyen re-defined the different comfort zones on the building psychrometric chart in accordance to actual adaptation and response to climate of Vietnamese people (see figure 2). The comfort level was indicated by the comfort temperature (T_{comf}) which was given in the relation with mean outdoor air temperature (T_{out}) through equation (1):

$$T_{comf} = 0.341 \times T_{out} + 18.83 \text{ (°C)} \quad (1)$$

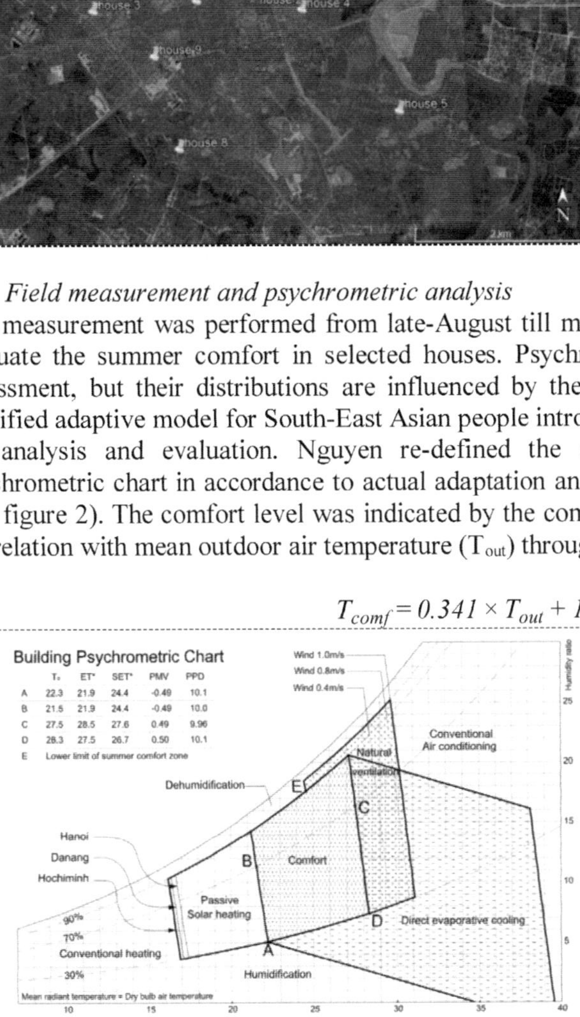

Figure 2. Building psychrometric chart with localized comfort zone dedicated to Vietnamese people [10].

The value of T_{out} can be taken as the arithmetic average of the mean daily minimum and the mean daily maximum outdoor (dry bulb) temperature of a month; however, the values of T_{out} which have been used for this research were actually taken from logged data in individual local climate zones, rather than the ones collected from local weather station. Besides the value of T_{comf} as comfort indication, the assessment was further performed using the Building Psychrometric chart analysis with basic comfort zone and extended ones (by application of natural ventilation).

The measurement was designed and performed with temperature and humidity data loggers (Velleman DVM171THD loggers) to obtain data on operative temperature during summer at locations inside and around selected houses. The accuracy of the measurement for air temperature and relative humidity can be found in Table 3.

Table 3. Characteristics of Velleman DVM171THD loggers.

	Overall range	Accuracy
Relative humidity measurement	0 - 100 % ± 5.0 %	(0 % - 20 % and 80 % - 100 %): ± 5.0 % (20 % - 40 % and 60 % - 80 %): ± 3.5 % (40 % - 60 %): ± 3.0 %
Dry-bulb temperature measurement	-40 °C to 70 °C	(-40 °C to -10 °C) and (+40 °C to +70 °C): ± 2 °C (-10 °C to +40 °C): ± 1 °C

For outdoor measurements, the sensors were protected against direct sunshine by radiation shields as recommended in the Guide to Meteorological Instruments and Methods of Observation [11]. Inside each house, various measurements were simultaneously conducted in most of the rooms during 4 days, see example in Figure 3 with all measurement locations in house 7.

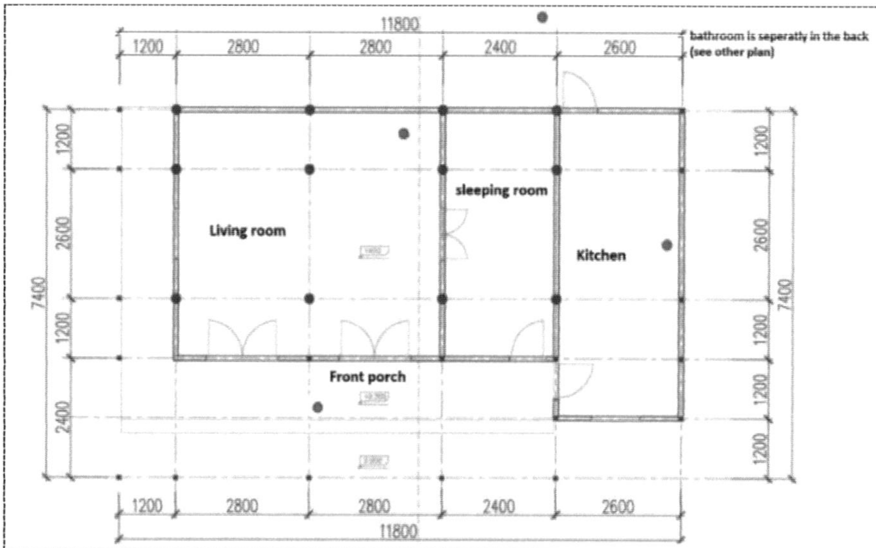

Figure 3. Plan layout of house 7: the measurement locations are indicated by red dots.

However, due to the limited availability of loggers, only two among nine selected houses could be measured in the same period. Unfortunately, the planned measurement in house 5 could not be realized due to the occurrence of a severe storm. The 'as-performed schedule' is shown in table 4.

Table 4. Schedule of on-site measurement.

	24-30/8/2017	30/8-5/9/2017	6-12/9/2017	21-28/9/2017	28/9-5/10/2017
Outdoor temperature (°C) [13]	28.5-33.8	32.8-36.5	33.4-37.4	29.4-32.9	26.4-32.9
Houses	1&2	3	4	6&7	8&9

2.3. Survey

In addition to the measurements, the evaluation of indoor comfort in the houses was complemented by surveys of thermal comfort perception by dwellers. The survey was already conducted as a part of previous study [6]. The questionnaires were handed out to 120 families in 4 different neighborhoods of the city (shown in Figure 4), and covered questions related to how people experienced the thermal environment inside the houses in both summer and winter. There was one representative of each family to answer the survey questions. To help the participants determine their satisfaction in two different seasons, a thermal comfort scale was derived from the psychological comfort model of Fanger [12] and represented by a specific indicator ranging from -2 to +2 in correspondence with resulting feeling from "very cold" to "very hot".

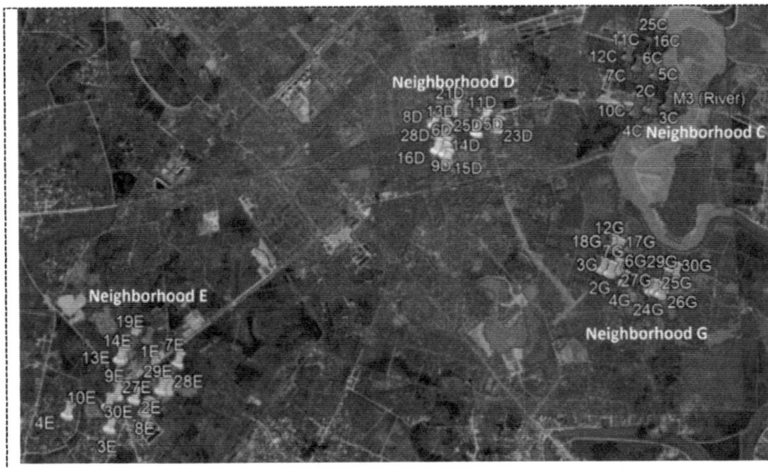

Neighborhood C- Situated along the Rao Cai river and susceptible to floods
Neighborhood D- Old center of Ha Tinh city, surrounded by a canal, high density of buildings (shops, houses, the combination of a shop and a house and government buildings)
Neighborhood E- Peri-urban neighborhood with a low building density and a lot of vegetation
Neighborhood G- Surrounded by rice fields, houses are smaller, more houses made out of loam, few cars, more animals, more agriculture activities ...

Figure 4. Satellite map of Ha Tinh with LCZs [9] and locations of neighborhoods for the survey in 2016, [7].

3. Results

In this section data are discussed on occupants' comfort level in summer and winter which were obtained from the survey in 2016, as well as results of the measurement of air temperature and relative humidity in selected houses in summer of 2017.

3.1. Results of survey of thermal comfort perception

Figures 5 to 8 show the thermal comfort perception by dwellers in summer and winter season. The results indicate that in general, indoor temperature provided a majority of inhabitants with satisfaction: 55 to 62% of the questionnaire participants classify the mean indoor temperature as in neutral situation in both winter and summer. However, almost one third (37%) of the inhabitants perceived the thermal condition in winter as too cold (-1 to -3) and 25 to 30% perceived it as too hot in summer (+1 to +3). If a reduced scale of comfort (-2/+2) was applied, a similar perception was provided by occupants.

The seasonal mean values were determined separately for both comfort scales. For the reduced scale, median-based seasonal values in winter and summer were 0.6 and 0.5 respectively. With the normal scale (-3/+3), the value for summer was 0.8 while that of winter was 0.9. This confirms the conclusion that the majority of the inhabitants are satisfied with the indoor temperature during summer and winter.

The thermal comfort perception was further examined in relation to the location of the houses in the local climate zones. In general, uniform pattern of occupants' perception revealed not a feasible co-relation between LCZ and thermal perception. Only a more visibly neutral state in winter perceived in LCZ3 was shown in Figure 8 when using the normal scale (-3/+3), [12]. It was an interesting result because LCZ3 is most densely paved with high density of low-rise houses and other buildings, and it could have been thought to bring less comfort than neighborhoods of other LCZs.

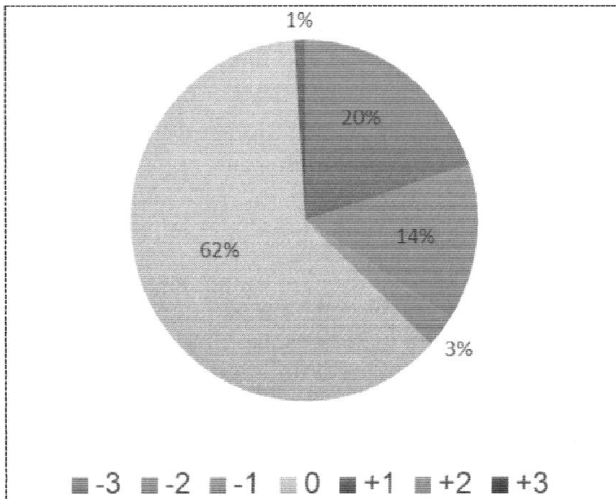

Figure 5. Perception temperature in winter of 120 surveyed participants with full comfort scale (-3/+3)

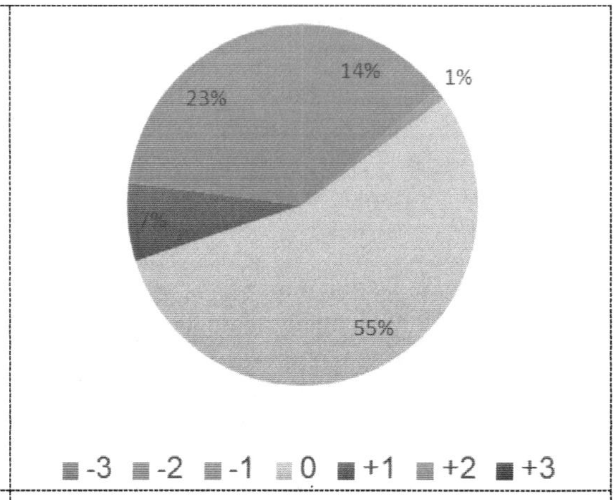

Figure 6. Perception temperature in summer of 120 surveyed participants with full comfort scale (-3/+3)

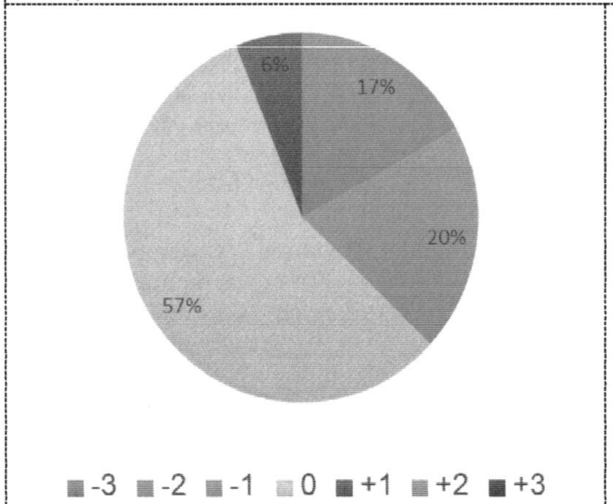

Figure 7. Perception temperature in winter of 120 surveyed participants with reduced comfort scale (-2/+2)

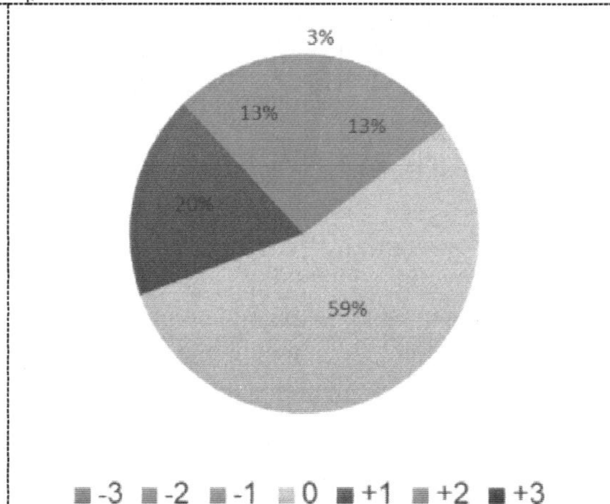

Figure 8. Perception temperature in summer of 120 surveyed participants with reduced comfort scale (-2/+2)

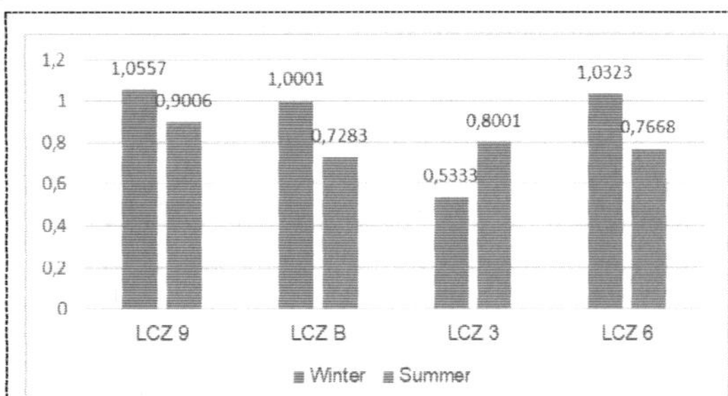

Figure 9. Comparison of thermal perception (average value of scale -3/+3) between location in different LCZ

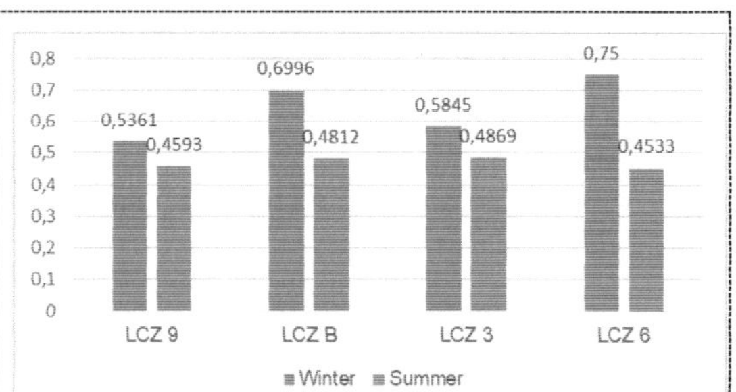

Figure 10. Comparison of thermal perception (average value of scale -2/+2) between location in different LCZ

3.2. Results of temperature and relative humidity measurement

Figure 11 shows the measurement results of the actual temperature in house 1 as an example together with the threshold values of comfort temperature T_{comf}.

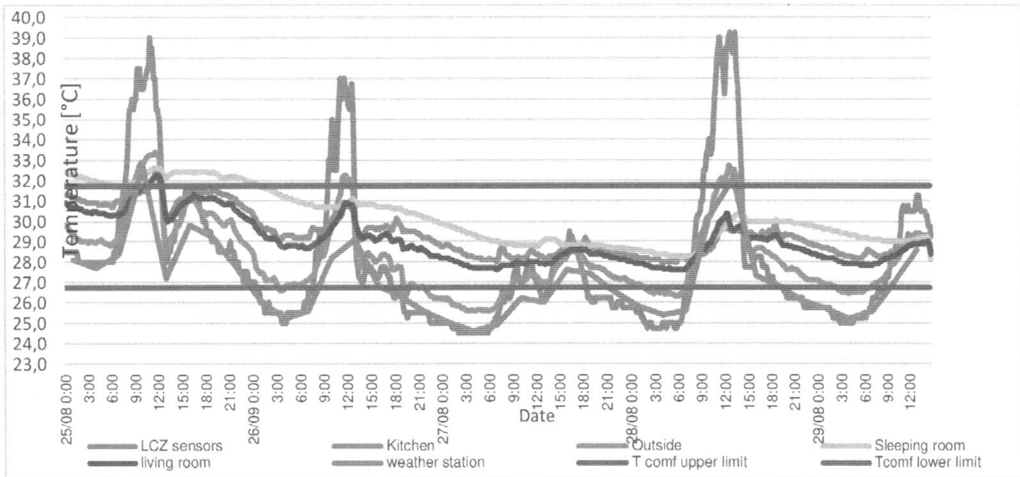

Figure 11. The statistic of air temperature variation in locations of house 1 during measurement time in comparison with values of comfort temperature

Psychrometric analysis for house 1 was also performed, and it is presented in Figure 12. The figure shows that indoor comfort of the house varied with different outdoor condition temperature, and duration when rooms of the house were exposed to discomfort was longer than it was as shown in figure 11 in which only air temperature was taken into account. The difference was due to the combining influence of relative humidity with temperature, causing unexpected comfort to the occupants. The analysis also indicates that the bedroom of house 1 was exposed to more discomfort than the other rooms. This can be explained by the difference in positions of the rooms in measured house: the living room is on the ground floor where least solar heat is transferred while the bedroom is on the first floor and closed to the roof structure, which gains high amount of solar heat during the daytime.

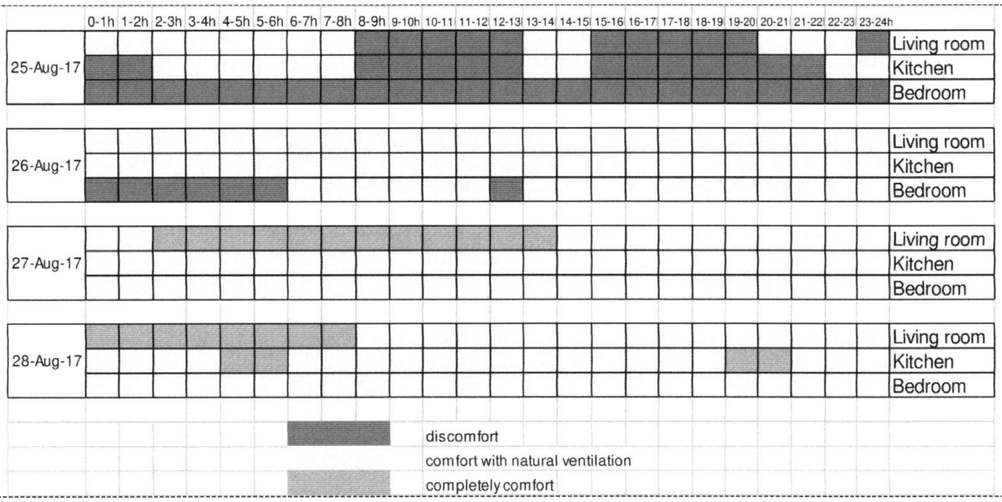

Figure 12. Result of psychrometric analysis for house 1 during four days of measurement.

Similar analyses were conducted with the measurement data in all the selected houses (except for house 5). A summarized statistic of comfort levels in all houses is given in Figure 13. Despite the fact that not all measurements were conducted at the same time, and therefore were influenced by unequal outside weather conditions, the comparison forms a basis for the evaluation of indoor thermal comfort level. The figure shows there was a high level of discomfort in houses 1, 3 and 4 while indoor climate in the other houses was significantly less out of comfort range.

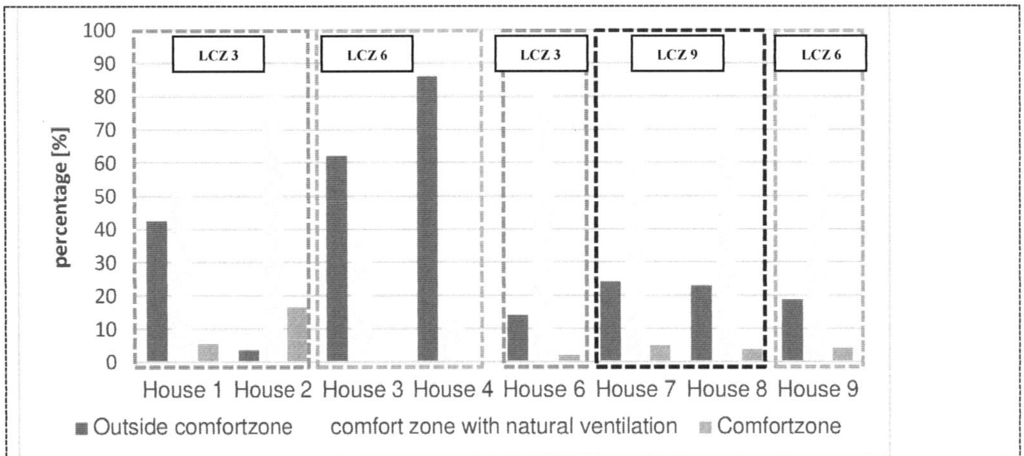

Figure 13. The comparison of results of psychrometric analyses for the measured houses

The difference can be explained by the fact that houses with more openings and surrounded by less buildings (houses 2, 6, 7 and 9), or the ones structured with double-layered roof as well as ventilated roof (houses 8 and 9) enable higher comfort percentage, while most discomfort houses (1, 3 and 4) are located in more populated quarters or closely surrounded by neighboring houses, which may obstruct ventilation cooling during summer time.

An unexpected result was detected in the measurement for house 3 where discomfort state was dominating. In fact, the house design was studied with sustainable climatic approach in expectation to achieve high level of comfort, and the real comfort in the house was also recognized by its occupants. The counter-intuition can be argued by the influence of harsh outdoor condition while conducting measurement, the inefficiency of novel vegetated roof, and large amount of solar heat penetrated through glass vanes on its main facade. One more reason for the discomfort situation is related to Southwest orientation of the house in which unpleasant Foehn hot-dry wind flow was prevailing in the summer and probably during the whole measurement time.

4. Conclusion

This paper presents the results of measurements and surveys on thermal comfort in residential buildings in Ha Tinh (Vietnam). According to what the survey found, the majority of local people was satisfied with the operative temperature in both seasons. The survey also presented various perception of inhabitants from different LCZs of the city, though the deviation was not significant. Besides, it gave an unexpectedly positive results in actual satisfaction of dwellers in most compact LCZ (LCZ3).

The results from measurements of operative temperature and relative humidity provided more detail about comfort level achieved in different housing typologies in the city, and indicated the dependence of thermal condition on properties of building envelope and natural ventilation efficiency. In the majority of measured houses, the comfort state could be achieved in most of the rooms by cross ventilation while natural comfort feeling could only occur in a very short duration of the measurement.

The study provides a clear insight on thermal condition of houses in Ha Tinh city and perception of local inhabitants from different parts of the city, and its results could be used as starting points in further studies for improvement of houses in terms of thermal condition and energy efficiency. For research at larger scales of neighborhood and urban areas, the study can contribute to the development and/or improvement of neighborhood typologies in complementation to typologies of buildings for a better climatic adaptation and thermal environment.

References

[1] Portal of Ha Tinh city *http://www.hatinh.gov.vn/vi/topicIntro/topic/2993*

[2] Kottek, M., J. Grieser, C. Beck, B. Rudolf, and F. Rubel. (2006). *World Map of the Köppen-Geiger climate classification updated. Meteorol. Z., 15, 259-263. DOI: 10.1127/0941-2948/2006/0130*

[3] Vietnam building code QCVN 02:2009. (2009) *Natural physical and climatic data for construction*

[4] Vietnam Aacademy for Water Resources (2017) *Technical consultancy of hydrological/hydraulic modelling of Rao Cai River basin and drainage modelling for Ha Tinh City, Ha Tinh Province, in relation to Climate Change*

[5] People's committee of Ha Tinh provine (2015 *Explanation of the modification to the master plan of Ha Tinh city to 2030 with the vision toward 2050*)

[6] WAter MAnagement and urban DEvelopment in Ha Tinh in relation to climate change (WAMADE) *https://www.vliruos.be/en/projects/project/22?pid=3187*

[7] Leyman, R. (2017). *Microclimate Simulations on Neighborhood Typologies in Ha Tinh Province , Vietnam.*

[8] Stewart, I. D., & Oke, T. R. (2012). Local climate zones for urban temperature studies. *Bulletin of the American Meteorological Society*, *93*(12), 1879–1900. http://doi.org/10.1175/BAMS-D-11-00019.1

[9] García, Á. G. (2017). *Development of a typology for green and urban structures using high resolution satellite imagery in Ha Tinh, Vietnam.*

[10] Nguyen, A.-T. (2013). *Sustainable housing in Vietnam: climate responsive design strategies to optimize thermal comfort, 305.*

[11] World Meteorological Organization (2008). *Guide to Meteorological Instruments and Methods of Observation*

[12] Ole Fanger, P., & Toftum, J. (2002). Extension of the PMV model to non-air-conditioned buildings in warm climates. *Energy and Buildings*, *34*(6), 533–536. http://doi.org/10.1016/S0378-7788(02)00003-8

[13] Statistic weather data of 2017 collected from Ha Tinh weather station (Ha Tinh province)

Climate implication and adaptation measures for energy use in buildings – a scoping review

A E Stagrum*[1], T Kvande[2], A Engebø[1], E Andenæs[1], J Lohne[3]

1. Ph.D. candidate, Department of Civil and Environmental Engineering, Norwegian University of Science and Technology, Trondheim, Norway
2. Professor, Department of Civil and Environmental Engineering, Norwegian University of Science and Technology, Trondheim, Norway
3. Research scientist, dr. art. Department of Civil and Environmental Engineering, Norwegian University of Science and Technology, Trondheim, Norway

Abstract. The purpose of the study is to investigate data on climate implication and adaptation measures for energy use in buildings. It is based on a scoping literature review, concerned mainly with the main journals operating in the field of climate adaptation of the built environment. Research documents that significant changes are taking place due to the implications of climatic change. Studies concerning climate change impact on energy use in buildings in warm climates represent the majority of the findings. The volume of research within the consequences for the built environment in cold regions is found to be surprisingly low, especially concerning the pecuniary stakes involved. However, significant regional differences are observed. Though further research is of essence, policy/regulatory measures ought already to be taken, based on climate scenarios.

Keywords: Climate change, building, adaptation, impact, energy use

1. Introduction

The global climate is constantly fluctuating to some degree. Data collected over the recent decades, however, show unusually rapid change in a historic context, caused primarily by anthropogenic activities. Increased concentrations of greenhouse gases in the atmosphere causes the global average temperature to rise, with wide-ranging consequences for the atmospheric climate [1]. Among other things, the increased global temperature causes weather patterns to shift and sea level to rise. This will have severe consequences for a built environment designed under the assumption of steady conditions. Representative Concentration Pathway (RCP) describes greenhouse gas concentration in the atmosphere for future climate scenarios, utilized to predict the climate change consequences [2]. The four RCPs used for climate modelling are RCP2.6, RCP4.5, RCP6, RCP8.5, where RCP2.6 represents the lowest, and RCP8.5 the highest concentration projection [3].

In Norway, climate change is expected to manifest itself in the form of higher temperatures and increased levels of precipitation. The national average air temperature has risen by 1 °C between 1900 and 2014, and precipitation has increased by 18 % over the same period [4]. The trends are expected to be prolonged over the next century. This leads to shorter and milder winters as well as more frequent and intense rainfall events. Sea level rise is a relatively minor concern in Norway, as it is largely counteracted by land rise. However, increased precipitation is expected to lead to costly damages to buildings and infrastructure by 2100 [5].

The amount of energy use in buildings depends on the building's location, outer factors, the building envelope and the intended use of the building [6]. Climate change will have consequences for energy requirements in buildings [7]. Since the winters will be warmer, the heating demand may be reduced. Rising summer temperatures cause higher demand for cooling to assure thermal comfort. Evaluation of adaptation measures for buildings, especially in urban areas, is therefore of high importance [8]. To examine this matter, the following research questions are addressed:

- What is known from existing literature about climate implication and adaptation measures for energy use in buildings?
- What are the most important gaps in the research?

This study is part of a larger literature research which concerns climate implication and adaptation measures for buildings (as a whole). Elements omitted are: policy, sustainability, mitigation, health impact, air quality, design tools and guidelines, overheating, extreme weather impact, precipitation and adaptation of the building envelope.

2. Methodology

This study is based on findings from a scoping literature review, carried out during the period of November-December 2018/January 2019. A scoping study might "aim to map rapidly the key concepts underpinning a research area and the main sources and types of evidence available and can be undertaken as stand-alone projects in their own right" [9]. Thus, the objective for this study was to map the literature available on the subject of climate change impact and adaptation for energy use in buildings, with the purpose of categorize and analyze the findings according to the proposed research questions. The method is based on the framework described in [9]. This involves a six-step procedure; 1) identifying the research question, 2) identifying relevant studies, 3) study selection, 4) charting data, 5) collating, summarizing, and reporting the results, 6) consultation.

The procedure also made it possible to identify the research gaps. In this manner, the study can provide a thorough and valid method for mapping the research area, to discover the measure and the characteristics of the research done in the field [10].

Given the extent of the material identified, certain limitations needed being outlined. The articles investigated concern mainly questions of building physics. Furthermore, as according to the guidelines provided by Arksey & O'Malley we did not assess research quality of the articles included in the review. The practical research procedure can be illustrated in the following figure:

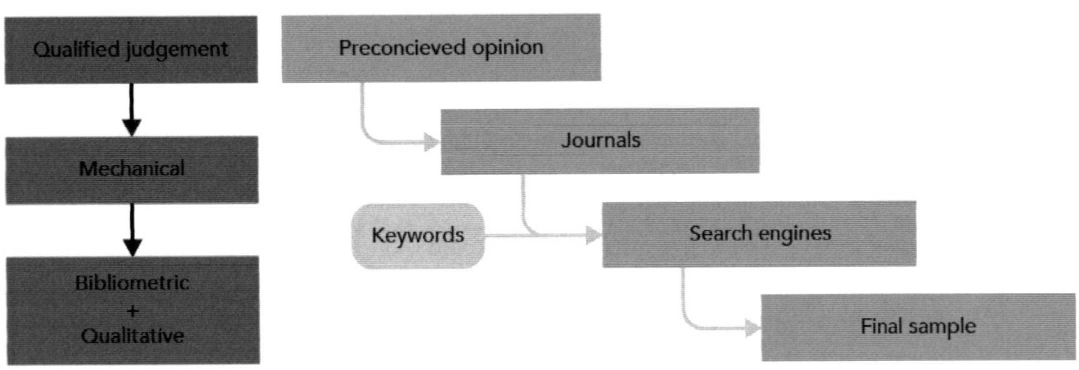

Figure 1. Research procedure

2.1 Identifying and selecting studies

To be as comprehensive as possible identifying studies, searching for research evidence should involve different sources [9]. Given the extensiveness of the field, two main search processes were employed. First, hand-searching of key-journals as described in 2.1.1. Second, a comprehensive search through carefully selected databases and search engines was conducted.

As part of the broader study, over 20 000 article titles and/or abstracts were examined in order to identify studies related to climate implication and adaptation measure for buildings. 163 articles were identified. To narrowing the scope, this study presents a set of 67 articles that all met inclusion criteria's regarding the topic of implications and measures for energy use in buildings.

2.2 Hand-searching of key-journals

Hand-searching of key-journals was carried out to identify research pertinent to the field of inquiry, as well as obtaining an overview of the content of the key journals within the field. Based on the experience of research co-workers and pre-conceived opinion, 12 journals were selected to examine. This selection consists of Building and Environment, Climate Services, Energy and Buildings, Building Research & Information, Journal of Climate Change, Buildings, Journal of Building & Physics, Climate Sustainable Cities and Society, Energy Policy, International Journal of Climate Change Strategies and Management, and Advances in Energy Building Research. The publications in Construction and Building Materials were examined only for 2018. It was discovered that this journal is very extensive. Since no relevant articles were found among the 40 volumes, the journal was discarded. The articles published in the selected journals are related to building science and/or climate change related issues. Articles newer than five years were selected through qualitative assessment of titles, keywords and abstract. 74 articles were found in this way.

2.2.1 Identification of search words. The search terms were decided based on the keywords in the articles that were found by hand-searching the key-journals. As figure 2 shows, the keywords in the selected articles were listed and counted. The final keywords were selected based on frequency and qualitative judgement. Thus, creating a consistent basis for the final search strategy as described in the following paragraph.

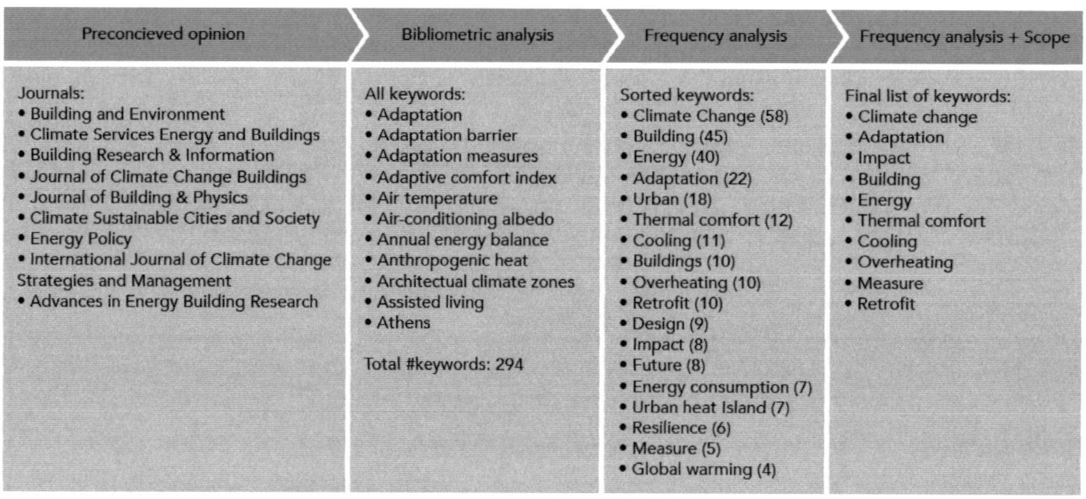

Figure 2: Procedure for identifying search words

2.3 Search through databases and search engines

The systematic search was carried out using combinations of search words identified through the key-journals, showed in figure 2. Three reference databases were then searched through, using the search strategy explained in table 1. The three major electronic databases included Google Scholar, Oria (Norwegian university library search engine), and ScienceDirect. Electronic databases provide several tools for narrowing the search, therefore, the search was limited to English scientific research and review articles, excluding patents and conference papers, published the past five years. Because databases operate with diverse filters and algorithms, some variation in the search strategy were employed for use in each of the databases. Regardless of variations in the structure of databases, all search words and combinations were consistently used in all databases. The filters for each database/search engine, filter explanation and the number of unique and double identified publications are listed below in table 1. Duplicate publications were later filtered out as from the final sample.

Table 1. Filters used for search through databases and search engines, filter explanation and number of unique identified publications.

Search engine	Filter	Filter explanation	Unique identified publications (doubles)
Google Scholar	Title and topics	All field-search gave an unmanageable number of hits	2 (13)
Oria	Title	All field-search gave an unmanageable number of hits	36 (70)
ScienceDirect	Title, keywords and abstract	Search results could be examined manually	50 (65)

2.4 Sorting of articles

The search-process produced a large number of hits in total, creating the need for an extensive screening. The screening process worked as follows: first, articles with obviously irrelevant titles were excluded. Second, if the title seemed relevant, abstract was examined. Third, if abstract seemed relevant as well, a throughout examination of the article was conducted. The screening was conducted primarily by the main author, but articles that were not irrelevant according to the inclusion criteria's set was discussed in weekly meetings between all of the authors.

2.5 Charting and reporting the results

All articles accepted through the screening process were retained for analysis. This step included "charting" of data, described as "a technique for synthesizing and interpreting qualitative data by sifting, charting and sorting material according to key issue and themes" [9]. As for analyzing the articles, a database was created using spreadsheets. The database collected the article's title, author, keywords, year of publication, study location, purpose, methodology, and highlights from the study. The articles were then categorized, as illustrated in [11]. It must be noted that some of the categories are overlapping. Some articles not included in this study, mentions energy use, without being the primary focus. As for reporting the results, the final sample consisting of 68 articles were divided into nine sub-categories; general aspects, electricity consumption, heat demand, cooling, heating and cooling demand, building envelope, passive adaptation measures, retrofit measures and costs. There is significant overlap between several categories; for instance, heating and cooling loads will be impacted by energy adaptation measures. It was also noted whether each article primarily concerns climate change impact, measures, or both.

The synthesis includes a qualitative analysis of the final sample. The qualitative analysis is primarily descriptive meaning that the results are described according to each category in the results section, mainly focusing on presenting the research purpose of each article. The discussion provides a synthesis of what is known about climate implication and adaptation measures for energy use in buildings and the current knowledge gap in the research.

3. Results

3.1 General overview of the material

Given the magnitude of data obtained, this chapter only reports briefly on each article. Furthermore, the results are organized around the nine sub-categories as described in the methodology. The material includes articles from 26 countries (first author affiliation), spread across all inhabited continents. The studied articles are sorted by categories and impact/measure in figure 3.

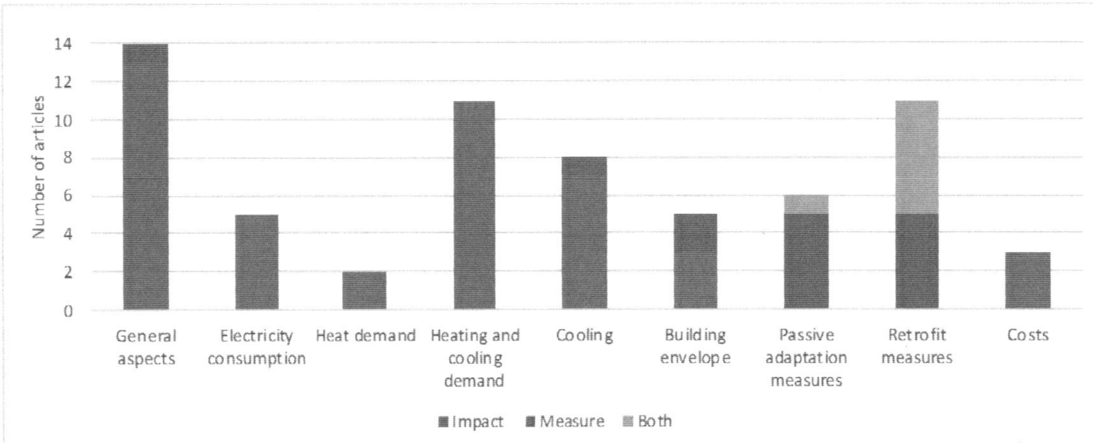

Figure 3. Number of articles sorted by categories and impact/measure.

3.2 General aspects

Variations in energy demand and consumption due to climate change and different impact factors, are investigated by several authors. Zhou et. al [12] investigate the impact of climate change on energy use in the US, using a sensitivity analysis. The results indicate that the impact varies greatly between states and between different energy carriers (i.e. electricity, oil and gas). Similar results are found in a different sensitivity analysis by Huang and Gurney [13], who also evaluate methods for estimating future energy consumption. The same authors [14] quantify how the relationship between climate change and building energy consumption varies across a range of building types in 925 U.S. locations. Changes in annual energy consumption range from −17% to +21%. Even within climate zones, changes in building energy consumption show strong variation. A similar study in the US was done by Dirks et. al [15], who simulated the impact of climate change on peak energy loads as well as annual energy consumption in buildings. It is found that peak demand can increase significantly faster than annual demand.

Shen [16] investigate how building energy use patterns in the US would change in response to climate change in the years of 2040-2069, using case studies in four cities. Future building performance regulations and codes are drafted. The change in annual energy use is predicted to range from −1.64% to 14.07% for residential buildings and from −3.27% to −0.12% for office buildings. Similar work was carried out by Kikumoto et. al [17], who constructed prototype future (2030s) weather data for locations in Japan. Building thermal loads were also simulated using these data, finding a predicted increase in sensible heat load of 15% for a house in Tokyo.

To address climate change issues, Cox et. al [18] investigate if a weather file constructed from a coarse (annual) estimate of future air temperature change can provide useful estimates of future energy demand of a building, if high-resolution data is not available. Future weather files are constructed based on annual, monthly and hourly temperature change estimates. The constructed weather files are found to give similar energy demand as high temporal resolution files. In a literature review, Fazeli et. al [19] analyze models of residential energy demand in response to temperature changes. The potential of the models to evaluate the adaptation responses is discussed. This study can be seen in relation with Pyrgou et. al [20] who compares typical weather files with locally collected weather data for a city in central Italy. Through analysis, a discrepancy between the data sets is identified. This may cause inaccuracies in building energy consumption estimates.

Downscaled future hourly weather data from the Global Climate Models (GCM) is used by Shen and Lior [21] to predict future performance of renewable energy systems for residential zero energy buildings in the U.S. The analysis shows that present systems in half the climate zones will not achieve net zero energy use.

Dodoo and Gustavsson [22] explore to what extent different climate scenarios influence overheating risk, energy use and peak loads for space conditioning of district heated multi-storey buildings in

Sweden. The thermal performance of the buildings changed notably under future climate scenarios, depending on the buildings' energy efficiency level.
A literature review is provided by Jenkins et. al [23], where some of the risks associated with energy use in buildings are discussed in the face of climate change. As climate change will affect the performance of energy systems, as well as building energy demand, it is proposed that climate change is taken into account during the energy assessment of buildings.
The thermal and energy performance of social housing projects in Brazil are investigated in Triana et. al [24], considering climate change. Energy efficiency measures for adaptation to climate change are also assessed.
Guan [25], on the other hand, investigates the influence of internal load densities on energy and thermal performance of office buildings in Australia. It is found through building simulation that reducing internal load densities has a high impact on the energy use of the building, and it is suggested that this could be an effective adaptation strategy in the face of climate change. Also in light of climate change, the impact of urban location on energy use of office buildings in Vienna is studied by Berger et. al [26]. It is shown that buildings in central urban areas yield lowest energy demand because of the urban heat island effect. Despite the lower need for heating, older buildings will still have a high energy demand, and may require cooling devices to remain habitable under certain future climate scenarios. Modern buildings are more robust in the face of climate change. A similar study was carried out by Wang and Chen [27], who found that the effect in energy use depends significantly on geographical location and building type. Natural ventilation availability will be reduced in some cities.

3.3 Electricity consumption
Analysis to relate the electricity consumption with outdoor climate during a 12-year period (2005–2016) is performed by Li [28]. Results show that electricity consumption may increase as much as 47% under the worst climate change scenario. The consumption increase is supported by Santamouris et. al [29], who compared existing studies on the impact of ambient temperature increase on electricity consumption. It was shown that the change in electricity demand per degree of temperature increase varies between 0.5% and 8.5%. In this case a more specific study by Braun et. al [30] is of interest. They investigated possible changes in gas and electricity consumption in supermarkets throughout Great Britain for the 2030s. Changes in temperature are expected to lead to a 2% increase in average electricity consumption and an average drop in the gas usage of 10%. Similar conclusions were shown in a 2014 article by the same authors [31]. A consequence of the increased electricity need is shown by Tarroja et. al [32], who indicate that up to a 31.6% increase in future grid capacity is needed, as heating electrification loads not align with future grid renewable generation.

3.4 Heat demand
The main scope of Andrić et. al [33] is to present a methodology to assess the possible long-term effects of climate change on building heat demand on a large scale. Through building simulation, it was discovered that heat demand density could decrease 22.3-52.4% in 2050. In another study performed by the same authors [34], reduction in heating hours was found to be insignificant in cold climates, while heating hours rate was substantial in warmer climates considered.

3.5 Cooling
A relational study of correlating cooling energy use with local weather station and apartment price in Seoul is presented by Yi and Peng [35], using simple bivariate regression models. The largest predicted increase rate is 96.1% in one city district (year 2047). The result may be seen in line with Li et. al [36] who investigated the climate change impact on cooling energy consumption of office buildings in China. Their building simulations show that cooling energy consumption is affected by different climate factors in different climate conditions but did not rise with rising temperatures in hot climate conditions. It is noted that humidity effect rather than single temperature effect should be fully considered to improve building energy efficiency, and humidity change should be used to predict future energy cooling consumption. Another impact study on office building is given by Yau and Peng [37] who studied the effect of weather variation on air conditioning in a building located in Malaysia.

The overall results show that every 1 °C causes the COP and the total cooling capacity to be lowered by 2%. Impact on residential buildings is given by Hwang et. al [38] who investigates the impact of climate change on current and future residential cooling energy use in central Taiwan, as influenced by the urban heat island (UHI) effect. Further, Arima et. al [39] assess the impact of climate change on energy consumption of a two-story detached house in Tokyo. A bias of the weather data was corrected using the statistical values of observations. The sensible heat load for the house was predicted to increase by 26% and latent heat load increased by 10%. These studies are of interest in the context of the building thermal simulation carried out by Zoras et. al [40]. They estimated the impact of outdoor bioclimatic measures on cooling demand during a summer day over a large city area. The main result of their study was the capability to synchronously evaluate building energy performance due to exterior micro- climate improvement for about 200 buildings.

Another example by Oliveira Panão [41] discusses benchmarks for cooling energy requirements for residential buildings in 30 regions in Portugal. Benchmarks for annual cooling energy needs from 5.5 to 37.6 kWh/m^2 are recommended. Finally, Santamouris [42] discusses a framework which defines the present and future cooling energy consumption of the building sector. The global cooling consumption of the residential sector is expected to increase up to 34% in 2050 and 61% in 2100.

3.6 Heating and cooling demand

Heating and cooling demand has been investigated by several authors and in many different locations in the world. The energy consumption of a university school campus for three typical future years (2039, 2069, 2099) is modeled by Mauree et. al [43]. Simulations predict, there will be a constant decrease in the heating demand in the future, while the cooling demand will increase substantially. This trend was also noted by Kalvelage et. al [44], who pointed out that peak heating load will decrease and the peak cooling load increase. Chen et. al [45] came to the same conclusions after similar investigations on Chinese cities. Similar results were also suggested by Orehounig et. al [46], who used weather data from expected climate change scenarios in Vienna, to evaluate different building design.

Spandagos and Ng [47] used a model building in Hong Kong, Seoul and Tokyo for energy simulations. The study shows an expected increase in energy usage for heating and cooling over the next few decades, even with future stabilization and reduction of GHG emissions (RCP-4.5). An increase in annual energy demand was also proved by Invidata and Ghisi [48], who investigated the impact of climate change on thermal comfort conditions and on heating and cooling energy demand in dwellings in three cities in Brazil. These studies are supported by a literature review performed by Santamouris [49], concerning the impact of the urban heat island effect and global warming on building energy demand. In the period 1970-2010, the average increase of the cooling demand is found to be 23% and the corresponding average reduction of heating demand is 19%. For typical buildings, the average energy consumption for heating and cooling purposes increased by 11% for the same period.

In colder climates, reduction in heating energy demand may be more substantial than the increased need for cooling. Jylhä et. al [50] assessed the energy demand of a typical detached house in Finland under observed recent and anticipated future climatic conditions. It is expected that the increased demand for cooling will be compensated by the high decrease in heating demand, and the net annual energy demand will therefore be reduced. Possible reduction of energy consumption was also found by Shibuya and Croxford [51]. They analyzed the impact of climate change on cooling and heating loads in office buildings in three sites in Japan, and over three periods (1981–2000, 2031–2050, and 2081–2100). Similar results were found by Damm et. al [52], who simulated changes in the heating and cooling electricity demand for 26 European countries, based on +2°C global warming. Electricity demand is found to decrease in most European countries, with the highest relative decrease found in Scandinavia. Similar trends were noted by Berger et. Al [53] who applied data from future climate scenarios in a simulation of office buildings in Vienna, Austria. Older office buildings are found to require less heating, which will slightly counteract the considerable increased need for cooling. Modern buildings, on the other hand, were found to have a higher cooling demand today which will increase further in the future, resulting in a higher net energy demand.

3.7 Building envelope

The influence of façade surfaces on the energy balance in buildings and thermal conditions in urban environments is investigated by Alonso et. al [54]. The findings suggest that the impact of the surface finish is significant. Similarly, Fontanini et. al [55] investigate how different attic forms are impacted by current and predicted future climates in multiple locations across the U.S. Relative changes in cooling, heating, and overall thermal loads are calculated for future climate scenarios. Further, Karimpour et. al [56] analyses both the external wall and roof construction on a case building in Adelaide, Australia. Climate change was found to shift the buildings energy demand from heating to cooling dominated by 2070.

Rubio-Bellido et. al [57] evaluates the impact of different elements of building design on energy demand in office buildings in Chile, under a projected climate change scenario. Energy demand is found to increase in all the 9 investigated climate zones in Chile. Further, Sajjadian et. al [58] presents a method to assess the effect of phase change materials (PCMs) on thermal comfort and energy consumption in UK dwellings in summer months. PCMs are found to be beneficial in current times and in the future, as the impact of climate change on thermal comfort and cooling loads are considerable in the UK.

3.8 Passive adaptation measures

A literature review by Ascione et. al [59] is focusing on renewable technologies for the passive cooling of buildings. Passive adaptation measures that reduce the cooling need and improve thermal comfort during summer include greenery of facades and roofs, phase change materials, solar screens, cool colors, thermal mass, evaporative cooling and natural ventilation. Fahmy et. al [60] explored the use of greenery as a climate change adaptation measure in urban areas in Egypt. Simulations show that the greenery improved the energy efficiency of buildings by as much as 23.8% in 2080, compared to a street without greenery. Another study, Van Hoof et. al [61] analyses the effect of passive climate adaptation measures on the cooling and heating energy demand of a terraced house, under a future climate change scenario. External solar shading and natural ventilation show the best performance in limiting the required cooling energy.

An adaptive comfort model was used in Huang and Hwang [62] to identify the air-conditioning operation status of a residential building to determine cooling energy use, in the past and the 2020s, 2050s, and 2080s. The cooling energy demand was found to increase by 31%, 59%, and 82% respectively. Five passive design strategies for reducing cooling energy demand are discussed. Pierangioli et. al [63] focuses on the effectiveness of passive adaptation measures in central Italy in the medium (2036–2065) and long term (2066–2095). In the medium term, the reduction of energy needed for heating could be bigger than the increase for cooling in case of poorly insulated buildings. This results in a total annual net energy need decrease; however, in the long term the opposite happens. Mitigation and adaptation strategies for sample office buildings in Vienna are simulated by Berger et. al [64], applying weather data from future climate scenarios. Internal heat gains are found to influence cooling demand more than climate change. Reduced internal heat gains and additional external insulation minimize cooling and heating demand. It is shown that winter savings due to external insulation will continue to outweigh increased cooling demand even in the future.

3.9 Retrofit measures

The energy performance for residential buildings in Stockholm from 1961-2100 for five climate scenarios is simulated by Nik et. al [65]. A method to assess and quantify the relative robustness of retrofitting measures on long term is presented, taking into account climate variations in different time scales, extreme conditions and uncertainties of climate change. Interesting in this context is also Cao et. al [66] who offers an overview of building energy-consumption situations, relevant energy-saving approaches, and the influence of global climate change. Improvements on building envelope and ventilation are found to play an essential role in reducing space heating and cooling consumption levels. They also provide some suggestions for further developing Zero Energy, while Pagliano et. al [67] discuss a retrofit of a building in Milan (Italy) on the basis of future weather scenarios. The analysis suggests that climate change might require the installation of active cooling systems to

compensate for harsher summer conditions in the long term. However, in the mid-term, passive cooling strategies may still guarantee thermal comfort. Strategies for reducing HVAC demand in US cities presented in Wang et. al [68] builds up on this statement.
Further aspects such as energy poverty and decision making were considered by Flippín et. al [69] who propose retrofit strategies for adapting to a climate change scenario in Argentina. The strategies might be seen in context with the model called CESAR (Combined Energy Simulation And Retrofitting) [70]. This model is used to generate future demand and emission projections of districts, including key economic indicators. Building energy performance under climate change projections is also evaluated by [70], while Waddicor et. al [71] investigate the impact of predicted climate change and building ageing on the energy performance of a library in Turin, Italy. The two phenomena are found to greatly influence the energy demand of the building. Mitigation measures against climate change are assessed.
Energy retrofitting measures for residential buildings in Sweden for five scenarios of future climatic conditions are compared by Nik et. al [72]. Nine energy retrofit measures are compared, as well as four combinations of these. Tuning the indoor set-point temperature to 20 °C can contribute to significant energy savings. A combination of improved thermal insulation of the building envelope with energy efficient windows is found to be the most effective and robust retrofitting measure. However, this seems to be of less importance in Australia. Here Daly et. al [73] have simulated the possible impact of climate change on the energy consumption of an office building in five Australian cities. The impact of climate change is compared to other possible changes to building energy consumption. For the Australian case, climate change is found to be relatively unimportant in selection of optimal retrofit strategies. Pérez-Andreu et. al [74] model a range of passive and active energy improvement measures applied to a typical Mediterranean residential building under various climate-change scenarios. Temperatures are projected under two Global Circulation Models for 2050 and 2100. Heating energy demand decreases significantly and cooling energy demand increases. It is shown that total energy demand is the most affected by thermal insulation and infiltration.

3.10 Costs
Potential future implications of climate change on building energy expenditures around the globe is explored by Clarke et. al [75]. Net expenditures decrease in some regions where heating demands currently dominate. Net expenditures increase the most in areas with higher need for space cooling. Similarly, based on the impacts of climate change on U.S. building energy consumption, Huang and Gurney [76] quantify the financial implications to consumers and suppliers. For the U.S. as a whole, building energy costs decrease by ~7 billion $/year in the residential sector, while costs increase by ~2.2 billion $/year in the commercial sector. For cold-weather states, residential energy savings are expected, while warmer states see increased residential energy costs. The latest is supported by Zachariadis and Hadjinicolaou [77] who assess the cost of additional electricity requirements because of projected climate change in Cyprus by the mid-21st century. Annual electricity demand is projected to rise by 6% by 2050. Additionally, welfare losses of more than 150 million Euros (at 2010 prices) are expected per year.

4. Discussion
In this paper, we set out to address the following research questions: What is known from existing literature about climate implication and adaptation measures for energy use in buildings, and what are the most important gaps in the research? Just from the past five years, there is a substantial body of literature on climate adaptation measures for energy use in buildings. As this scoping review indicates, the literature covers a wide array of topics. To provide clarity, the results are discussed according to the following nine sub-categories: building envelope, cooling, costs, heat demand, heating and cooling demand, passive adaptation measures, retrofit measures, and "others".
A general theme noted in a majority of the articles is that climate change will manifest in the form of higher temperatures, which will reduce the energy demand for heating and increase the demand for cooling. In warmer climates the overall energy demand for heating and cooling will increase, while it will decrease in colder climates. Highly insulated buildings may experience substantially higher

cooling loads to prevent overheating. Poorly insulated buildings may benefit comparably more from reduced heating demand. However, they will also require more cooling in high temperatures. Peak loads will be affected more than annual consumption.

Adaptation measures can be active or passive. Active measures include more efficient use of heating, ventilation and air-condition (HVAC)- systems, as well as active solar shading. Passive measures include building insulation, greenery of facades and roofs, phase change materials, solar screens, cool colors, thermal mass, evaporative cooling and natural ventilation. Many articles emphasize that use of measures may compensate for increased energy demands from climate change, and even lower energy usage in individual buildings compared to the current practice.

There are notably fewer articles about adaptation measures than the impact of climate change. Few studies examine building energy use in cold climates, although this might be because warming climate will be a more serious challenge in warmer regions. As building designs in colder regions are dominated by the need for heating, a changing climate may have a significant impact on construction practices. For instance, the implications of a reduction of freezing index, including impact on slab-on-ground insulation have not been addressed in the studied literature, Lower demand for heating may lead to a loosening of the stringent insulation requirements in the Norwegian building code.

5. Conclusions

This scoping literature review was first step by the authors toward exploring what is known about climate implication and adaptation measures for energy use in buildings. In sum, it is shown that impact of climate change in energy use in buildings has been thoroughly studied for climate regions all over the globe. The material includes articles from 26 countries, spread across all inhabited continents. However, there is a certain lack of material concerning the implication of climate change and relevant adaptation measures in cold climates. Climate change is expected to manifest in the form of higher temperatures, which will increase cooling demand and reduce heating demand. This has been a focus in the majority of the examined literature.

Further work on the subject will examine the implications of climate change in other fields than energy use, as well as different forms of climate change adaptation measures. Research is required on wider effects of climate change on buildings, including resilience, design guidelines, impacts of extreme weather events and more.

Acknowledgments

The authors gratefully acknowledge the financial support by the Research Council of Norway and several partners through the Centre for Research-based Innovation "Klima 2050" (see www.klima2050.no). A special thanks to Remy Eik at SINTEF for help with making the images.

References

[1] Pachauri RK and Mayer L 2015 Intergovernmental Panel on Climate Change (eds). *Climate change 2014: synthesis report.* Geneva, Switzerland: Intergovernmental Panel on Climate Change

[2] Stocker TF et al. 2013 Climate Change 2013. The Physical Science Basis. Working Group I Contribution to the Fifth Assessment Report of the Intergovernmental Panel on Climate Change Cambridge University Press, Cambridge, United Kingdom and New York, NY, USA, 1535 http://www.ipcc.ch/pdf/ assessment-report/ar5/wg1/WG1AR5_ALL_FINAL.pdf

[3] van Vuuren DP, Edmonds J, Kainuma M, et al. 2011 *Clim Change* **109** 5–31

[4] Hanssen-Bauer I, Førland EJ, Haddeland I, et al. 2017 *Climate in Norway 2100.* Oslo, Norway: The Norwegian Centre for Climate Services (NCCS)

[5] Lisø KR, Kvande T, Time B 2017 *Energy Procedia* **132** 628–633

[6] Novakovic V, Hanssen SO, Thue J, et al. 2016 *ENØK i bygninger - effektiv energibruk.* 3rd ed. Oslo: Gyldendal 22-24

[7] de Wilde P and Coley D 2012 *Build Environ* **55** 1–7

[8] Hughes S and Sarzynski A 2015 *Urban Clim* **14** 1–3

[9] Arksey H and O'Malley L 2005 *Int J Soc Res Methodol* **8** 19–32.

[10] Levac D, Colquhoun H and O'Brien KK 2010 *Implement Sci* **5** 69.
[11] Corbin JM and Strauss A 1990 *Qual Sociol* **13** 3–21
[12] Zhou Y, Clarke L, Eom J, et al. 2014 *Appl Energy* **113** 1077–1088
[13] Huang J and Gurney KR 2016 *Clim Change* **137** 171–185
[14] Huang J and Gurney KR 2016 *Energy* **111** 137–153
[15] Dirks JA, Gorrissen WJ, Hathaway JH, et al. 2015 *Energy* **79** 20–32
[16] Shen P 2017 *Energy Build* **134** 61–70
[17] Kikumoto H, Ooka R, Arima Y, et al. 2015 *Sustain Cities Soc* **14** 404–413
[18] Cox RA, Drews M, Rode C, et al. 2015 *Build Environ* **83** 104–114
[19] Fazeli R, Ruth M and Davidsdottir B. 2016 *Urban Clim* **15** 45–59
[20] Pyrgou A, Castaldo VL, Pisello AL, et al. 2017 *Sol Energy* **153** 224–237
[21] Shen P and Lior N 2016 *Energy* **114** 1288–1305
[22] Dodoo A and Gustavsson L 2016 *Energy* **97** 534–548
[23] Jenkins DP, Patidar S and Simpson SA 2015 *Buildings* **5** 985–1002
[24] Triana MA, Lamberts R, Sassi P 2018 *Energy Build* **158** 1379–1392
[25] Guan L 2015 *Archit Sci Rev* **58** 162–173
[26] Berger T, Amann C, Formayer H, et al. 2014 *Build Environ* **81** 258–269
[27] Wang H and Chen Q 2014 *Energy Build* **82** 428–436
[28] Li X-X 2018 *Energy* **157** 734–743
[29] Santamouris M, Cartalis C, Synnefa A, et al. 2015 *Energy Build* **98** 119–124
[30] Braun MR, Beck SBM, Walton P, et al. 2016 *Energy Build* **111** 109–119
[31] Braun MR, Altan H and Beck SBM 2014 *Appl Energy* **130** 305–313
[32] Tarroja B, Chiang F, AghaKouchak A, et al. 2018 *Appl Energy* **225** 522–534
[33] Andrić I, Gomes N, Pina A, et al. 2018 *Energy Build* **126** 77–93
[34] Andrić I, Pina A, Ferrão P, et al. 2017 *Energy Build* **149** 225–234
[35] Yi CY and Peng C. 2017 *Sustain Cities Soc* **35** 645–659
[36] Li M, Cao J, Xiong M, et al. 2018 *Energy Build* **173** 38–44
[37] Yau YH and Pean HL 2014 *Energy Build* **72** 1–7
[38] Hwang R-L, Lin C-Y and Huang K-T 2017 *Energy Build* **152** 804–812
[39] Arima Y, Ooka R, Kikumoto H, et al. 2016 *Energy Build* **114** 123–129
[40] Zoras S, Veranoudis S and Dimoudi A. 2017 *Energy Build* **150** 81–89
[41] Oliveira Panão MJN 2014 *Energy Build* **76** 354–362
[42] Santamouris M 2016 *Energy Build* **128** 617–638
[43] Mauree D, Coccolo S, Perera ATD, et al. 2018 *Sustainability* **10** 1134
[44] Kalvelage K, Passe U, Rabideau S, et al. 2014 *Energy Build* 76: 373–380
[45] Chen Y, Li M, Xiong M, et al. 2018 *Pol J Environ Stud* **27** 45–53
[46] Orehounig K, Mahdavi A, Doppelbauer E-M, et al. 2014 *Sustain Cities Soc* **12** 92–101
[47] Spandagos C and Ng TL 2017 *Appl Energy* **189** 352–368
[48] Invidiata A and Ghisi E 2016 *Energy Build* **130** 20–32
[49] Santamouris M. 2014 *Energy Build* **82** 100–113
[50] Jylhä K, Jokisalo J, Ruosteenoja K, et al. 2015 *Energy Build* **99** 104–116
[51] Shibuya T and Croxford B 2016 *Energy Build* **117** 149–159
[52] Damm A, Köberl J, Prettenthaler F, et al. 2017 *Clim Serv* **7** 12–30
[53] Berger T, Amann C, Formayer H, et al. 2014 *Energy Build* **80** 517–530
[54] Alonso C, Martín-Consuegra F, Oteiza I, et al. 2017 *Sol Energy* **146** 470–483
[55] Fontanini AD, Pr'Out KM, Kosny J, et al. 2016 *Energy Build* **129** 32–45
[56] Karimpour M, Belusko M, Xing K, et al. 2015 *Energy Build* **87** 142–154
[57] Rubio-Bellido C, Pérez-Fargallo A, Pulido-Arcas JA. 2016 *Energy* **114** 569–585.
[58] Sajjadian SM, Lewis J and Sharples S 2015 *Energy Build* **93** 83–89
[59] Ascione F 2017 *Sol Energy* **154** 34–100
[60] Fahmy M, El-Hady H, Mahdy M, et al. 2017 *Energy Build* **153** 241–261
[61] van Hooff T, Blocken B, Timmermans HJP, et al. 2016 *Energy* **94** 811–820
[62] Huang K-T and Hwang R-L 2016 *Appl Energy* **184** 1230–1240

[63] Pierangioli L, Cellai G, Ferrise R, et al. 2017 *Build Simul* **10** 459–479
[64] Berger T, Amann C, Formayer H, et al. 2016 *J Build Eng* **5** 86–95
[65] Nik VM, Mata É and Sasic Kalagasidis A 2015 *Energy Build* **88** 262–275
[66] Cao X, Dai X, Liu J 2016 *Energy Build* **128** 198–213
[67] Pagliano L, Carlucci S, Causone F, et al. 2016 *Energy Build* **127** 1117–1132
[68] Wang L, Liu X and Brown H 2017 *Energy Build* **157** 218–226
[69] Filippín C, Flores Larsen S, Ricard F 2018 *Energy Build* **165** 399–415
[70] Wang D, Landolt J, Mavromatidis G, et al. 2018 *Energy Build* **169** 9–26
[71] Waddicor DA, Fuentes E, Sisó L, et al. 2016 *Build Environ* **102** 13–25
[72] Nik VM, Mata E, Sasic Kalagasidis A, et al. 2016 *Energy Build* **121** 176–187
[73] Daly D, Cooper P and Ma Z 2014 *Build Environ* **74** 86–95
[74] Pérez-Andreu V, Aparicio-Fernández C, Martínez-Ibernón A, et al. 2018 *Energy* **165** 63–74
[75] Clarke L, Eom J, Marten EH, et al. 2018 *Energy Econ* **72** 667–677
[76] Huang J and Gurney KR 2017 *Energy Build* **139** 747–754
[77] Zachariadis T and Hadjinicolaou P 2014 *Energy* **76** 899–910

Do Town Planners in England feel a professional responsibility for a climate-resilient built environment?

N Murtagh[1], ND Odeleye[2] and C Maidment[2]

[1]The Bartlett School of Construction and Project Management, University College London (UCL), London, UK
[2]School of Engineering and the Built Environment, Anglia Ruskin University, Chelmsford, UK

Email: n.murtagh@ucl.ac.uk

Abstract.
With global warming already underway and the sense of urgency around climate breakdown growing, delivery of a resilient built environment is critical. Town planners are crucial intermediaries in this work but there is little recent work in the UK exploring their role in climate change resilience. In addressing the gap, the objective of the study was to examine resilience and professional identity of town planners. Interviews were conducted with 19 planners working in England who had at least seven years' experience. Thematic analysis of the data found a strong professional and public service identity but little awareness of resilience. Many participants were unfamiliar with the concept, with the exception of specific actions such as planning for flood protection, and almost none were aware of resilience thinking incorporating conceptualisation of places as complex, interconnected, dynamic systems. We conclude that the built environment in England is not being prepared for imminent and future climate risks, and that the gaps on climate-resilience pose a threat to the profession's jurisdiction.

1. Introduction

The latest report of the IPCC, detailing the difference between a global mean surface temperature rise of 1.5°C versus 2°C above pre-industrial levels, carries stark evidence for a world changed dramatically by global warming [1]. The term 'climate breakdown' has been suggested as an appropriate description of the degree of change underway. The level of risk depends, in part, on the extent of mitigation and adaptation that are pursued and both adaptation and mitigation are crucial. Even if global commitments to greenhouse gas emission reductions made as part of COP21 are achieved, global warming is still likely to surpass a 2°C threshold [2]. In the UK, risks have been identified from flooding, higher temperatures and extreme weather to people, business, infrastructure and buildings [3]. A more resilient built environment has a key role to play in reducing vulnerability to risk and capacity to recover in these domains [4]. Amongst the range of professionals involved in the delivery of the built environment, town (urban) planners are key [5]. We examined the question of whether planners in England feel a sense of professional responsibility to improve resilience to climate breakdown in the built environment.

Resilience has been seen as an aspect of sustainability, in that a sustainable system will also be resilient [6]. The two terms however have taken somewhat different paths in academic and policy discourse (ibid.). Definitions of resilience of the built environment are still evolving. Earlier conceptualisations referred to capacity of an urban system to maintain its functions under disturbance [7]. More recent thought has extended the definition to "a whole spectrum of global environmental change" [8], including social, economic and political crises, in addition to climatic [9]. Quigley et al. [6] suggest that definitions to date can be viewed as dichotomous. Definitions with what they term as an engineering focus define urban resilience in terms of robustness, ability to predict and plan, capacity to withstand disturbance and to return to the status quo in an efficient manner. In contrast, socio-ecological approaches to resilience emphasise capacity to adapt, to maintain function after disturbance while evolving through feedback and learning. Acknowledging resilience as a disputed term, Davoudi et al. [10] proposed a four-dimensional framework comprising learning capacity, robustness, innovation and flexibility. Of importance in these and other recent papers is the notion of resilience as an approach which recognises the essential interconnectedness of social and ecological systems, and of urban places as non-linear, dynamic, complex systems [11]. While resilience is seen as a useful concept for climate change adaptation by scholars, there is evidence that implementation of resilience thinking is not yet happening in practice [12, 13]. While a number of studies have explored practical advances in Ghana, other African cities, Asian countries and the Netherlands, there is little current empirical work on progress in the UK. The current study aimed to address this gap,

In England, local authorities are responsible for determining what development may take place. In compliance with national policy, planners employed by local authorities gather evidence to develop local policy and manage its introduction, offer guidance on compliance, and provide judgement on applications for development. While they have 'delegated powers' to decide on typically small developments, their work is set within a wider context of local planning committees comprised of elected officials and a planning inspectorate which oversees appeals against planning judgements. The Royal Town Planning Institute (RTPI) grants the professional qualification of Chartered Town Planner. To attain this status, planners are required to complete an accredited, planning-related postgraduate degree and to have achieved relevant experience. The RTPI also requires its chartered members to continue to update their knowledge through systematic and recorded Continuing Professional Development (CPD).

The role of town planner thus meets the definition of a 'profession' by Freidson [14], Evetts [15] and others, at least in part. Writers on the subject of professions have listed attributes of professions including: a representative body which accredits qualifications and determines membership; an agreed code of conduct; a specialised body of knowledge; and complex or specialised work, typically with exclusive jurisdiction. Most also specify an overarching aim of working for social good. This latter point is emphasised in a current and comprehensive work on professionals in construction which argued that "the wider public benefit…[is] an essential characteristic of all the professions [in the built environment]" (Foxell, 2019, p. 6). Foxell [16] explicitly links a sustainable and climate-resilient built environment to the notion of public benefit.

Much previous work on the concept of the profession, such as Abbott's [17] 'system of professions' and Freidson's (2013) 'third logic', has been sociological in approach, exploring the how 'professions' have developed in societies over time. Less attention has been paid to the lived experience of a profession, in particular how internalisation of the profession, or professional identity, may motivate behaviour. Additionally, whilst the importance of thinking about resilience in the built environment has been outlined, this leaves open the question of how it becomes part of the profession's body of knowledge.

Identity theorists have argued that social structures, such as a profession, shape identities and that an individual identity represents subjective processes or responses to such external institutions (Burke & Stryker 2016). Thus to identify as a particular type of professional is to act in a manner consistent with that identity. As the professionals who enact the planning regime, through application of national policy, formation of local policy and judgements on development, planners have the potential to be critical mediators of resilience and to influence the extent to which it is seen as key part of the profession's body of knowledge. We sought to explore whether planners in England saw themselves as professionals, and to what extent their overall aim was that of public good. Positioning a climate-resilient built environment as a public good, and applying the theoretical lens of professional identity, the research question was: To what extent do town planners in England feel a professional responsibility for a climate-resilient built environment?

2. Method

Experienced urban planners were recruited through communications with alumni of accredited programmes in planning at the authors' institutions, and through authors' professional networks. The recruitment advertisement sought planners with more than seven years' experience working in England. Semi-structured interviews were conducted during 2018 with a total of 19 planners. Their experience ranged from 7 to 19 years. All but one worked in local authorities and eight were women. Job titles encompassed Senior Planning Officer or equivalent (4), Principal (5) and Manager (5) as well as employer-specific titles such as Commissioner for Planning. Four participants worked primarily in development management (control), seven worked in policy and six combined both (this categorisation was not applicable for two participants). Interviews lasted approximately 1 hour, were audio recorded and transcribed verbatim, and had prior ethical approval. The interview schedule included sections on meanings of professionalism and of sustainability and resilience, which are analysed here. Using NVivo, thematic analysis was conducted [18]. The accounts were read and re-read, and phrases, sentence or short sections relating to resilience and professionalism were coded. The codes were clustered into more general themes and the most relevant of these to the research question are outlined below. All names are pseudonyms to protect confidentiality.

3. Findings

In this section, the theme of planning as a profession is first presented, before considering evidence for professional identity. Evidence on the theme of meanings of resilience is then presented, including planners' perceptions of responsibility in this domain.

3.1 Planning as a profession

Participants were asked whether they considered the role of planner to be a profession or a job. Although a few participants felt that planning may be treated as a job by a small number of colleagues in their experience, the majority stated definitively that they viewed planning as a profession: "*Definitely a profession. Absolutely.*" [Heidi]. Given what Freidson [19][19] (2004:17) refers to the "gentlemanly status of the traditional… professions", it is unsurprising to find near-unanimous claims to membership of a profession. However, when asked to elaborate on their reasons for this belief, the participants described a number of features of their occupation and career path that accorded with characteristics of professions defined by scholars of professionalism such as Freidson (2013), Evetts (2012), Foxell (2019) and others. All of the participants had complete university education, most to post-graduate level. Many referred to the professional body (RTPI) and membership or chartership as indicators of professional status.

In exploring whether planners applied a unique body of knowledge, participants were asked what expertise planners bring to their job. The participants referred to the legislative framework, and beyond that, to the breadth of knowledge they require:

> *Planning is a broad church, isn't it? It covers a whole array of different fields of expertise and knowledge, ranging from anything like highways and traffic, all the way through to dormice and flooding issues.* [Harry].

Some acknowledged the overlap of their expertise with other professionals, describing themselves as having some but not extensive knowledge of many areas. Nevertheless, they also described how they brought a possibly unique perspective in considering how places interact and how cities develop over time: *"I think when [an engineer and an architect] do it, they're doing it on specific projects or a bridge or a building, but we're doing it for a whole place."* [Beth].

They went on to outline skills in integration of varied and often technical or complex information, "*form[ing] a judgement based upon a balanced view of those*" [Jean], communicating with multiple stakeholders and negotiating an optimal outcome. Thus they referenced many of the attributes of a profession including a representative body, accredited qualifications, arguably a specialised body of knowledge, and complex work, primarily within planners' exclusive jurisdiction

An additional professional characteristic described by several was the need for continuous learning. This related to the constantly-changing political landscape, but was also evident in the narratives of participants of developing additional specialisms and broadening their expertise along their career trajectory. This appeared to relate to internalised professional identity as well as being an institutional requirement:

> *You do look to develop as a planner, and develop your knowledge, because you just have an interest in it. ...Obviously, most people I know, even though they didn't have [CPD requirements] hanging over them, would still want to develop their knowledge.* [Gail]

The broadening and deepening of knowledge as they progressed in their careers contributed to perceptions by others of their experience and authority, and enhanced their perception of being treated as a professional in their daily work.

3.2 Professional identity

Further aspects of professional identity were in evidence in the participants' accounts of their day-to-day dealings with stakeholders ranging from developers, to elected local politicians to members of the public. Transparency, fairness and balance were seen as important. There was a sense of a shared set of values, evidenced in references to the RTPI's Code of Conduct, to ethics and to impartiality. Of particular interest with regards to the research question was the extent to which reference was made to the social good of planning. The notion of planning as contributing to public benefit emerged spontaneously in most of the interviews. One participant said: *"There's a sense of your professional principles as well in terms of is what you're doing ultimately going to be for the public benefit."* [Jack]. Some referred also to an identity as a public servant: *"I see myself as a public sector, a public servant first and a planner second"* [Harry]. Another positioned social good as a driving motivation for planners in general:

> *Most people I know who have been successful as planners all share certain characteristics. It is that kind of dogged determination to get on with things, and oh, it sounds really hippyish, but make the world a better place.* [Gail]

Participants referred to design quality, happiness and well-being in the community, protection of the green belt[1] and provision of housing. Several linked public good to sustainability in the built environment in broad terms although only one referred directly to resilience: *"if you're a planner and you're not thinking about longer term issues of resilience, you're really not a very good planner"* [Jack].

3.3 Responses to resilience

When asked about the applicability of resilience in their work, there was no initial recognition of the concept by almost half of the participants and in a few cases, there was acknowledgement of lack of knowledge on the topic:

> *Researcher: I was wondering is resilience something that you do talk about and what it might mean for the built environment.*
> *Participant: Not really. As a word we don't really use that, so I'm not sure how I would apply it to planning. Yes, it's not a word that appears in any of our policies, that I know of, and it's not something that really crops up* [Ella]

However, more than half of the participants went on to provide some associations with the concept. Some confounded it with climate change mitigation and others referred to a wide range of factors including accessibility of homes, an ageing population, economic outcomes and biodiversity. Just over half connected the concept with a response to climate change, referring to planning for flooding, water stress and heatwaves. One talked about emergency planning and a number interpreted the concept in terms of future proofing, coping with adverse conditions and an ability to adapt: *"The ability for our developments to be able to mitigate, and adapt to, some of the issues is going to become more important."* [Debbie]. Of note here is positioning the issues around resilience in the future tense ("is going to become"), in the same way as [Kevin] says: *"we don't yet know what it means properly...But it's early days."* Understanding of resilience was markedly different to knowledge of sustainability, which all participants readily discussed and could explain how it was embedded in policy.

A small number of participants saw resilience as a concept with a broader meaning than preparing for climate breakdown, and incorporating flexibility to deal with change in market conditions and political contexts. One had attended RTPI-supported information sessions on resilient cities and another saw it as a goal:

> *Coming back to resilience, it's a big part of what we're trying to achieve and it's a very useful way to view some of those issues and a way to group things together and get people thinking about that longer term* [Jack].

With the exception of this participant [Jack], there was little evidence of activity relating to resilience in planning work currently. Participants noted the absence of national policy and the corresponding gaps in local plans. Some also pointed to the challenge in development management of applying the overarching aims of resilience on a single development: *"We think very development specific, and the short term, in terms of from the point of receiving an application to granting consent, that's one of our thought processes really"* [Beth]. Where they offered a view on where responsibility lay for increasing resilience in the built environment, participants considered that planners had a role to play but that responsibility extended to all stakeholders:

---

[1] [1] Green belt land is designated for the explicit purpose of preventing urban sprawl by severely restricting development on previously undeveloped 'greenfield sites' beyond existing settlements.

> *I think it's a shared responsibility. I don't think that it resides in planners, though we have it. I don't think it's just a political thing though certainly politicians have a responsibility. I don't think there's anybody who's not touched by the issues of resilience.* [Jack]

4. Discussion and conclusions

In the current study, 19 planners in England were interviewed, each with at least seven years' experience. There was near unanimous agreement that urban planning is a profession. In deeming it as such, the participants made reference to attributes of professions described in the academic literature, including academic qualification, a professional body and a shared code of conduct. Departing somewhat from previous definitions, not all participants were members of the professional institution as membership is not a mandatory for practice as a planner in the public sector, although eligibility for chartered membership is expected for advancement. While the body of knowledge was not exclusive as some scholars have implied [17], the planners argued that their holistic perspective which integrates knowledge and information across multiple sources and relates solutions to place and over time is specific to their profession. Further, they stressed the requirement for continuous learning and development over the course of the careers. They saw this as a common motivation amongst the planning community which contributed to how others recognised their authority and professional status.

Of particular interest for our research question was the evidence for social good as a motivator for most of the participants. The interviews showed a common sense of purpose among the planners involved, to provide an improved living and working environment for the local community. A particular theme in the participants' responses was a sustainable built and natural environment and this supports Foxell's (2019) argument for sustainability as an important public benefit to be delivered by professionals associated with the built environment. The shared sense of purpose represented one facet of professional identity, which was also expressed through reference to standards of impartiality, transparency and fairness. The evidence here then demonstrates an internalised professional identity in which the purpose of delivering social benefit is a central tenet.

The notion of resilience was not familiar to nearly half of participants, aligning with recent work with planners in Ghana (Poku-Boansi & Cobbinah, 2018). When invited to consider the concept in interview, most could make some associations, more or less tenuously related to the term. However, the level of familiarity was markedly different from that of sustainability. There was little evidence of understanding of underlying principles of complex, non-linear, dynamic urban systems and of the relationship between social and ecological systems. The approach of 'resilience thinking', heralded as having the potential for a paradigm shift in planning [20], was not evident. In a small number of cases, participants used the future tense, positioning resilience as something that will have to be addressed in the future. With strong current evidence for weather patterns affected by anthropogenic influence and the long-term impact of planners' decisions, as referenced by the participants themselves, this is a worrying finding. If planners today are not developing local policy and making judgements based on enhancing resilience, the impacts of the changing climate will have greater adverse impact into the future than could be the case.

In the UK, there has historically been a wide separation between town planning and health and emergency/disaster preparedness, with national policy being focused on civil emergencies (whether arising from accidents, natural hazards or human threats). Although the

Planning and Compulsory Purchase Act of 2004 was meant to re-integrate health and wider concerns back into a new spatial planning approach, it would appear from the evidence here that this has been limited. It is possible that planning for climate change resilience may be being conducted within other functions of local authorities such as environmental health or emergency planning, and that there is greater knowledge and understanding in such areas. However, such an approach would indicate an engineering approach to resilience, that is, a reactive approach, planning to protect against change and aimed at returning a system to its original state with greatest efficiency. The potential benefits of an evolutionary approach, of developing learning, robustness, innovation and flexibility, are seemingly not being pursued. Resilient thinking, of necessity, should involve town planners, given the necessarily future-orientation of local development plans, which typically look 20-25 years ahead.

With few exceptions, there was little evidence of reference to urban resilience in national or local planning policies. However, the participants considered responsibility for sustainable and resilient environment to be shared, and including planners, government and other stakeholders in the built environment. Although the evidence demonstrated lack of detailed knowledge and action, the theoretical framing of professional identity showed an underlying motivation is there.

5. Conclusions

The conclusion from the analysis is that while planners' professional identity drives a strong purpose of delivering social good, there is too little knowledge, awareness or requirement around climate-resilience for the concept to inform day-to-day planning policy and decisions. This holds important implications for the profession, its professional body and policy makers.

Development of the built environment is guided by planners' expertise and long-term view. If planners nationally are unaware of the predicted impacts of climate breakdown on urban and other development, then the built environment now and into the future is not being prepared for future risks.

As the planning profession derives part of its body of knowledge from its perspective over time as well as space, the gaps on climate-resilience could pose a fundamental threat to the profession. If the profession does not fully cover its claimed jurisdiction, other professions may encroach on its traditional domain.

Led by the professional body (RTPI), and drawing on experts within relevant economic sectors and from academe, it is imperative that the predicted risks to the built environment from climate breakdown and principles around urban resilience are disseminated. Resilience thinking must become embedded in day-to-day work of planners.

When knowledge of principles and approaches to resilience becomes professionally disseminated, the evidence from this study suggests that planners have the motivation, through their professional identity to serve the public good, and the skills, of integrating knowledge from disparate disciplines, to seek strategic national and local policy that incorporates requirements to ensure greater resilience.

National planning policy requires urgent updating but policy changes are not only top-down. Without a bottom-up drive from planners and their representatives, national policy will not develop in the direction needed.

Further questions to consider might be: given the political context of planning, are professional knowledge and motivation alone sufficient to ensure that best practice is pursued and implemented? Is the current structure of planning departments and processes in itself part of the problem? What would resilience planning mean, and look like, if emergency planning coordination was brought within UK town planning departments? And would such a shift strengthen or erode the professional identity of town planners?

References

[1] IPCC, "Global warming of 1.5'C," Intergovernmental Panel on Climate Change, 2018. Accessed: 12.10.2018. [Online]. Available: http://www.ipcc.ch/report/sr15/
[2] J. Rogelj, M. den Elzen, N. Hohne, T. Fransen, H. Fekete, and e. al., "Paris Agreement climate proposals need a boost to keep warming well below 2'C," *Nature,* vol. 534, no. June, pp. 631-639, 2016.
[3] DEFRA, "UK climate change risk assessment 2017," UK Government, Online, 2017. Accessed: 12.6.2017. [Online]. Available: https://www.gov.uk/government/publications/uk-climate-change-risk-assessment-2017
[4] O. Lucon, A. Urge-Vorsatz, H. Zain Ahmed, P. Akbari, L. F. Bertoldi, and e. al., "Buildings. In Climate Change 2014: Mitigation of Climate Change. Contribution of Working Group III to the Fifth Assessment Report of the IPCC," in "5th Assessment," Intergovernmental Panel on Climate Change, Cambridge, 2014.
[5] Y. Jabareen, "Planning the resilient city: concepts and strategies for coping with climate change and environmental risks," *Cities,* vol. 31, no. 0, pp. 220-229, 2013.
[6] M. Quigley, N. Blair, and K. Davison, "Articulating a social-ecological resilience agenda for urban design," *Journal of urban design,* vol. 23, no. 4, pp. 581-602, 2018.
[7] L. Gunderson and C. S. Holling, *Panarchy: understanding transformations in human and natural systems*. Washington, DC: Island Press, 2001.
[8] C. Folke, A. Jansson, J. Rockstrom, P. Olsson, S. R. Carpenter, and e. al., "Reconnecting to the biosphere," *AMBIO: A journal of the human environment,* vol. 40, no. 7, pp. 719-738, 2011.
[9] A. Mehmood, "Of resilient places: planning for urban resilience," *European planning studies,* vol. 24, no. 2, pp. 407-419, 2016.
[10] S. Davoudi, E. Brooks, and A. Mehmood, "Evolutionary resilience and strategies for climate adaptation.," *Planning practice and research,* vol. 28, no. 3, pp. 307-322, 2013.
[11] L. Porter and S. Davoudi, "The politics of resilience for planning: a cautionary note," *Planning theory and practice,* vol. 13, no. 2, pp. 329-333, 2013.
[12] H. Funfgeld and D. McEvoy, "Resilience as a useful concept for climate change adaptation?," *Planning theory and practice,* vol. 13, no. 2, pp. 324-328, 2013.
[13] M. Poku-Boansi and P. B. Cobbinah, "Are we planning for resilient cities in Ghana? An analysis of policy and planners' perspectives," *Cities,* vol. 72, pp. 252-260, 2018.
[14] E. Freidson, *Professionalism: the third logic*. Oxford: Wiley, 2013.
[15] J. Evetts, "A new professionalism: challenges and opportunities," *Current sociology,* vol. 59, no. 4, pp. 406-422, 2011.
[16] S. Foxell, *Professionalism for the Built Environment*. Abingdon, Oxford: Earthscan/Routledge, 2018.
[17] A. Abbott, *The system of professions*. Chicago: University of Chicago Press, 1988.
[18] V. Braun and V. Clarke, "Using thematic analysis in psychology," *Qualitative Research in Psychology,* vol. 3, no. 2, pp. 77-101, 2006, doi: 10.1191/1478088706qp063oa.
[19] E. Freidson, *Professionalism reborn: theory, prophecy, and policy*. Cambridge: Polity Press, 2004.
[20] K. Shaw, ""Reframing" resilience: challenges for planning theory and practice," *Planning theory and practice,* vol. 13, no. 2, pp. 308-318, 2013.

IOP Conf. Series: Earth and Environmental Science **297** (2019) 012037 doi:10.1088/1755-1315/297/1/012037

BIM and Energy Efficiency training requirement for the construction industry

Sunil Suwal[1], Maaria Laukkanen[1], Päivi Jäväjä[1], Tarja Häkkinen[2], Sylvain Kubicki[3]

1. Metropolia University of Applied Sciences, Espoo, Finland.
2. VTT Technical Research Center of Finland, Espoo, Finland.
3. Luxembourg Institute of Science and Technology (LIST), Luxembourg.

Abstract. Construction industry has a wide impact to our built environment. It plays a vital role for governments and supports in both developed and developing economies. The industry supports economic growth of a nation through various trades and activities as well as provides new jobs. It is one of the largest industrial sectors that has a projected output of 13.5% by 2025. Climate change, resource depletion and rapid urbanization are the most severe amongst the challenges we face today. The construction industry alone is responsible for 20% of the global energy consumption and approximately one-third of energy-related CO2 emissions. The industry entails the high energy saving potentials and provides various opportunities to implement sustainable solutions to decrease the environmental impacts and thus lower greenhouse gas emissions. In the construction projects today, we can implement different energy efficiency strategies and products virtually during the planning phases of the construction projects using model based environment and tools commonly known today as virtual design and construction (VDC) as well as building information modeling (BIM). We can accurately simulate and calculate the impacts and thus support the increasing demand of complex high-performance buildings in order to effectively address energy and carbon reduction targets. However, one of the key challenges, the industry faces is the availability of skilled workforce and experts that have enough BIM and energy efficiency skills. The paper addresses such needs and presents the current status of a H2020 project with a focus on defining roles and responsibilities construction project stakeholders have for the energy efficiency measures along the different project phases. It also provides an outlook for the development of the learning outcomes based on knowledge, skills and competence (KSC) framework as well as systematically presents the base for harmonization of the learning outcomes at the EU level.

1. Introduction

Construction is important and the impact of the construction industry is huge. It strongly affects our society, economy as well as our environment. It is a determinant of where and how almost everyone lives, works and plays. The built environment heavily influences quality of life. The construction industry alone amounts for a total of 6% of global GDP and plays a vital role for governments and

supports economic development both in both developed and developing economies. It accounts for about 5% of total GDP in developed countries, while in developing countries it tends to account for more than 8% of GDP. While the construction industry has a direct impact on economic growth of a nation, it is also the single largest global consumer of resources and raw materials.
Climate change, resource depletion and rapid urbanization are the most severe amongst the challenges we face today. The construction industry alone is responsible for 20% of the global energy consumption and approximately one-third of energy-related CO2 emissions. Whereas our built environment utilizes the world's 40% of energy, 50% of natural resources and contributes to 40% of carbon dioxide emissions.[1]

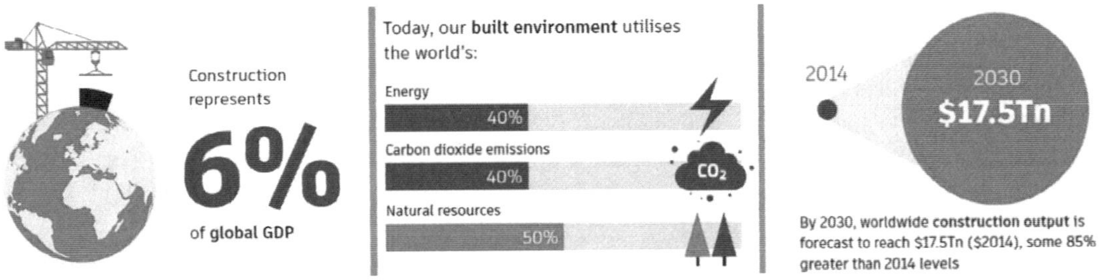

Figure 1: Impacts of the construction industry (Left: World Economic Forum, Middle and Right: Global construction 2030 as seen in Constructing with the power of digital, Autodesk)[2]

The industry entails the high energy saving potentials and provides various opportunities to implement sustainable solutions and processes to result in better and sustainable built environments. By 2030, worldwide construction output is forecast to increase by 85% greater than that of 2014 levels. This directs to the growing demand of both skilled and unskilled workforce at very high levels. Furthermore, the technological advancements today have reshaped out way of working in most of the industries. Construction industry professionals need to embrace the knowledge, the skills and the competencies of various tools and processes incorporated for enhancing the construction projects. More specifically, the construction industry has a great inclination today towards the use of the model based tools and processes commonly known today as virtual design and construction (VDC) as well as building information modeling (BIM). VDC and BIM can be actively implemented for the construction projects to accurately simulate and analyze the impacts of different sustainable solutions. BIM and Energy Efficiency competence today is considered very important.

This paper presents the ongoing results of BIMEET (BIM-based EU -wide Standardized Qualification Framework for achieving Energy Efficiency) research program funded under H2020 program. BIMEET project aims to leverage the take-up of ICT and BIM through a significant upgrade of the skills and capacities of the EU construction workforce. This paper provides an overview of the BIMEET project and discusses the current development of learning outcomes for different roles required for different stakeholder training to support BIM based energy efficiency.

2. Performance based buildings

The ideas of performance-based building formulate an important basis for the formulation of the learning outcomes with regard to energy-efficiency and BIM. Performance approach is concerned with what the building is required to do, and not with describing the technical solutions i.e. how it is constructed [3]. A prescriptive approach describes an acceptable solution while a performance approach describes the required performance. The performance approach is the practice of thinking and working in terms of ends rather than means. Performance requirements do not say anything about the ways and means of buildings, e.g. the types of materials, the dimensions, and sizes of building parts, or the methods of construction, but they define the required end result.

Performance based approach is important to consider in customer-driven processes to enable owners and users to present what the building is required to do instead of describing how it is constructed. Performance based and qualitative requirements give a better basis for innovative design solutions than a prescriptive approach [4 5].

Building performance requirement is the minimum acceptable level of a critical property (ISO 6707-1 2014). The overall performance of a building is based on diverse aspects such as energy performance, fire safety, structural safety, acoustics, thermal comfort, and the quality of indoor air, among others. Requirements on sustainability increase the number of aspects and also environmental concerns need to be considered. There have been many attempts to outline building performance and describe different performance aspects with the help of indicators [6 7 8 9] ISO 21929-1 [10] presents indicators of a sustainable building with the help of 14 aspects of sustainable building: (1) Emissions to air, (2) Amount of non-renewable resources consumption by type, (3) Amount of fresh water consumption, (4) Amount of waste generation by type, (5) Change of land use, (6) Access to services by type, (7) Accessibility, (8) Indoor conditions and air quality, (9) Adaptability, (10) Life Cycle Costs, (11) Maintainability, (12) Safety, (13) Serviceability and (14) Aesthetic quality. A standard indicator of building energy efficiency is annual energy consumption (kWh/m2) as a function of climate, envelope design, heating ventilation and air-conditioning (HVAC) systems, and occupant behaviour, among other parameters[11].

3. Learning outcomes

Learning outcomes are the explicit statements of what a learner is expected to know, understand and is able to do after the completion of a learning activity. Learning outcomes discussed refer to the intended leaning outcomes (ILOs) rather than achieved learning outcomes (ALOs). Learning-outcomes-based frameworks enable the comparison of qualifications across different types of institutions and stakeholders. By providing a common language makes it possible to compare qualifications over national borders.

> *"Learning outcomes are attributed to individual educational components and to programmes at a whole. Learning outcomes are specified in three categories – as knowledge, skills and competence. This signals that qualifications – in different combinations – capture a broad scope of learning outcomes, including theoretical knowledge, practical and technical skills, and social competences where the ability to work with others will be crucial."*[12]

Learning outcomes are valuable. Bloom's taxonomy and revised Bloom's taxonomy is the most often used frequent tool while developing learning outcomes. The cognitive domain of the Taxonomy comprises of six hierarchical levels of learning. The categories are ordered form simple to complex and from concrete to abstract[13] with a focus towards the level of cognitive processing required in the levels of learning particularly termed as low level thinking skills (LOTS) and high level thinking skills (HOTS) as shown in Figure 2. Carefully established learning outcomes prioritize the application of knowledge and equally play an important role as their use within a unit of instruction primarily guides learning and assessment activities that enhances student engagement and learning. [14] The approach adopted to describe and develop EU wide learning outcomes is based on the principles of the European Qualification Framework (EQF) particularly focusing on the KSC framework and the use of action verbs in relation to what a learner should know, have skills and be competent on.

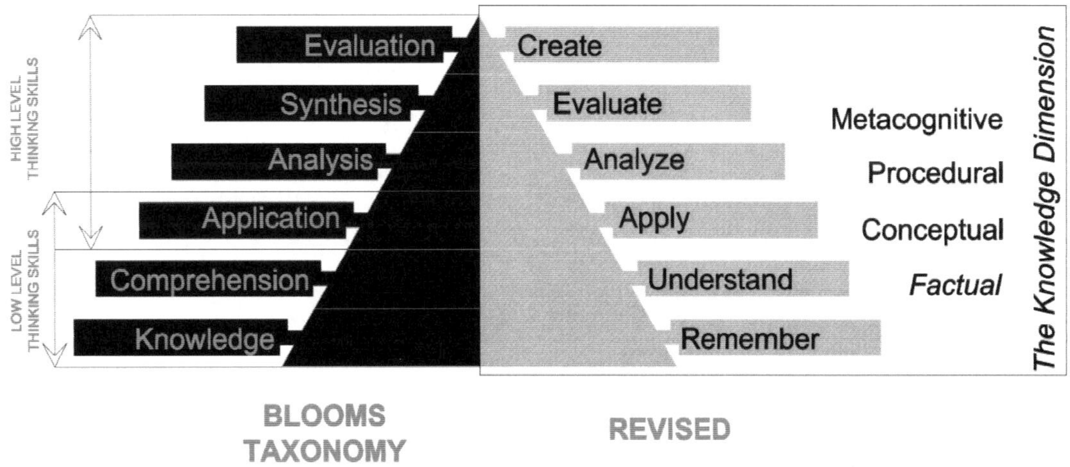

Figure 2: Bloom's taxonomy and revised Bloom's taxonomy hierarchical levels

The process used to develop the learning outcomes furthermore also relates to the case based approach. It focuses on mapping the standard RIBA Plan of Work stages and identified stakeholders followed by the process to defining learning outcomes for the role based courses. The three categories of KSC should be collectively perceived and should not be read in isolation from each other. Figure 3 highlights some of the action verbs used to define the levels of taxonomy in the cognitive domain. An example showing the basic structure of learning outcomes are presented in the Figure 3.

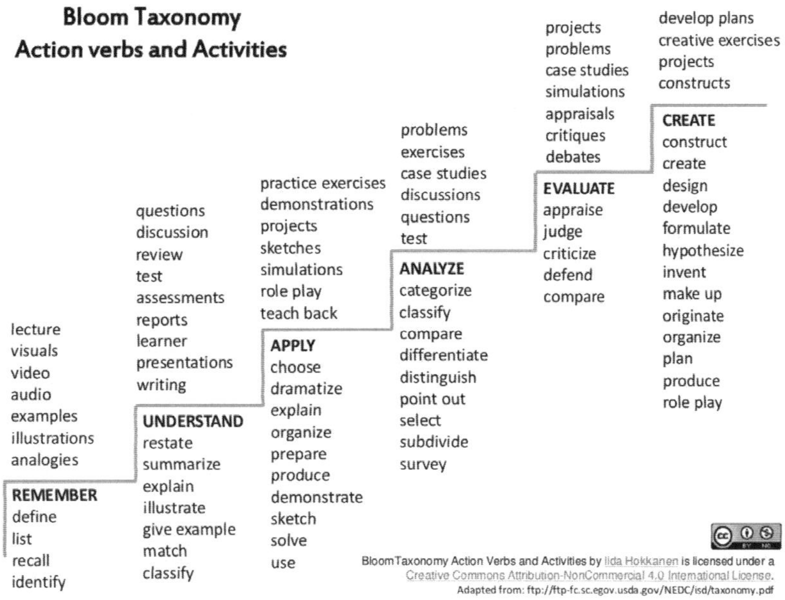

Figure 3 Bloom taxonomy action verbs and activities (Iida Hokkanen 2015) [15]

4. European wide learning outcomes

BIMEET collects, creates and validates the required knowledge, skills and competencies for different stakeholders of a construction project with respect to BIM and energy efficiency. The basic method for defining role based skills, knowledge, competence and finally learning outcomes was based on the following procedure:

Step1	• Defining the main roles in design, building and maintenance processes
Step 2	• Identification of the main process and phases of construction project
Step 3	• Identification of tasks/responsibilities related to energy performance management and BIM for different roles in the main phases
Step 4	• Defining the required skills, knowledge and competence to manage these tasks in different phases
Step 5	• Summarizing the results for role-specific requirements
Step 6	• Concluding the required learning outcomes because of the summarized required knowledge, skills, and competencies.

Figure 4: Procedure for defining KSC and LO

During this procedure, national guides for plans of works for different roles and national guides for common BIM requirements were made use of in defining phases, tasks and roles. For example in Finland, guides for plan-of-work have been formulated for architectural design, structural design, HVAC/MEP design, and management of building projects. The following Table 1 presents the results for one selected role - Chief designer - as an example of the results.

Table 1: An example of Skills, Knowledge, Competence and Learning outcomes for Chief designer in Finland

Skills	• Skills for BIM-based collaboration such as with the help of CAVE (computer aided virtual environment) and Big Room working • Basic skills in using assessment and optimisation tools • Skills in using visualisation tools to enable good information sharing in the design team and with the construction site management • Skills in using model checking tools - such as Solibri and NavisWorks - for quality assurance • Skills in using project data management systems (file based or model based) • Skills in using electronic services provided by the building permission authorities • Skills in Presentation of animated models, gaming models, and virtual models EQF

	Levels 6 - 7
Knowledge	• Excellent knowledge about national guidelines for building information modelling (COBIM) • Comprehensive knowledge about BIM terminology and definitions • Good understanding about the use of BIM authoring software • Excellent knowledge about the aspects of building performance and impacts (financial and environmental) and related indicators and benchmarks, labels and certificates • Understanding about needs of initial data and the potentials of different surveys and building inventions in refurbishment projects • Understanding about the feasibility with the help of different kinds of concept solutions • Understanding about the importance of systematic modelling and data management • Knowledge about the effect of orientation, volume and space design, and main product type selections on energy performance and building performance and knowledge about alternative potential solutions to fulfil the set targets • Knowledge about the potentials of different assessment, calculation and optimisation tools for parametric iteration and design of performance aspects • Knowledge about electronic application process and its data format requirements • Understanding about input and output data requirements of different design domains in general and especially in detailed technical design and understanding about data needs in specific tasks (such as bill of quantities and environmental assessment), and procurement processes. • Knowledge about the information requirements in quality control and assurance process in construction • Knowledge about data and information requirements of sustainable care and maintenance processes EQF Levels 6 - 7
Competence	Competence to use knowledge and skills • to understand owner's targets and feasibility with the help of different kinds of concept solutions • to lead the design team to prepare, compare and improve alternative concepts • to direct the design towards set targets utilizing the capacity of different kinds of digital assessment and simulation methods. • in collaborative BIM-based working together with the owner/project manager, users, building authorities, and other stakeholders. • to lead the preparation of the presentation materials and presentation agenda to support owner's effective decision-making and opinion formation of other stakeholders

	• to take care and lead the tasks related to the electronic building permission process including communication with the building authorities. • to verify the achievement of the targets on the basis of the results received with the help of different kinds of assessment, simulation and labelling tools. • to lead the process resulting in the publication of the merged model (as designed) together with all needed information to support sustainable procurement, construction and maintenance EQF Levels 6 - 7
Learning outcomes	Proven ability based on exams • about good theoretical in overall design principles for sustainable and energy-efficient buildings Proven ability based on exams and demonstration • about design management supervising the overall design quality of each design domain taking care of the compatibility and the achievement of set targets Proven ability based on exams and demonstration • about good understanding and skills in applying collaborative processes and data management systems

Comparison between gained learning outcomes within countries involved in the project (Luxemburg, France, Finland, Greece and United Kingdom) were then collected, compared and studied. Learning outcomes from other standards and EU projects were studied and made use of in order to formulate first draft of European wide stakeholder specific learning outcomes. The identified other projects that had relevance to the subject matter are as follows:

- BIMcert: Project learning tools, focus on testing the BIM approaches to green and passive building design https://energybimcert.eu/ (Partners from: UK, Ireland, Portugal, Macedonia, Croatia)
- BIMplement: BIM and quality control approach to achieve energy efficient buildings www.bimplement-project.eu (Partners from: France, Netherlands Lithuania, Spain and Poland)
- NET-UBIEP: Aims at increasing energy performance of buildings by wide spreading the use of BIM www.net-ubiep.eu (Partners from: Croatia, Spain, Estonia, Italy, Slovak, Denmark, Lithuania)

The information thus generated and developed were compiled and tracked into a "Super matrix" as and information storage and to map the existing learning outcomes for the specified roles. The process of formulating the EU level learning outcomes and the structure of "Super matrix" is explained in Figure 5 and Figure 6 respectively.

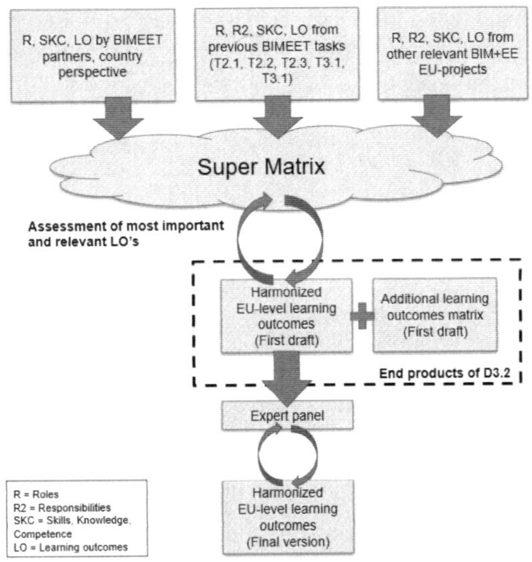

Figure 5: Process and content development of Super matrix

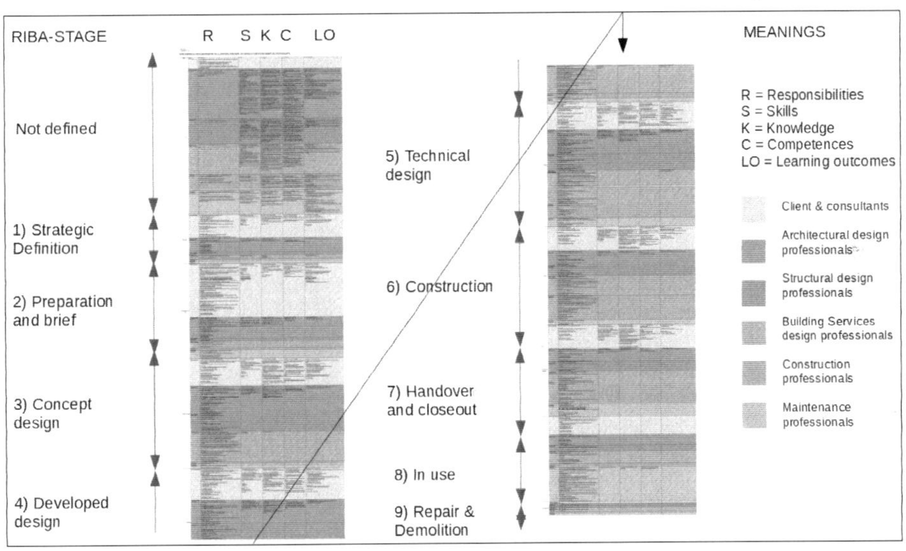

Figure 6: Data collection and structure of the Super matrix

Construction industry and building projects have several roles and stakeholders. The four main role categories were identified in earlier deliverables of BIMEET to generalize the roles during a constructin project. The roles idenfied are that of Client, Designers/Design Consultant, Contractor/Sub-contractor, Facility/Asset management[16]. The existing roles (as well as emerging roles like BIM coordinator, BIM manager) have been mapped at a general level to the four sectoral groups. However when defining and detailing out the EU wide learning outcomes, design consultant: Architect, Structural engineer, Building services engineer are mentioned by all partners and the generalized roles based upon the information related with BIM and EE KSC. The selected roles and their EU learning outcomes for BIM and EE trainings are presented in the Results and Conclusion section.

5. Results and Conclusions

The work resulted in proposing 6 - 8 specified groups of learning outcomes for the each selected main category role. Each of the groups consists of 4 - 14 learning outcomes that clarify and supplement the required qualifications for the selected roles. The categories of roles that were chosen for the EU-wide approach are presented in Table 2.

Table 2: Role groups and selected roles

Role Group	Roles
Client & Clients advisors	Client, Project manager, BIM manager, BIM coordinator, Briefing consultant
Architectural design	Architectural Design and BIM Coordinator, Chief Designer, Architect, Assistant designer
Structural design	Structural design and BIM coordinator (structural), Assistant designer
Building services design	HVAC and Energy design and BIM coordinator (HVAC), Assistant designer
Construction work	Site manager, construction site workers and installers
Maintenance work	Maintenance operator, property manager, care taker

Thus developed European wide learning outcomes for different stakeholders are grouped into a generalized focus that they provide. An example of such groupings for architectural design role group is presented in Table 3 and Table 4 presents EU learning outcomes (LO1) and (LO2) for the same role group mapped with the EQF levels as an example.

Table 3: Learning outcome groups and their main focus for architectural design role group

Role Group	Roles
Architectural design	• (LO1) Fundamentals of BIM and principles of its uses with respect to building life-cycle • (LO2) Fundamentals of sustainable and energy-efficient buildings and building performance • (LO3) Leading of design process, supporting the client and other stakeholders in decision making • (LO4) Implementation of energy performance, building performance and sustainability targets into design process. • (LO5) Production of BIM models with accurate and required information content for the different uses and phases of a building project. • (LO6) Collaboration, communication and visualization with help of BIM • (LO7) Implementation of target and quality management procedures in the building project. • (LO8) Skills for relevant software and interfaces between software.

Table 4: EU wide learning outcomes 1 and 2 for architectural role group (CD = Chief Designer; ARCH = Architectural design and BIM Coordinator; ASS = Assistant designer)

	Learning outcomes	EQF Levels		
		CD	ARCH	ASS
LO1	**Learner is able to explain the fundamentals of BIM and the underlying principles of uses with respect to building life-cycle.**	**6**	**6**	**3**
1.1	Recall essential contents, summarize and give examples of BIM terminologies, definitions and standards.	6	6	3
1.2	Recall essential contents, summarize and give examples of overall BIM process for a building's life cycle.	6	6	3
1.3	Explain and use standard information exchange processes for different design domains in general and especially in detailed technical design.	5	5	2
1.4	Explain the essential issues related to information management, data transfer and sharing.	5	5	2
1.5	Explain the added value of using open file formats (i.e. IFC) to ensure interoperability.	5	5	2
1.6	Recall, summarize and explain essential contents and relevant parts of national BIM guidelines.	6	6	3
LO2	**Learner is able to explain the fundamentals of sustainable and energy-efficient buildings and building performance.**	**4**	**6**	**2**
2.1	Explain and give examples of aspects and terminologies of energy and building performance.	6	6	2
2.2	Describe the financial and environmental aspects and related indicators, benchmarks and certification systems of energy and building performance.	2	6	2
2.3	Explain the issues that affect energy performance of buildings and demonstrate competence in domain specific solutions.	4	6	1
2.4	Explain relations between life-cycle costs, energy performance and building performance.	4	6	2
2.5	List and explain the core concepts of sustainable building rating and certification systems.	4	4	2
2.6	Summarize and give examples about the potentials of renewable energy sources applicable to buildings including district-scale solutions.	4	4	1
2.7	Point out legislation and regulations related to energy performance, thermal comfort and air quality.	6	6	3

The EU learning outcomes developed are further to be validated for their use in developing BIM and EE trainings with an input from BIM EE expert panel in coming work deliverables. These would then further be used for defining national level learning outcomes for the project partner countries in the project. The validated learning outcomes will thus be used as the base for specifying the BIM and EE related training outcomes for the EU.

Acknowledgement
The authors acknowledge financial support from the European Commission (H2020) to the BIMEET project, grant reference: 753994.

References

[1] World Economic Forum. (2016). Shaping the Future of Construction: A Breakthrough in Mindset and Technology. Cologny, Switzerland: World Economic Forum.
[2] Autodesk. Construction with the power of digital. https://damassets.autodesk.net/content/dam/autodesk/www/AEC-Solutions/pdf/fy18-aec-construction-manifesto-single-page_EN-US.pdf [Accessed: Feb 1, 2019]
[3] CIB 1982. CIB Report. Working with the Performance Approach in Building. CIB Publication no. 64. Rotterdam, Netherlands, 1982, 30 p.
[4] Eriksson, P. and Westerberg, M. (2011). Effects of cooperative procurement procedures on construction project performance: A conceptual framework, International journal of project management 29: 197 208
[5] Sexton, M. and Barrett, P. (2005). Performance-based building and innovation: balancing client and industry needs, Building Research & Information 33(2): 142 148.
[6] Alwaer, H. and Clements-Croome, D. J. (2010). Key performance indicators (KPIs) and priority setting in using the multiattribute approach for assessing sustainable intelligent buildings, Building and Environment 45(4): 799 807.
[7] Augenbroe, G. and Park, C.-S. (2005). Quantification methods of technical building performance, Building Research & Information 33(2): 159 172.
[8] Loomans, M. G. L. C., Huovila, A., Lefebvre, P., Porkka, J., Huovila, P., Desmyter, J. and Vaturi, A. (2011). Key Performance Indicators for the Indoor Environment, In Proceedings World Sustainable Building Conference SB11, Helsinki, pp. 1666 - 675.
[9] Frandsen, A. K., Bertelsen, N. H., Haugbølle, K., Porkka, J., Huovila, P., Hietanen, P., Widén, K., Hansson, B. and Karud, O. J. (2010). Performance Indicator Framework, In V. Huovila, P., Karvonen, A. & Raasakka (Ed.), Proceeding of Sustainable Building Conference SB10, Espoo: Finnish Association of Civil Engineers.
[10] ISO 21929-1. (2011). Sustainability in building construction - Sustainability indicators - Part 1: Framework for the development of indicators and a core set of indicators for buildings. International Standard ISO 21929-1.
[11] IEA ECBCS Annex 53. (2013). Annex 53 Total Energy Use in Buildings: Analysis & Evaluation Methods
[12] Users'Guide, E. C. T. S. "Luxembourg: Publications Office of the European Union." DOI 10 (2015): 87192. https://europass.cedefop.europa.eu/sites/default/files/ects-users-guide_en.pdf
[13] Krathwohl, D. R. (2002). A revision of Bloom's taxonomy: An overview. Theory into practice, 41(4), 212-218.
[14] University of Toronto. Centre for Teaching Support and Innovation, and Emily Gregor-Greenleaf. Developing Learning Outcomes: A Guide for Faculty. Centre for Teaching Support and Innovation, University of Toronto, 2008. https://teaching.utoronto.ca/wp-content/uploads/2015/08/Developing-Learning-Outcomes-Guide-Aug-2014.pdf
[15] Honkkannen, I. (2015). Bloom taxonomy action verbs and activities. https://www.slideshare.net/IidaHokkanen/bloom-taxonomy-action-verbs-and-activities [Accessed: December 10,2018]
[16] BIM for Energy Efficiency Training (BIMEET). (2018). Deliverable D3.1

IOP Conf. Series: Earth and Environmental Science **297** (2019) 012038 doi:10.1088/1755-1315/297/1/012038

Multi-criteria material selection for buildings in challenging environments

L Domingos[1] and V Rato[1]

[1] ISCTE-Instituto Universitário de Lisboa, ISTAR-IUL, Lisboa, Portugal

leonordomingosarq@gmail.com

Abstract. Climate change and future weather conditions are likely to challenge the way buildings are designed because there will be an increase in extreme climatic conditions. What should architects change in their design process to produce human habitats able to withstand those extreme conditions to ensure adequate comfort conditions? This paper presents preliminary results within the scope of an on-going research that addresses one single key issue: what materials will be most suitable in extreme temperature conditions. A set of 52 materials is analysed through a multi-criteria decision process that includes thermal conductivity, thermal diffusivity, thermal effusivity, linear thermal expansion, service temperature, fracture toughness, recycle potential and embodied carbon as criteria. The goals are to find the best-fit materials for each climate scenario within the scope of contradictory objectives and to develop a methodology for the selection of construction materials for buildings in challenging environments. Results show that the best possible material for extreme temperatures, whether it would be a very cold or a very hot environment, is one that could combine the properties of polymers with a very low environmental impact (at the level of the impact from materials such as natural fibres, wood or wood derivatives). The results thus suggest that further research may be directed at biomaterials development.

1. Introduction

Climate change and future weather conditions are likely to challenge the way buildings are designed because there will be an increase in extreme climatic conditions. What should architects change in their design process to produce human habitats able to withstand those extreme conditions assuring adequate comfort conditions? The answer to this question requires comprehensive interdisciplinary research. This paper presents preliminary results within the scope of an on-going research that addresses one single key issue: what materials will be most suitable in extreme temperature conditions.

1.1. Challenging Environments

The Astrobiology Institute of NASA (NAI) defines an extreme environment as one characterized by extreme physical conditions, that are outside the boundaries in which humans habit and dwell comfortably [1]. These are conditions such as temperature, desiccation, air quality, pressure, salinity, radiation, and acidity (or alkalinity, pH). An extreme environment is nowadays defined mostly by its location, but due to global warming and environmental disasters, researchers now understand that it is possible for these environments to develop in different and unexpected areas. Examples of extreme environments include the geographical poles of Earth, deserts, very high mountains, the interior of volcanoes, deep oceans, upper atmosphere, and other planets or outer space [1].

Although extreme environments have always existed on Earth, now, thanks to climate change, it is highly likely that other places will face extreme climate scenarios. So, the real question is how will we be able to provide people with the tools to adapt to these new situations? How can architecture help people habit comfortably while facing the possibilities of lack of water and high temperatures, or even the opposite, floods, blizzards or polar deserts? Organizations such as NASA and ESA have been studying self-sufficient human habitats for decades to colonize other planets, and deal with challenges like these. To better understand this topic, it is possible to look at the work of architects such as Neil Leach [2], who guest-edited the edition of Architectural Digest magazine, Space Architecture: The New Frontier for Design Research, or even Lydia Kallipoliti [3], more specifically, her exhibition "Closed Worlds", which deals with the various concepts of closed systems, experiences that have been conducted since the 20th century, with the purpose of experimenting autonomous living. Manuel Kretzer's research in his platform "Materiability" [4] also offers new insight in terms of materials and architectural dynamics. He approached these environments in 2015, on his workshop "Dynamics in Extreme Environments". He states that architecture will have to respond to extreme weather conditions, and that materials will play a critical role in the architectural process of facing these new challenges, on a global scale.

1.2. Climate and Architecture

Bioclimatic Architecture is a concept that encompasses design strategies and the use of materials with sustainable criteria, aiming at assuring comfortable and healthy internal environments respecting the limits of the natural systems [5]. It represents the optimal energy management for buildings, through collection and distribution of renewable energies, in an active or in a passive way [6]. This should be done using proper (and preferably native) materials, with ecological and eco-construction criteria. This turns up as an inseparable concept when researching for construction materials for challenging environments, the relationship between climate and architecture being therefore essential [7]. This research aims to understand how professionals can use and design with materials that fulfil the demands environments put on them, while being at the same time as effective as possible, and how can the impact they have on the environment be minimized. The climate of a region, however, is defined by a variety of factors, being the group of meteorological phenomena that defines the state of the atmosphere, and it varies according to various climate factors. Regardless of these, however, buildings must adapt and change, according to necessity [7]. For example, if one studies the traditional architecture of Aït-Ben-Haddou, in Marroco, which has a desertic climate and hot steppe, it will be drastically different from the architecture of an Igloo, which can be located on the icy tundras of northern Canada (fig. 1). On the first case, the city consists of fortresses and of a defensive architecture, created due to tribal and monarchic wars. The extreme climate conditions these constructions must handle conditions their structure, therefore they form compact group buildings, with narrow interior roads, which offers less exposure to the external climate factors, managing the thermal leaps and offering wind protection. They are build using two wooden barriers, which they fill up with stone first, to avoid rising humidity, and then with rammed earth, which is later coated with a mixture of native clay, sand, lime and sometimes plaster, so that the walls would be water-resistant. When it comes to the igloo, these are subjected to temperatures close to -50° and very strong winds, and they are used by Inuits, which base their economic activities in hunting, and therefore are consistent of nomadic groups, who use the igloo as temporary housing. The base material for its construction is snow which has been compacted by the wind, which is cut into 90cm wide blocks and placed in a spiral to form the igloo's domes; it is always built from the inside. The hemispherical shape of the igloo ensures minimal surface exposed to the winds, and a large interior, which is warmed with an oil lamp. The interior insulation is ensured thanks to animal fur that lines the interior, creating small pockets of air between the fur and the ice. The igloo gains solidity with time, making it a monolithic and very solid structure. With these examples in mind one can begin to understand how buildings differ depending of the environment they exist in; however, most of the examples of bioclimatic architecture are of vernacular origin, and with the existence of air-conditioned units and generally used materials, a sort of generalization of building assemblies happened, where wherever the place, the construction technologies do not vary significantly. What this research aims to achieve is to offer a way of selecting materials for nowadays buildings, that will face new climate-based challenges, that haven't been faced before. As vernacular and ancient architecture responds to the demands of populations from ancient times, so must contemporary architecture do so [7].

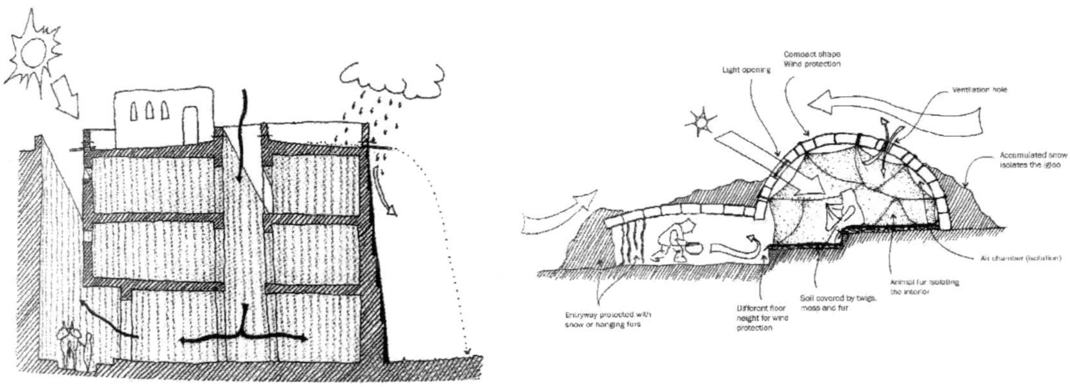

Figure 1. Drawn sections of the traditional architecture of the city of Aït-Ben-Haddou in Marroco and a typical Igloo from the Canadian Tundra (Adapted Illustration [7])

1.3. *Spatial Configuration and Materials*

When it comes to buildings needing to adapt to new challenges and surroundings, Robert Kronenburg studies lead him [8] to believe that human beings and buildings can and will be flexible and that the latter should adapt to changing situations, depending on place, culture or use. He gives examples of spatially flexible buildings, such as Toyo Ito's Matsumoto Performing Arts Centre, in Japan, which has transformable elements and movable parts in its architecture, such as the ceiling, stage and seating, that can move and adapt according to necessity. It also features fluid spaces, with no determined used, and encourages visitors to interact with each other. He believes this sort of scheme allows buildings to be more responsive, instead of remaining standardized and static [9]. On the other hand, psychologist Sally Augustin studied the attributes of well-designed spaces and their comforting qualities [10], which she states are often under looked in challenging environments, both because of budget limitations and lack of designer involvement. When it comes to feeling well in these sort of spaces, researchers Yan and England, who focused on artic environments [11], concluded that the physical built environment becomes one of the few resources available to help inhabitants cope and adapt to their extreme climate surrounding conditions. When a building is so isolated from other stimuli, the smallest detail can make a difference in one's perception of it. More than adapting to exterior climate factors, spatial configuration and materials in buildings influence greatly those who experience it, and this becomes even more important when we deal with environments that are not ideal, and even may be aggressive, to the human condition; within this context, the best choice of materials and good interior spatial configuration can make the experience as comfortable and ecologically responsible as possible, for both humans and the environment. Having an efficient tool to uncover the best materials for each environment will certainly make the task of designing and building much simpler.

2. Methodology

2.1. Choosing Climate Conditions

To select what would be the temperature conditions that could be defined as 'extreme', a simple criterion was used: to refer to the most extreme temperature values ever recorded on permanently inhabited places on Earth. These values are -62 °C and +57 °C, corresponding, respectively, to Oymyakon (Siberia, Russia, registered in 2018) [12] and to Furnace Creek (Death Valley, USA, registered in 1913) [13].

2.2. Material Library and Selection Criteria

Next step in this research methodology was to identify the set of materials that would build the library for multicriteria selection based on different climate scenarios. The complete set has a total of 52 materials organized in twelve categories: metals, glass, ceramics, stone, concrete, elastomers, polymers, polymer foams, biopolymers, wood, wood derivatives and natural fibres. Besides materials commonly used in construction (e.g. aluminium, steel, soda glass, bricks, stone, concrete, PVC, PS and PU foams, wood), other materials were included to further extend the base of comparison and find if any of the non-common materials would have suitable properties for extreme climates. The information regarding material properties was retrieved from two main sources, both from the same author, to ensure coherence among the set of data [14] [15].

As previously mentioned, extreme low and extreme high temperatures were chosen as climate scenarios for this analysis (section 2.1.). Eight material properties were used as criteria for analysing the suitability of each material for the two climate scenarios, categorized as thermal, mechanical and eco properties. Thermal category includes thermal conductivity, thermal diffusivity, thermal effusivity, linear thermal expansion and service temperature. Therefore, heat transfer, thermal inertia and thermal stability were considered. The mechanical behaviour was characterized by the fracture toughness because this property was considered as an effective way of dealing with different types of materials having quite dissimilar mechanical performance (e.g. fibres and concrete). Eco properties include recycle potential and embodied carbon (accounting for climate change potential), thus considering the environmental impact of each material.

Service temperature was used to assign to each material a qualitative characterization about its resistance to extreme low, and high, temperatures. By comparing the minimum and maximum service temperature values with the temperature threshold considered for climate conditions (section 2.1.), materials were classified as having high, medium or low resistance to extreme temperatures. It was found that maximum service temperature is higher than +57 °C for every material. Therefore, this is not a critical property for very hot climates and therefore it was excluded from the models. In what concerns extreme low temperatures, materials having minimum service temperature below -62 °C were classified as high resistant, the ones having minimum service temperature between -62 °C and -30 °C were considered as medium resistant, and those having minimum service temperatures above -30 °C were classified as having low resistance to extreme low temperatures.

The set of 52 materials, characterized by the above-mentioned eight criteria were the base to build the multi-criteria decision model.

2.3. Multi-criteria decision analysis (MCDA)

To select the best suitable material, or group of materials, for a given climate condition, considering, simultaneously, the eight properties mentioned above (used as criteria), requires that these criteria are prioritized and put together in a single analysis tool. Together with the complexity of simultaneously analysing eight criteria, the fact that some of these are contradictory (e.g. thermal inertia and environmental impact) [16] is the reason why MCDA is an adequate option in this case. MCDA is a general term for systematic approaches that are used to support the analysis of various alternatives in complex problems involving multiple criteria. Various multi-criteria methods have been developed, usually supported by software tools built to help create the evaluation and analysis model [17]. The method used in this research is MACBETH (Measuring Attractiveness by a Categorical Based

Evaluation Technique), supported by the software tool M-MACBETH. It is a measurement approach from the early 90's that uses non-numerical judgments and weights for criteria in MCDA [17]. The non-numerical judgments implemented by the user are based on a qualitative difference of attractiveness attributed to each pair of criteria. A numerical scale is then generated according to the decision maker's judgments [18]. The MACBETH method was chosen for this research because its efficacy has been proved in various fields, such as urban strategies, environmental planning and eco-system managing. Moreover, the method can include a large number of options and criteria and has a high degree of flexibility in the way options are characterized and judgments are performed. This flexibility allows for a straightforward way of representing different scenarios by changing the difference of attractiveness between criteria and/or between levels of a given criteria to further fine tune the analysis. Furthermore, the decision maker may adjust the relative weights of the criteria within a certain range keeping the consistency of the overall set of judgements.

To represent the two climate scenarios – extreme low and extreme high temperatures, two base models were used. For the extreme cold climate scenario, thermal conductivity and resistance to extreme low temperatures were the first priority properties to consider, having the same relative weight. Environmental impact was considered as the second priority and thus recycle potential and embodied carbon were equally weighted with a weak difference of attractiveness in relation to thermal conductivity and resistance to extreme low temperatures. The third level of priority was given to fracture toughness and linear thermal expansion to which a moderate difference of attractiveness was attributed in relation to the top priority properties. Finally, thermal diffusivity and thermal effusivity were considered as not significant for extreme cold climates because thermal inertia is not a determinant requirement for comfort in these conditions; therefore, these two properties were judged in the model as having an extreme difference of attractiveness to all the others. For the extreme hot climate scenario, thermal conductivity, thermal diffusivity and thermal effusivity were considered as the top priorities because, besides the need to reduce heat transfer, thermal inertia generally plays a crucial role in obtaining comfort in hot climates. The second and third levels of priority are the same as for the previous model, with the same differences of attractiveness. Finally, resistance to extreme low temperatures is not a priority in this case.

In both models, as for the resistance to extreme low temperatures, recycle potential was also characterized by qualitative levels: high, moderate and low. All the other properties were quantified for each material option. In all these quantified properties but thermal conductivity, the levels used for judging and weighting correspond to the full interval of the set of individual values for each material. In the case of thermal conductivity, and considering the scope of this specific research, a limit was established for a material to be considered as suitable in this particular property. In this case, a lower reference of 0.51 W/m.K was considered so that materials with higher thermal conductivity would have a negative rate for heat transfer.

The research further evolved to consider a second set of analysis scenarios in which climate conditions are the same as previously described but where a higher importance was given to the materials environmental impact. In the MACBETH models, this new priority was considered by changing the difference of attractiveness between the levels of recycle potential and of embodied carbon. In the former, levels 'medium' and 'low' were considered as having an extreme difference of attractiveness to 'high', whereas in the base models' variation amongst criterion levels is linear. In the latter, an upper threshold was introduced at 1.30 $kgCO_2/kg$ together with a lower reference of 1.95 $kgCO_2/kg$ judged as having an extreme difference of attractiveness to the upper threshold value. This procedure was the way to model an analysis where materials having a value for embodied carbon up to that threshold would be highly rated while above that threshold the rate would rapidly decrease to zero and, above the lower reference value, have a negative rate. The threshold of 1.30 $kgCO_2/kg$ corresponds to the upper value of a set of materials including stone, simple concrete, ceramics, glass, wood, wood derivatives and some natural fibres.

The four models are designated as 'cold', 'hot', 'cold-eco' and 'hot-eco'.

IOP Conf. Series: Earth and Environmental Science **297** (2019) 012038 doi:10.1088/1755-1315/297/1/012038

3. Results and Discussion

As previously mentioned, this paper reports a preliminary phase of an on-going research about selecting materials for buildings in extreme environments. A set of conclusions may be drawn that gives valuable information to direct next steps.

In the base scenario for extreme low temperatures ('cold'), best ranked materials are polymers, polymer foams, glass, biopolymers, wood and wood derivatives. Ceramics, concrete, stone and metals are very poorly ranked, with negative scores for the cases of stone and metals. The highest ranked material is Polytetrafluoroethylene (PTFE) with a weighted score of 66.48 (out of 100); this modest score shows that there is no material that performs very well in all the criteria within the model structure that was adopted for this research. This means that the ranking of each material is a compromise between the several types of performance considered. For instance, soda-lime glass is the fifth best ranked material in the 'cold' scenario although its thermal conductivity is high within the purpose of this study (1.0 W/m.K). However, thermal stability, resistance to extreme low temperatures and recycle potential are strong properties for glass; on the other hand, fracture toughness for glass is very low. These first observations show that the overall ranking is useful for a general analysis, but a more detailed consideration of each individual property is necessary to assure a good match to specific performance requirements of each design case.

In the case of the base scenario for extreme high temperatures ('hot'), best ranked materials are polymers, biopolymers, polymer foams and glass. The above conclusion for the 'cold' scenario about a compromise between the different criteria is again relevant although some nuances may be observed. The best ranked material for this 'hot' scenario is polythylene terephthalate (PET) with a very modest score of 48.04. On the other hand, there is now more consistency in the ranking of each individual criterion. The materials that are on the top of the ranking list have a relatively low thermal conductivity and a relatively low environmental impact if assessed by recycle potential and embodied carbon. Wood is an interesting case in this scenario analysis: its rank (24th) is a result of a relatively higher thermal conductivity and a low recycle potential. This observation makes the case for the need of further developing and detailing the way of analysing the environmental impact of materials. Another critical observation in the 'hot' scenario is related to thermal inertia. The two material properties to assess this type of performance – thermal diffusivity and thermal effusivity – were given priority for extreme high temperatures. In the MCDA model, these two properties have the same weighting judgements as thermal conductivity. However, the materials that have a good thermal inertia performance (concrete, stone and metals) have a very poor thermal insulation performance as assessed by the thermal conductivity. The weighted balance between these three properties privileges thermal conductivity to the detriment of thermal inertia due to the relative differences between the values of these properties in each material. Therefore, there is in this preliminary phase a clear indication that the way to consider thermal inertia for very hot climates needs further development and research.

When a higher importance is given to the materials environmental impact (scenarios 'cold-eco' and 'hot-eco'), generally the overall tendencies observed in the previous models are maintained, with two significant changes. The first is related to the consistency of the ranking of each individual criterion (above mentioned for the 'hot' scenario); the higher relative importance attributed to recycle potential and embodied carbon in the 'hot-eco' scenario reduces the weight of thermal conductivity and there are good ranking materials that have a poor thermal insulation performance. The second significant change is that the best ranking materials, both for extreme high and low temperatures, are now natural fibres, wood and wood derivatives.

Two additional overall features may be observed from the results, related to contradictory objectives between criteria. There is a conflict between thermal insulation and mechanical strength because the materials with a lower thermal conductivity are the ones with a lower fracture toughness. Another conflict exists between thermal storage and thermal insulation; this observation is not at all surprising because materials that effectively share heat with the surrounding environment must have a moderate to high thermal conductivity. Finally, it is worthwhile to note that almost all the materials have a good thermal stability as assessed by the linear thermal expansion; therefore, this property does not represent

a challenge in material selection for the purpose of extreme temperatures and may thus be removed from the analysis.
The results seem to show that the best material for extreme climate conditions would be one combining the properties of polymers with a very low environmental impact. Specifically, for very hot climates, the ability of storing heat (thermal inertia) would be crucial. It may be said that biopolymers and natural materials as mycelium [19] seem to be good opportunities for further research.

4. Conclusion and future directions

The selection of building materials for buildings in the context of extreme temperatures and challenging climates is an issue that requires special attention and constant research, not only due to the effects of climate change but also due to an increasing global environmental concern. Because of this, designers and builders need to be able to ensure that buildings respond to the needs of their users, but also ensure that this can be done without further damaging the environment and thus contributing to the aggravation of the fragile environmental conditions of our world. This process can be made faster and more effective with the use of multi-criteria decision models, for a selection process that has into account various options and criteria in a flexible way. This process allowed to perform a series of analysis within the set of chosen materials considering two base scenarios of extreme temperatures, a very cold and a very hot scenario. With these as references, two other scenarios were introduced, through preference levels, which offered more environmentally positive alternative results to the base ones. Preliminary results show that the best possible material for extreme temperatures, whether it would be a very cold or a very hot environment, is one that could combine the properties of polymers with a very low environmental impact (at the level of the impact from materials such as natural fibres, wood or wood derivatives). The results thus suggest that further research may be directed at biomaterials development to combine high thermal insulation, high mechanical strength and very low environmental impact; for hot climates, thermal storage is an additional requirement. Material performance has a critical role in the search of the best possible ways to design and build in ever-changing and ever-challenging environments.

Acknowledgments
The authors wish to thank the support of ISTAR-IUL through funding from the Portuguese Foundation for Science and Technology (Project UID/Multi/04466/2019).

References

[1] Bannova O 2014 *Extreme environments - Design and human factors considerations* (Gothenburg: Chalmers University of Technology)
[2] Leach N 2014 Space Architecture: The New Frontier for Design Research *Architectural Design*, **232** pp 8-15
[3] Kallipoliti L 2017 *Closed Worlds* [Online] 20 December 2017. http://www.closedworlds.net/.
[4] Kretzer M 2018 *Dynamics in Extreme Environments* Materiability [Online] 12 February 2018. http://materiability.com/portfolio/dynamics-in-extreme-environments/.
[5] Kibert C 2008 *Sustainable construction. Green building design and delivery* (New Jersey: John Wiley & Sons)
[6] Hegger M, Fuchs M, Stark T and Zeumer M 2008 *Energy Manual – Sustainable Architecture* (Basel: Birkhäuser Architecture)
[7] González J 2004 *Arquitectura Bioclimática en un entorno sostenible* (Madrid: Editorial Munilla-Lería)
[8] Kronenburg R 2007 *Flexible: Architecture that Responds to Change* (London: Laurence King)
[9] Kretzer M 2017 *Information Materials: Smart Materials for Adaptive Architecture* (New York: Springer)
[10] Augustin S 2009 *Place Advantage: Applied Psychology for Interior Architecture* (New Jersey:

John Wiley & Sons, Inc.)
[11] Yan W and England M 2001 Design Evaluation of an Artic research Station From a User Perspective. *Environment and Behaviour* **33** pp 457
[12] Wright M 2018 Thermometer in world's coldest village breaks as temperatures plunge to -62C *The Telegraph* [Online]16 January 2018. https://www.telegraph.co.uk/news/2018/01/16/thermometer-worlds-coldest-village-breaks-temperatures-plunge/.
[13] Sample I 2012 Death Valley gains the title of hottest place ever recorded on earth *The Guardian* [Online] 13 September 2012. http://www.theguardian.com/science/2012/sep/13/el-azizia-libya-title-hottest-place-recorded.
[14] Ashby M 2013 *Materials and the Environment - Eco-Informed Material Choice* (Nashville: BH Publishing)
[15] Ashby M and Johnson K 2014 *Materials and Design - The art and science of material selection in product design* (Nashville: BH Publishing)
[16] Veludo S and Rato V 2016 Performance-based Selection of Sustainable Construction Soluction for Extenal Walls *Oikonet 3rd Internation Conference - Global Dwelling: Sustainability - Design – Participation* (Manchester: Oikonet & University of Central Lancashire) pp 95-103
[17] Costa C, De Corte J and Vansnick J 2003 MACBETH (Overview of MACBETH multicriteria decision analysis approach) *Internation Journal of Information Technology and Decision Making* **11** pp 359-387
[18] Mustajoki J and Marttunen M 2013 *Comparison of Multi-Criteria Decision Analytical Software* (Helsinki: Finnish Environment Institute)
[19] Heisel F, Schlesier K, Lee J, Rippmann M, Saeidi N, Javadian A, Nugroho A, Hebel D and Block P 2017 Design of a load-bearing mycelium structure through informed structural engineering: The MycoTree at the 2017 Seoul Biennale of Architecture and Urbanism *World Congress on Sustainable Technologies* (Cambridge: University of Cambridge)

IOP Conf. Series: Earth and Environmental Science **297** (2019) 012039 doi:10.1088/1755-1315/297/1/012039

A sustainable built environment for healthcare: generative mechanisms for future-proofing

I Krystallis[1, 2], B Rickaby[1] and N Murtagh[1]

[1] The Bartlett Faculty, School of Construction and Project Management, 1-19 Torrington Place, London, WC1E7HB, UK

2 email: i.krystallis@ucl.ac.uk

Abstract. Construction projects are planned from a set of deterministic objectives and constraints that dictates what designers must accomplish. In the case of healthcare facilities, where the policy environment is characterized by change on several fronts, fixed requirements could constrain their evolvability and sustainability. Future-proofing is a response to uncertainty whereby a physical structure is designed to respond to future changes in requirements, change of use, strategic perspectives, business drivers, new policies and changing climate. The paper investigates the causal powers that explain how and why future-proofing decisions achieve a sustainable outcome. A critical realist lens is adopted to develop a configurational perspective of future-proofing design evolution. Critical Realism's open systems ontology of social reality can account for the fact that outcomes are not predictable and can better explain the nature of causation in complex social interactions such as construction projects. A case study research design was conducted comprising in-depth interviews with healthcare construction professionals in a UK setting. The study contributes to the sustainability literature by offering four generative mechanisms of future-proofing decision making and how they contingently lead – or fail to lead - to future-proof outcomes. Two change mechanisms and two problem mechanisms are presented, together with their configuration under the right conditions and context to lead to design decisions for construction projects capable of adapting to a range of possible futures.

1. Introduction

Large construction projects are complex systems at the intersection of engineering, management and social sciences [1]. Developing such complex systems requires not only skills such as architectural and engineering design, but also knowledge of policy issues, societal norms and trends [2]. In addition, these systems include challenges that need to be dealt with due to their size [3], complexity, and technological change [4]. Further, sustainability objectives must be met, with national legislation and strategies, such as the UK's Climate Change Act (2008) and Clean Growth Strategy [5], paving the way for businesses and industries including the construction sector to invest in and deliver sustainable developments.

One type of large and complex construction system is the healthcare estate, which has become less flexible compared to a few decades ago [6]. Healthcare and its executive teams face a policy environment characterized by change on several fronts (e.g. reforms in health services, medical technology [7]. Healthcare building assets exist in a complex operational and technological environment and consequently identifying the requirements from a whole-life management perspective is crucial for ensuring these assets will meet their design life [8]. Additionally, for a physical asset to achieve sustainability, designers and investors alike recognize that it has to be also future-proofed [9].

In this paper, the concept of future-proofing (FP) is considered as a 'shield' against uncertainty. Recent studies suggest that FP is a phenomenon that can only exist in an open system [2], [10]. However, the traditional approach is to procure and deliver healthcare facilities as 'complete' projects, in other words, projects with fixed requirements or closed systems. Standard practice proceeds from a set of deterministic objectives and constraints that defines what designers must accomplish, thus standard practice does not deal with rapid change [2]. FP may be considered as an essential feature of "true sustainability" [11] – the notion of persistence over time without input of limited or non-renewable resources. In terms of construction, a project in which FP has influenced decision making is expected to remain in operational use beyond its initial span. This results, at least, in minimization of additional resources to reconfigure the development and, at best, avoidance of demolition waste and embodied carbon of materials for a new building. Our research question is: "Which mechanisms cause future-proof outcomes?". The study aims to contribute by identifying of generative mechanisms within a configurational approach. The adopted approach advances current knowledge about why some complex systems are future-proofed while other systems are not.

2. Literature review

Planning is a process that involves making decisions about future actions [12], often characterized by uncertainty due to incomplete knowledge. Numerous factors contribute to uncertainty in construction projects, such as, but not limited to, climate change and sustainability challenges, digitalization of the built environment and change in regulations. Planning for projects under conditions of uncertainty is challenging, as many of the decisions are interconnected. Strategic planning decisions are actions that pave the way and are irreversible. In situations where they are not compatible with future circumstances, subsequent interdependent decisions are also incompatible [12]. This poses significant challenges for practitioners. The flexibility perspective of planning adopts approaches that accommodate future changes. The intellectual basis of flexibility includes contingency theory [13], and evolutionary theory [14], based on the notion that practices need to be adaptable and compatible with a changing and uncertain future. This approach recognizes the need for being future ready – as the project evolves, unexpected challenges, requirements or opportunities, such as new technology, could emerge. Flexible design approaches are decisions that can either accommodate flexibility during project implementation or after completion, for example, embedding options, modularity and late design freeze [15], [16]. Planning for flexible approaches, such as flexible designs, implementation tends to enable smoother and adaptive working practices, which reduces the risk of time and cost overruns. Approaches that contribute to flexibility in the long-term and after project implementation (are referred to as FP, which enable projects to accommodate future needs [10].

Within the flexibility perspective, real-options theory examines the role of embedded options [2]. The term real-options refers to the "right, but not an obligation, to take some future specified action at a specified cost" related to a physical asset [17]. The theory originates from the financial options logic. Embedded options increase the flexibility of construction projects and can result in less disruption during project operations [18]. FP of a project can be conceptualized as embedding options that could be used by the users/owners in the later life-cycle of projects [16]. These options could be dealing with downside situations (e.g. change of use, due to shift in customer requirements) or upside opportunities (e.g. exploiting new technologies) [19].

3. Research Methodology

Considering the above, our research question is: *"Which mechanisms contingently cause future-proof outcomes?".* We addressed the question by adopting case study research and applied previously established methodological principles [20]. A total of 26 interviews of between 60 and 240 minutes were conducted with senior managers from procurers and supply chain organizations (suppliers) engaged in health estate projects for the UK National Health Service (NHS) who had experience in future-proofing.

IOP Conf. Series: Earth and Environmental Science **297** (2019) 012039 doi:10.1088/1755-1315/297/1/012039

Secondary data were also used, including direct reports and papers from the NHS and the Department of Health (DH), and third-party reports which reviewed NHS estate investment strategies and processes. Methodologically, we used Wynn and Williams [20] four-step approach to report the findings, as summarized in Table 1. In analyzing the transcripts and secondary data, the NVivo qualitative package software was used.

Table 1. Methodological stages of Critical Realism (adapted from Wynn and Williams, 2012)

Methodological stage	Outputs and evaluation criteria
Explication of Events	• Basic description of case study, setting the scene • Summary and streamlined version of events as they occurred. Use of direct quotes to show perception in the empirical domain • Description of events includes details of key actions and outcomes • Documentation of the effects (or not) of change
Explication of Structures and Context	• Identification of social and physical structures (actors, rules), and relationships among them • Identification of context and necessary conditions (tendencies) • Description of causal tendencies that caused the events • Identification of changes (or not) to the structure • Description of emergent properties
Explication of mechanisms	• Identification of proposed mechanisms that emerged in the real domain • Use of abduction reasoning to logically support the mechanisms • Identify *problem* mechanisms and *change* mechanisms
Alternative mechanisms	• Identification of alternative mechanisms that could exist • Validation of proposed mechanisms over alternative mechanisms

Configurational approach

Working within a stratified CR framework, a configurational approach [21], distils the philosophical stance into a methodological framework. Key terms are mechanisms, objects, context and conditions, which will now be explained. This approach enables outcomes to be explained by analyzing configurations of possible mechanisms and context-variations. A generative mechanism may be *active* or *passive* in a system - its activation can lead to a desired outcome if that mechanism operates within an appropriate context. Context is a general concept that encapsulates the 'relevant circumstances. The mechanisms exist in a given context and under the right conditions become enacted. If the context/conditions are not in place the mechanism's causal powers have no impact.

The context of this study is the construction contract. The entities include both agents such as procurers and supply chain consultants and contractors, and the relationships between them. Both agent types, procurers and suppliers have powers. In this context, a supply chain consultant can take actions in the best interests of the procurer, and the procurer can make decisions based on the recommendations made by the consultant and what is exchanged has the power to influence both parties. Future-proofing decision-making is enacted within this context. conditions that directly or indirectly lead to the success of a construction project are deemed *success factors* [22].

4. Case Study Findings

4.1 Background

The setting of this case study was a large-scale public service organization, the UK NHS which delivers (predominantly) free healthcare nationally. The service, to meet demand and become proactive in terms of responding to change, has developed capital plans which aim to maximize value for money. All proposed plans must be supported via business cases which are compliant with Department of Health guidelines. However, the consensus is that the current capital investment is insufficient to fund transformation and maintain the current and future estate business.

4.2. Explication of events

In the first step of analysis (see Table 1), a series of events is described, derived from the participants' accounts of experiences on healthcare projects procured around the UK.

4.2.1 Initiating the Business case

For NHS England, business case development is triggered by the need for a new service and estate. A business case must comply with a series of guiding principles. National data are used for the development of the business case which will cover current healthcare and new treatments. The appraisal of business case options must demonstrate that various options have been informed by stakeholder consultation, lessons learnt from past projects, and international comparative cases.

4.2.2 Establishing the Procurement route

As part of the tender process, the briefing document outlines the technical requirements the design must meet. The brief will describe the current needs including any aspects of future-proofing and sustainability goals the design should address. If NHS England is procuring with its own capital, then the supply chain interviewees claim it will go for the cheapest solution.: "To put it simply, the Trusts decide … the design with the lowest price wins" [S-HDiv]. If on the other hand, it is some form of public–private partnerships (PPP) project, then, potential investors look for the most economical solution over the whole life of the investment. In this case, the investors will be interested to explore what flexibility could offer in terms of return on investment (RO):

"The drivers for PFI [Private Finance Initiative] are totally different than NHS Capital Procurement, the PFI is driven by the banks and how much return on investment they can get. And if they are prepared to put some additional money up front to save a lot of money in the long term" [S-MD]

4.2.3 Preparing the Supply Chain Response

Once the tender documents are released, the supply chains form teams to respond to tender requirements with a proposal. During this stage, the team will formulate decisions around technical and commercial requirements to win the project. In going through this short term but demanding process, the main issues can be grouped into three broad categories. The first category is around prejudices on the concept of FP - stakeholders in both the procurers and supply chain involved in the project can see FP as a simple tick box exercise. The second category is around existing processes. The supply chain must convey their thinking against existing processes. For example, a supply side Design Director identified the constraints due to the brief content as limiting the ability of the technical response to accommodate options thinking: "The briefing is the main barrier, it constrains the approach. This is the primary concern… we would include flexibility in its full form [if not constrained by the brief]" [S-DirDes]. The third category is around financial issues. The supply chain struggles to justify offerings of value against offerings that are built on price. All participants highlighted financial issues that exist in incorporating options thinking in the technical solution, specifically the potential for higher construction costs. A development director noted: "The drivers of course are always cost related. It is cost, it is time! Why would we do anything like [future-proofing]?" [S-DirDev]. Despite these issues, technical solutions for FP are sought, and the teams respond in the best way possible in what they hope will be the winning solution. A sector director added: "you will win the work because you are the cheapest. So, you are not looking to duplicate anything when you are bidding" [S-SDir].

4.2.4 Initiating the Project

NHS England assesses all submissions and evaluates the value propositions based on two key categories: technical criteria and commercial criteria. The proposal which achieves the highest marks will be awarded the project. The supply chain team will try to incorporate FP after understanding where FP is needed. However, challenges emerge as additional funding for FP, improving quality or meeting sustainability standards is difficult to secure. Some of the participants felt that the only way that additional funding could be achieved was on the grounds of increased patient capacity and so FP becomes very difficult to justify. Changing old ways of thinking can be a challenging task for the supply chain team and as a result these conversations often fail. As a result, these complex capital investments in most cases do not have FP embedded in them and face risks of becoming obsolete before the end of their assumed design life.

4.3 Explication of Structure and Context
Having unpacked the sequence of events, we now turn to the explication of structure and context, the relationships of actors and identification of necessary conditions (Table 1, stage 2).
With regards to objects (e.g. actors) and the relationships amongst them, the supply chain actors act on behalf of the procurers' team to produce design solutions according to the specifications of the brief requirements. At first glance, this relationship may be perceived as a peer to peer relationship. However, the relationship between the two agents is directive. The structural relationship between those two actors is summarized by a managing director of a major UK contractor where he puts into perspective how the relational model may be formed:

"I think it [future-proofing] is very difficult …It is like one of those things where an editor can always tell you what you should have done previously rather [than] tell you what you need to do to go forward" [S-MD].

4.3.1 Contextual conditions
Having analyzed the context and structural relationships of objects, we turn to the critical success factors that actualized the FP configuration. Two conditions have been identified through our analysis.

Condition #1 Capital targets versus whole-life targets
This condition in the case study setting originates in the relationship between the Government's policy directives and the service's performance. The service's delivery performance was aimed to meet the government targets, hence reducing capital cost to meet set targets was more important than achieving whole-life targets. For the above reasons, justifying investment in the estate business, particularly in improving the future performance of the estate, received lower priority than performance of the service delivery, although arguably these are interrelated. On the other hand, having whole-life targets to identify where investment now would achieve savings later was identified as an enabling condition in the analysis.

Condition #2 Tight brief versus loosely-defined brief
This condition relates to the opposing tendencies of tight versus loosely-defined brief from the outset of the business case. In the case study, in general, client requirements for FP were superficial. Thus, when such solutions were proposed, they tended to be rejected, because the solutions were outside of the scope of the brief requirements. Instead, the brief was tightly focused on what was needed at that point in time. This tendency runs counter to a more loosely-defined approach, where the brief did not freeze prior to entering the design phase. An example of such approach was presented by the sector director of a consultancy firm:
"But FP should really come from designing a building such as the [Project N] Hospital which was allowed fast track construction. It was designed by not knowing were the departments will be, so it ended up with having huge floor plates, ducts on a regular basis and you could do what you like" [S-SDir].
During the development activity, a detailed account of solutions is developed. This detailed account in turn could feed back and redefine the requirements and thus lock-in of the baseline design would emerge

at a more developed stage. In the above example, the team challenged the brief and shifted its focus from being capital driven to being operations driven. This would not have been possible if the procurement followed a tight brief schedule.

4.4 Explication of generative mechanisms

In the next stage of analysis, we tested various generative mechanisms (Table 1, stage 3) against the events of the case study. We adopted a backward-chaining approach [23], looking at the two outcomes of interest (obsolete systems; future-proof systems) and the causes that drive towards these outcomes. In total, we identified four generative mechanisms. We further analyzed them into two categories depending on their powers to influence an outcome. These mechanisms are now described.

Change mechanism #1 – Iteration between problem, design and solution spaces

In the in-depth interviews of this case study, there existed an insightful optimism that a construction system could expand its service life and offer best value. Not freezing the brief (condition #2) before the design solution reaches the required maturity created a loop of iterations between problem space and solution space [24]. These iterations result in both spaces being continually informed as the project matures. As a result, the two spaces feed back into the design space, where solutions can mature and be checked against the latest set of requirements. The participants described how a conversation between principal and agent takes place to discuss possible options enacted by the interplay of repeatable standardized design elements, overengineering and flexible design elements. Through this mechanism, the teams implemented an FP approach which resulted in numerous design solutions being assembled.

Change mechanism #2 – Making the case for affordability

The outcome of the first mechanism is the solution space, in other words, the number of candidate solutions that could become plausible scenarios for implementation. Yet, it is not determined at this point that these options will be implemented. The outcome of the second mechanism is the ultimate decision whether a combination of plausible interconnected solutions is affordable. The participants' propositions for affordable solutions to the client were also informed by previous cases.

In this mechanism, the case study findings indicate that the design which incorporates FP need to be perceived to be affordable to be attractive. Furthermore, the solution should feature payment-by-results, whereby additional investment to justify the business case to include FP thinking is contingent on the independent verification of results. A participant added: "The concept of Payment-by-results … implies that the only way that you afford anything … by saying 'I am going to put more patients through' so [FP] becomes very difficult to justify" [S-MD].

The participants identified that financial issues could stall approval and funding of a potential development. The dominant view was that budgets on a potential development are often already under financial pressure and thus clients would not decide on design choices for the future unless these were perceived as affordable.

Problem mechanism #1 – Cognitively-bounded decision-making

Having identified two mechanisms that could enact future-proof outcomes, we now turn to the resident problem mechanisms that may prevent successful outcomes. The first problem mechanism is that of cognitively-bounded decision-making. This draws on the long-established notion of bounded rationality [25]. Bounded rationality acknowledges that much human decision-making is highly complex and inevitably involves uncertainty but that people's abilities to deal with uncertainty, to identify options, to foresee consequences and to prioritize competing goals are very limited. To cope, the problem is simplified to be manageable and a 'satisficing' or 'good enough' outcome is pursued.

The interviewees alluded to the complexity of the problem, as a Development Director argued:

"Everything needs to account together… so that we can use [the building] at any given point in time in the future. And it is that bit that people do not get and they do not think about… we just do not think about it enough as a process." [S-DirDev].

The speaker here recognizes the complex, interconnected nature of FP and the bounded nature of thinking, suggesting that decision makers on project teams do not even address the more complex requirements. A programme manager for NHS estate stated: "I do not think we do it in detail" [P-PgmM] – a reference to simplifying the problem.

The problem of uncertainty was addressed explicitly in the interviews– uncertainty of outcomes and uncertainty over the meaning and value of FP. There was recognition that a simplified, all-or-nothing understanding of FP is inadequate, that the actors in the given context have not thought about it enough, and the problem lies in limited cognition (an interviewee referred to "insufficient thinking" [S-SDir].

Problem mechanism #2 – Motivational gap

The second problem mechanism's causal powers stem from the lack of personal commitment which was in evidence in the interviews. The interviewees highlighted that actors in the social system of the construction project have different values. For instance, according to a managing director and owner of an architectural practice, the Tier 1 actors who are leading the process lack personal commitment, thus he and his team try to push different agendas such as sustainability and FP: "But the truth is, those end up being words in a presentation to win the job and then after that the contractor is not interested. It is very frustrating" [S-MD2].

5. Discussion and Conclusions

From this study, we sought to explain why some construction projects are future-proofed whilst others become obsolete. We described two change and two problem mechanisms behind future-proofed outcomes and the configurational setting and the context within which they operate.

We now turn to assess what alternative mechanisms could explain the occurrences of the observed events (Table 1, stage 4). From the literature, optimism bias emerged as an alternative mechanism. In the findings here, the interviewees did not underestimate how complex FP is, nor did they overestimate the beneficial outcomes of implementing FP. On the contrary, they criticized the current level of thinking on FP as inadequate. This inadequate thinking went beyond a simple cognitive bias and in fact the participants recognized varied cognitive limitations with regards to taking actions to pursue FP.

The second alternative mechanism explored was that of strategic misrepresentation. Strategic misrepresentation assumes a deterministic view of the outcome created by the planner's conscious intent. However, our findings did not support a tendency by the supply chain deliberately to underestimate efforts required with regards to planning for FP.

In addition to the above, optimism bias and strategic representation can be rejected as potential alternative mechanisms because of their underlying assumption of a closed system approach. The two candidate mechanisms assume requirements fixation at pre-acquisition stage, whereas in complex projects such as healthcare requirement evolve over long time-spans of definition and development stages. Thus, whereas these projects should be treated as open systems, the above candidate mechanisms assume the planning activity takes place in a closed system.

We theorize that a construction system can be made sustainable as well as future-proof if the right conditions and the two change mechanisms identified above are enacted. Our findings suggest that decisions could lead to non-FP outcomes if the change mechanisms are insufficiently strong or if the necessary conditions are not in place. With regards to implication for policy, our findings have shown that policymakers need to allow for budget flexibility in the way funds are released, in order construction capital investments to achieve sustainability targets and be benchmarked beyond time, cost and quality. Allowing for a loose-fitting brief will contribute towards project success. Shifting assurance procedures towards whole-life targets instead of capital targets should be considered. Awareness of cognitive boundaries and personal commitment in FP agendas can enhance policy makers' understanding of how procurers and supply chain partners will develop future-proof solutions and avert uncertainty.

Future research could broaden this study's empirical and theoretical approach. Empirically, settings other than healthcare projects could be examined from a CR perspective. In addition, there is more work to be done with regards to CR and the literature on uncertainty in decision-making. Underlying each mechanism and condition there may lie other powers waiting to be discovered [26]. Investigating further the mechanisms outlined in this study using CR would offer valuable insights to rethink about how FP might be applied in the built environment.

References

[1] N. Gil and S. Beckman, "Introduction: Infrastructure meets business: Building new bridges, mending old ones," *Calif. Manage. Rev.*, vol. 51, no. 2, pp. 6–29, 2009.
[2] R. De Neufville and S. Scholtes, *Flexibility in engineering design*. MIT Press, 2011.
[3] B. Flyvbjerg, "What you should know about megaprojects and why: An overview," *Proj. Manag. J.*, vol. 45, no. 2, pp. 6–19, 2014.
[4] M. Dodgson and D. Gann, "Technological innovation and complex systems in cities," *J. Urban Technol.*, vol. 18, no. 3, pp. 101–113, 2011.
[5] BEIS, "Clean Growth Strategy," 2018.
[6] P.-Y. Crémieux, P. Ouellette, F. Rimbaud, and S. Vigeant, "Hospital cost flexibility in the presence of many outputs: A public-private comparison," *Health Care Manag. Sci.*, vol. 8, no. 2, pp. 111–120, 2005.
[7] P. M. Ginter, W. J. Duncan, and L. E. Swayne, *The strategic management of health care organizations*. John Wiley & Sons, 2018.
[8] I. Krystallis, "A study of the concept of future-proofing in healthcare building asset management and the role of BIM in its delivery," {\copyright} Ilias Krystallis, 2016.
[9] E. Krygiel and B. Nies, *Green BIM: successful sustainable design with building information modeling*. John Wiley & Sons, 2008.
[10] O. U. Rehman and M. J. Ryan, "A framework for design for sustainable future-proofing," *J. Clean. Prod.*, vol. 170, pp. 715–726, 2018.
[11] M. Pagell and Z. Wu, "Building a more complete theory of sustainable supply chain management using case studies of 10 exemplars," *J. supply Chain Manag.*, vol. 45, no. 2, pp. 37–56, 2009.
[12] R. Ackoff, "A concept of corporate planning," *Long Range Plann.*, vol. 3, no. 1, pp. 2–8, 1970.
[13] K. M. Eisenhardt and B. N. Tabrizi, "Accelerating adaptive processes: Product innovation in the global computer industry," *Adm. Sci. Q.*, pp. 84–110, 1995.
[14] H. Aldrich, *Organizations and environments*. NJ: Prentice Hall: Englewood Cliffs, 1979.
[15] N. Gil and B. S. Tether, "Project risk management and design flexibility: Analysing a case and conditions of complementarity," *Res. Policy*, vol. 40, no. 3, pp. 415–428, 2011.
[16] N. Gil, "On the value of project safeguards: Embedding real options in complex products and systems," *Res. Policy*, vol. 36, no. 7, pp. 980–999, 2007.
[17] L. Trigeorgis and J. J. Reuer, "Real options theory in strategic management," *Strateg. Manag. J.*, vol. 38, no. 1, pp. 42–63, 2017.
[18] I. Krystallis, P. Demian, and A. D. F. Price, "Using BIM to integrate and achieve holistic future-proofing objectives in healthcare projects," *Constr. Manag. Econ.*, vol. 33, no. 11, 2015.
[19] M.-A. Cardin, G. L. Kolfschoten, D. D. Frey, R. de Neufville, O. L. de Weck, and D. M. Geltner, "Empirical evaluation of procedures to generate flexibility in engineering systems and improve lifecycle performance," *Res. Eng. Des.*, vol. 24, no. 3, pp. 277–295, 2013.
[20] D. Wynn Jr and C. K. Williams, "Principles for conducting critical realist case study research in information systems," *MIS Q.*, pp. 787–810, 2012.
[21] R. Pawson and N. Tilley, *Realistic evaluation*. Sage, 1997.
[22] A. De Wit, "Measurement of project success," *Int. J. Proj. Manag.*, vol. 6, no. 3, pp. 164–170, 1988.
[23] A. M. Pettigrew, "Contextualist research and the study of organizational change processes," *Res. methods Inf. Syst.*, vol. 1, pp. 53–78, 1985.
[24] I. Krystallis, P. Demian, and A. D. F. Price, "Design of flexible and adaptable healthcare

IOP Conf. Series: Earth and Environmental Science **297** (2019) 012039 doi:10.1088/1755-1315/297/1/012039

buildings of the future - a BIM approach," *Proc. First UK Acad. Conf. BIM, Newcastle Bus. Sch. Sch. Law Build. Northumbria Univ. 5-7 Sept. 2012*, pp. 222–232, 2012.
[25] H. A. Simon, "A behavioral model of rational choice," *Q. J. Econ.*, vol. 69, no. 1, pp. 99–118, 1955.
[26] A. Collier, "Critical realism: an introduction to Roy Bhaskar's philosophy," 1994.

IOP Conf. Series: Earth and Environmental Science **297** (2019) 012040 doi:10.1088/1755-1315/297/1/012040

Assessment of sustainability of the built environment in Latin American neighbourhoods: five cases in southern Chile.

A Zumelzu[1], D Gruehn[2] and J Hosni[3]

[1] Instituto de Arquitectura y Urbanismo, Universidad Austral de Chile, Chile.
[2] School of Spatial Planning, TU Dortmund, Germany.
[3] Institute of Urban Planning and Design, University of Stuttgart, Germany.

E-mail: antonio.zumelzu@uach.cl

Abstract. This article aims to show the results of the evaluation of the sustainable potential of the built environment in five neighborhoods in Southern Chile, to deal with the complex agenda of urban transformations in intermediate cities. Five neighborhoods in Isla Teja district, in the city of Valdivia, are evaluated in this research. The methodology considers five dimensions widely discussed in scientific literature: Accessibility, Connectivity, Density, Diversity and Vitality. The results show that certain neighborhoods have greater degree of adaptability in their morphology to generate diversity of uses, which in turn promote greater sustainability of the urban form. Mixed-use development and graded densities in urban blocks show best response to promote social life and collective activity in urban space. The article concludes with respect to future challenges to promote the development of more sustainable urban conditions in Latin American cities.

1. Introduction

Lack of planning has especially placed South American cities at the center of 21st Century development (UN-Habitat, 2017). In Latin America, evidences indicate that cities are following a pattern of change similar to that observed in European and North American contexts. This mainly refers to changes in two basic aspects of the city: form and function. For Salinas and Pérez (2011), this transformation is characterized by a morphological modification in relation to attributes of contiguity, compactness and limits of the historical model of the city, by those attributes of discontinuity, fragmentation or diffusion of the contemporary urban development, which subsequently produces the dispersion of the city. The literature also suggests that one predominant element is the massification of private residential projects, which have transformed the landscape and urban lifestyles (Borsdorf and Hidalgo, 2013; López-Morales et al., 2015). Other studies suggest that these same processes have generated new forms whose main features are fragmentation and privatization of the space, reducing social polarization and increasing segregation on a reduced scale (Inzulza and Gatica, 2018). In this way, urban units such as the neighborhood -bases of a city microstructure- where the community articulates the private space of the family with the public space, gradually lose their structure, identity and vitality (Zumelzu, 2016). There is very little evidence of these processes in intermediate cities, especially in southern Chile, which today represents a great range of potential and interest to promote sustainable urban development (UN-Habitat, 2016).

The principal objective of this article is to apply a methodology to explore and evaluate the sustainable potential of the urban form in five neighborhoods in Southern Chile. Sustainability is used as an operational concept to face urban problems, where the strength of the concept consists of integration of different aspects – economic, social, mobility, accessibility, land use, environment, among other. The methodology establishes the application of five dimensions of sustainable urban form -widely discussed in scientific literature- to evaluate the potential of some places to be catalysts for an improved -more sustainable- urban form: accessibility, connectivity, density, diversity and vitality. To do so, it conducts a critical review of the concept of urban form, sets out its key dimensions and understands their claimed relationship with sustainability. Building on the comprehensive study of Isla Teja, a district located in Valdivia in Southern Chile, five neighborhoods are identified and evaluated. Both quantitative and qualitative research methods are adopted to test the associated sustainability outcomes. Finally, we use our results to highlight critical aspects and future challenges to promote the development of more sustainable urban conditions in Latin American cities.

2. Sustainability as an operational concept

The concept of sustainability in urbanism is broad, complex and even ambiguous, for which there is no single definition, leading to a variety of perspectives and visions. Due to its complexity, the term "sustainability" is often used unsustainably. The stigma of "sustainable city" or "sustainable development" is a metaphysical ideal and a utopian myth -powerful but problematic (Davidson, 2010). There is no pragmatic model, only partial visions (Frey and Yaneske, 2007). While there is evidence of progress, it has been at the level of policymaking. The concept has shown its vagueness in terms of action and implementation, in which the relationship with the purpose of planning is lost (Zumelzu and Doevendans, 2016, p.2).

Application of sustainability in the urban form has been widely discussed during last two decades, in which debates have focused mainly on the search for "ideal" urban models which contribute more to sustainability than others (Jabareen, 2006; Ehlers, 2011). These views suggest that sustainable urban form must be compact from the center to the edges (Breheny *et al.*, 1996; Burton, 2002; Artmann *et al.*, 2019). Other authors suggest that the city could consist of different compact units which are decentralized but connected by a public transport networks (Jenks and Dempsey, 2005; Neuman, 2005). The debate has been, in discourse and in practice, confusing, inconclusive, in which researchers have focused only on a limited number of aspects (Naess, 2014; Jenks and Jones, 2010; Sakamoto *et al.*, 2018). However, urban models can play a determining role because they contain interesting elements which can be applied in an existing city. But often, these models are theoretical constructs, and, in practice, they are still a long way from solving the real problems of cities -mainly related to high levels of energy consumption and urban areas dependent on cars, social and spatial segregation, high levels of pollution, among others.

We believe that discussions and debates on a sustainable urban form should follow a more heuristic trajectory, broadening and interrelating the debate towards the development of methods, instruments and tools instead of producing optimal solutions or models, such as the compact city for example (Kärrholm, 2011; Zumelzu, 2016). It is very important to study the causes of city problems and the development of methods and tools for the practical application of sustainability (Säynäjoki et al., 2015).

3. Dimensions of sustainable urban form

The definition and measurement of sustainable urban form -sometimes referred to as "sustainable urban neighborhoods" or "sustainable urbanism"- has advanced and evolved significantly in the last two decades (Frey, 1999; Song and Knaap, 2004; Karimi, 2013; Luederitz *et al.*, 2013; Zumelzu and Doevendans, 2016; Hemani *et al.*, 2017; Mouraditis, 2018). The application has demonstrated the importance of the concept of sustainability as a tool for integrating economic, environmental, social and morphological issues to deal with urban problems. In spite of this, its direct use in developing countries is debated, due to a lack of clarity regarding how to sensitize these aspects to local problems, not to mention the lack of a database with the magnitudes and specificities required to apply them. It should be mentioned, however, that there are many other needs, aspirations and dimensions to promote a place that are more specific in terms of sustainable development. This challenge has prompted planners, architects, and local and international governments to propose new frameworks for restructuring and redesigning urban areas to achieve sustainability at different spatial levels.

In scientific literature, discussions of urban form related to sustainability encompass a number of physical and intangible characteristics, which are related not only to morphological elements but also to the configuration of the social environment and its interaction within neighborhoods and districts (Dempsey *et al.*, 2010). Accordingly, we propose five key dimensions, broad and interrelated, that have been identified -usually in scientific literature- and that can be measured and evaluated. The scientific evidence shows that these dimensions are most demanded to influence sustainability and human behavior in the city.

3.1. Accessibility

Accessibility is a long-standing theme among the theories of "good" urban form (Lynch, 1981; Sternberg, 2000). The sustainable urban form is defined by the degree to which it is compatible with the needs of pedestrians and cyclists over car drivers. This has been motivated by a concern about negative effects generated by the built environment on the physical activity and health of inhabitants. It is believed that pedestrian-oriented streets have a positive effect not only on the quality of the place, but also on the degree to which people are willing to walk (Talen and Koschinsky, 2013). A pattern of sustainable settlement should increase access among residents, their workplaces and required services in an equitable manner, considering inhabitants with displacement problems -older adults, disabled, pregnant women and households without car (Barton *et al.*, 2010; Sigh, 2016). Accessibility measures have been used in recent times as part of an effort to assess built environment for health effects (Frey and Bagaeen, 2010; Jun and Hur, 2015; Kowaleski-Jones *et al.,* 2018).Walkable access to services and equipment is an essential part of urban sustainability, because people who live in places with services -at pedestrian reach- tend to reduce car-dependent mobility and generate a smaller carbon footprint.

3.2. Connectivity

Urban form plays an important role in promoting or limiting connectivity. Connectivity refers to the degree to which local environments offer connection and contact points -to people and resources- in a variety of scales and for multiple purposes. This quality promotes sustainability in which the highest connectivity leads to higher levels of interaction between people, the environment, society, economic and cultural activities, all of which are believed to improve the long-term stability and community of the neighborhood (Talen, 2011). The social connection at neighborhood scale is seen as pedestrian phenomenon, in which networks of "neighborhood relationships" are related to interconnected pedestrian streets and to the internal access to the neighborhood generated by these street networks (He *et al.*, 2019). The importance of maximizing connectivity in the urban space has been a vitally important aspect in studies of urban form, wherein the focus has been to maximize opportunities for interaction and exchange, and increase the number of routes -streets, sidewalks, and other throughways. There is a common agreement in the literature that urban block of great length or scale, cul-de-sacs, and dendritic street systems are less likely to provide good connectivity (Sevtsuk et al., 2016).

3.3. Density

Density is an essential component of sustainable urban form. This criterion has been one of the main factors in maintaining pedestrian access to the basic services and equipment needed in a neighborhood. It is known that an increase in carbon emissions causes declines in density and mixed use (Cervero, 2013). However, no specific rules exist about how density form or the mixed-use level should vary given different regions and contexts (Zumelzu and Doevendans, 2016). In the literature, in relation to neighborhoods, values of total population size are generally established between 4,000 to

10,000 inhabitants, gross population densities between 50 and 58 inhab/he, and net dwelling densities between ranges of 43 and 85 dw/he (Frey and Bagaeen, 2010; Dempsey *et al*., 2010). The average density of housing must reflect, and is conditioning, to local situations. These may vary depending on the local conditions, either near the edge or the center of the neighborhood. The objective is to support the development of local services and equipment within the neighborhood.

3.4. Vitality
Vitality, also referred as "neighborhood vitality" or "nodality" in urban form, is understood as fundamental for community sustainability. Some authors characterize urban vitality at the neighborhood scale according to three main factors: the continuous presence of people on streets and in public spaces, their activities and opportunities, and the environment in which these activities are carried out (Zarin *et al*., 2015; Zeng *et al*, 2018). In urban form, the idea of nodality suggests that urban development should be organized around nodes of various levels and sizes (Frey, 1999; Kärrholm, 2011; Talen, 2011). While expansion tends to segregate the territory, the urban sustainable form tends towards a distinguishable hierarchy: from regional nodes of development to neighborhood centers or even public spaces at urban block level. At neighborhood level, nodes of activity promote sustainability of the urban form, providing there are public spaces around which the buildings are organized, where shops, local services and facilities, and social interaction can take place. Centralized nodes of activity at neighborhood scale might initiate physical interaction among the community, by giving the surrounding residents a common destination. These types of spaces support other aspects of sustainable urbanism, such as an increase in the density of surroundings, mixtures of types of housing anchored by a centralized space, or the commerce viability at neighborhood scale (Ellin, 2006).

3.5. Diversity
Diversity, as a dimension of sustainable urban form, relates to two aspects. In the first, the diversity of land use -understood as balances in the development of residential and non-residential land- is related to promoting a number of benefits: economic vitality, social interaction between individuals, and walkable provision of various services and equipment that a neighborhood requires (Talen, 2008; Tarbatt, 2012). In the second, socially diverse neighborhoods, it continues to be seen as essential for the well-being of a community and the goals of social equity, where the relationship with sustainability is made through a combination of income as well as races and ethnic groups which are believed to form the basis of an "authentic" sustainable community (Dempsey et al., 2011; Landry, 2016). The diversity and mixture of types of housing units is also important, from single family detached, semi-detached, etc. Studies of socially mixed neighborhoods -either through individuals and typologies- have consequently identified the urban form as a key factor in sustainable diversity (Talen, 2008; Zumelzu, 2015). .

4. Materials and methods

4.1. Study area
To study the sustainable potential of the urban form, we carried out a case study in the district of Isla Teja in the city of Valdivia, Southern Chile. which represents an area of important spatial transformations from the beginning of 20th Century. Isla Teja represents one of the pioneering districts of industrial principles during the 20th century in Chile. Isla Teja is a residential and service district, which covers approximately 15 km^2 and contains several residential areas. The growth and urban development of Isla Teja has been generated in a dispersed and fragmented way, given the natural geomorphology of the territory composed of natural systems such as wetlands, rivers, lagoons, forests and hills, generating "residential islands" within the same island, adapted to the natural borders (figure 1).

IOP Conf. Series: Earth and Environmental Science **297** (2019) 012040 doi:10.1088/1755-1315/297/1/012040

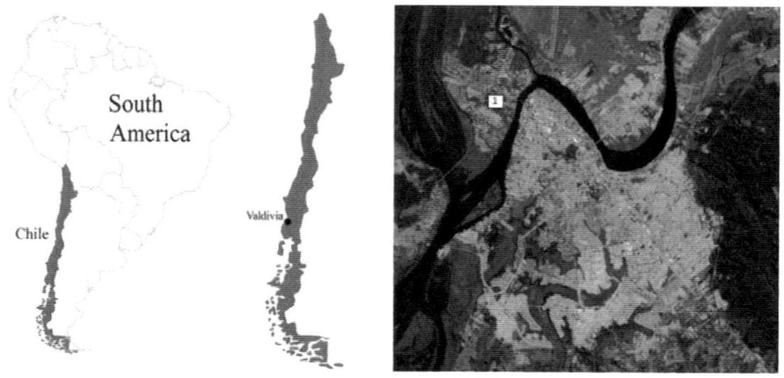

Figure 1. Location of Isla Teja district (1) within the city of Valdivia, Chile.

Despite its early occupation, the real urban development of Valdivia began with the arrival of German settlers in the late 19th century, bringing their entrepreneurial spirit and extensive knowledge of industrial processes and labor division. Industrial development and the subdivision and specialization of work creates the need for large labor forces, and therefore, the development of neighborhoods for workers came about in the early 1930s. These neighborhoods were developed in Isla Teja in close proximity to the industries, taking advantage of the connected areas.

Despite the efforts made by families descended from German settlers, the great earthquake of the year 1960 abruptly ended the industrial development of the island and may have implied an end to the urban development of Isla Teja. The 1960 earthquake not only altered the economic development of the island, but also redirected its productive role to a residential and service role. These changes were accompanied by morphological variations, including the descent of extensive areas which currently make up wetlands considered nature sanctuaries. The original types of housing with social character also stopped developing along with the end of industry, being replaced by residential complexes for the middle class and later, the upper class. In this sense, due to the varied construction phases and applications of planning concepts, substantial differences in urban form can be noted among the neighbourhoods of Isla Teja.

4.2. Methods

Table 1 summarizes the methodological structure to evaluate and measure sustainability. To study the sustainability of urban form in Isla Teja, the identification of activity nodes that provide neighborhood vitality is realized, to select which places and types of nodes could provide "nodality" dimension. To measure neighborhood vitality, the intensity of occupation is calculated through surveys to measure the count of pedestrians in an area, by using Gate method from Space Syntax theory (Vaughan, 2001).

Specifically, density is a measured by calculating gross population and net dwelling density, considering groups of urban blocks close to the node of activity. Accessibility is measured based on distance between residential lots and retail. In addition, the method of *Following People* (Vaughan, 2001) was incorporated to assess two aspects: the relationship between one movement route and another one within the system; and the distance that a person walks between their points of interest and their route choice. Connectivity is measured by counting the number of intersections per area unit by using Depthmap analysis from Space Syntax Theory. Finally, diversity is measured by Simpson´s diversity index for housing units' types and socio-economic group. According to Talen (2008), greater mix of housing units from single-family detached to apartment buildings would have an urban form more supportive of social diversity, a key feature of sustainability. It expected to obtain varying levels about places that could potentially be more sustainable than others in terms of urban form.

Table 1. Dimensions of sustainable urban form and how these are explored and measured. Source: authors, based on Talen (2011).

Dimension	How is measured	Spatial unit	Result
Vitality	Intensity of pedestrian occupation and movement in the urban environment.	street, lots, squares	To qualify and localize nodalities
Density	Population per hectare and dwellings per hectare.	Block group	To distinguish and characterize variations in households' size and factors of land uses.
Accessibility	Count of residential lots within 400 mts of services and facilities, divided by area of tract.	Unit area or tract	To evaluate the pedestrian access to daily-basic needs.
Connectivity	Number of intersections per area of tract.	Unit area or tract	To evaluate the degree in which the environment offers points of connection to people and resources.
Diversity	Diversity of housing types and socioeconomic groups.	Unit area or tract	To quantify the diversity of typologies and social mix of an area.

5. Analysis of results

5.1. Identification of potential nodes

Figure 2 shows the potential nodes that provide vitality, together with the neighborhoods of Isla Teja that had highest levels of intensity of pedestrian occupation and movement. Through the "Gates" method, the intensity of occupation was measured (Vaughan, 2011). Several captures were taken in the field for three days in the morning, three days in the afternoon, and one in the evening. Moreover, the captures were taken during two days of the week and one of the weekend.

The identified nodes correspond to neighborhoods Prochelle, Santa Ines, Teja, Raulies and Teja Sur. These areas reveal a mixture of urban occupation and grain types, from a predominantly urban fabric to those that are largely culs-de-sac. Several activity nodes were identified, ranging from parking areas and public squares to supermarkets and green open spaces. Figure 2 shows five types of places that could provide nodes of greater vitality in the neighborhoods of Isla Teja. These nodes serve as the "focal point" of some neighborhoods, presenting diverse uses and activities. Criteria for selection including places that provide a mixture of uses, high density of housing and a meeting place with pedestrian activity.

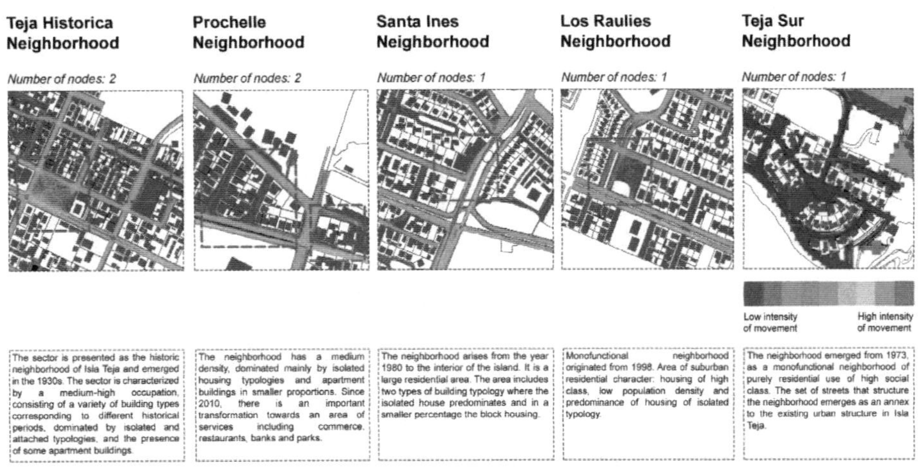

Figure 2. Potential nodes of activity within neighbourhood areas that provide high levels of vitality.

5.2. Results

Table 2 presents the results numerically. Accordingly, several observations about these data can be made. First, the results show that the most sustainable areas do not seem to have much relation to potential nodes which have greater loads and high values of intensity of use. But rather, they relate to places with a higher mix of building use and higher diversity of activities in their nodes, which are oriented towards open spaces such as squares and green areas, and not towards streets with high intensity of use. Second, it is observed that in neighborhoods with gross housing densities between 35 and 65 houses per hectare, and gross population densities between 50 and 65 inhabitants per hectare, they show a greater graduality of density between their blocks, offering a higher variety of types and size of housing. This promotes greater functional exchanges in the nodes, increasing pedestrian access to opportunities and services. From the net values, it is observed the development of two types of density: density of vertical type and density of horizontal type. The density of vertical type generates higher positive effects within the neighborhoods, especially in Teja and Santa Inés neighborhoods. The high horizontal density, on the other hand, due to the growth of rental housing in residential plots, decreases the pedestrian accessibility in the inner streets.

Table 2. Five measures of sustainable urban form in neighborhoods in Isla Teja. Source: authors.

Measure	Isla Teja Neighborhoods				
	Teja	**Prochelle**	**Santa Ines**	**Raulies**	**Teja Sur**
Vitality (intensity of pedestrian use, average per week)	7129	6370	9344	1540	1430
Population density (mean block group density)	155	62	81	46	16
Dwelling density (mean block group density)	63	35	59	21	25
Accessibility	0.35	0.25	0.34	0.11	0.07
Connectivity	5	7	4.4	3.7	3
Diversity	0.55	0.46	0.3	0.15	0.12

A high accessibility is observed in neighborhoods with a high vertical housing density and a greater balance between their land uses, such as the Teja and Santa Inés neighborhoods. The pedestrian accessibility decreases in places with greater horizontal density of housing due to the expansion of single-family house that generates effects on the public space. Elements such as blind walls, the lack of front yard, the low distance between street fronts, the absence of green areas, and the decrease in the size of the sidewalks in the inner streets, they negatively affect the choice of routes, and consequently pedestrian accessibility. Moreover, the routes with higher pedestrian accessibility are observed in streets conformed by shorter blocks in their façade dimension, which do not exceed 100 m on average. These routes have a high quality of built environment in all its categories, from wide sidewalks, presence of green areas and good quality of infrastructure of shops and services, as it happens in the nodes of Teja and Prochelle neighborhoods.

Regarding diversity, most of the diverse areas are generally located in areas of higher connectivity and in nodes that show a greater diversity of uses, as in the case of the nodes in Teja, Prochelle, and Santa Inés neighborhoods. These neighborhoods show more balanced average values of diversity among the four categories analyzed: land use, dwelling size, typology of housing and socio-economic group. Furthermore, these "diverse" neighborhoods are positive because of the functional exchanges they offer, promoting greater effective access to opportunities, goods, and services. In addition, it is observed a greater diversity of socio-economic groups (per block), especially in places of greater mixture of building use, through high-rise housing with services on the first floor or the adaptability of an existing building (transformation of existing typology) to a new use. The mix of housing types and the combination of different shapes and sizes provide multiple opportunities for access in the neighborhoods. In Teja and Prochelle neighborhoods, for instance, the mix of housing types has generated opportunities for the arrival of immigrants and university students, which has led to the appearance of new services and local economies. In these neighborhoods, the commercial activities coexist with residential areas (mixed use), with rental properties that not only provide greater purchase options, but are also essential to retain low-income people in the neighborhoods.

6. Discussion

The maps in figure 3 show the most and least sustainable areas within neighborhoods that show a better balance between their values. Some of the most interesting aspects observed, according to the results, is that the nodes with the highest load of use are not related to places that provide a greater diversity of activities, and in turn, greater social interaction.

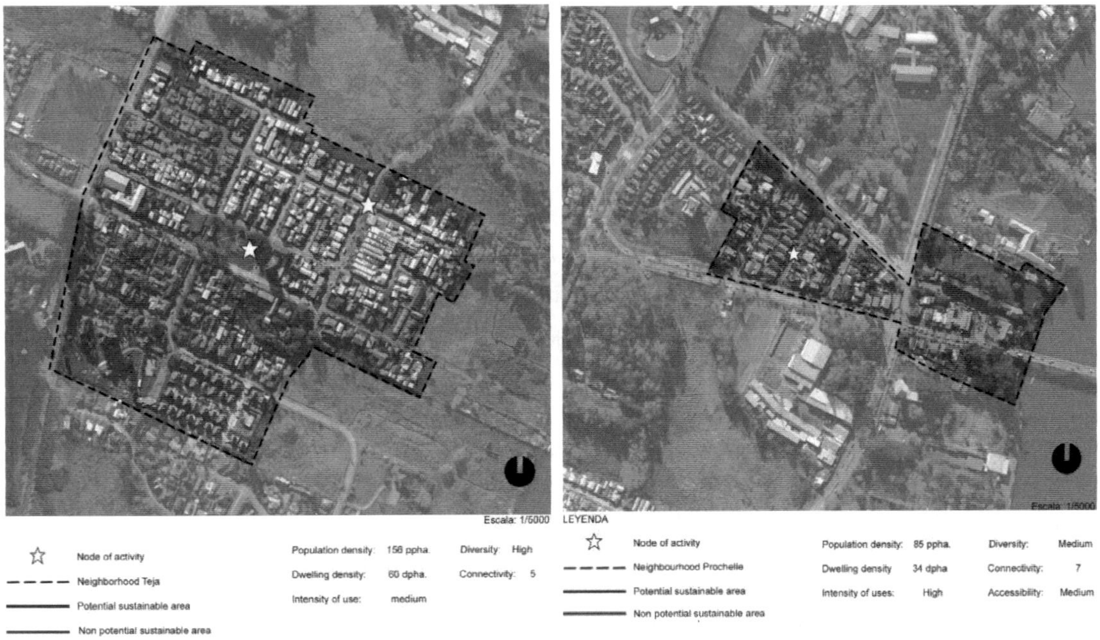

Figure 3. areas scoring high in three dimensions of sustainable urban form, in neighborhoods Teja (left) and Prochelle (right).

For instance, in Santa Ines neighborhood the node is generated due to the high topological accessibility of the street, which, in addition to be the main connector, has a preferably vehicular use intensity. Thus, this has generated the appearance of new uses such as a supermarket and a fire station, which have modified the size of the original plots and have transformed the scale of the neighborhood. Despite being the node with the highest recorded load, there are no activities associated with social interaction and collective life in the space. These constraints do not promote the generation of vibrant nodes with a greater diversity of activities. On the contrary, what is observed in the Prochelle and Teja neighborhoods is interesting in terms of the grouping of activities. In the node of Prochelle neighborhood, it is observed a high pedestrian overlay within the neighborhood. This is mainly given by the average size of their blocks, 98 meters, which are very convenient to promote pedestrian activity. These neighborhoods are those that show a higher degree of adaptability in the morphology of their nodes in order to generate diversity of uses, which in turn promotes greater sustainability of the urban form (figure 4). The presence of front yards, transformed into terraces with commercial use and services, generate a greater diversity of activities, being the place of greatest pedestrian attraction of Isla Teja. The places of high sustainable potential occur in general around nodes of activity that promote greater diversity of activities and greater adaptability in the building morphology to adapt new uses. In this places, local activities are promoted in the structure of the neighborhoods, such as, grocery stores, bakeries, greengrocers, small restaurants, local commerce, and green areas.

These nodes are generated mainly in open places with greater visibility, such as green areas or squares (figure 4). This condition allows to increase the opportunities for social interaction between people, since there is a bigger temporality of the meetings, and in turn, increases access to opportunities and resources within the units. That condition is strongly observed in Teja, Prochelle and Santa Ines neighborhoods. Besides that, the places that tend towards an "unsustainability" or low sustainability according to the analyzed dimensions are places with low diversity of socio-economic groups, and low diversity of typology and size of housing. Furthermore, they are places which tend to be mono-functional areas with low pedestrian accessibility, high horizontal density, and sizes of blocks much larger, as is the case of Teja Sur and Los Raulies neighborhoods.

Figure 4. Morphological adaptability of the urban fabric in potential node in Prochelle neighborhood (left); Teja Neighborhood square, potential node that provide high levels of pedestrian vitality (right).

7. Conclusions

In general, it is concluded that there is no single neighborhood that is more sustainable than the other in all its categories or dimensions analyzed. Rather, there are specific places within the analyzed units that show values that indicate a greater potential for sustainability given by particular situations of urban development. The places with the highest sustainable potential are mainly related to two factors: one, to units that present a higher mix of building use in their neighborhoods and a greater diversity of activities in their nodes, oriented towards open spaces and not towards streets with a high load and intensity of use. Second, the vertical mixed use and the high variation of densities in blocks continue to be the best response to promote social life and high collective activity in the space.

The five dimensions measured and evaluated are shown to promote greater sustainability in this type of context. Sustainability implies the ability to endure. In this sense, the scope of urban design is particularly broad because it includes not only the capacity of the natural environment to endure, but also the built environment: the place, its community and economy. This parallel focus, having to simultaneously consider maintenance of nature and the economy, is a source of tension when it comes to interpreting what sustainability means in practice. It has also been noted that the neighborhoods in transformation in medium-scale cities of southern Chile still have the human scale, which justifies taking measures to avoid repeating the errors of big cities. Continuing in this line, one challenge in urban morphology is to strengthen the physical-social dimension, which can serve as a strategy for sustainable planning in the 21st century. This dimension complements scalar hierarchy related to the forms of spatial organization of the city (home, block, neighborhood, district) with the forms of social-civil organization (from families to urban communities). Furthermore, this association between spatial and social organization positions the neighborhood as the basic "sustainable unit" of the city.

This research recommends that the initiatives associated with urban design in this context should be oriented, on one hand, to improve the morphological conditions of neighborhoods, understanding their potential and weaknesses in terms of vitality; and on the other, that they promote human development and the generation of necessary conditions for sustainable social and economic development. The sustainable city of the 21st century should not be conceived as a neutral machine with zero carbon emissions, but rather, as the maximum expression of complex social, economic and cultural fabrics, in which the constant interaction of these is what determines, to a large extent, the success or failure of a sustainable urban experience.

References

[1] Artmann M *et al* 2019 *J. Ecolind* **96** pp 10-22
[2] Borsdorf A and Hidalgo R 2013 *Cities* **31** pp 96-104
[3] Breheny M *et al* 1996 *The Compact City: A Sustainable Urban Form?* (London: E&FN Spon)
[4] Burton E 2002 *Environ. Plan. B- Plan. Des.* **29** b2713
[5] Cervero R 2013 *J. Transp. Land Use.* **6** pp 7-24
[6] Davidson M 2010 *City* **14** pp 390-405
[7] Dempsey N *et al* 2010 *Elements of Urban Form* (Dimensions of the Sustainable City 2) ed M Jenks and C Jones (Dordrecht: Springer) chapter 2 pp 21-52.
[8] Dempsey N *et al* 2011 *Sust. Dev.* **19** pp 289-300
[9] Ehlers E 2011 *Urban. Morph.* **15** pp 97-119.
[10] Ellin N 2006 *Integral Urbanism* (New York: Routledge)
[11] Frey H 1999 *Designing the city: Towards a More Sustainable Urban Form* (London: Spon Press)
[12] Frey H and Bagaeen S 2010 *Adapting the City* (Dimensions of the Sustainable City 2) ed M Jenks and C Jones (Dordrecht: Springer) chapter 8 pp 163-182.
[13] He S *et al* 2018 *Sust. Cities. Soc.* **44** pp 880-887
[14] Hemani S *et al* 2017 *Urban Des Int* **22** pp 168-194
[15] Inzulza-Contardo J and Gatica-Araya P 2018 *Urban Studies* **55** 0042098018787967
[16] Jabareen Y 2006 *J. Plan. Edu. Res.* **26** pp 38-52.
[17] Jackson R *et al* 2013 *Am. J. Public Health* **103** pp 1542-1544
[18] Jenks M and Dempsey N 2005 *Future forms and design for sustainable cities* (Oxford, UK: Architectural Press)
[19] Jenks M and Jones C 2010 *Dimensions of the sustainable city* (Dordrecht: Springer)
[20] Jun H and Hur M 2015 *Appl. Geogr.* **62** pp 115-124
[21] Kärrholm M 2011 *Eur. Plan. Stud.* **19** pp 97-112
[22] Karimi K 2013 *J. Urban Des.* **18** pp 168-170
[23] Kowaleski-Jones L *et al* 2018 *Popul. Health.* **6** pp 9-15
[24] Landry B 2016 *Race, gender and class: Theory and methods of analysis* (New York: Routledge)
[25] López-Morales E *et al* 2015 *Eure* **41** pp 45-67
[26] Luederit, C *et al* 2013 *Landsc. Urban Plan.* **118** pp 40-52.
[27] Lynch K 1981 *Good City Form* (Cambridge: MIT Press)
[28] Mouratidis K 2018 *Cities* **74** pp 7-20
[29] Naess P 2014 *Eur. Plan. Stud* **22** pp 1524-1543
[30] Neuman M 2005 *J. Plan. Edu. Res* **25** pp 11-26
[31] Sakamoto K *et al* 2018 Cities **81** pp 230-241
[32] Salinas, E and Pérez L 2011 *Rev.geogr.Norte Gd.* **49** pp 79-97.
[33] Säynäjoki E-S *et al* 2015 *J.Urban. Plann. Dev.* **141** 04014011
[34] Sevtsuk A *et al* 2016 *Urban. Morpho.* **20** pp 89-106
[35] Singh R 2016 *Procedia Soc. Behav. Sci.* **216** pp 643-654
[36] Song Y and Knaap G 2004 *J. Am. Plan.Assoc.* **70** pp 210-225
[37] Sternberg E 2000 *J. Am. Plan.Assoc.* **66** pp 265-278

[38] Talen E 2008 *Design for Diversity: Exploring Socially Mixed Neighborhoods* (London: Architectural Press)
[39] Talen E 2011. *Environ. Plan. B- Plan. Des* **38** pp 952-978.
[40] Talen E and Koschinsky J 2013 *Int. J. Sust. Land Use. Urban. Plan.* **1** pp 42-63.
[41] Tarbatt J 2012 *The Plot: Designing diversity in the Built Environment: a Manual for Architects and Urban Designers* (London: RIBA Publishing)
[42] UN-Habitat 2016 *World cities report 2016: urbanization and development-emerging futures* (Nairobi: UN-Habitat)
[43] UN-Habitat 2017 *Panorama multidimensional del desarrollo urbano en América Latina y El Caribe* (Santiago de Chile: Naciones Unidas CEPAL)
[44] Vaughan L 2001 *Space Syntax Observation Manual* (London: UCL)
[45] Zarin S *et al* 2015 *Procedia Soc. Behav. Sci.* **170** pp 659-668
[46] Zeng C *et al* 2018 *Sust. Cities. Soc.* **40** pp 296-306
[47] Zumelzu A 2015 *Sustainable transformation of cities: urban design pragmatics to achieve a sustainable city* (Eindhoven: Bouwstenen series)
[48] Zumelzu A and Doevendans K 2016 *Urban Des Int* **21** pp 1-18
[49] Zumelzu A 2016 *AUS* **20** pp 77-85

Acknowledgements
Our thanks to CONICYT Chile for funding this research under the project FONDECYT number 11160096, and to Universidad Austral de Chile.

The role of prosumers in supporting renewable energies sources

Sara Torabi Moghadam[1,3], Maria Valentina Di Nicoli[1], Alessia Giacomini[2], Patrizia Lombardi[1], Jacopo Toniolo[1]

[1] Interuniversity Department of Regional and Urban Studies and Planning (DIST), Politecnico di Torino, Viale Mattioli 39, 10125, Torino, Italy.
[2] Department of Energy (DENERG), Politecnico di Torino, Corso Duca degli Abruzzi 24, 10129, Torino, Italy.
[3] Author to whom any correspondence should be addressed.

sara.torabi@polito.it

Abstract. The present study is a part of an on-going Horizon 2020 project, named "SCORE" (Supporting Consumer Co-Ownership in Renewable Energies), which focuses on sustainable cities and communities developments goal. Particularly, this project aims at (1) overcoming usage of energy from fossil sources in favour of renewable sources, (2) increasing the energy efficiency and (3) reducing the energy consumption. Since the first project's task was completed, i.e. identification and description of different case studies, the next one is to propose for each of them several retrofit alternatives, in order to address the above mentioned "SCORE" purposes. In this framework, the main goal of the present study is to select and rank the relevant evaluation criteria, with the aim at building an evaluative matrix, which later makes possible to analyse the feasibility of the different case studies and choice the best retrofit alternative through a Multi-Criteria analysis (MCA). The criteria were pre-selected through a literature reviews, while the final selection took place by organizing a specific working group composed by real stakeholders. The results show how the role of the working groups (one composed by energy experts and the other by evaluation experts) were fundamental since the application of the playing card method allowed to (i) select and rank the set of relevant evaluation criteria and (ii) associate a weight for each evaluation criteria. Finally, it was decided not to aggregate the results of the two expert groups in order to show and consider the difference in stakeholders' point of view.

1. Introduction

In the current context, the implementation of the Agenda 2030 and the achievement of the 17 Sustainable Development Goals (SDGs), contributes to global development, promoting human wellbeing and protecting the environment. In this framework, the on-going Horizon 2020 project, named "SCORE" (Supporting Consumer co-Ownership in Renewable Energies), is focused mainly on the Goal 11 "Sustainable cities and communities", which makes cities and human settlements inclusive, safe, resilient and sustainable [1]. Specifically, "SCORE" aims at (1) overcoming the usage of energy coming from fossil sources by taking advantage of energy from renewable sources, (2) increasing energy efficiency of the building systems (e.g. envelope or energy system) and (3) reducing energy consumption related to building/neighbourhood users' behaviour. Consequently, local and decentralized renewable energy sources (RES) production leads to positive implications, such as

community self-sufficiency and better air quality, related to CO_2 emissions reduction. In addition, the project lends attention to the dynamics that allow or not the birth of the specific communities, analysing the role of citizens. The intention is to shift the attention from the individual to the community, and consequently, from the consumer to the prosumers. The term "prosumer" is the synthesis of two words: "producer" and "consumer" and indicates an individual strongly independent from the classical economy. In fact, a prosumer is a user with a more active role in the phases of production, distribution and consumption of energy, and for which monitoring, energy saving, and accumulation take on an ever-increasing significance. This leads to numerous prosumers of renewable energy (first of all taking advantage of photovoltaic) and the concept of "self-produced energy sharing" is spreading, directing the market towards a decentralized, more democratic and efficient model; a new way of producing and distributing energy, near to consumption places, is emerging. From this, it is clear that another purpose of "SCORE" project is encouraging the consumer to play an active role, contributing himself to the production of energy in the community. Therefore, the identification of segments of the population that are interested or would like to be part of the project, but do not have the possibility (for different reasons such as economic, social, etc.), is the core of the research. The project aims at implementing these communities in three pilot regions: Italy, Poland and Czech Republic. These pilot projects are at the core of "SCORE" as they have to demonstrate the practical feasibility of optimized joint prosumer investments with local municipalities. Once the first task of the project, concerns the identification and the characteristics description of different case studies in each pilot region, is completed, the next step regards their implementation. For each case study, several retrofit alternatives will be proposed in order to address, with different extents, the "SCORE" purposes; but, only one of them will be carried out: the most feasible alternative, evaluated under different perspectives. Accordingly, the feasibility analysis, among different Italian pilot case studies (Susa Valley, in Piedmont region), will be performed through a Multi-Criteria Analysis (MCA), which allows to support the choice of the best project through an analysis of a complex system in which different aspects (social, technical, economic, environmental and administrative) are assessed. In particular, the present study aims at identifying the evaluation criteria and determine the respective weights. This represents an important phase because it allows to analyse and compare different projects through an evaluative matrix, assessing and considering the preferences of different stakeholders' point of views.
The paper is divided as follows. Section 2 illustrates the methodological approach through which the evaluation criteria were defined and ranked. In section 3, the results will be discussed in detail. The paper lasts giving a few perspectives for the current work (section 4).

2. Methodology and material

This section demonstrates the methodology employed to create a hierarchy rank among the different selected criteria, which will analyse the pilot projects, according to a convenience and feasibility priority. In this regard, a selection process is defined and it consists of five phases. The process begins with the detection of the case study (diagnosis phase) and continues with the selection of the evaluation criteria, taking into account the stakeholders' objectives and preferences. This last part is started from the pre-selection step and ended by the final selection of a set of criteria. The selection process has been carried out in the following order as shown in figure 1.

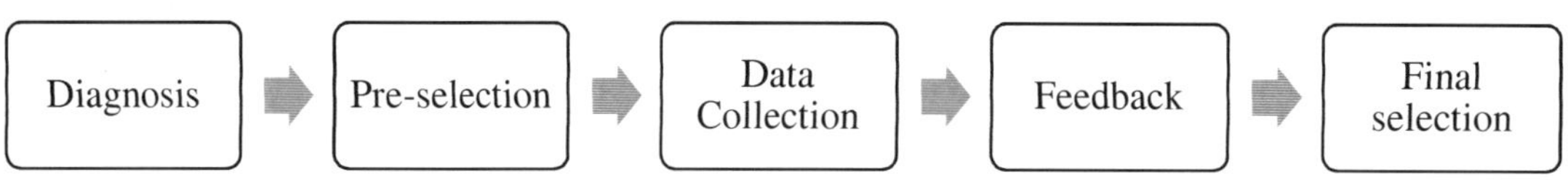

Figure 1. The steps of the selection process

As mentioned above, the main output of the present work allows to later perform the feasibility analysis of different Italian case studies. For this purpose, after having collected and elaborated all the needed information and data, the decision criteria need to be carefully identified and selected in order to rank the different retrofit alternatives.

2.1. Diagnosis

The first step concerns the identification of the different case studies pilots in an Italian selected territory: the Susa Valley. Specifically, the pilots are small municipalities in which some buildings, mostly of public property, have been identified since they are characterized by inefficient and polluting heating plants system, fuelled by diesel or natural gas. In addition, the proximity of the buildings allows their connection to a new plant system, a district heating network supplied by biomass. This project strategy leads to overcome the use of energy from fossil sources in favour of the use of wood chip, coming from the local territory in a controlled and certified perspective of sustainable development. Table 1 shows the different selected case studies indicating the existing heating system and the planned one.

Table 1: List of the Italian case studies

N°	Case study	Existing conventional energy sources for heating	Planned RES sources project
1	Oulx	Oil and natural gas boiler	District heating network (biomass)
2	Novalesa	Oil and natural gas boiler	District heating network (biomass)
3	Salbertrand	Natural gas boiler	Biomass plant
4	Bruzolo	Natural gas boiler	Biomass plant
5	Almese	Natural gas boiler	District heating network (biomass) and centralized solar thermal collectors
6	Perosa Argentina	District heating network (natural gas)	District heating network (biomass)
7	Ala di Stura	Oil boiler	District heating network (biomass)
8	Rueglio	Oil boiler	District heating network (biomass)
9	Caprie	Biomass plant	New biomass plant and solar thermal collectors
10	Villardora	Natural gas boiler	District heating network (biomass) and solar thermal collectors

2.2. Pre-selection

The selection process of the criteria begins with the pre-selection phase. The different criteria are important to identify the most feasible and sustainable project, not only from a technical and economic point of view, but also considering environmental, social and administrative aspects.

The pre-selection step is based primary on two aspects. The first aspect is an existing literature, specially, according to [2], [3], [4]; these studies showed that the efficiency, investment cost, operation and maintenance cost, NO_x emission, CO_2 emission, land use, social acceptability and job creation are the most widely used evaluation criteria in energy planning purposes. The second aspect is setting up five internal teamwork discussions with different expertise in energy engineering, multi-criteria analyst, plant system designer and socio-energetic planner. Considering both above-mentioned aspects, we have classified the criteria into five main categories, so called environmental, economic, technical, social and administrative [5]. Afterward, the first repository of criteria was built up by considering the high frequency used evaluation criteria in the literature [2] and, especially, the criteria which should be affected by energy retrofitting measures. Furthermore, the set of criteria were modified through five internal teamwork meetings and based on the specific particularities of the project (see table 2).

Table 2: Description of the considered pre-selected criteria for "SCORE" project.

	Criteria	Description	Unit
Environmental	Environmental constraints	Environmental restrictions (park or protected area) and constraints such hydrogeological, seismic, etc.	-
	Land use	Surface occupied by the plant.	m^2
	Primary energy saving	Primary energy that would be saved if the new plant was built (it is linked to the renewable nature of the investment and to the interventions on the building envelope).	$kWh_{primary\ energy}$
	Global emissions CO_2	Reduction of CO_2 emissions guaranteed by the project plant compared to the current one	kg
	Local emissions NO_X, PM_{10}	Reduction of NO_X and PM_{10} emissions guaranteed by the project plant compared to the current one.	kg
Economic	Payback period (PBP)	Time in which negative and positive cash flows are equal. It represents the moment after which the expenses are amortized and there is the actual gain.	years
	Investment cost	Investment costs related to refurbishment of the building (efficiency investment) and/or new heating system (infrastructure investment).	euro
	Public incentives	Percentage of savings linked to the share of investment cost covered by administrative incentives	%
	Savings on energy expenditure	Savings on annual expenditure	euro/year
	Economic impact - installation	Money that remains on the territory because of installation.	euro
	Economic impact – operation and maintenance	Money that remains on the territory because of operation and maintenance.	euro/year
Technical	The increase of plant system efficiency	The increase in the efficiency of the new system plant compared to the existing one.	%
	Operational difficulty of installation	Presence of physical constraints or impediments that make difficult the installation of the system. It takes into account difficulties related to the size of the components or particular work for a buried plant, etc.	-
Social	Number of users	Number of people who use the structure.	-
	Ownership	The property of the building can be public, private or mixed. Depending on the ownership, it may be more or less easy to obtain consent to proceed with the refurbishments work.	-
	Architectural impact	The visual and architectural impact of refurbishments in the existing built environment.	-
Administrative	Interest of public administration and opportunities	Level of interest project and participation for the project. Opportunities are linked to the proximity of the elections, historical situation, citizens' interest, etc.	-

2.3. Data collection

The data on the ten case study pilots (municipalities showed in table 1), in the Susa Valley, is collected using two pre-defined questionnaires provided by project partners. The first survey regards the investments identification of renewable energy sources; the second one regards the energy costs and tariffs for the actual situation, for the use of non-renewable energy sources.

The first survey is composed by five parts which allow a general description of the buildings considered for each case study, describing the current situation (geometry and energy plant system) and the design one (planned project in terms of RES and financial aspects). The first part identifies the

building characteristics (building ownership, building construction year, year of the last refurbishment, gas/energy/heat and DHW distribution system operator, average of gas/energy/heat and DHW expenses, the total number of dwellings or offices, the total official number of inhabitants/employees, number of floors, total usable area and total roof area). In the second part, the existing conventional energy sources or external supplier is investigated (type of energy sources, installed power or purchased power if the district network is present). The third part identifies the existing renewable energy sources (type of energy sources, installed power and active surface if photovoltaics (PV) and solar thermal panels are present). In the fourth part, the planned renewable energy sources are investigated (type of energy sources, installed power and active surface if PV and solar thermal panels are present). Finally, the fifth part identifies the planned structured of financial sources for the renewable energy sources investment (type of financial sources and percentage of overall costs).
The aim of the second survey is to collect information about the use of non-renewable energy sources; specifically, in this questionnaire, the average consumption fee [€/GJ] (annual consumption [GJ] and historical data (from 2013 to 2017) for oil and natural gas cost [€/GJ]) and the average fixed fee [€/month] are investigated. According to the data collected through these two surveys, it will be possible to fill in the evaluative matrix containing the criteria identified in the pre-selection phase.

2.4. *Feedback*
The final list of the criteria was established through a workshop including stakeholders and experts. In this workshop, the author played a role of analyst who aids Decision Makers (DMs) in making better their decision without expressing any personal preferences [6]. The first half day workshop was set up on 29th January 2019 at Politecnico di Torino (Italy). As said above, the purpose of the workshop was to select and rank the most important criteria to further assess the feasibility analysis of different pilot case studies. Initially, an official email has been sent to the stakeholders, introducing them the material and the structure of workshop and their role. During this first contact, the aims of the workshop were explained to each individual stakeholder. The invited stakeholders were included in two main backgrounds: the energy engineers and plant system designers and the evaluation experts.

2.5. *Final selection*
To rank and define the importance of the criteria during this research project, it was decided to use the "Playing Cards" method, which is a semi-structured participative procedure proposed by Simos [7]. The "Playing Cards" method is appropriate to support group discussions; this is one of the important reasons why this method has been selected. In fact, one of its specific aims is to stimulate the interactive and constructive discussion between different stakeholders with different backgrounds. The methodology helps the stakeholders to express the way in which they wish to rank the different criteria in a specific context. This method is very easy to use and it consists in associating a "card" with each criterion. Moreover, the stakeholders have a set of "white cards" available, the use of which depends on specific needs. The application of the procedure is very simple: first, the stakeholders should order the "cards" according to the importance of the criteria (from the less important to the most important one) and, if some criteria have the same importance, the stakeholders should build a subset of cards clipping them together; second, according to the fact that the importance of two successive criteria in the ranking can be more or less close, the stakeholders are asked to insert as many "white cards" as much is the distance between two successive criteria, according to the number meaning of white card showed in table 3, and, then, providing a final ranking of the importance; third, the final ranking of criteria is transformed into weights according to Simos' algorithms [7].

Table 3: Description of the meaning of the white cards.

Number of with cards	Importance
0	two following criteria do not have the same weights, but there is a minimal difference
1	two times the minimal difference

2	three times the minimal difference
N	n+1 times the minimal difference

3. Results and discussion

First, an official mail, in which the description and the schedule of the workshop were explicit, has been sent to nine stakeholders with different background and field of interest. Only six stakeholders actually took part in the workshop and they were divided into two main groups according to their backgrounds and interests. The first group (G1) is composed by the energy experts: an external expert in the building energy efficiency field and two academic professors from the energy department. The second group (G2) is composed majority by the evaluation experts: two academic professors who deals with economic evaluation of built environment issues from the urban planning department, and finally, one energy professor in order to guarantee to meet the project energy issues aiding the stakeholders in the specific energy aspects.

The work done by the workshop has been organized in two main steps, according to the following structure: (i) firstly, the authors played the role of moderator and the analyst provided to the two groups of stakeholders two set of colored cards (one for each group) in which were written the pre-selected criteria (described in table 2). Then the analyst asked to discuss the importance of each criteria in order to select them and define a rank, according to their opinion. (ii) Secondly, the main task was the addition of the white cards with the purpose of inserting them (if necessary) between two consecutive cards [8] (according to table 3); in this way, each group was able to define the level of importance of each ranked criterion, determining the weights of each criterion.

Finally, the two performed ranks were illustrated in a plenary session and each group of stakeholders explained the reasons of their selection. Then, according to the Simos' algorithm, the final ranking of criteria was converted into weights and the results are shown in table 4 and table 5.

Table 4: Final results coming from the Playing Cards method - energy experts (G1)

Rank	Subset of Ex-Equo	Number of Cards	Positions*	Non-normalize weights	Normalized Weights	Total**
1	Architectural impact, land use, number of users	3	1, 2, 3	2	0.97	2.91
2	White card	1	[4]	-	-	-
3	Economic impact – operation and maintenance, economic impact – installation	2	5, 6	5.5	2.67	5.34
4	Local emissions, global emissions CO_2, primary energy saving, increase of plant system efficiency	4	7, 8, 9, 10	8.5	4.13	16.50
5	Interest of public administration and opportunities, ownership	2	11, 12	11.5	5.58	11.17
6	Investment cost, public incentives, savings on energy expenditure	3	13, 14, 15	14	6.8	20.39
7	Payback period (PBP), duration of the intervention contract***	2	16, 17	16.5	8.01	16.02

8	Environmental constraints, availability of primary resource****, operational difficulty of installation	3	18, 19, 20	19	9.22	27.67
SUM		**20**	**206**			**100**

*This sum does not include the positions of the white cards (in brackets).
** The total column reports the normalized weights multiplied for the number of cards of each position.
*** This criterion has been recorded between the economic aspects by the energy experts.
**** This criterion has been recorded between the technical aspects by the energy experts.

Table 5: Final results coming from the Playing Cards method - evaluation experts (G2)

Rank	Subset of Ex-Equo	Number of Cards	Positions*	Non-Normalize weights	Normalize d Weights	Total**
1	Environmental constraints, land use	2	1, 2	1.5	0.84	1.68
2	Architectural impact	1	3	3	1.68	1.68
3	White card	2	[4, 5]	-	-	-
4	Economic impact – installation	1	6	6	3.35	3.35
5	Operational difficulty of installation	1	7	7	3.91	3.91
6	White card	1	[8]	-	-	-
7	Investment cost	1	9	9	5.03	5.03
8	Economic impact – operation and maintenance	1	10	10	5.59	5.59
9	Savings on energy expenditure	1	11	11	6.15	6.15
10	White card	2	[12, 13]	-	-	-
11	Payback period (PBP), public incentives	2	14, 15	14.5	8.10	16.20
12	White card	1	[16]	-	-	-
13	Ownership	1	17	17	9.5	9.5
14	Interest of public administration and opportunities	1	18	18	10.06	10.06
15	White card	2	[19, 20]	-	-	-
16	Primary energy saving	1	21	21	11.73	11.73
17	Global emissions CO_2, local emissions	2	22, 23	22.5	12.57	25.14
SUM		**23**	**179**			**100**

*This sum does not include the positions of the white cards (in brackets).
** The total column reports the normalized weights multiplied for the number of cards of each position.

As it is shown in tables 4 and 5, the criteria were ordered from the less important to the most important; in addition, some of the initially considered criteria (table 2) have been removed or added from the stakeholders during the discussion. In detail, the G1 have considered the process in terms of technical aspects. Indeed, they stated that if some criteria such as "environmental constraints", "operational difficulty of installation" and "availability of primary resource" were not meet the targets, the project is not able to proceed. According to this, they defined these three criteria the most important and ranked the criteria (table 4) according to their technical process importance. In addition, two main criteria, which are "availability of primary resource" (included among the technical aspects) and "duration of the intervention contract" (included among the economic ones), were added by G1 into the final rank. The second rank (table 5) demonstrates the selected criteria by the G2. In particular, this group have removed two aspects "number of users" and "increase of plant system efficiency" since they believed that these criteria were redundant; the first one with respect to the

"interest of public administration and opportunities" and the second one respect to the "primary energy saving". Instead, all the criteria related to the global and local "emissions" have been considered as fundamental ones.
Finally, there are similarities among the two rank for the less important criteria. It is notable to say that, even if the literature suggests taking into account the social criteria, the practice neglects this important evidence. Probably one of the reasons why the social aspects are partially ignored such as "architectural impact" is due to their complex and qualitative approach of assessment. Similarly, also the "land use" (in the environmental category) has been considered not essential from the stakeholders.

4. Conclusions and future developments

This study has demonstrated how the evaluation criteria have been selected and ranked through a workshop attended by experts in energy and evaluation. The selected criteria will be used later to analyze the feasibility of different Italian case study within the "SCORE" project. The two groups of experts worked in a parallel way to define the relevant criteria. For this purpose, the playing cards method has been employed; the latter allowed stakeholders to order the criteria in a very intuitive way from the least important to the most important one. Consequently, from the subsequent application of the playing cards, which applies the Simos' algorithm, it was possible to obtain the weights associated to each criterion. In this way, two different sets of ranks and weights were obtained. It was decided not to aggregate the two results in order to have two different stakeholders' point of view. Furthermore, the obtained weights are especially suited to this specific projects where the energy component is predominant and this is evident from the composition of the two groups of experts. The methodology used is robust and usable in wider and different contexts, but it is important that the interest participants of the workshops should meet the project purposes in question. One of the interesting future developments is the use of PROMETHEE method, which is an outranking Multi Criteria Analysis (MCA) in order to rank a feasibility analysis of ten case studies pilots, taking into account the criteria weights defined by this study. This will allow defining the most convenient pilot project in a sustainable perspective.

Acknowledgments

This research project has been financially supported by the H2020 project, entitled Supporting Consumer Co-Ownership in Renewable Energies – "SCORE". Grant Agreement number: 784960-SCORE-H2020-EE-2016-2017/H2020-EE-2017-CSA-PPI.

References

[1] https://sustainabledevelopment.un.org/sdg11. Last access: April 2019
[2] J. J. Wang, Y. Y. Jing, C. F. Zhang, and J. H. Zhao, "Review on multi-criteria decision analysis aid in sustainable energy decision-making," *Renewable and Sustainable Energy Reviews*, vol. 13, no. 9. pp. 2263–2278, 2009.
[3] E. Strantzali and K. Aravossis, "Decision making in renewable energy investments: A review," *Renewable and Sustainable Energy Reviews*, vol. 55, pp. 885–898, 2016.
[4] P. Lombardi, F. Abastante, and S. T. Moghadam, "Multicriteria Spatial Decision Support Systems for Future Urban Energy Retrofitting Scenarios," 2017.
[5] T. Kaya and C. Kahraman, "Multicriteria renewable energy planning using an integrated fuzzy VIKOR & AHP methodology: The case of Istanbul," *Energy*, vol. 35, no. 6, pp. 2517–2527, 2010.
[6] E. Løken, "Multi-Criteria Planning of Local Energy Systems with Multiple Energy Carriers," 2007.
[7] J. Simos, "L'evaluation environnementale: Un processus cognitif negocie. These de doctorat," 1990.
[8] J. Figueira and B. Roy, "Determining the weights of criteria in the ELECTRE type methods with a revised Simos' procedure," *European Journal of Operational Research*, vol. 139, no. 2, pp. 317–326, 2002.

IOP Conf. Series: Earth and Environmental Science **297** (2019) 012042 doi:10.1088/1755-1315/297/1/012042

HOUSING: Dense, high-rise buildings with green, low carbon solutions. Case study: Santiago, Chile.

E Pollock[1], S, Romano[2], J Kareco[2]
[1]Lecturer, ConREM, Metropolia UAS, Helsinki, Finland
[2] Master students, ConREM, Metropolia UAS, Helsinki, Finland

E-mail: eric.pollock@metropolia.fi, scarlet.arq@gmail.com

Abstract. Considering the international relevance of sustainability, Chile has implemented several regulations in the housing area during the last years through the Ministry of Housing and Urbanism. The goal of these regulations is to improve the performance of Chilean cities regarding the three pillars of sustainability (social, economic and environmental). Santiago, as the capital of Chile and the focus of this research, has defined different methods to apply these regulations. However, this city still has challenges to implement them, such as environmental pollution, social segregation and the lack of economic and natural sources. There are different alternatives to overcome the challenges to implement the new sustainable housing guidelines in Santiago, one of them is to rethink the current density, high and sustainable solutions that the buildings have. The main objective of this paper is to describe and analyse the situation of the housing in Santiago considering the density, high-rise buildings with green guidelines and low carbon solutions as analysis parameters. The methodology established to achieve this objective is exploratory-descriptive with the literature review technique. The partial results illustrate that despite the benefits that represent the incorporation of new guidelines related to the increase in density and high-rise and low carbon solutions in housing projects of Santiago, it is important to consider that not all of these are applicable to this city. Each territory has its features (weather, culture, resources and norms), which determine the requirements that the projects need and the real possibilities to implement changes.

Keywords: high-rise building, housing project, low carbon solution, green guidelines, Santiago - Chile.

1. Introduction

Sustainability is a significant concept in the Architecture, Engineering, and Construction (AEC) industry because it defines the guidelines that this business implements in projects. The Brundtland Commission defined sustainability as a "*development that meets the needs of the present without compromising the ability of future generations to meet their own needs*" [1]. This definition was complemented by the Agenda for Development established by the United Nations (UN) in 1997 with the description of three pillars of this concept: social development, economic development and environmental protection.

Internationally, there are two facts which illustrate the relevance of sustainability. First, the agreements established to define the sustainable goals for the future of the world. For example, the 17 Sustainable Development Goals (SDGs) of the UN [2] and the Paris Agreement signed in 2015 [3]. Second, the new green certifications for building and materials. For instance, Energy Start, Leadership in Energy and Environmental Design (LEED), and Green Globes [4].

Considering the international relevance of sustainability and the guidelines and certifications related to this topic, Chile has implemented three regulations in the housing area during the last years through the Ministry of Housing and Urbanism: The Sustainable Construction Code, The National Sustainable Construction Strategy and The Energy Rating of Housing [5]. The objective of these regulations is to improve the performance of Chilean cities regarding three pillars of sustainability. Santiago, as the capital of Chile, has defined different methods to apply these regulations. However, this city still has challenges to implement them [6]. For example, environmental pollution, social segregation and the lack of economic and natural sources.

There are different alternatives to overcome the challenges to implement the new sustainable housing rules in Santiago. One of them is to rethink the current density and the high and sustainable solutions that the buildings have. The main objective of this work "is to describe and analyses the situation of the housing in Santiago considering the density, high-rise buildings with green guidelines and low carbon solutions as analysis parameters". This research is descriptive exploratory, and the methodology involves a bibliographic review.

The justification of this work is based on three reasons. First, exploring different guidelines possible to apply to Santiago as a possible alternative to improve the current situation of this city. Second, learning about two successful examples of sustainable high-rise housing. And third, studying different strategies to achieve at least 4 of the 17 SDGs of the United Nations (UN) [2], Goal 7: Affordable and clean energy, Goal 9: Industry, innovation and infrastructure, Goal 10: Reduce inequality and Goal 11: Sustainable cities and communities.

This research is organized into 5 sections. First, the introduction. Second, the methodology. Third, the description of the housing topic related to density, high-rise buildings with green guidelines and low carbon solutions and the description of two international successful examples. Fourth, the analysis of Santiago. Fifth, the exploration of possible guidelines to apply to Santiago. And sixth, the conclusion.

2. Methodology

In addition to the main objectives of this research, this work has four specific objectives:

1. To define the key concepts of this research: density, high-rise buildings with green guidelines and low carbon solutions.
2. To describe two successful examples of dense, high-rise buildings with green, low carbon solutions for housing considering the features of the city of Santiago.
3. To describe the case study: Santiago.
4. To explore new guidelines, from the case studies analysed, possible to apply in Santiago to improve the current situation of housing.

This research is exploratory-descriptive and uses the method of a literature review, which is focused on analysing information from secondary sources connected to the themes of this research, such as papers of relevant authors; books with peer editorial review; and reports from architecture and construction companies associated with the case studies.

The three criteria to define the two successful examples analysed in this paper were the typology of the project (residential), the climate condition (similar to Santiago) and the population (30% more or less of the inhabitants of Santiago).

3. Housing: Dense, high-rise buildings with green, low carbon solutions

3.1. Definition of concepts:

High rise buildings

Farouk defines high rise buildings based on the criteria of height as buildings which fit at least one of two criteria: height over 36m or height over 12 floors [7]. Another factor is the structural perspective, where a high-rise building can be defined as 'the building that its height will be affected by lateral forces resulting from earthquakes & wind forces to the extent that such forces will play a major role in the process of design' [7]. The uses of high-rise building vary between residential, commercial, administrative, office buildings, hotels and mixed-use buildings.

Green or sustainable low carbon buildings
A building can be named sustainable when it has the scarcest troublesome impact on the environment [8]. The sustainability can be represented through three perspectives environmental, social and financial responsibility, on the other hand, services or products that have less impact on environment and health can be named green. In the case of buildings that have fewer carbon emissions, they are called low carbon buildings [9].

The attractive indoor environment of buildings, higher quality of life, renewable energy use, and reuse and recycling of materials are substantial benefits of the high rise, dense, green and low carbon buildings. Considering these advantages, these typologies of buildings are relevant for the future environment, because of the limited natural resources and the necessity of new methods and techniques in construction project management in order to minimize the effects of construction.

3.2. Successful international buildings
Two international projects were chosen to be analyzed considering the criteria explained in the methodology.

One Central Park is located in Sydney, Australia. This project has three main similarities to Santiago. First, the monthly average temperature which varies in Sydney between 13°C and 23°C and in Santiago between 9°C and 21°C. Second, the average humidity which in Sydney is 71% whereas in Santiago it is 70%. Finally, the population of Sydney in 2017 was about 5.07 million [10], while the population of Santiago in 2016 was about 6.5 million [11].

One Central Park is a mixed-use type of building. It integrates residential, commercial, cultural, institutional and entertainment applications and it provides four distinguishing sustainable features. A tri-generation energy plant which is run by natural gas and provides space heating and cooling energy by producing low-carbon thermal energy. This tri-generation energy plant reduces the greenhouse gas emissions by190,000 tons over 25 years [12]. An on-site water recycling plant creates the opportunity for inhabitants to use 40-50 per cent less drinking water leading to saving money and drinking water supplies. The facility has hanging gardens that incorporate 23 green walls including over 35,000 green wall plants with 350 different species, occupying 1,200 sqm area of the façade. These vertical gardens give a unique view to the building and they purify the air [13]. In addition, the cantilevered heliostat made by a series of motorized mirrors reflects the daylight to the cantilever. There are 220 reflecting panels fixed in the cantilever and they bounce the light back to below spaces [13] (figure 1).

This project has received numerous awards that have for its structural ingenuity and sustainability measures, for example, Winner of the High-Density Development by 2015 UDIA National Awards and Winner of the Best Innovative Green Building (global) from MIPIM Awards 2015 [12].

One Central Park, through the use of different sustainable systems, provides advantages by producing energy through low carbon emissions, lowering water consumption by water recycling and maximizing natural daylight through a reflecting system. Nevertheless, it has disadvantages, as for example the high accuracy and order regarding plant pests and diseases and the expensive maintenance cost for the vertical garden.

The Solaire is located in New York, USA. The project has two main features similar to Santiago. First, the monthly average temperature which displays values in New York between 1°C and 25°C and in Santiago between 9°C and 21°C. Second, the population, which in New York in 2017 was about 8.6 million [14], while the population of Santiago in 2016 was about 6.5 million [11].

Figure 1: One Central Park building in Sydney [25]

Figure 2: Solar cells on The Solaire façade [15].

This project is a 250-unit apartment building and it implemented sustainability guidelines through four different methods. First, resident education about the benefits and daily components of sustainability, and behavior reinforcing communications and operations protocols. Energy consumption by residents is 24% less than the New York State code [15]. Second, onsite wastewater treatment, storage and reuse system located in the building's basement. After a process of recycling and removing color and odor, the water is placed in the storage tanks. The treated water is used as flush water, make-up water for the cooling towers, and for irrigation. Third the cooling and heating system, implemented in all the apartments includes programmable thermostats that efficiently control four-pipe fan coil units providing heating and cooling all year. High-efficiency, variable-speed pumps, fans, and motors circulate ventilation air and water throughout the building, and the high-efficiency absorption chillers run on natural gas instead of electricity [15]. Finally, solar cells made by panels that become part of the façade structure, replacing the use of brick or glass. The material used is recycled silicon. The solar electric systems are sized to meet 5% of the base building's non-residential electric load [15] (figure 2).

This Solaire provides advantages such as; well-insulated and sealed building envelope, centralized air filtration and seasonal humidity adjustment, integrated photovoltaic panels, on-site water treatment and 85% of construction waste recycled. However, it has disadvantages as higher initial and maintenance costs.

4. Case study: Santiago

The metropolitan area of Santiago is configurated by 32 municipalities and accommodates approximately 35% of the population of the entire country (INE, 2012). It is the biggest city by population (6.5 million inhabitants), density (393 ha/k2) and provides up to 50% of Chile's total Gross Domestic Product (GDP) of the country (figure 3 and 4).

From an environmental point of view, Santiago is one of Latin America's most polluted cities due to thermal inversion [16]. Considering this situation, the government of Chile has generated different alternatives to overcome this situation, for example, the creation of Chile's National Environmental Commission (CONAMA), controlled agriculture burning, eliminating lead gasoline, prohibiting open fires, the introduction of tradable emissions permits, the introduction of a new transport system.

Figure 3: Santiago, the capital of Chile [26].

The municipalities of Santiago are aligned towards high rise buildings. This is due to the funding provided by the private sector. The north-east part of Santiago is an example of high-rise buildings, with high population and green areas density, which provide social and ecological value for their inhabitants. The parking system of the city is based on underground parking areas, limiting the above area necessary to provide this service [17]. These standards are not achieved in the other municipalities of Santiago. The lack of required green areas density and housing standards provided in these municipalities produce limited high-rise buildings which also do not meet the sustainable objectives.

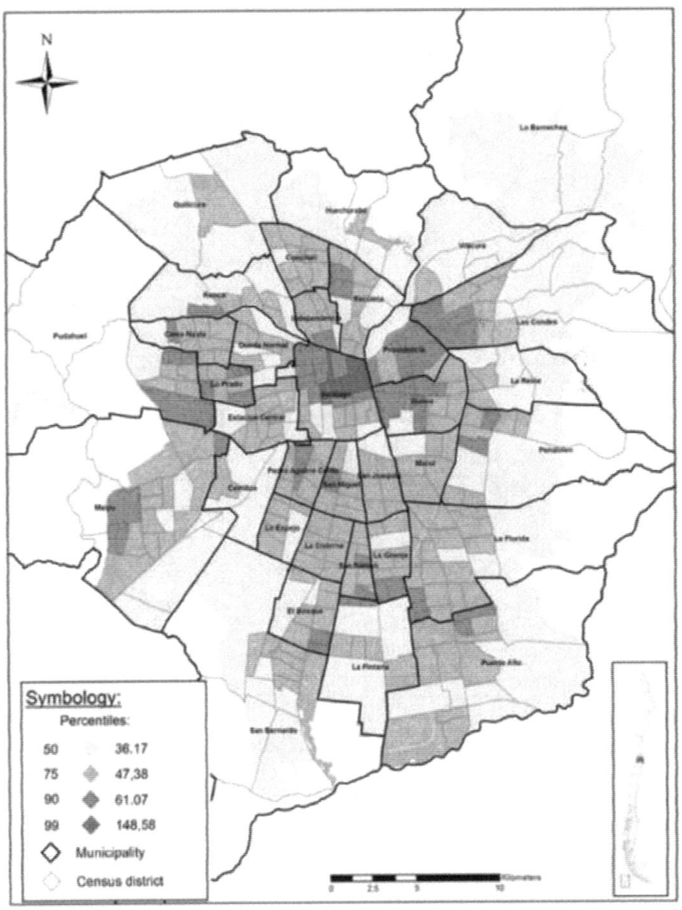

Figure 4: Dwelling density at the census district level, 2011 [26].

The social and economic polarization of this city affects the urban landscape and the density distribution. The standard of housing such as dimension, localization, type of calefaction and quality of the design depends on the social and economic status of the inhabitants. The infrastructure is also unevenly distributed in the different areas of the city. For example, the green area for a person is also affected varying from 9 square meters per person in the north-east area (rich sectors) to below less than 3 square meters per person in other areas (poor sectors) [18] (figure 5, 6 and 7).

The distribution of housing in the peri-urban areas of Santiago indicates the low density of middle-class residences in these areas. The lack of middle-class inhabitants contributes to accentuate the social differences in these areas [19]. Some municipalities allow the construction of high-rise building and others prioritizing low density. The criteria to define the density and the height of the buildings depends on the goal of the municipalities and the number of resources that they have because these topics are related to available facilities that the territory has.

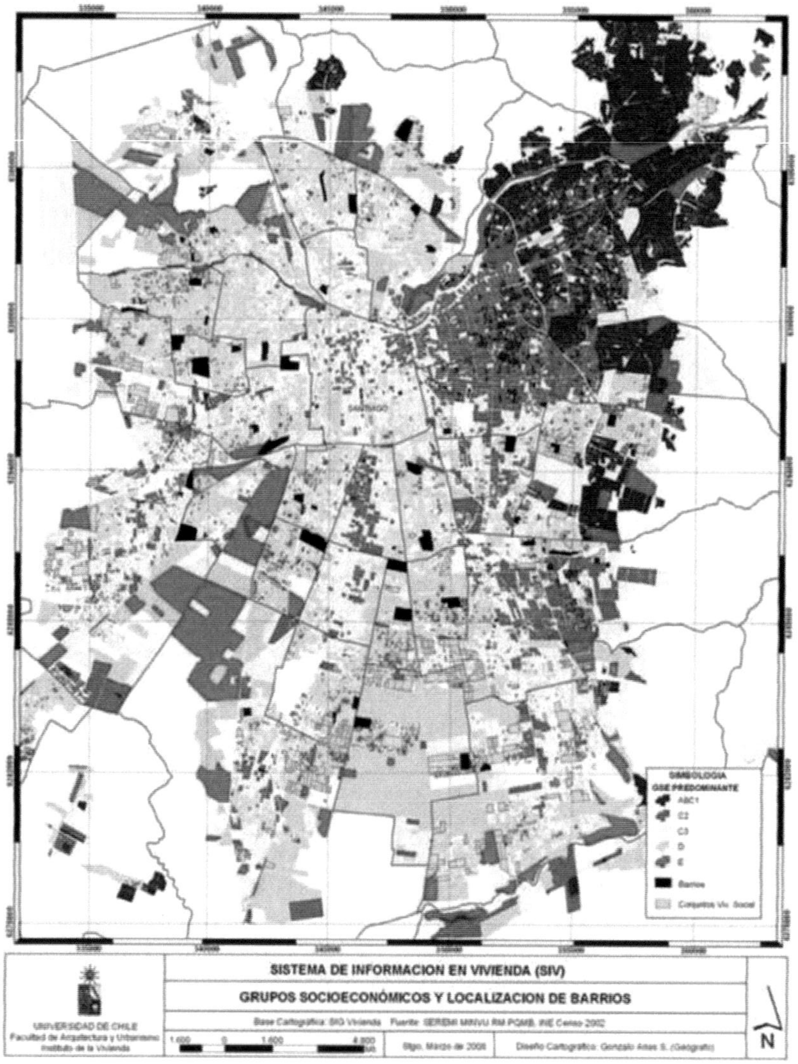

Figure 5: Socio-economic groups and location of neighborhoods. Blue and green areas are upper classes, red areas lower class. Green is park areas [17].

Figure 6: High-income neighborhood in Santiago [26].

Figure 7: Low-income neighborhood in Santiago [27].

The concepts of high-rise, green guidelines and low carbons emission were and are considered in the develop of housing projects in Santiago, for example the project Exequiel González Cortés Housing complex: built in the 1960s by the government is considered relevant because it increased the density of the housing projects considering the accessibility to facilities and green areas.

Nowadays, projects, as for example Amanda which is a 5-story residential building with the goal to improve the performance of the building and reduce the negative impacts in the environment implemented green guidelines and low carbon solutions thought the super insulation, excellent environmental quality and clean climatization [20].

Despite the existence of good examples of a high-rise building with green guideless and low carbons emissions, there is a new trend of construction in Santiago which is defined as Vertical ghettos. The main features of these buildings are the lack of green and common areas, the limited natural sunlight and fresh air, high energy consumption product of the limited access to natural sources (for example, light and ventilation), restricted entrances and exits considering the number of residents, and the poor acoustic conditions (figure 8).

Figure 8: Non-Green Buildings in Santiago [28].

In summary, the inequality of distribution of the population based on social and economic status is a relevant issue in Santiago. The development of the dense areas populated by inhabitants with middle to low incomes in high-density areas, which provide better infrastructure, services and green areas for their population is a challenge faced to achieve sustainable goals for the city. High standard buildings and infrastructure to the less wealthy areas of Great Santiago can only be provided through legislation and government initiatives. This would raise the well-being state of the city but is connected to political actions.

Based on these existing issues, the government of Chile has provided guidelines defined by the National Policy of Urban Development [21]. The focus of these guidelines is shifting towards providing residential buildings to citizens with all the required necessities. The goal is to implement a strategy that will contribute to produce a cohesive housing situation for all the inhabitants of Santiago and to achieve the sustainable housing objectives.

Table 1: SWOT analysis. Source: Own elaboration.

Strengths	Weaknesses	Opportunities	Threats
Economic and development centre for the country.	Social and economic polarization.	Build denser urban areas.	Construction propensity in peri-urban areas.
Historical, cultural and architectural values of existing buildings.	Distribution of Green areas in sectors.	Provide better infrastructure and services.	Private sector construction tendency.
Existing areas with high rise buildings with social and ecological value.	Standard of housing in different sectors.	Existing National Policy of Urban Development.	
Existing parking areas below the street level.	Lack of infrastructure in different sectors.	Neighbourhood unit concept.	
	Low middle-class population in peri-urban areas.		

5. Analysis: New guidelines to Santiago

Considering the international cases analyzed and the current situation of Santiago, this research defined two typologies of guidelines possible to apply to this city. The first typology considers the design of the housing projects from an architecture and engineering perspective and the second, establishes relevant urban policies.

5.1. Design (architecture and engineering)

New technologies Followings are significant and appropriate technologies for applying in a high-rise building in Santiago that create sustainability:

On-site tri-generation thermal system: Tri-generation technology produces clean power and energy for buildings. This technology works with natural gas. It is compatible in Santiago because of the existing of natural gas resources. If the tri-generation system is installed on site, it will supply thermal and electrical energy for residents; therefore, the electrical energy will not transfer from hundreds of kilometers away. Tri-generation technology is environmentally friendly because it reduces greenhouse gases and increases air quality.

On-site water recycling plant: According to the weather and climate information of Santiago, the water recycling system is suitable to use, because of the considerable amount of humidity. The on-site water recycling plant is a significant environmentally sustainable technology. It provides the opportunity to use rainwater, groundwater, sewage water and irrigation water in a more effective way. In this way, it grants a solution for the lack of water resources in these days and also avoided a potential critical crisis in the future.

Solar panels: The weather and daylight statistics of Santiago justify the use of solar panels. These panels can be installed on the roof or on the facade of the building. Moreover, solar panels can also provide shade and shelter from the rain.

Rooftop Gardens: They increase the green area of Santiago. This technology reduces heat effect and water runoff. It also helps thermal insulation in the building and gives some social benefits for residents. Energy efficiency: In this part, for increasing energy efficiency different alternative methods can be used. These methods reduce non-renewable resources usage and provide more efficient and environmental solutions. The use of solar and gas energy is used to provide electricity and also improve the efficiency of the HVAC system by using less energy. Using waste heat through different systems to achieve heating and cooling efficiency. Moreover, using the appropriate sensors in different spaces to reduce the amount of energy by inhabitants.

Design Specification: Achieving sustainable goals is an international practice that has been followed by different states. The states have developed different sustainable guidelines for construction purposes. The implementation of guidelines such as LEED, BREEAM, iiSBE, Green Globes, and CASBEE have achieved ranged success in sustainable goals. These guidelines can be used as the main base to create the national specification for Chile.

Building life-cycle consideration: Considering the life cycle of the building to reduce the perception of the big initial investment of money necessary to implement new guidelines and the reduction of the energy during the complete life of the building. One of the main problems with the life cycle cost of buildings is the emphasizes giving to the planning and construction process over the operation and maintenance process. The maintenance cost is at least three times more than the initial cost (design and planning) [23].

Utilization of specific international standards for life cycle assessment like ISO 14040 and ISO 14044 offers an opportunity to establish life cycle cost as a requirement for construction projects.

Mixed-use building design: Implementing high rise mixed-use buildings as areas which provide not only residential spaces for the residents, but also commercial, cultural, and entertainment spaces. This enhances social interaction and provides a better social environment.

5.2. Urban policies

People education: Sustainability is a concept that is reaching a widespread in the last few years. As a relatively new idea, it should be incorporated in the Educational system as a social concept. Students can understand it and later implement in society. Also, different educational programs within private companies and public institutions can provide lectures to develop the knowledge of the people.

New policies implementation: Social awareness is a key factor in understanding the benefits of achieving sustainable goals. The new guidelines can be realized, only if people have a complete understanding of the manner how the implementation of sustainable policies will improve their lives. The government should conceive extensive advertising campaigns to provide the population with an understanding of the impact of sustainability.

Private sector incentives: While the government produces conceives policies related to construction requirements, the partnership with construction companies is essential to implementing them. Construction companies, as private businesses are income oriented. A partnership can be reached if different incentives are given for achieving sustainable goals. These incentives can be in the form of beneficial loans, tax discounts or subsidies for implementing sustainable technologies.

Professionals knowledge: The development of new technologies provides tools to achieve sustainable goals. However, their usage is confined to the possibilities and abilities to implement them. The process of implementing and maintaining these new technologies over the life cycle of the buildings can be feasible, only if there are specialists, capable of implementing the new technological systems and are capable of maintaining them.

Society integration through the design of the cities: The main problem in Santiago that affects its capabilities of achieving sustainable goals is segregation. This problem can be challenged, through striving towards integration and social cooperation. High rise buildings can provide social apartments. This helps integrate people with different social and economic capabilities or background with each other. Also offering more common spaces in buildings for residents.

6. Conclusion

During the last years, AEC industry has included several sustainable guidelines in housings projects. Nowadays, these new strategies have improved the design of the housings, their sustainability performance and the quality of life of the inhabitants. The conclusions are structured according to the four specific objectives defined in the methodology.

First objective: *"To define the key concepts of this research: density, high-rise buildings with green guidelines and low carbon solutions"*. Nowadays, there are several innovations and new material which allow the construction of high-rise and dense buildings decreasing their cost. These new alternatives represent a real possibility to re-think about the design of buildings, specifically housing projects because there are more opportunities to plan and develop cities. Another important parameter is the current advance of green solutions for buildings which not only contribute to reducing the environmental impacts but also to improve the health of the inhabitants and limit of economic expenses during the life circle of the buildings. In summary to discuss and define the appropriate density, high-rise and low carbon solutions for the housing projects of a city is crucial to analyse the current performance of the buildings and the quality of life of the people from this city and the sources available to make changes.

Second objective: *"To describe successful examples of dense, high-rise buildings with green, low carbon solutions for housing considering the features of the city of Santiago"*. Considering the examples analysed, the possible guidelines to apply to Santiago are diverse. These contemplate technological innovations, vertical gardens, mix-uses, energy efficiency according to the local characteristics and bioclimate design. Nevertheless, all these projects show disadvantages such as difficult and expensive maintenance and investment, errors in the design and execution and ignorance of the inhabitants on how to use the new technologies. These weaknesses are important to overcome if the goal is to reply to the guidelines of these buildings to other housings complex. For example, it is required to improve the knowledge of the professional and inhabitants in these areas to avoid mistakes during the design and maintenance, incentivize the private sector to invest in housing projects, strengthen the local norm and regulation and incorporate the sustainability guidelines and behaviours as part of the daily life.

Third objective: *"To describe the social, economic and environmental areas of the Santiago"*. The city chosen has a strong economy and a variety of landscapes and natural resources and several sustainable construction laws and urban regulations, however, the distribution of wealth is unequal, a situation that is reflected in housing projects. For this reason, this city is considered a suitable case of study because the current situation of the guidelines to build and design housings can be improved through the analysis of the performance of other residential buildings.

Fourth objective: *"To explore new guidelines, from the case studies analysed, possible to apply in Santiago to improve the current situation of housing"*. The guidelines (architecture and engineering) examined contribute to understanding the existing variety of possibilities to design better housing complexes and also the necessity to discuss the policies that support these guidelines. The political, social, an economic context of Santiago determines the guidelines possible to apply, since the current law allows for new technologies and architectural strategies; the perception and the education of the inhabitant contribute to increasing the demand for environmentally friendly housing projects, and the resources of the city define the investment that it can do.

These four specific objectives contributed to achieving the general objective of this research: *"To describe and analyse the situation of the housing in Santiago considering the density, high-rise buildings with green guidelines and low carbon solutions as analysis parameters"*.

In summary, despite the benefits from the incorporation of new guidelines for increasing density and high-rise and low carbon solutions in housing projects of Santiago, it is important to consider that not all of these are applicable to this city. Each territory has its features (weather, culture, resources and norms), which determine the requirements that the projects need and the real possibilities to implement changes.

Considering the conclusion, the future research topics are the methods to implement and evaluate the new guidelines proposed to Santiago because it is relevant to study the feasibility to modify the current situation of this city and the availability of resources to do it. Without these studies, these guidelines only contribute to generating an academic discussion but not to change and overcome the housings deficiencies of Santiago in a practical manner. Another topic to deepen is the effects that the new regulations of Santiago such as The Sustainable Construction Code, The National Sustainable Construction Strategy and The Energy Rating of Housing, can produce improvement on the configuration of the city and the quality of housing.

Acknowledgments
The authors wish to acknowledge encouragement from our editor Eric Pollock, and our classmates Ali Amiri, Babak Shahmehr, Maan Singh, Hatef Hajian, and Gbolahan Ola for their cooperation in the development of this paper which is based on the book *Sustainable Cities*, Metropolia Publications, developed for the Master´s Programme in Construction and Real Estate Management (ConREM).

References
[1] Environment, UWCO, & Development 1987 Our common future pp 89 doi:10.4324/9781912281220
[2] Clark H and Wu H 2016 The sustainable development goals: 17 goals to transform our world Furthering the Work of the United Nations 36–54 doi:10.18356/69725e5a-en
[3] UN n.d. Paris climate agreement (Access Science) Retrieved October 20, 2017, from http://unfccc.int/paris_agreement/items/9485.php doi:10.1036/1097-8542.br0612171
[4] Principles n.d. Green Building Certification Systems. Retrieved October 12, 2017, from https://www.wbdg.org/resources/green-building-standards-and-certification-systems doi:10.11129/detail.9783955531683.8
[5] Ministerio de Vivienda y Urbanismo n.d. Construcción Sustentable. Retrieved January 20, 2018, from http://csustentable.minvu.gob.cl/
[6] Santiago municipality 2015. Municipio de Santiago camino a convertirse en una comuna sustentable, Retrieved January 20, 2018, from http://www.munistgo.cl/municipio-de-santiago-camino-a-convertirse-en-una-comuna-sustentable/

[7] Farouk A 2011 High Rise Buildings and How They Affect Countries Progression pp. 2
[8] Du Plessis C 2007 A strategic framework for sustainable construction in developing countries *Cons. Man. and Ec.* *25*(1) 67–76 doi:10.1080/01446190600601313
[9] Wong J K W and Zhou J 2015 Enhancing environmental sustainability over building life cycles through green BIM: *a review Autom. in Cons.* 57 156–65 doi:10.1016/j.autcon.2015.06.003
[10] Deacon M. 2017 Forecasting the future of NSW: inner city population rises as Sydney builds upwards, The Population experts, Jan. 27, 2018 https://blog.id.com.au/tag/sydney-population-2017/
[11] INE, 2017 Retrieved January 20, 2018, https://www.ine.cl/estadisticas/demograficas-y-vitales
[12] Central Park Sydney 2013 Retrieved January 20, 2018, from Central Park: https://www.centralparksydney.com/
[13] Nouvel J, Beissel B 2014. Case Study: One Central Park, Sydney. *CTBUH Journal* 4, 12-18 Retrieved February 09, 2019, from http://global.ctbuh.org/resources/papers/download/1836-case-study-one-central-park-sydney.pdf
[14] World Population Review 2019, February 09, 2019, http://worldpopulationreview.com/
[15] Epstein K 2008 NYC's Living Lesson. High Performing Buildings, pp. 56-65.
[16] Garcia-Chevesich P, Alvarado S, Neary D, Valdes R., Valdes J, Aguirre J, Olivares C. 2014 Respiratory disease and particulate air pollution in Santiago Chile: Contribution of erosion particles from fine sediments. *Environmental Pollution*, 187, 202–205. doi:10.1016/j.envpol.2013.12.028
[17] Ocampo R, Rooms J, Barroso V, Waldron B, González J, 2009 *Bicentenario: oportunidad de repensar las politicas urbano-habitacionales en chile.* Revista INVI, **24**(67). doi:10.4067/s0718-83582009000300002
[18] Weiland U, Kindler A, Banzhaf E, Ebert A, Reyes-Paecke S. 2011 *Indicators for sustainable land use management in Santiago de Chile. Ecological Indicators*, **11**(5), 1074–1083. doi:10.1016/j.ecolind.2010.12.007
[19] Cáceres C 2014 Exploring the peri-urban residential areas in Santiago de Chile: The asymmetric residential model between Corporate and State residential areas in Lampa 2000-2010. Hafen City Universität Hamburg. Thesis. Retrieved February 09, 2019, from http://edoc.sub.uni-hamburg.de/hcu/volltexte/2015/177/pdf/Exploring_the_periurban_residential_areas_in_Santiago_de_Chile.The_asymmetric_residential_model_between_Corporate_and_State_residential_areas_in_Lampa_2000_2010.pdf
[20] URBES 2016 Retrieved February 09, 2019, from https://urbes.cl/proyecto/dona-amanda-nunoa/
[21] Ministerio de Vivienda y Urbanismo 2014 Politica nacional de desarrollo urbano: Ciudades Sustentables y Calidad de Vida Retrieved February 09, 2019, from http://cndu.gob.cl/wp-content/uploads/2014/10/L4-Politica-Nacional-Urbana.pdf
[22] Fuller S 2016 Life-Cycle Cost Analysis (LCCA). National Institute of Standards and Technology (NIST) Retrieved February 09, 2019, from https://www.wbdg.org/resources/life-cycle-cost-analysis-lcca
[23] Atelier Jean Nouvel n.d. February 09, 2019, http://www.jeannouvel.com/en/projects/one-central-park/
[24] Sportstour n.d. Retrieved February 09, 2019, from http://www.sportstour.cl/destinations-chile/santiago-surroundings/
[25] Aquino F and Gainza X 2014 *Understanding density in an uneven city, Santiago de Chile: implications for social and environmental sustainability*. Sustainability **6**(9) 5876–97 doi:10.3390/su6095876
[26] Viajeros n.d Retrieved February 09, 2019, from http://www.viajeros.com/destinos/santiago-de-chile/3-que-hacer/barrio-paris-londres
[27] Labbé G 2017 ¿Por qué en mi barrio no? Las "Zonas Rojas" como forma de retracción institucional en Santiago de Chile. February 09, 2019, from https://invi.uchilefau.cl/por-que-en-mi-barrio-no-las-zonas-rojas-como-forma-de-retraccion-institucional-en-santiago-de-chile/
[28] Sustainable Built Enviromental Journal 2017 Retrieved February 09, 2019, from https://sbejournal.wordpress.com/2017/04/11/compact-cities-a-vertical-ghetto/#jp-carousel-780

Interaction of obligatory regulations and voluntary ecolabels for climate change mitigation in the Finnish building sector

R Holopainen[1], **M Airaksinen**[2], **K Bergbom**[1], **H Vaarala**[1]

[1]Ecolabelling Finland, Urho Kekkosen katu 4-6 E, Helsinki, 00100, Finland
[2]Finnish Association of Civil Engineers RIL, Lapinlahdenkatu 1 B, Helsinki, 00180, Finland

riikka.holopainen@ecolabel.fi

Abstract. This paper compares the existing Nordic Swan Ecolabel criteria for new buildings with the National Building Code of Finland. The added value of the ecolabelling is presented and the role of the ecolabelling as a test bed for upcoming building regulations is introduced.

1. Introduction

Three types of voluntary environmental labels have been identified by the International Organization for Standardisation (ISO). Type III labels are voluntary programmes providing quantified environmental data of a product, e.g. the Environmental Product Declarations (EPDs). Type II label is based on self-declaration of environmental predication, not certified by an impartial third party. Type I is a voluntary, multiple-criteria based, third party programme. It awards a license that authorises the use of the environmental label on a product or service that is proven environmentally preferable within a specific product or service category [1, 2].

One example of Type I labels is the official ecolabel of the Nordic countries, the Nordic Swan Ecolabel, which was established in 1989 by the Nordic Council of Ministers as a voluntary ecolabelling scheme for Denmark, Finland, Iceland, Norway and Sweden. Currently the Nordic Swan Ecolabel has over 60 different product groups with more than 200 different product types and more than 25.000 different products.

The criteria development process (Fig. 1) of the Nordic Swan Ecolabel is based on preliminary studies highlighting areas where the need for requirements is greatest. Development of new criteria and revision of existing criteria takes place in close contact with businesses and other stakeholders with relevant knowledge, using available literature from academic institutions and experts. The final proposal is passed by the Nordic Ecolabelling Boards. The final proposal is then uploaded to both Nordic and national websites, together with background documents and comments from the public consultation.

Figure 1. Nordic Swan criteria development process.

2. Nordic Swan criteria for new buildings

Nordic Swan includes criteria for both new and renovated buildings. The criteria have requirements for the entire building process, from cutting of trees to recycling of waste. This paper concentrates on new buildings, where the criteria for new built small houses, apartment buildings and pre-school buildings have been revised in 2015 [3].

The criteria are valid for following newly constructed building types: small houses, apartment buildings, buildings for pre-schools and schools/educational buildings, extensions to existing buildings and homes for the elderly. The extension to an existing building must be a residential building, pre-school or school, and only the extension part can be ecolabelled. The homes for the elderly must be classified as homes and not as care facilities. This also applies to residential institutions for persons with physical or mental functional impairment. Shared areas for the home's residents and staff areas can also be covered by the Nordic Swan Ecolabel.

The criteria consist of 41 obligatory requirements and 14 point-score requirements. The 41 obligatory requirements consist of

a) general requirements (O1-O3)
b) requirements for resource efficiency (O4-O14) divided into energy and climate (O4-O6), waste (O7 and O8) and indoor environment (O9-O14)
c) requirements for chemical products, construction products, construction goods and materials (O15-O29) divided into general requirements (O15), chemical products (O16-O21), construction products, construction goods and materials (O22-O26) and timber, bamboo and fibre materials (O27-O29).
d) quality management of the construction process (O30-O34)
e) quality and regulatory requirements (O35-O40) and
f) instructions for residents and property managers (O41).

The obligatory requirement O3 sets the number of point-score requirements to be achieved. Point-score requirements P1-P5 refer to the requirements for resource efficiency, especially to energy and climate. Point-score requirements P6-P13 refer to the requirements for chemical products, construction products, construction goods and materials. Point-score requirement P14 deals with innovative measures taken in the construction process such as ecosystem services, environmentally friendly transports or energy-related measures.

3. Comparison between Swan label criteria and the National Building Code of Finland

The Swan label criteria and the National Building Code of Finland [4] are compared in this chapter. Here the obligatory (O) and voluntary point-score (P) ecolabelling requirements are shortly presented and followed by the possible corresponding requirements set by the National Building Code of Finland.

3.1. General requirements of the Nordic Swan Ecolabel

The Overall description of the building (O1) includes details of the building dimensions, structures, systems, materials, supplementary buildings. It also demands a statement of individual electricity metering. Responsibility for Nordic Swan Ecolabelling (O2) states the licence holder's responsibility for all requirements in the criteria document and for the fulfilment of the requirements, no matter who the work is performed by, until the building is ready for occupation. The number of points achieved (O3) defines the minimum number of filled point-score requirements in order to be granted a Nordic Swan Ecolabel licence. For Finnish buildings this minimum number is 16 for apartment buildings, 15 for small houses and 14 for pre-school and school buildings.

3.2. Energy consumption and production

The overall energy consumption O4 must fulfill the energy class A according to the Ministry of the Environment's regulation for buildings' energy performance (1010/ 2017). For pre-schools and schools, the maximum energy consumption is 85% of the regulations limit of 100 kWh/m^2. Cottages and holiday homes are treated as small houses for permanent residence and holiday apartments as apartment buildings. The energy calculation must be performed in accordance with Ministry of the Environment's regulation for buildings' energy performance or equivalent in Finland. For extensions to existing buildings, the energy requirement and energy calculation concern only the extension part of the building.

The point-score requirement P1 includes energy contributions from local energy sources or energy recovery. Here maximum 6 points can be achieved by the utilization of local energy sources (solar collectors, solar cell panels or heat recovery from domestic hot water consumption).

National Building Code of Finland sets the minimum requirement for the calculated energy performance reference value (E-value), which is a building's calculated annual consumption of delivered energy weighted by energy carrier factors per the building's net heated area. The E-value calculated in accordance with the building's intended use category and shall not exceed the limit values set by the building code. National Building Code of Finland calculates the energy use by using standardized building occupancy and equipment usage schedules.

One point-score can be achieved by individual metering and display of domestic hot water consumption (P2). Also, one point can be achieved by calculating the hot water circulation (HWC) energy losses for building types with requirements on maximum waiting time for hot water in the tap by the national building regulations (P3).

National Building Code of Finland assumes the domestic hot water consumption calculated by standard heat load which is a function of square meters.

The lighting management requirement O5 requires automatic demand control for outdoor and indoor lighting. Exceptions are lighting inside apartments, workplace lighting, worktop lighting, lighting fitted into technical installations and equipment, lighting of lifts and emergency lighting.

National Building Code of Finland by default uses a standard load by using standardized occupancy/load periods, but there is an option to calculate also in a detailed level energy consumption if the building has an advanced lighting management system.

Energy-efficient white goods must fulfill specific energy class requirements: class A++ for washing machines, fridges, freezers and tumble dryers; class A+ for dishwashers; class A for ovens and combined washing machine and tumble dryers and class B for extractor hoods (O6). Maximum 3 point-score requirements can be collected from products classified in the best energy class within a product type/category (P4). Maximum 3 point-score requirements can also be collected if all products within one type/category of energy-labelled sanitary tapware fulfill the required energy class of that type/category (P5).

National Building Code of Finland assumes all white goods as a standard load and does not specify the loads based on different classifications.

3.3. Climate impact of materials

The point-score requirement P6 recognizes reducing the climate impact of cement and concrete use with two alternative ways. First alternative: at least 50 % of a certain product type is covered by cement products that contain cement clinker at maximum 70% by weight. Points can be achieved from max 2 of following product types (1 point per product type): foundations, bearing systems and ground/floor decks, roof elements, wall elements, façade elements, balcony elements, terraces and verandas. Second alternative is to collect max 2 points of following measures (1 point per measure): deliberate work to differentiate concrete quality as required for the building (i.e. different concrete qualities for different construction details) or deliberate work with slim/lean concrete structures by taking various measures (high-performance concrete, reinforcement material, reinforcement technique, edge foundations, etc.).

National Building Code of Finland takes the material properties into account as thermal properties and calculates the energy consumption based on these. Currently Finland is creating a method to calculate embodied energy and CO_2 emissions of building materials during the whole life cycle. The method is assumed to be in line with European standards.

3.4. Green initiatives

A maximum of 3 points can be achieved for innovative measures taken in the construction process (P14). These innovative measures cover ecosystem services (green roofs/façades, local disposal of surface water, opportunities created for urban cultivation, creation of gardens for biological diversity, creation of habitats for insects, birds, and bats), environmentally friendly transports (indoor bicycle workshops for residents, at least 1.5 bicycle parking spaces per flat provided and equipped with access to frame locks, at least 50% of bicycle parking weather protected, at least one parking space equipped with access to charge electric vehicles) and energy-related measures (fixed or movable exterior solar protection on all south-facing windows, intelligent monitoring and display of the residential unit/preschool/school's energy consumption, possibility of energy storage in the building for flexibility, all white goods within a product group connected directly to the district heating network or the domestic hot water network).

National Building Code of Finland does not recognize green indicatives, but those are voluntary bases.

3.5. Requirements for waste sorting and recycling

Possibility for sorting at source O7 requires sorting vessels with at least four waste fractions in residential units and at least five waste fractions in pre-school or school kitchens or adjacent spaces. O9: Waste sorting stations with at least six waste fractions are obligatory for apartment buildings with more than eight flats, pre-schools and schools (O9). Maximum 3 points can be achieved by recycling of building waste (P13). Here 3 points are achieved if the ratio of building waste from the construction process that is sorted for recovery or recycling of materials is 70 % or higher. For 60-70 % 2 points are given and for 50-60 % 1 point. National Building Code of Finland does not set yet the requirement of building waste recycling. However, many construction companies have already long time developed their own system. The household waste is well recycled or used as energy, but in building material waste new investments are needed in order to increase building site waste recycling.

National Building Code of Finland does not set minimum requirement for waste sorting, but at city level there are requirements for waste sorting.

3.6. Requirements for indoor environment

Radon prevention measures should comply with national official requirements and radon level limits (O10). If the building is not built radon-proof, the risk analysis of the presence of radon from both the ground and filling materials must be conducted and support the level of radon prevention measures.

National Building Code of Finland sets requirement levels for radon concentration.

A moisture prevention plan or a description for moisture prevention work is required (O11). A competent moisture expert technician with required experience must be nominated to monitor adherence to the moisture prevention plan. The moisture prevention work should at least include the selection of damp sensitive materials and technology, relevant for the origin of harmful moisture, weather protection and other handling of materials and building/construction elements at the construction site, measures taken to ensure that the building dries out sufficiently and statement of the time this is estimated to take, determination of the highest permitted moisture status of various materials (critical moisture state) and control of moisture protection by calculating or measuring the moisture in concrete. If measures are conducted, borehole measurements must be undertaken as surface moisture measurement is not sufficient to determine whether the concrete slab is dry enough. National guidelines must be followed.

National Building Code of Finland sets requirement for moisture prevention and prevention of risks of moisture damage.

A function-control must be performed for ventilation systems before they are taken into use (O12). For apartment buildings the control must be performed for at least 10% of the total number of flats, but minimum one flat. Pre-schools and schools must have automatic controls for demand management of air volume/ventilation.

National Building Code of Finland sets requirement for ventilation rate and system performance. The function control is done by the system providers.

Noise environment planning/projecting must be performed for pre-schools and schools (O13). Here reverberation and one additional parameter must fulfil noise class B and other parameters noise class C. Temporarily occupied rooms, such as entrance halls, photocopying rooms, changing rooms and WCs, are exempted from this requirement. The P8: Residential buildings can achieve 1 point if the building fulfills recommended supplementary requirements for low frequencies concerning impact noise and/or airborne noise according to the national acoustic environment standard, in combination with sound class C. Residential buildings can achieve 3 points if the building fulfills the sound class B for two optional acoustic environment parameters.

National Building Code of Finland set requirements for maximum noise levels in different types of occupation spaces and at day and night time. In addition, the health authorities have recommendations for noise levels.

Daylight requirements are set by the obligatory requirement O14. Daylight factors (DF) for pre-schools and schools must be at least 2.5% in common rooms/playrooms or classrooms as the average value. For residential buildings the required DF of the national building code must be fulfilled in at least one occupiable room per residential unit. The DFs must be calculated using a specified computer program. If a DF limit is not stated in the relevant country's building code, compliance can be demonstrated by using a DF requirement from another Nordic country of choice.

National Building Code of Finland sets requirements for daylight indirectly by setting the minimum requirement of window area in living rooms.

Requirements for formaldehyde emissions are set by the obligatory requirement O15. All wood-based panels containing more than 3% by weight formaldehyde-based additives have specific requirements for formaldehyde emissions except panels solely marketed as façade panels.

National Building Code of Finland sets requirements for impurities in indoor air.

3.7. Requirements for chemical products, construction products, construction goods and materials

All construction products, construction goods, materials and chemical products to be used to construct the Nordic Swan Ecolabelled building must be listed in the product list and logbook of the building (O15) except products that are used to a very limited extent or has limited impact on health and the environment.

National Building Code of Finland does not set requirements for labelling, that is for voluntary bases.

The classification of chemical products (O16) must be in line with current legislation (CLP Regulation (EC) No 1272/2008. Carcinogenic (Carc), mutagenic (Muta.) or reprotoxic (Rep) classified chemical substances according to CLP Regulation 1272/2008, may not be used in the production of Nordic Swan Ecolabelled buildings (O17). Few exemptions are listed in the criteria documents. The criteria contain concentration limitations to preservatives or combination of preservatives in indoor paints and varnishes (O18) and other chemical products for indoor use (O19). The criteria have prohibitions or restrictions to various individual substances or substance groups (O20) including e.g. the substances on the ECHA Candidate List, substances evaluated by the EU to be PBT substances (persistent, bioaccumulative and toxic) or vPvB substances (very persistent and very bioaccumulative).

National Building Code of Finland only sets requirements for indoor air concentration levels and other issues must be done according to law.

Nanoparticles from nanomaterial may not be constituent in chemical products, excluding pigments, naturally occurring inorganic fillers, synthetic amorphous silica and calcium carbonate and polymer dispersions (O21). The criteria have restrictions (O23) for active adding of nanoparticles from nanomaterial to the glass on balconies or the outer glass surface of windowpanes on windows, window doors and exterior doors. There are also restrictions to the surfaces where chemicals or additives added to provide an antibacterial or disinfecting surface can be used.

National Building Code of Finland do not give the criteria for this, but the issue is handled in other parts of the Finnish legislation.

Several construction products, goods and materials are excluded from the set requirements, such as thermal, acoustic and technical insulation, sealing products, drainage pipes and electricity cables (O22). However, also these groups have restrictions to used substances such as substances on EU Candidate list etc.

National Building Code of Finland is not giving requirements but the law.

The use of chlorinated plastics (PVC) including polyvinylidene chloride PVDC is not permitted on interior surface layers on floors, roofs and walls (O24). However, service areas where unauthorized persons do not have access, are excepted from this rule.

National Building Code of Finland only sets requirements for indoor air concentration.

3.8. Requirements for recycled material, copper, wood products and ecolabelled products

A certain proportion of recycled material must be used in windows and exterior doors made from non-renewable materials (O25). Recycled plastic resources may not contain lead or cadmium in levels exceeding 100 ppm. According to the point-score requirement P12 one point can be achieved for each category of construction product and construction material outside the vapor barrier consisting of at least 25% of recycled or reused raw material. Maximum point score is 3.

National Building Code of Finland does not set requirements for the use of recycled material.

Obligatory requirement O26 concerns copper in domestic water pipes and as façade and roofing material. Domestic water pipes made of copper are not permitted. The maximum copper content of the cladding for roofs and façades and products for roofs and façades is 10% by weight. The requirement excludes closed water piping systems and service areas.

National Building Code of Finland does not set requirements for the use of copper.

Nordic Ecolabelling maintains a list of prohibited tree species (www.nordic-ecolabel.org/wood/), which are not permitted to be used in Nordic Swan Ecolabelled buildings (O27). This restriction is also valid for timber and wood products used in construction but not incorporated in the building (e.g. wood in casing and mold). The used wood raw material must be stated, and the supplier of the material must be Chain of Custody certified by the FSC/PEFC schemes (O28). Point-score requirement P11 gives one point if at least half of the need for wooden mouldings (mouldings, skirtings, baseboards, level drops and similar) is covered by certified timber products and 2 points if the entire requirement is fulfilled by certified timber products. Same requirements for the supplier as in O28 apply also for P11. Of all wood raw material used in Nordic Swan Ecolabelled buildings a minimum of 70% by weight must origin from forestry certified under the FSC or PEFC schemes or be recycled material. The remaining proportion of wood raw material must be covered by the FSC/PEFC control schemes or be recycled material. Point-score requirements P7 gives maximum 2 points can be achieved by fulfilling 1-2 of following measures (1 point per measure): 1) bearing systems/joists of wood or other renewable material, 2) bearing walls or bearing roof structures of wood or other renewable material or 3) façades for which more than 50% of the façade area is of maintenance-free wood or another maintenance-free renewable material. Obligatory requirement O29 concerns durable/resistant wood for outdoor use. Use of timber impregnated with heavy metals and/or biocides is not allowed with the exception of timber in class B according to the Nordiska Träskyddsrådet (Nordic Wood Preservation Council) classification or corresponding classification. There can also be exceptions for wood in direct contact with water or ground or with specific demands of strength.

National Building Code of Finland does not set requirements for the use of wood products.

Points can be achieved from using ecolabelled construction products or conscious product choice. Maximum 10 points can be achieved by using Ecolabelled (Nordic Swan Ecolabel or EU Ecolabel) construction products (P9). For each product category 1 point is given if at least 10% of the product requirement is covered by ecolabelled products and 3 points if at least 50% of the product requirement is covered by ecolabelled products. P10 gives maximum 2 points can be achieved by following measures: 1) 2 points if all expansion joints with the highest demand on flexibility are phthalate free, 2) 2 points if all conduits are PVC-free and contain a maximum of 0.05 weight% bromine and chlorine respectively and c) 1 point if all products within one of the following product categories are PVC-free: drainage pipes, main cables and plastic pipes for central vacuum cleaners.

National Building Code of Finland does not set requirements for the use of ecolabelled products.

3.9. Quality management of the construction process

Requirement concerning air permeability (O30): The air-tightness of the whole building envelope must be measured for small houses, pre-schools and school buildings and it must fulfill the requirements set by the construction project. For apartment buildings and extensions to apartment buildings, the air-tightness must be measured for a representative selection. This selection must constitute at least 10% of the total number of flats, but at least one flat. Both flatwise and stairwise measurement of air permeability are approved.

National Building Code of Finland sets requirement for air tightness value.

Requirement concerning management of requirements on products and materials (O31): The license holder must ensure that the requirements O15 - O29 are fulfilled. If the license holder uses subcontractors, a documentation is needed that the subcontractor is aware of and adheres to the materials requirement. Also new products and materials, added after the license is granted, must be approved by Nordic Ecolabelling if they are subject to the requirements in chapter. All parties involved in the construction process must have the relevant knowledge to be able to ensure fulfilment of the Nordic Swan requirements (O32). This includes employees, supervisors, site managers, sub suppliers and subcontractors. Information on how chemical products must be handled to avoid risks to people must be translated if necessary. The self-monitoring of the contractor is obligatory, and this must be documented during the construction (O33). The necessary routines to be monitored are waste handling at construction sites, moisture/damp prevention, secure execution of water installations, air permeability and testing, electrical installations, ventilation, heating system and routines for the performance of pre-inspection of the building before the independent third-party control or final inspection. An independent third party with relevant expertise must inspect the completed building (O34).

Here the National Building Code of Finland refers to the Finnish law.

The documentation that is sent in with the application must be archived by the license holder (O33). The license holder must document a list of the constructed Nordic Swan Ecolabelled buildings and store the documentation for at least five years after occupation (O36). If the license holder is planning to make product and market changes (O37) or unforeseen non-conformities (O38) that affect Nordic Ecolabelling's requirements, a written notice of these must be submitted to Nordic Ecolabelling. The license holder must have routines for the documentation, reporting and handling of any complaints/claims concerning Nordic Swan Ecolabelled buildings. It must also be clearly stated that the license applicant is responsible for the customer and is the party that the customer must contact concerning any complaints and claims (O39). The license holder must ensure that all relevant applicable laws and provisions regard to safety, working environment, environmental legislation and site-specific terms/concessions apply at all production sites for Nordic Swan Ecolabelled buildings (O40).

National Building Code of Finland has no requirements here but refers to the Finnish law.

3.10. Instructions for residents and property managers

The Nordic Swan Ecolabelled building must contain overall general information and operation and maintenance instructions (O41) including a maintenance record book. The maintenance record book guides the property manager and the inhabitant how to use and maintain the building and the technical systems considering also the environmental viewpoint.

National Building Code of Finland refers to the Finnish law.

4. Discussion

The Finnish Building Code sets the minimum level for construction. The mandate of the regulation is not able to include all relevant issues in the National Building Code, therefore both the Ministry of the Environment and the Ministry of the Interior have regulations concerning e.g. indoor environment.

The comparison between the criteria of the Nordic Swan Ecolabel and the National Building Code of Finland clearly shows how the ecolabelling criteria has a wider perspective to safe and environmentally conscious construction, especially regarding the use of chemical substances and material. The tighter requirements for materials, energy-efficiency and utilization of natural daylight enable a more comprehensive inspection of the indoor environment.

5. Conclusions

As a conclusion of the comparison between the Nordic Swan ecolabelling of buildings and the Finnish Building Code, it can be concluded that they both have an important role in the climate change mitigation of the Finnish building sector. The role of the obligatory Building Code is to state the minimum level for construction. The role of the voluntary ecolabelling of buildings is to enhance the construction industry into continuous improvement and to speed up the technological development.

Ecolabelling can therefore act as a test bench for upcoming building regulations. Here the forerunners can first test fulfilling the requirements of the ecolabel and after this piloting period specific issues from the ecolabelling criteria can be introduced in the National Building Code.

In line with European standards, the Ministry of the Environment is introducing a method to calculate embodied energy and CO_2 emissions of building materials during the whole life cycle. The Nordic Swan Ecolabel should therefore also consider introducing the calculation of the carbon footprint in the upcoming criteria revision.

6. References

[1] GEN 2004. Introduction to ecolabelling. Global ecolabelling network (gen) Information paper. https://globalecolabelling.net/assets/Uploads/intro-to-ecolabelling.pdf
[2] ISO 14024:2018 Environmental labels and declarations. Type I environmental labelling, Principles and procedures.
[3] Nordic Ecolabelling 2016 Small houses, apartment buildings and buildings for schools and pre-schools. *Criteria document.*
[4] The National Building Code of Finland. http://www.ym.fi/en-US/Land_use_and_building/Legislation_and_instructions/The_National_Building_Code_of_Finland

IOP Conf. Series: Earth and Environmental Science **297** (2019) 012044 doi:10.1088/1755-1315/297/1/012044

Residential movements in connection to renovation of rented multi-residential housing: A pilot study

P Femenías[1,6], L Jonsdotter[1], J Forsemalm[2], E Punzi[3], E Bogdanova[4], C Thodelius[1], and K Granath[1,]

[1] Department of Architecture and Civil Engineering, Chalmers University of Technology, Sven Hultins gata 6, SE-412 96, Gothenburg, Sweden
[2] Radar Arkitektur och planering, Surbrunnsgatan 6, SE-411 19 Gothenburg, Sweden
[3] Department of Psychology, University of Gothenburg, PO Box 500, SE-405 30 Gothenburg, Sweden
[4] Department of Sociology and Work Science, University of Gothenburg, PO Box 750, SE-405 30 Gothenburg, Sweden
[6] Corresponding author.

Abstract. Residential movement and displacement as an effect of renovation has earned attention and also affected renovation practices in Sweden. While statistical studies have linked deep renovation to residential movement and displacement, there are no recent studies that investigate why people move or remain in housing areas that are renovated, and if and how the relocation is determined by the renovation. A pilot study was initiated as a means to develop a methodology to study residential movement in connection to renovation. In this paper, methodological considerations are discussed based on 31 interviews (face-to-face and telephone) with movers related to 34 municipally owned rented housing areas about to undergo renovation, as well results from a questionnaire sent to two finalised renovation projects (N=113). So far, the pilot study indicate that few relocations can be linked to the up-coming or finalised renovation in the studies cases. The questionnaire that was sent out to remaining tenants had a low response rate of 29%, and the efficiency of using questionnaires is discussed.

1. Introduction

The European Energy Efficiency Directive (EED) (2012/27/EU) has identified the existing building stock as the single biggest potential sector for energy savings and reductions of green-house emissions. As a result, European policy has encouraged member countries to promote and facilitate deep renovation [2]. Deep renovation has been defined as leading to up to 60% of energy savings through energy renovation [3]. Although a range of European policies has been set in place, the rate of deep energy renovation is far below expected [4]. Barriers for deep energy renovation has been found to be of technical, financial, and regulatory character but also linked to a lack of knowledge and awareness [4]. Even when technical solutions are found viable, the property owner still needs to make decisions on *"where to start, and what measure to implement in which order"* [5].

In parallel to the energy debate, the negative social consequences of deep (energy) renovation has been highlighted. Deep renovation is often combined with standard improvements, and costs to finance the investments are often pushed over to the residents [6]. Recent studies have highlighted the social worries among tenants when their home is to be renovated, and in special the risk for a forced relocation when rents are increased [7],[8]. Furthermore, statistical studies have shown that residential movements as a result of deep renovation of large stocks of housing could lead to a more segregated society at large [8] a phenomenon which would be reinforced by the fact that the socio-economic most vulnerable residents live in areas with the largest need for deep energy renovation [9]. While the argument can be made that actually few of the renovations that are carried out are of the magnitude of deep renovation [4],[10], also more limited renovations and rent increases would be a problem to many households in rented apartments as they have limited incomes [12].

The mentioned studies have some limitations. The statistical study [9] does not provide any insights to why people move in relation to a renovation. The motive for the move could be raised rents, but it could also be other kinds of nuisance in relation to a renovation. It could also be that the tenants have thought of moving earlier and get a push when the renovation is happening. Furthermore, renovation is likely to affect residents in different ways, and elderly people might be especially vulnerable to cope with the nuisance and worries that a renovation brings on to them [13].

Several of the referred studies are also projections of what could happen in the case of deep renovation, but do not provide any evidence of what actually happens [8],[10],[12]. Furthermore, several studies are limited to a few case studies owned by a limited number of property owners and it is not certain that the reported incidents [7],[8] are representative for all rented housing. Renovation strategies also change over time due to changing financial and policy structures [11], and it is not sure that the findings would be the same if the studies were carried out today. Finally, while Swedish studies state a tendency of people moving to more deprived areas after a relocation in relation to a renovation [9],[14], Dutch studies show that forced relocation can lead to housing improvements. A longitudinal study showed that forced relocation or displacement, in the long-term, can lead to improvements for individuals in relation to their housing situation [15]. A more recent study, using individual-level population registry data, showed that forced relocates from the demolition of three housing areas, live in less deprived neighbourhoods after the move [16]. However, the results were not conclusive in that this upgrade in housing leads to more socio-economic opportunities for these people in other areas.

1.1. Aim and scope

The main aim of this study is to investigate why tenants in rented housing move in relation to a renovation. A longer study, that will run for at least three years, is planned together with the Mother Company of three municipally owned Housing Companies in Gothenburg, the second largest city in Sweden. The local authorities in the city, has decided that no tenants should be forced to relocate as a result of rent increases. While the main aim is to investigate the reasons behind a move and the eventual connection to the renovation, the study also has a broader aim of investigating nuisances of renovation among tenants. The question of relocation and other kinds of disturbances among residents during and after a renovation needs further research in order to design feasible and sustainable renovation strategies.

The aim of the study is to understand why people move and how they experience a renovation, consequently, statistical studies were ruled out in favour for interviews and questionnaires. The interviews and questionnaires should answer the following questions: How many tenants move in relation to a renovation? What are their main causes for moving? To what extent is the renovation the cause for relocation? Where do they move after the renovation? What are the reasons to stay or move in after a renovation? How did the residents experience nuisance in relation to the renovation? Is there a relation between the renovation strategies, the disturbances during the renovation, and the tenants' decisions to relocate? Can we see some differences in the demographics of the 'movers' vs the 'stayers'? Is there a relation between the quality, standard and rent of the housing before and after the renovation and the decision to stay/relocate?

In order to develop and test a methodology, a one-year pilot project was initiated. In this paper, early results from the pilot project, 31 interviews with movers and the results from one questionnaire among stayers sent out to two competed renovation projects (N=113) are used for a methodological discussion. We compare our experiences to earlier studies of residential movement patterns.

2. Earlier studies

A short overview is here provided of the methodology and data collection used in a few earlier qualitative studies of residential movements in connection to urban regeneration and renovation.

The social effects of regeneration and renovation in connection to investments in the housing stock in 1960s to the 1980s have been the subject for a larger number of studies in Sweden [17]. While many of these studies were limited to unique case studies, some were thoroughly made. In a famous sociological study by Laval et al. [18], interviews were made with a larger population in an area that was to be demolished, before, during and after the intervention, in order to capture the residents experiences. The results that were brought forward showed that the residence appreciated their run-down housing and neighborhood came as a surprise to those working with urban clearance at the time.

Recent Swedish studies have highlighted negative social effects of residential movement in connection to renovation. Baeten et al., [7] draw conclusions from an non-specified number of interviews with residents, housing company representatives (not indicated if private or public), representatives of the Rent Tribunal and representatives of the Swedish Tenants' Association. The authors declare that the interviews were semi-structured, recorded and transcribed. Residents and housing companies were from three different housing areas in one city and the representative from the Rent Tribal form another. The recruitment of interviewees is not précised. The residents were interviewed at two occasions before and after the renovation in order to *"capture the experience of displacement and displacement pressure at different points in time"* [7] (p. 633).

Polanska and Richard [8] studied how tenants experience fear of forced displacement prior to a renovation of their housing area which is owned by a property owner that has become known for making standard improvement with larger rent increases. The authors refer to an ethnographic study spanning from 2011 to 2018 based on 32 semi-structured interviews with tenants and observation studies (not specified). The interviews encircled *"collective representations of the Swedish society, the tenants' own analysis of the situation and how they have decided to act and make resistance against the up-coming renovation"* [Authors translation from Swedish] [8] p 417. When the study was made, the rents in the area were to be increases by 43%, but they finally settled upon 18-34% [8] , p 432.

Some studies from other countries present results from longitudinal studies on the effect of regeneration and renovation. Egan et al., [19] followed 12 households during the regeneration of a neighbourhood that involved clearance and relocation. The households were recruited through local housing associations, community groups, the church and snowballing through these first contacts. In the Netherlands, residential movements have been studied using focus groups and interviews with 65 movers [15], but also longitudinal studies using a combination of face-to-face interviews and telephone interviews [20] . In the latter study, the previous and the current addresses of movers where found in a data base of a social housing and a random sample were interviewed in 2001 (N=243) and 2004 (N=225) using a structured interview guide and at times other languages than Dutch. Later on, a third round of interviews were made adding to the long-term perspective [21]. Finally, 144 in-depth interviews was used in a study with displaced tenants surveys [22] which were recruited from the previous studies [20],[21].

3. Methodology

The study was initiated by researchers with special expertise in architecture, renovation and housing at Chalmers University of Technology. In order to bring in a broader spectrum of expertise in order to develop an appropriate methodology, an inter-disciplinary group was formed that includes expertise from the fields of ethnography, psychology, sociology, and criminology.

The study targets tenants living in municipally owned housing in the city of Gothenburg. The project is planned together with three independent municipally owned Housing Companies that together owns, manages and let out almost 70.000 apartments. For the study, 34, on-going, planned or recently completed renovation projects were identified by these Municipal Housing Companies, and for which there is a planned rent increase of at least € 10 per month, or other potential disturbances for the tenants during the renovation.

Firstly, the study addresses all tenants that decide to move before, during and up to two years after the completed renovation in the selected housing areas. All these movers are to be interviewed about their decision to move and the eventual connection to the renovation. Second, the remaining tenants and the newcomers that move into the studied housing areas are asked why they decided to remain in the area, or move there, and how they were affected by the renovation before, during and after a renovation. The stayers and the newcomers are addressed through a questionnaire which is sent out six months after the completion of the renovation.

3.1. The interviews

During the first weeks of the pilot study, four interviews (three face-to-face and one over the telephone) were carried out as a means to get a better idea of how to approach the interviewees and to get a hold on what kind of subjects and experiences that the interviewees would bring forward. Based on these first interviews, an interview guide and a questionnaire where designed.

Information about tenants, in hold of the lease contract, and who have decided to move (here called movers) are transferred to the project leader at Chalmers, and in accordance with GDPR regulation. The project leader or an assistant then contact the movers to ask for a personal interview. These interviews are preferably held in the mover's home, but on their request, they can be held in a public place or at their work place. If the mover does not wish to meet for an interview they are asked if they can accept a short telephone interview.

The face-to-face interviews are focused around a few questions in an open interview guide: Why have you decided to terminate the lease of the apartment? Why do you decide to move now? Is there any relation between your move and the renovation? Where do you move? Furthermore, questions are posed about the move, preferences in the old/new home, costs for the old/new home and experiences from the renovation (disturbances, information provided, possibilities to influence the renovation).

During the interview the interviewee is also approached with a short paper-based questionnaire that include questions on personal demographic: their gender, age, and primary occupation, the size of the household, and the households' collective income. These same demographic questions are included in the questionnaire (section C) which is sent out to those who remain after the renovation (see 3.2.).

Regarding the telephone interviews, the same open questions as in the face-to-face interviews are posed. In the first telephone interviews the short questionnaire with the demographics was also filled in over the phone, but this was later abandoned, for reasons of personal integrity.

The interviews have been carried out either by the project leader and an assistant or the assistant alone. At this stage, the project leader has been present at 12 out of 31 interviews. So far, all interviewees for the face-to-face interviews have accepted to be recorded. The interviews are then transcribed. For the telephone interviews, these are not recorded. Notes are taken during the interview and documented in a short report directly after the interview.

Two kinds of analysis are planned regarding the interviews. First, addressing the main questions for the study, why the tenant has decided to move and the eventual relation to the renovation, is analysed using results from both face-to-face interviews and the telephone interviews. Secondly, an in-depth analysis is made on only on the results from the face-to-face interviews in order to get a more holistic understanding of the circumstances around the decision to move.

3.2. The questionnaire

The questionnaire was designed so that it could be used for all different renovation project in the study. This demanded some effort in order to guide the respondent through all questions non-respect to weather they had been evacuated or remained in the apartment during the renovation or moved in after the renovation. The respondents that moved in after the renovation were not supposed to answer questions about the renovation process.

The questionnaire was divided in three separate sections: Section A with 28 questions about the satisfaction with the apartment, the area and the services of both these, including question of why the tenants choose to move here or stay after the renovation; Section B, with 18 question about the renovation process, the satisfaction with the process in itself, their possibilities to influence the plan of the renovation and their satisfaction with the final results; and finally, Section C with 15 questions about demographics (the same as the interviews, see section 3.1.). In connections to some, but not all questions there were possibilities to add free comments.

The questionnaire was prepared in paper format and digitally. The paper version included a link to the on-line version. No reward was provided for filling in the questionnaire. The anonymity of the respondents was guaranteed by coding the answers only to staircases not to single apartments.

Results from the questionnaire has been analysed both using descriptive statistics and multivariate statistical analysis to find correlations. Free text comments have also been coded and analysed.

4. Results

At this stage, only preliminary results are presented. The material is rich and more in-depth analysis are on-going.

4.1. Preliminary results from the interviews

So far, 63 tenants that have decided to move, have been contacted for an interview. Table 1, shows the number of movers that have accepted to meet for a personal interview, those who have accepted a telephone interview and those who have declined to be part of the study.

Table 1. Number of interviews, telephone interviews and rejected interviews.

Category of interview	Number	Comment
Face to face	13	10 held in the home, 3 at café/work place
Telephone	18	
Subtotal	**31**	
Rejected interview	32	23 did not answer or had already moved[a], 5 declined the interview, 2 did not show up for a booked interview, and in 2 cases the telephone number was incorrect.
Total	**63**	

[a] *Form two of the Housing Companies we got the information too late after some tenants had already moved*

The face-to-face interviews hold qualitative information about the move, the reasons behind moving, and qualities that they appreciate in their home and its' environment. Regarding demographics of the interviewees, data from all of the interviews have not been collected. In the face-to-face interviews, the short questionnaire was left for the interviewee to fill in after the interview and sent by post. Unfortunately, not all of them have done so. In addition, as already said, these demographics questionnaires were not filled in during all telephone interviews. The data we have, show that the interviewees are either in age group 25-34 or 55-64, of these 19 were women and 13 men, and all were employed. More demographics data could probably be retrieved from an analysis of the transcribed record which has not be done so far.

Table 2. Main reason to relocate according to interviewees (face-to-face and telephone).

Main reason to relocate	Number	Comment	Connection to renovation
Family reasons	7	Move in with partner, want to move closer to family and friends or closer to the job or the childrens' schools, in two cases up-sizing is mentioned.	No
Wanted to move/live elsewhere	6	Two tenants want to live in countryside, one wants to up-size, one has got a push to move in relation to the renovation	No (one partly)
Have bought a new home	5	Enter a housing career, want new production, live on the country side or up-size	No
Does not live in the apartment	4	The apartment has been rented out second hand, one of the tenants was evicted due to illegal sub-letting	No (except eviction case)
Raised rent after renovation	4	One directly due to this, two have decided to downsize in the area, one got a push to relocate	Yes
Lack of quality in old apartment	3	One tenant experience high levels of noise in the area, two refer to badly maintained apartments	No
Lack of accessibility	1	No elevator	No
Want to downsize	1	The rent increase is limited but they want to live cheaper	No
Nuisance caused by the up-coming renovation	1	Work night time, cannot plan up-coming travel plans due to lack of precision in renovation plan	Yes
Total	**31**		

In Table 2, a first categorisation of the main reasons behind their relocation is presented. Five of the thirty-one relocations have been judged as caused by the renovation. Four of these five tenants mentioned the rent increase as the reason behind moving and one tenants' relocation is motivated by the expected nuisance of the up-coming renovation. In the category that mention that they want to live somewhere else, as a main reason for relocating, we also find one relocation that is partly due to the renovation. In the category of tenants that sub-let their apartment, one tenant was evicted due to illegal sub-letting which could also be judged as being provoked by the renovation. A majority (14) of the movers, move to another rented public apartment which they either have searched for through housing ques or been proposed by the property owner in connection to the renovation. Five movers have exchanged their publicly-owned rental apartment for a privately-owned rental apartment. Four of the movers have bought a new home (condominium or villa). For the four movers which did not live in the apartment that was to be renovated, now either live abroad or has moved in with family for which the tenure of the home is not known.

4.2. Preliminary results from the questionnaire

For the pilot study, the questionnaire was sent out to the completed phases of two separate renovation projects. Case A is a housing area from the 1950s with more than 500 apartments in lamella and tower blocks. The buildings undergo a medium intense renovation including new facades, the installation of double flux ventilation, and complete renovation of bathrooms. The tenants stayed in the apartment during the renovation. The tenants were not able to use parts of their apartment between two to six weeks, and the water was cut off for a few weeks. The property owner provided evacuation apartments for those with special needs (e.g. working shift, disabilities). The rent increase after the renovation is about 23% compared to the earlier rent level.

Case B are four higher slab blocks with about 400 apartments constructed in the late 1960s. The questionnaire was sent out to phase two in the renovation project. The buildings underwent a façade renovation some years ago. In this new intense renovation, the buildings are equipped with double flux ventilation and the apartments are completely refurbished with new bathrooms and kitchens. Two floors with additional apartments are added on the top of the buildings. All tenants were evacuated for at least 7 months. In the first phase, the rent increase was up to 45% but the standard of the internal renovation and consequently the rent raise will be lower in the next three phases.

Table 3. Respondents of the questionnaire for the two case studies.

Case	Number of apartments	Number of respondents	Percentage
Case A	270	65	24%
Case B	122 (100 originally and 22 new apartments on top)	42	34%
Total	**392**	**113**	29%

The questionnaire was sent on twice and the total number of respondents for each case, combined paper and digital answers, is shown in Table 3. The choice of coding the answers per staircase meant that we had to distribute the reminder to all tenants, also those who had already answered, to the irritation of some that contacted us.

The respondents are about 52% women and 48% men in both case areas. About 50% are employed and about 20% are retired. In Case A, for more than 30% of the responding households, the yearly income is approx. 3000 to 4000 €. For Case B, for more than 30% of the responding household, the yearly income is approx. between 3500 and 4500 €.

In Case A, the property owner had already sent out a questionnaire prior to the start of this research project which could explain the low response rate in that case. In Case B, actually very few of the original tenants moved back after the evacuation according to data provided by the property owner. Originally, 75 tenants had permanent lease contracts and were evacuated prior to the renovation. Of these 33 chose not to move back but relocated to an apartment in the neighbourhood provided by the property owner. Another 31 tenants chose to move directly to the building in the property that was already renovated (phase one), eight of those chose a smaller apartment in the move and seven a larger apartment. In total, seven, of the original tenants moved back to the same building (phase 2), four to an apartment of the same size, two to a smaller seize, and one to an apartment on a lower level. Two tenants moved permanently and are no longer renting a home from this property owner.

Without this valuable information about the movement patterns, the results from the questionnaire would have been difficult to understand. Only one out of the 42 respondents answered that they had lived in the building more than one year, and at first, we thought that all respondents in Case B had misunderstood the thoroughly prepared question on when they moved in.

Regarding the evacuation, the questionnaire also provided ambiguous answers. In Case B, where all tenants were evacuated according to the property owner, 50%, or 21, of the tenants answered that they belonged to this group something which does not correspond to the information given, that only seven tenants were evacuated and moved back. It might be that some of these tenants were temporarily evacuated from another phase (phase 3) prior to moving into phase two.

In Case A, where there was no common evacuation of tenants, only for those with special needs. Still as many as 75% answered that they were evacuated. Of these, 22% answered that they had lived with family temporarily with a comment that they found it difficult to stay in the apartment when the kitchen or bathroom were disabled for a couple of weeks and workmen present during daytime. In the commentary field, some tenants claim that they were able to get an evacuation apartment through repeated complains.

5. Discussion and conclusions

Recent studies of residential movement in Sweden in relation to renovation of rented housing are based on the tenants experiences prior a renovation [8], and in relation to a few areas [23]. The aim of this study is thus to complement these earlier studies with insights from actual movers and stayers and in relation to a larger number of renovations and including limited renovations and deep renovation. A limitation of the presented study is that only municipally owned housing companies are studied and in one city. If possible, future studies could include other types of property owners and locations.

So far, 31 movers have been interviewed, and of these, six could be linked to the renovation. In several cases, the interviewee did not recognise that their house should be renovated, this was for example the case for those who rented out their apartment but also true for others. Of these six movers that could be linked to the renovation, five mentioned the rent increase as a result of the planned renovation as one main motivation. In one case, the nuisance before and during the renovation was the reason to move. A majority of the movers seemed to be content with the prospect of their new home which they in most cases had found on their own, and in a few cases, with the help of their housing company.

Earlier studies have indicated that income levels is determining whether tenants are forced to relocate in connection to a renovation [9],[12]. Further analysis of the transcriptions of the interviews in our study must be done before the movers can be established as having a more or less advantageous economic situation than the stayers. That few movers indicate that the rent increase, as a result of the renovation, is the primarily cause for their relocation, the question can be raised if the stayers are more negatively affected then the movers. Unfortunately, the questionnaire had a low response rate which probably can be linked to several factors such as the length of the questionnaire, its design or formulation of questions [24; 25]. For example, the information provided on the income levels might not be accurate, but the property owners have purchased economic information for their housing areas which can be used for verification.

For the continuation, the questionnaire as a method has to be reconsidered. Besides reformulating some questions, a reward for answering the questionnaire could be installed. The many comments left by the respondents, even though the answer they gave in the comments was actually defined as an answer, could be an indication that the respondents wish to give more qualitative answers. Focus groups as a replacement or complement to a questionnaire will be considered.

Acknowledgments

The authors wish to thank project assistant Michelle Lundin who worked with the study in the initial months, and Alexander Knutsson who contributed to the design of the questionnaire and also made the statistical analysis of the data. Furthermore, we wish to thank Liane Thuvander, Chalmers University of technology and Özum Durgun from Research Institutes of Sweden who initially were part of the inter-disciplinary working group and active in giving feed-back to the development of the questionnaire. We also want to acknowledge the Centre for Management in the Building Sector and Framtiden Förvaltning AB for financing the study.

References

[1] EED 2012 *Directive 2012/27/EU of the European Parliament and of the Council of 25 October 2012 on energy efficiency*

[2] Saheb Y, Ódis KB, Szabó S, Ossenbrink H, Panev S 2015 *Energy Renovation: The Trump Card for the New Start for Europe* (JRC Science and Policy Reports, European Commission, Brussels);

[3] Castellazzi L, Zangheri P, Paci D 2016 *Synthesis Report on the assessment of Member States' building renovation strategies* (JRC Science for Policy Report, European Comission, Brussels);

[4] Artola I, Rademaekers K, Williams R, Yearwood J 2016 *Boosting Building Renovation: What potentials and Value for Europe?* (Study for the ITRE Commission, European Parliament, Brussels);

[5] Fabbri M, De Groote M, Rapf O 2016 *Building Renovation Passports; Customised Roadmaps towards Deep Renovation and Better Homes* (Buildings Performance Institute Europe, Brussels);

[6] Ástmarsson B, Jensen PA, Maslesa E 2013 *Sustainable renovation of residential buildings and the landlord/tenant dilemma* Energy Policy [Internet]. 63(0):355–62.Available from: http://www.sciencedirect.com/science/article/pii/S0301421513008501

[7] Baeten G, Westin S, Pull E, Molina I 2017 *Pressure and violence: Housing renovation and displacement in Sweden* Environ Plan A Econ Sp. 49(3):631–51.

[8] Polanska D V, Richard Å 2018 *Bortträngning pågår: Renovering som kulturellt trauma* Sociol Forsk. 55(2–3):415–39.

[9] Boverket 2014 *Flyttmönster till följd av omfattande renoveringar* Rapport 2014:34 Regeringsuppdrag. (Karlskrona: Boverket)

[10] Mangold M, Österbring M, Wallbaum H, Thuvander L, Femenias P 2016 *Socio-economic impact of renovation and energy retrofitting of the Gothenburg building stock* Energy Build. 123.

[11] Femenías P, Mjörnell K, Thuvander L 2018 *Rethinking deep renovation: The perspective of rental housing in Sweden* J Clean Prod. 195.

[12] Bo-Analys 2017 *Var tredje kan tvingas flytta: En rapport om effekterna av hyreshöjningar i samband med standardhöjande åtgärder i Göteborg* (Göteborg, Sweden: Vostra konsulter);

[13] Serrano-Jiménez A, Barrios-Padura Á, Molina-Huelva M 2018 *Sustainable building renovation for an ageing population: Decision support system through an integral assessment method of architectural interventions* Sustain Cities Soc. 39:144–54.

[14] Bråmå Å 2011 *Går det att vända utvecklingen i utsatta bostadsområden? Exemplet Navestad/Ringdansen i Norrköping* (Linköpings Universitet, Centrum för kommunstrategiska studier);

[15] Kleinhans R 2003 *Displaced but still moving upwards in the housing career? Implications of*

forced residential relocation in the Netherlands Hous Stud. 18(4):473–99.

[16] Miltenburg EM (Emily., van de Werfhorst HG (Herman., Musterd S (Sako), Tieskens K (Koen) 2018 *Consequences of Forced Residential Relocation: Early Impacts of Urban Renewal Strategies on Forced Relocatees' Housing Opportunities and Socioeconomic Outcomes* Hous Policy Debate. 28(4):609–34.

[17] Femenías P 2015 *Det sociala och renovering – en översikt av tidigare erfarenheter* In: Lind H, editor. Antologi om social hållbarhet och renovering. (Stockholm; SIRen), p. 65–76.

[18] De Laval G, Egerö B, Lindskoug K, Sjöstrand P 1965 *En stadsdel försvinner: Saneringen av Stigberget i sociologisk belysning.* (Stockholm: Riksbyggen).

[19] Egan M, Lawson L, Kearns A, Conway E, Neary J 2015 *Neighbourhood demolition, relocation and health. A qualitative longitudinal study of housing-led urban regeneration in Glasgow, UK* Heal Place. 33:101–8.

[20] Kleinhans R, Van der Laan Bouma-Doff W 2008 *On priority and progress: Forced residential relocation and housing chances in Haaglanden, the Netherlands* Hous Stud. 23(4):565–87.

[21] Doff W, Kleinhans R 2011 *Residential outcomes of forced relocation: Lifting a corner of the veil on neighbourhood selection* Urban Stud. 48(4):661–80.

[22] Posthumus H, Kleinhans R 2014 *Choice within limits: how the institutional context of forced relocation affects tenants' housing searches and choice strategies* J Hous Built Environ. 29(1):105–22.

[23] Baeten G, Listerborn C 2015 *Renewing urban renewal in Landskrona, Sweden: Pursuing displacement through housing policies* Geografiska Annaler. Series B: Human Geography 97(3):249–61.

[24] Bradburn, N. M., Sudman, S., Blair, E., Locander, W., Miles, C., Singer, E., & Stocking, C. 1979 *Improving interview method and questionnaire design: Respone effects to threatening questions in survey research.* (San Fransicso: Jossey-Bass)

[25] Kaplowitz, M D., Hadlock, T. D. & Levine, R. 2004 A comparison of web and mail survey response rates. *Public opinion quarterly,* 68(1):94-101

IOP Conf. Series: Earth and Environmental Science **297** (2019) 012045 doi:10.1088/1755-1315/297/1/012045

Classification of Indoor Environment 2018 and Updated Criteria of RTS Environmental Classification

Laura Sariola
The Building Information Foundation RTS sr, Po.Box 1004, 00100 Helsinki, Finland, Europe, laura.sariola@rakennustieto.fi

Jorma Säteri, Helsinki Metropolia University of Applied Sciences, Finland, jorma.sateri@metropolia.fi

Timo Rintala, Green Building Partners Oy, Kutomotie 16, 00380 Helsinki, Finland, timo.rintala@gbp.fi

Abstract. The Finnish Society of Indoor Air Quality and Climate (FiSIAQ) introduced over twenty years ago in 1995 a Classification of Indoor Climate, Construction Cleanliness, and Finishing Materials. After two revised versions, the fourth edition Classification of Indoor Environment was published on May 2018. Based on the criteria set in the classifications, The Building Information Foundation RTS sr started M1-labelling of building products in 1996, named Emission Classification of Building Materials and Cleanliness Classification of Air-handling Components.

Indoor air quality and instruments in improving all the sectors to concern are in great importance in Finland. First part of The Classification of Indoor Environment 2018 (CIE 2018) includes new Requirements for Technical Target Values. The other part, Guidance for design and Construction, includes surface structural design, construction site design, Classification of Construction and Construction Work Cleanliness (P1), Classification of Building Materials (M1), Cleanliness Class of Air-handling components (M1) and instructions for designing mechanical systems for buildings.

The main changes in the 2018 edition are revised target values for indoor operative temperature, inclusion of a PM2.5 criteria, revision of some acoustical target values and revision of recommended outdoor air flows. Regarding thermal comfort, the target values of the individual indoor climate, S1, will be slightly more demanding than the targets of the S2, good indoor climate, class. The criteria for particulate matter, PM2.5, has been included due to the significant health impact of fine particles. The targets for acoustic environment in offices have been revised to reflect the importance of speech transmission. The recommended outdoor air flows follow the principles of EN15251-2007, considering the common usage of very low-emitting M1-class materials in Finland. The outdoor air flow rates have also been aligned with the recently changed building code values.

Companies have developed low-emitting building materials, fixtures, furniture and HVAC-devices during the years to meet the requirements. There are almost 5200 M1-classified low emitting building materials, fixture, furniture and HVAC-devices at the moment (May 2019). The Classification of Indoor Environment 2018 (classes S1, S2) and sub-classifications (P1) and (M1) are integrated part of the new Finnish environmental classification system for building processes called RTS Environmental Classification launched in the spring 2017 and updated in September 2018.

The Classification of Indoor Environment has been used in multiple office and residential construction projects during the years. By integrating the CIE 2018 requirements to the RTS 2018, third-party audit and certification of the whole construction project is possible. Therefore, the continuous improvement will be one considerable option.

1. Introduction

High indoor air quality is recognized as a crucial issue of both national health and economy in Finland. There are several projects and programs going on to ensure and excite to improve good indoor air quality and moisture management in building projects.

The Finnish Society of Indoor Air Quality and Climate (FiSIAQ) introduced twenty years ago in 1995 a Classification of Indoor Climate, Construction Cleanliness, and Finishing Materials. The first revised edition, Classification of Indoor Climate 2000, was published in 2001 and the second edition Classification of Indoor Climate 2008 was published in 2008 and the third edition was published in 2018. Based on the criteria set in the classifications, The Building Information Foundation RTS started the M1-labelling of building products twenty years ago in 1996, named Emission Classification of Building Materials and Cleanliness Classification of Air-handling Components. The testing protocol is updated continuously (RTS).

The purpose of the Finnish Classification for Indoor Environment is to help the user, building owner, developer, and designer of the building to set the target levels for indoor environment. Based on current knowledge of health and comfort, the target levels specified in the classification describe safe indoor environment conditions, the quality of which exceeds the requirements set by authorities. Indoor climate issues are considered in the conventional construction project in the same way as other functional requirements.

2. Classification for Indoor Environment 2018

The Finnish Society for Indoor Air Quality, FISIAQ, has a leading role in indoor air improvement and especially in developing the Finnish Classification of Indoor Environment 2018, even though the publication is a result of co-operation of many organizations. The targets, requirements, and instructions of the classification shall be taken into consideration at all stages of the construction project.

The developer or subscriber chooses the target values for the indoor environment in collaboration with the design team. The target values are chosen for each construction project from the values presented in the section 'Target values for indoor environment', either by choosing all the values from the selected category or by setting individually considered values for various parameters.

In order to reach the desired result, the developer shall control design by explicitly recording the desired indoor environment targets and making them known to all of the designers. Each member of the design team shall take care that the decisions made concerning the selected indoor environment class are presented in the following documents: drawings, work specifications, the supplement to the construction contract, and the quality control plan for the construction site.

The Classification of Indoor Environment has three parts:

PART 1: The classification of indoor climate gives target and design values for thermal conditions, odour intensity, noise levels, ventilation and indoor air pollutants.

PART 2: Guidelines for design and construction including classification of cleanliness in construction, principles and procedures for the main stages of construction works.

PART 3: The Emission Classification of Building Materials contains target values for odours and emissions of the materials, fixtures and furniture and recommended maximum surface area of materials based on their emissions.

The content must be considered when setting targets in the Green Leadership Tool.

2.1. PART 1: Target values for thermal environment and operative temperature in the CIE 2018

The following table (Table1) present the technical target values for thermal environment in the indoor air classes S1 and S2 which shall be used to specify the target level for indoor environment during the design phase of the construction project.

Table 1 Target values for thermal environment, new and old criteria represented

	New criteria S1 class	**Old criteria S1class**	**New criteria S2 class**	**Old criteria S2 class**
Operative temperature t_{op} [°C]				
$t_u \leq 0$ °C	21,5 [1)]	21,5 [1)]	21,5	21,5
$0 < t_u \leq 20$ °C	21,5+0,15xt_u [1)]	21,5+0,3xt_u [1)]	21,5+0,2xt_u	21,5+0,3xt_u
$t_u > 20$ °C	24,5 [1)]	24,5 [1)]	25,5	24,5
Deviation allowed from set value [°C]		±0,5		±1,0
Deviation upward				
$t_u \leq 0$ °C	<22,5		<23	
$0 < t_u \leq 15$ °C	22,5+0,166xt_u		23+0,2xt_u	
$t_u > 15$ °C	<25		<26	
Deviation downward				
$t_u \leq 0$ °C	>20,5		>20,5	
$0 < t_u \leq 20$ °C	20,5+0,075xt_u		20,5+0,025xt_u	
$t_u > 20$ °C	>22		>21	
Maximum value of operative		t_{op} +1,5		$t_u \leq 10$°C:t_{op} +1,5
temperature [°C]	<23		<23	$10 < t_u \leq 20$ °C: 23 +
$t_u \leq 0$ °C	23+0,2xt_u		23+0,2xt_u	0,4 * (t_u -10)
$0 < t_u \leq 20$ °C	<27		<27	$t_u >$ 20° :27
$t_u > 15$ °C				
Minimum value of operative temperature [°C]	20	20	20	20

A temperature control is essential for the indoor environment. In the updated classes S1 and S2, the temperature must stay within target value range 90% of the operating time. There is an exception in the class S2 for residences where the temperature must stay 80% of the time within target value range. The content must be considered when setting targets in the RTS Environmental Classification. One third of the points in the tool can be gathered from indoor air related criteria.

2.2. PART 1: Indoor environment quality in the CIE 2018

Technical target values of the indoor environment during the use phase apply to the occupied zone of a room (with respect to the speed of air, 0.1 m from the workstation at an elevation of 1.1 m). It usually extends from the floor surface up to 1.8 metres and begins at 0.6 metres from the walls.

The carbon dioxide concentration target applies to carbon dioxide from humans. The stability of the conditions is observed with the fluctuating one-hour average for carbon dioxide concentration. The target concentration of the particulates is the average concentration in indoor air during 24-hour measurement. The radon concentration is determined by a measurement method approved by Radiation and Nuclear Safety Authority.

$PM_{2.5}$ particulate concentration refers to airborne dust in indoor air with particles having an aerodynamic diameter of less than 2.5 μm. The particulate concentration is measured according to standard SFS-EN 12341 during normal use of the building and the measurement lasts for 24 hours.

The relative humidity of the air has not been given a target value because it can be very low during freezing temperatures. When using local humidification, the relative humidity of the air should be below 60 % and humidification should not cause the risk of moisture and microbial damage. In humidifying the air, attention shall be paid to the humidifiers not increasing impurities in the air. The relative humidity of the air is measured with, for example, a psychrometer or capacitive sensor in accordance with the standard SFS-EN 12599.

Table 2 Target values for indoor environment quality

	S1	*S2*	*S3*
Additional carbon dioxide concentration [ppm]*	*<350*	*<550*	*<800*
Radon concentration [Bq/m³]	*<100*	*<100*	*<200*
$PM_{2.5}$ [μg/m³]	*<10*	*<10*	*<25*
$PM_{2.5}$ indoors / outdoors	*<0.5*	*<0.7*	–
Relative humidity [% RH]	–	–	–
Stability of environment [% of operating time]			
• *office and school spaces*	*90%*	*90%*	
• *residential spaces*	*90%*	*80%*	

* Higher than the ambient carbon dioxide concentration.

2.3. PART 2: Guidance for design and construction in the CIE 2018

During the project design phase of the construction project, the developer /client / construction manager shall in co-operation with the design team determine and choose the target values related to indoor environment, construction cleanliness, and the emissions of building materials at the beginning of the construction project and shall make them known to all designers involved in the construction project.

In the project design stage, the indoor environment categories (S1, S2, S3) of space types central to the project programme are defined. In the proposed design, the targets are specified as individual target values. The chosen target values of the indoor environment and the related technical specifications shall be explicitly presented in all documents. The most crucial document from the standpoint of indoor environment is the contract limits appendix.

More detailed procedures are presented in the work specifications and instructions and obligations are presented in the contract documents. The contract limits appendix shall specify who is responsible for:

- schedules concerning (the drying out of structures before application of finishing materials the testing, adjustment and commissioning of mechanical systems; the required verification of cleanliness before the operation tests; protection of structures from the harmful effects of the weather)
- proper sealing of pass-throughs and joints in the structures
- on-site insulation of vibration and sound for vibrating mechanical units and pipes supported by the structures
- removal of water during construction phase so that the structures remain dry
- quality control plan of the construction site
- inclusion of the moisture control plan in the quality control plan of the construction site

- provision of clean and dry storage facilities for classified building materials and air-handling equipment, or corresponding arrangements
- premises for operational testing
- training the designers and contractors in mastery of the requirements of the classification

2.4. PART 3: New requirements of the Emission Classification of Building Materials

The classification of building materials has three emission classes, where emission class M1 corresponds to the highest quality and emission class M3 includes materials that don't meet the requirements of class M2. The classification presents emission requirements for the building materials, fixture and furniture used in ordinary work spaces and residences with respect to good indoor air quality.

Table 3 Requirements for the classes M1 and M2

Examined qualities	**M1-class [mg/m²h]**	**M2-class [mg/m²h]**	**Upholstered office chairs M1-class [µg/m³]**
The emission of total volatile organic compounds (TVOC). A minimum of 70% of the compounds shall be identified.	< 0,2	< 0,4	≤ 20
single VOC (µg/m³)	≤ EU-LCI	≤ EU-LCI	≤ EU-LCI or ≤ 10
The emission of formaldehyde (HCOH)	< 0,05	< 0,125	≤ 10
The emission of ammonia (NH_3)	< 0,03	< 0,06	≤ 10
The emission of compounds (CMR) belonging to category 1A or 1B in Annex VI to Regulation (EC) No 1272/2008 [mg/m³][1*]	< 0,001	< 0,001	≤ 1
Odour [2*]	Is not odorous	Is not odorous	Is not odorous

1* does not apply to formaldehyde
2* The result of sensory evaluation shall be at least + 0,0

The committee of Indoor Air PT17 has an authorization to accept new criteria for the classes M1, M2 and M3. The PT17 accepted one new tighter requirement for M1-class for building materials on 5th December 2018. The new criterion for "The emission of compounds (CMR) belonging to category 1A or 1B in Annex VI to Regulation (EC) No 1272/2008 [mg/m3]" is 0,001, the former requirement was 0,005 mg/m^2h. In addition, there have been included single voc criterion where have been referred to the EU-LCI-list.

The new specific requirements and criteria of the classification of building materials concerns especially office chairs with paddings and textile coverings. The criteria vary from the common criteria of M1 because of the special character of upholstered furniture. When one aims for the best indoor environment category, S1, the use of materials with higher emissions (emission classes M2 and M3) shall be restricted. Upholstered office chairs can only be classified in class M1.

The new criteria for Emission Classification of Building Materials mentioned in the third part of the Classification of Indoor Environment has been set in the classes M1 and M2. The content must be considered when setting targets in the Green Leadership Tool. All the standards behind the Classification of Building Materials can be found from the protocol: M1 Emission Classification of Building Materials: Protocol for Chemical and Sensory Testing of Building Materials. One of the accepted testing laboratories is in the USA and the other in Europe, two of them in Finland.

Figure 1. Classification of Building Materials and Classification of Air-Handling Components

3. Finnish Classification of Indoor Environment 2018 in the RTS Environmental Classification

RTS Environmental Classification is a system for managing construction process in a sustainable way. The system recognizes best national (Finnish) building strategies and practices. While there are globally accepted certification schemes, a need for national tool with local perspective and integration to local best practices like the Classification of Indoor Environment 2018, has been recognized especially in non-profit real estate sector as municipalities renovating and building schools, day care centers, health care centers and other social property. The environmental classification tool makes it easier to agree upon matters such as the energy efficiency targets of buildings and solution development.

The tool has been developed for the Finnish public real estate owners as municipalities, but it is suitable for all property owner's despite of the background and business model. The classification was launched in the spring 2017 and there are 90 active projects and 140 users in the tool on May 2019. The RTS Environmental Classification follows the changes, developments and needs of the market, therefore the criteria will be updated continuously.

The Finnish Classification of Indoor Environment 2018 is mentioned in the following criteria:
P3.2 Site dust management, S1.1 Indoor temperature, S1.2 Air quality, S1.3 User controls, S1.4 Material emissions, S3.1 Space acoustics, S3.2 Sound insulation. Although there are not specific requirements for air-handling components, it is assumed that there will be used M1-classified products like ducts, fittings, air filters etc. in the projects to achieve class S1 or S2 indoor environment conditions.

3.1. An example of the criterion S1.4 Material emissions

Low-emitting building products and furniture are essential part in achieving good indoor air quality. The assessment will consider all materials inside the building's vapor barrier, except for technical facilities that do not have direct contact with indoor areas (district heating substation, AHU-rooms, etc.) and products that are connected to indoor air. The low-emission of products is reported at the site phase with a list showing the name of the product, the manufacturer, target of use, and note of the emission certificates. The following certification marks and levels are approved for material emissions certifications: Emission class M1 for building materials, GEV Emicode EC1 and EC1Plus, Blue Angel, GUT. The criteria for low-emitting products include:

Low-emission materials have been effective in significantly reducing the total concentration of harmful indoor pollutants.
Low-emission indoor materials, 50% of the weighted value:
1: Paints, adhesives, carpets and floor coating used inside the vapor barrier and wood panels meet the emission limits of materials.

- Designer: Recording of material requirements into contract documents.

- Developer: Summary of the used products and certificates
- User:

2: Fixed furniture is low-emitting or manufactured from low-emitted materials (adhesives and coatings)

- Designer: Requirements are recorded in the plans
- Developer: Fixed furniture certificate or sub-component list and certificates
- User:

3: The inorganic fibres must be shielded or encapsulated in all indoor spaces. In renovation, all existing unprotected fibre sources should be removed or encapsulated.

- Designer: Requirement to encapsulate the inorganic fibres
- Developer: Inspection lists for building components containing inorganic fibres
- User:

4: In renovation, contaminants (PAHs, creosote, asbestos) should be removed from the contract area

- Designer: Clarification of contaminants if construction year before 1990
- Developer: Inspection lists for removing contaminants
- User:

Room air quality is indicated by measurements, 50% of the weighted value:

5: Measurements made using approved measurement methods in the commissioning of a building indicate that total room air concentrations are on acceptable level before the use of the building.

- Designer: Measurement obligation is recorded in the contract material
- Developer: Measurement report
- User: Measurement report

Alternatively, low emissions can be demonstrated by verifying equivalence with the emission class M1 requirements for materials by an acceptable test method (EN 16516: 2017 + EN ISO 16000-9: 2006 + ISO 16000-28: 2012). Low-emission base materials such as concrete, natural stone, ceramic and crimped tiles, as well as untreated timber are approved directly and are not required to be certified. In renovation, it is assumed that all residual surface materials can be thought to be low-emitting in an existing situation.

Inorganic fibres include e.g. fibreglass (technical fibreglass) and mineral wool fibres. Mineral wool fibres are used in thermal insulation materials for exterior walls, ground and top floor structures, thermal, sound and fire insulation materials in ventilation ducts, in air filters and in room acoustic insulation materials such as acoustic boards and panels. As a protective measure is accepted e.g. coating, encapsulation, paint coating of open surfaces.

Requirements for indoor air total concentrations

Measurements are made as one-time measurements in type-specific spaces before users with ventilation already in use. The measured concentrations must be below the following limit values before use:

Formaldehyde $\leq$ 100 $\mu g/m^3$, 30 min average

TVOC $\leq$ 1000 $\mu g/m^3$, 30 min average

The concentrations measured during the operating phase must be below the following limits:

Formaldehyde $\leq$ 50 $\mu g/m^3$, 30 min average

TVOC $\leq$ 200 $\mu g/m^3$, 30 min average

Accepted measurement methods for indoor air concentrations

Measurements of concentrations should be made according to the following standards:

Formaldehyde: ISO 16000-3

TVOC: ISO 16000-5 and ISO 16000-6 or ISO 16017-1

IOP Conf. Series: Earth and Environmental Science **297** (2019) 012045 doi:10.1088/1755-1315/297/1/012045

4. Conclusions

Classification of Indoor Environment 2018 is intended for the setting of indoor air targets concerning usual work and occupied spaces as public buildings, schools, day-care centres and dwellings. The previous version Classification of Indoor Environment 2008 was widely used in construction industry and therefore the updated version is suitable for RTS Environmental Classification as well to help setting indoor air targets, controlling and helping design and material choices, helping water and moisture control planning and construction cleanliness as well as construction site design.

Companies have developed low-emitting building materials, fixtures, furniture and HVAC- devices during the years to meet the requirements for Emission Classification of Building Materials and Cleanliness Class for Air-handling Components. These classifications are included in the Classification for Indoor Environment 2018. There are almost 5200 classified low emitting building materials including 70 furniture and 370 HVAC-devices (May 2019).

There have been clear improvements also in construction technology and the use of buildings. Moisture management has improved in construction sites and designers have been committed to achieve the targets. Companies that have chosen to use RTS Environmental Classification for construction processes have noted that most important thing is that the tool concentrate on the most relevant issues, it considers Finnish legislation and northern climate and it will guide the process and help with material choices.

The RTS Environmental Classification will help construction industry in setting environmentally acceptable targets for construction projects when operating in Finland. The Finnish legislation and both indoor air and outdoor air circumstances have been considered. By including CIE 2018 requirements to the RTS the process and outcome can be certified after the project. We have a long tradition in Finland to cooperate and create together acceptable systems and the ability is even more important in the future because of the accelerating demand for sustainable methods and instruments.

5. References (some of them only in Finnish)

[1] CEN/TC 350 standards
[2] Classification of Indoor Environment 2018, Finnish Society of Indoor Air Quality and Climate, FISIAQ. Espoo, Finland (will be published on May 2019 in English)
[3] Cleanliness Classification of Air-handling components, The Building Information Foundation RTS sr, http://m1.rts.fi
[4] Emission Classification of Building Materials, The Building Information Foundation RTS sr, http://m1.rts.fi
[5] M1 Emission Classification of Building Materials: Protocol for Chemical and Sensory Testing of Building Materials, The Building Information Foundation RTS sr, http://m1.rts.fi
[6] HTP Values 2016: CONCENTRATIONS KNOWN TO BE HARMFUL, Publications of the Ministry of Social Affairs and Health 2016:8
[7] Green Leadership Tool Criteria 2018, The Building Information Foundation RTS sr, http://glt.rts.fi

Finnish legislation: (can be searched from www.finlex.fi by reference code (e.g 782/2017), some of them are described only in Finnish or Swedish)
[8] Decree of the Ministry of Social Affairs and Health on Health-related Conditions of Housing and Other Residential Buildings and Qualification Requirements for Third-party Experts (545/2015)
[9] Terveydensuojelulaki. Suomen säädöskokoelma 763/1994.
[10] Ympäristöministeriön asetus rakennuksen ääniympäristöstä 2017
[11] Ympäristöministeriön asetus rakennusten kosteusteknisestä toimivuudesta 782/2017

[12] Ympäristöministeriön asetus uuden rakennuksen energiatehokkuudesta 1010/2017
[13] Ympäristöministeriön asetus uuden rakennuksen sisäilmastosta ja ilmanvaihdosta 2017
[14] The National Building Code of Finland, Ministry of Environment, http://www.ym.fi/en-US/Land_use_and_building

Impact of hidden defects on the durability and reliability of gas pipelines in cities

A Romanovs[1], J Tihana[1] and A Borodinecs[1]

[1] Riga Tehnical University, FCE, 6A/6B Kipsala Street Riga LV-1048, Latvia
E-mail: *aleksandrs.romanovslg@gmail.com*

Abstract. The article describes the methods of improving the durability of new gas pipelines, as well as examines the need of rising the level of inspection security. Reliable gas pipelines is an important safety indicator of people's housing areas and the industrial and commercial organizations work areas. The verification of durability and tightness of gas pipelines is one of the most important inspection, which shows the existence of defects in compliance of the gas pipelines and debt construction technologies and instructions. Detected on time defects influence the number of uncontrolled natural gas leaks in the pipelines. When assessing the impact of hidden defects on the durability and reliability of the gas pipeline, the gas pipeline monitoring process was made, and the defects in the construction of gas pipelines (polyethylene, steel) and fittings were more often detected, the effect of hidden defects of gas pipelines on the safety of pipelines was analyzed. The experiments described in the article were carried out to evaluate the durability and tightness of gas pipelines by simulating the hidden defects on new installed gas pipeline in two sites: Site 1 and Site 2, both places are situated in Latvia. The obtained data was processed and reflected in tables, charts, graphs, and the results were analyzed and summarized in the conclusions, and also proposals are made to improve the problem. In the article is described a calculation methodology, which proves what needs to be improved in order to increase the safety of newly built and reconstructed gas pipelines.

*This work has been supported by the European Regional Development Fund within the Activity 1.1.1.2 "Post-doctoral Research Aid" of the Specific Aid Objective 1.1.1 "To increase the research and innovative capacity of scientific institutions of Latvia and the ability to attract external financing, investing in human resources and infrastructure" of the Operational Programme "Growth and Employment" **(No.1.1.1.2/VIAA/ /2/18/259 Efficiency of compact gas hybrid appliance in Latvian climate conditions**)*

1 Introduction

Currently there is an increase in the gas pipeline construction sector in Latvia, and methods of assembly and quality testing are also developing.

The monitoring organisations are involved in construction from the design of the project to the commissioning of the site. In Latvia, the technical supervision of gas pipeline construction is ensured by representatives of gas pipeline distribution network operator "Gaso".

The gas pipeline construction monitoring process consists of multiple operations to be tested by network operator engineers: inspection and registration of a building project, inspection of gas pipelines at the site, taking a trench and accepting other communications at the site, verification of the quality of the welding process technologies, inspection of welding and assembly at the site, closure of the pipeline filling and improvement at the installation, inspection of the internal cleanliness of the gas pipeline, removal at the site. Inspection of the strength and tightness of the gas pipeline at the installation, inspection of the project execution documentation and acceptance of the gas pipeline in operation [7]

[8], [9]. Checks on pipeline networks are necessary to prevent accidents from causing great harm and damage to people (financial, economic, health and life, etc).

2 Gas pipeline strength tests

The stress test for gas pipelines is one of the most important. It shall demonstrate the safety and operational readiness of the projects envisaged. It is the stress and tightness tests of the gas pipeline that show that there were errors in the construction of gas pipelines and that construction technology and instructions were not followed. The defects detected at the time prevent the uncontrolled number of natural gas leakages (discards) by putting gas pipelines into service.

It also has a big impact on the safe lifestyle of the satellite, not only the external gas pipeline systems, but also the internal gas pipeline systems. Where there are a series of their own factors [10].

Construction works are a complex technological process that starts with project calculations, matching materials and continues through the implementation of complex building physics processes in practical construction activities [1].

In Latvia, nitrogen has been used to check strength and tightness in recent years. The stress and tightness of gas systems shall be tested with manometers having an accuracy class of at least 0,6 and a maximum measurement range of 150% of the test pressure.

The duration of the strength test for external gas pipelines shall be 1 hour during which the pressure is not allowed to fall. [12], [2].

The leakage of underground gas pipelines shall be checked after their backfilling to the markings of the building project. The test pressure shall be maintained prior to the leakage test in the gas pipeline so that the temperature is aligned with the ground temperature. The minimum temperature adjustment duration shall be as follows:

- for a gas pipeline with a notional diameter (DN) of up to 300 mm (inclusive) for 6 hours;
- for a gas pipeline with a notional diameter (DN) above 300 mm to 500 mm (inclusive) for 12 hours;
- for a gas pipeline with a notional diameter (DN) above 500 mm - 24 hours. [10], [18].

During the monitoring process, the following data for the 2016-2018 year have been compiled for the assessment of the strength and tightness of the pipeline system for newly constructed sites (Figure 1):

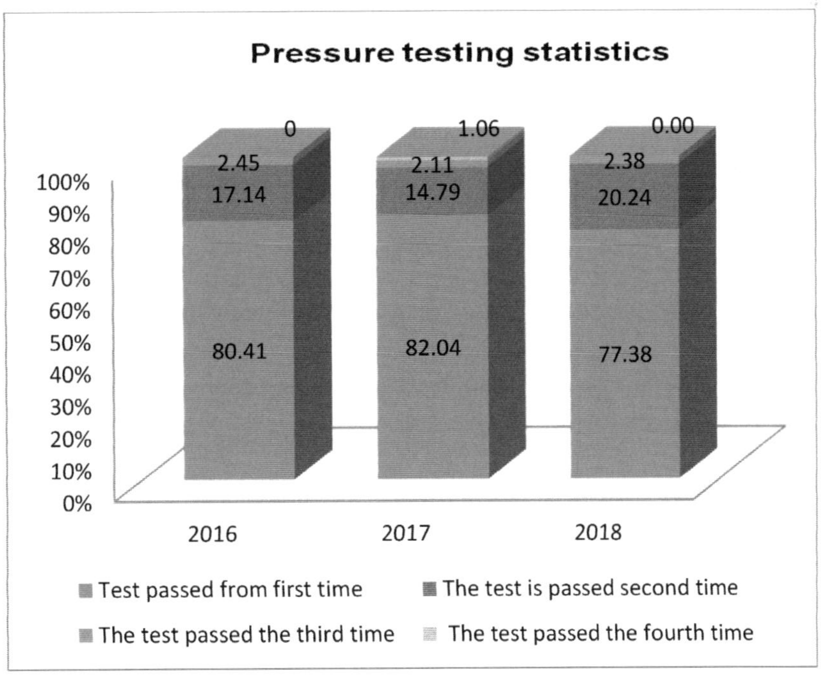

Figure 1. *Site strength and tightness testing 2016-2018 (source AS "Gaso")[13].*

On the basis of aggregated data, it can be concluded how on average 80% of the tested sites have passed the strength and leakage test from the first time, 17% from the second time, and 3% only from the third

time [13]. Dmitri Ohotski years 2015 has also complidet the results of calls by AS "Latvijas Gaze" (now AS "GASO") for his master's work (to 2012 by 2014) and has concluded that on average 20.84% of the cases of gas leaksges have been detected for operating underground gas pipelines [11].
Accident survey allows for analysis of the situation and improvement of emergency work by means such as updating emergency plans, high-quality training of personal and the setting up of modern equipment at sites [17].
This means how, before the new gas pipeline was taken into service, 20% of gas leakages were detected and eliminated.
During the leakage test of the underground gas pipeline, the actual drop in pressure must not exceed the calculated permissible value, which is determined if the pressure is measured in millimetres of the mercury column, using the following formula [12], [10]:

$$\Delta P = \frac{20T}{d} \qquad \left(\Delta P' = \frac{150T}{d}\right) \tag{1.}$$

where ΔP – alloweble pressure falls (KPa);
$\Delta P'$ – allowed pressure drop (mm Hg);
d – the internal diameter of the pipeline (mm);
T – test time (h).

The actual pressure on the pipeline ΔPf (kPa vai mm Hg) during the test shall be calculated using the following formula [10]:

$$\Delta P_f = (P_1 + B_1) - (P_2 + B_2), \tag{2.}$$

where P_1 and P_2 – indication of the manometerm at the beginning and end of the test (kPa or mm Hg);
B_1 or B_2 – Barometer readings at the beginning and end of the test (kPa or mm Hg).

If the pipeline cinsists of pipeline sections with different diameters (d_1, d_2, d_3...d_n), the mean internaldiameter shall be calculated using the following formula [10]:

$$d = \frac{d_1^2 l_1 + d_2^2 l_2 + d_3^2 l_3 + \cdots d_n^2 l_n}{d_1 l_1 + d_2 l_2 + d_3 l_3 + \cdots d_n l_n}, \tag{3.}$$

where d_1, d_2, d_3...d_n - internal diameter of the phases of the pipeline (mm);
l_1, l_2, l_3...l_n – length of the relevant diameter pipeline section (m).

The duration of the leakege test for underground gas pipeline shall be 24 hours. The leak test may be carried aut at the same time as the strength test (combined inspection) using the same anvironment and pressure [10], [12].
However, the required length and volume of the pipelines in metres of the calculation pipelines have not been determined and there is no appropriate methodology for calculating them.
The calculation methodology presented in the LBN 242-15 allows for the determination of the permissible loss reduction by calculating the ratio between the test time and the average diameter of the gas pipeline to be inspected. There may be a possibility of leakage that can be classified as "hidden defects" below the limit of allowable losses. In identifying the most important factors affecting the existence of "hidden defects", a set of the size, length and diameter of the pipelines.

3 Gas pipeline defects

Such defects, which are not detected during the strength and leak test, are called "hidden defects":

- poor quality of welded compounds [20];
- defects found during transportation and assembly of the pipeline;
- industrial defects of the pipeline;
- pollution which has a negative impact on the quality and integrity of the pipeline connections;
- corrosion of steel pipelines;
- defects in pipeline fittings;
- Failures resulting from third part activities [4], [5], [6].

If the pipe connection is welded to the technology, the joint should be as strong as the pipe [19].

3.1. Poor fusion interface

Cracking down of gas pipelines through the fusion interface which was formed when the oxidation layer from the external surface of a weldable pipeline has been removed unqualitatively (Figure 2a), there was an inappropriate temperature at the welding site (Figure 2b), no purification and degreasing methods [15], the article contains specific defects that resulted from non-application of the technology.

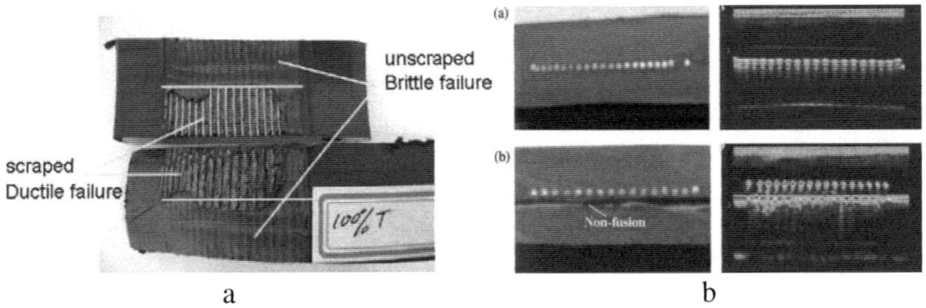

Figure 2. a - *Defect with unremoved oxidation layer, 4073* b - *Defect – Cold welding, 4072) [15]*

3.2. Welding tank formation

Cracking of gas pipelines during assembly - electrical welding compounds containing voids. A defect in emptiness can result in a structural rupture at the welds sites and cause a stress concentration that can further create a gap. The main causes of the voids are the choking air and no dew point was removed on the outer surface or there has been wet contamination, shown in (Figure 3a) [15].

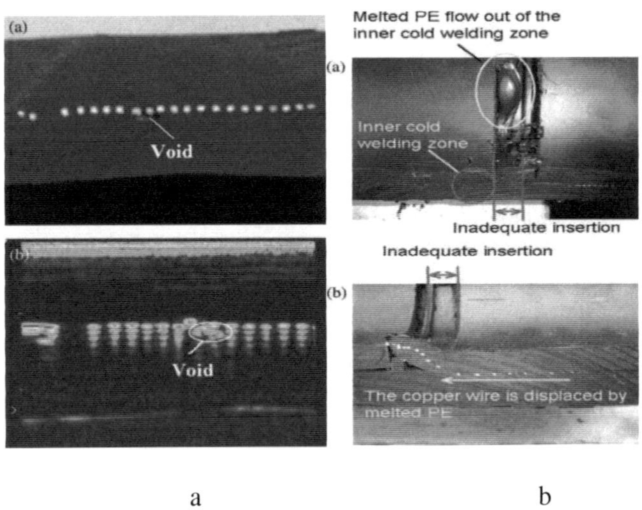

Figure 3. a - Welding tank formation, 4074 *b - Structural deformation, 4072 [15].*

3.3. Structural deformation

Cracking of gas pipelines through a built-in copper heating element wire. Failure of the electrical welding coupling to maintain the expected structure after welding when deviations from the correct pipeline fitting welding fitting fitting. Consequently, the gas pipeline contains bottlenecks filled with the mass of the melting process as a result of the welding process. This mass may pass through cold areas of the fitting outwards or inwards. If the mass is released outwards, it may be visually detected and, if inwards, cannot be detected. Position of the deformation heating element in the same process outwards or inwards running with the mass formed in the melting process (Figure 3b) [15], [14].

4 Hidden defects" in the object

Monitoring of newly constructed gas pipelines was carried out at the construction site in Latvia (hereinafter the Site1). The newly constructed pressure P ≤ 4 bar distribution pipeline and 38 privat house connection to the site. Length of the newly constructed gas pipeline 2844,51 m. Distribution pipeline: 1841,04 m: DN 50 mm - 1405,23 m and DN 130,8 mm - 435 m and leads DN 25 mm - 1003,47 m. Average diameter of the gas pipeline **DN** 76,11548 mm (formula 3). Decrease limit **ΔP** = 47.297 mmHg (formula 1). The volume of gas pipelines without pressure shall be 9,103142 m^3 (formula 4). Volume of the gas pipeline systems:

$$v_{syst} = \frac{\Pi \cdot Dn^2}{4} \cdot L \text{ (m}^3\text{)}, \quad (4.)$$

where ∏ - pi – 3.14 const;
Dn – internal diameter (m);
L – length og the pipeline section (m).

The pipeline inspection was carried out with combined strength and tightness testing (24 hours) and during the test after all obtained for measurements, a "Hidden Defect" was detected in the gas pipeline (Table 1) When reading the test data and making all calculations using the formulae in Chapter 2, the results obtained has not exceed the permissible pressure loss limit.

Table 1. Results of the Site 1 "Work Site".

Stages	Test time (hour)	Manometric pressure				Barometric pressure (mmHg)		Allowable loss limit	Actual drop in losses
		P1 (bar)	P2 (bar)	P1 (mmHg)	P2 (mmHg)	B1	B2	mmHg	mmHg
1	24	7.092	7.031	5319	5273.25	761	763	47.297	43.8
2	24	7.127	7.091	5345.25	5318.25	762	761	47.297	28.00

In Table 1, can be seen that after the results of Stage 1, it is apparent that the allowable loss limit is not exceeded, but in fact two gas leakage sites were found in the pipeline. The results of Stage 2 show that the actual margin of reductions (from 43.8 mmHg to 28.00 mmHg – 36.07%) has changed, preventing gas leaks in the pipeline system. Consequently, it can be concluded that there is a relationship between the diameters, lengths, system volumes that can be verified by formulas and conclusions drawn. An experiment was needed to test this relationship in real terms.
Stankevič N. has mentioned in his book that there should be a certain gas pipe diameter and maximum pipeline inspection length (up to DN 200mm - 12 km, DN 300mm - 8 km, DN 400mm - 6 km, DN 500mm - 5 km) but not clear from how it was calculated [21]. Also, the DVGW standard G 469 (2010) mentions the relationship between the diameter of the gas pipelines to be tested and the maximum length, taking in to account the geometric volume of the pipeline [2]. Also, the methodology for calculating these maximum diameters and lengths is not shown here. Lithuanian Standard Order for Approval of Distribution Pipeline Installation Regulations 2016 May 17 No. 1-162 (2016) takes the maximum length of the gas pipeline test, but the volume only to determine the required stabilization time of the gas pipelines when the gas pipeline is filled with nitrogen and is ready to test for strength and tightness (leakage) [16].

5 Experiment

The experiment was couducted on a new pipeline distribution grid and 11 privat house connection to the site with pressure P ≤ 100 mbar in Latvia (hereinafter Site 2). The length of the newly constructed gas pipeline, which has been studied, is 2171.23m. From it distribution gas pipeline - 1969.55m: DN 50mm - 188.83m, DN 73.6mm - 1229.44m, DN 90mm - 42.69m and DN 100mm - 508.59m. Drives - 201.68m: DN 25mm 144.34m and DN 32mm - 57.34m. The average diameter of gas pipelines is DN 79,40831mm (formula 3). Permissible drop limit ΔP = 45.34mmHg (formula 1). The volume of gas pipelines without pressure is 9,9821m3 (formula 4). The pipeline scheme can be seen in Figure 4.

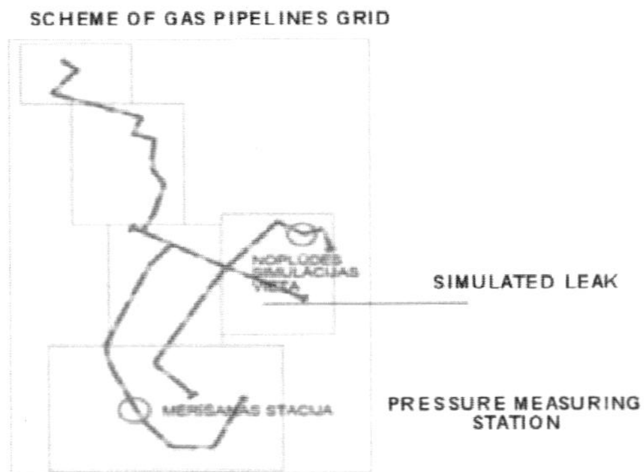

Figure 4 Pipeline diagram with the specified leakage simulation site and measuring station [13].

5.1 Preparing a Sie 2 for an experiment

The measuring station was integrated in the system (Figure 5) where the measuring devices (pressure gauge) are installed and the results of the pressure test are read. Inert gas — nitrogen (N2) has been used for pressure testing. The pipeline system is nitrogen-filled, not exceeding the permissible pressure test parameters.

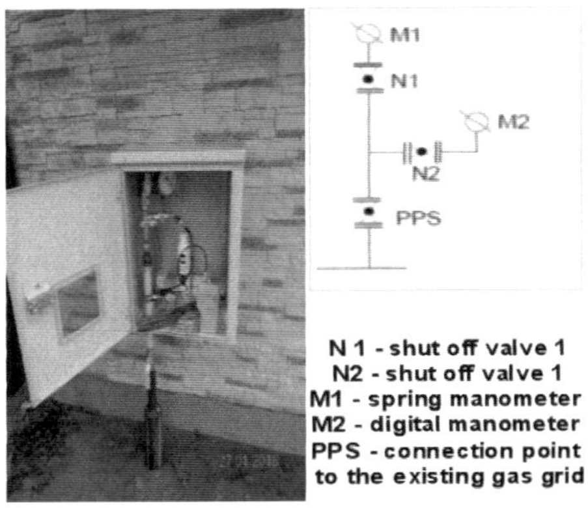

Figure 5. Scheme and photo of the measuring station.

Figure 5 shows the pressure regulator and gas meter and schematically describes the Measuring Station device. During the experiment, a gas leak simulation site was selected. The simulation equipment consists of two shut off valves, two flow interrupters and manometers (Figure 6).

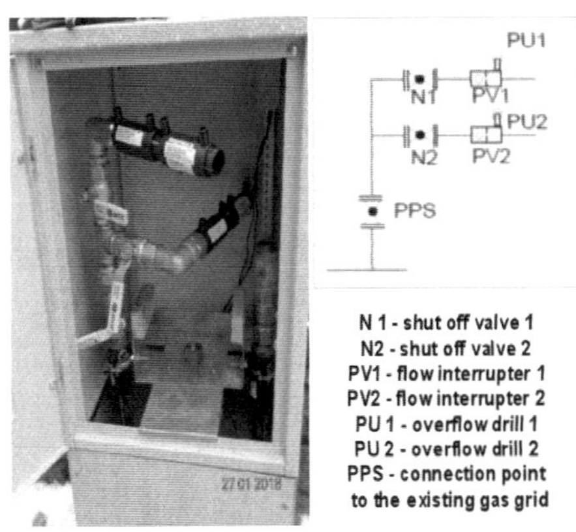

Figure 6. Gas leakage simulation site.

5.2. Description of the experiments progress

Experiment performed in 3 steps:

In step 1, the author reads the pressure data of the pipeline system from the site of the pressure measuring station, using manometers to establish that the pipeline system is airtight.

In step 2, the author opens the shut off valve No 1 and gives a verifiable system pressure to flow interruptor No 1. As a result of the operational principles of the flow interruption, the interruptor No 1 closes. (the flows were interrupted by the principle of action) [3].

Knowing the selected flows disrupted the industrial data in the V_L — (overflow well airflow — V_L l/h < 30 l @ 1bar per hour) and the pressure in the system, the estimated amount of released nitrogen in m3 for 1 hour and 24 hours was calculated [22]:

$$V_L=30*P_n\ l/h \tag{5}$$

where P_n – pressure system by measurement number (bar):

$$V_L=30*7{,}118=213{,}54 \text{ l/h or } V_L=213{,}54/1000=0{,}21354 \text{ m}^3\text{/h}.$$

Leakage through the overflow well No 1 simulates the "hidden defect" to be investigated in the system. Experimental time of the leak - 1 hour. After an hour, the pressure in the system at the location of the measuring station is re-measured. Experimental time of the leak - 24 hours. After 24 hours, the pressure in the system at the location of the measuring station is re-measured.

In step 3, the author shall open shut off valve No 1 and 2 and give the system pressure to be checked to the flow interruptors Nos 1 and 2. In the context of the operating principles of the flow transformer, interruptors Nos 1 and 2 shall be closed. Leakages simulating the "hidden defect" system to be investigated shall occur through overflow drills No 1 and No 2. Experimental time of the leak - 1 hour. After an hour, the pressure in the system at the location of the measuring station is re-measured. Experimental time of the leak - 24 hours. After 24 hours, the pressure in the system at the location of the measuring station is re-measured.

The author reads all the data and makes calculations based on the formulae mentioned above, compiling the results in Table 2 [13].

Table 2. Results of the experimental Site 2.

Stages	Test time (hour)	Manometric pressure						Barometric pressure (mmHg)		Allowable loss limit		Measurement error 0.1%		Expected drop in losses		Actual drop in losses	
		P1 (bar)		P2 (bar)		P1 mmHg	P2 mmHg	B1	B2	bar	mmHg	Robeža (+-)	bar	bar	mmHg	bar	mmHg
1	1	Nr.1	7.29	Nr.2	7.29			753	753	7.28	0	0.0073	7.28	0	0	0.00	0
	21.52	Nr.2	7.29	Nr.3	7.27	5466	5456	753	758	7.23	40.65	0.0073	7.28	0	0	0.01	5.5
2	1	Nr.4	7.12	Nr.5	7.11			764	764	7.11	0	0.0071	7.11	0.02	15.75	0.006	4.5
	24	Nr.8	7.10	Nr.9	6.91	5325	5185	764	753	7.04	45.34	0.0069	6.92	0.51	395	0.19	151
3	1	Nr.6	7.11	Nr.7	7.10			764	764	7.10	0	0.0071	7.10	0.04	32.25	0.01	6
	24	Nr.10	6.91	Nr.11	6.68	5183	5012	753	750	6.85	45.34	0.0067	6.69	0.99	748.5	0.23	174

In the analysis of the results obtained, which were summarised in Table 2, it can be concluded:

1- Phase results (steps): After reading the measurements of No.1 and No.2 from the gas pipeline hermetic system with digital and reflective pressure gauges with 1 hour difference the results showed that the pressure drop is not observed, the strength test is passed and can take No.3 measurement after 24 hours.

When calculating the actual pressure drop at the site after 24 hours, losses amounted to 5.5 mmHg. In the light of all standards, the permissible pressure reduction limit has not been exceeded, the leak check has been passed and the pipeline system can be accepted into service [13].

2- Phase results (steps): The expected gas leak at an installation with a started one flow interruptor after 1 hour and, after all calculations, was composed at 213,54 l/h (formula 5) or 7,078 bar according to pressure gauge figures.

The actual gas leak at an installation with a started one flow interruptor was 60 l/h (formula 5) or 7.11 bar after 1 hour, following pressure gauge parameters – 3.5 times less than expected – 8.4 l*1 bar/h (formula 5) and maximum drilling overflow (according to producers' data) was not achieved.

So then: After phase 1 and phase 2, the first hour pressure losses at the installation amount to 0.006 bar – do not exceed the error of the digital manometer measurement. This means that the experiment can be continued with 24 hours of combined tightness testing in phase 2.

In Figure 7, you can see an author's calculation schedule for the loss of gas pipeline pressure with a started one flow interruptor (PV1) after 1 hour of action (bar), where the expected loss of pressure exceeds the measurement error limit by 19 minutes, while the actual loss of pressure at 60 minutes does not reach the measurement error limit.

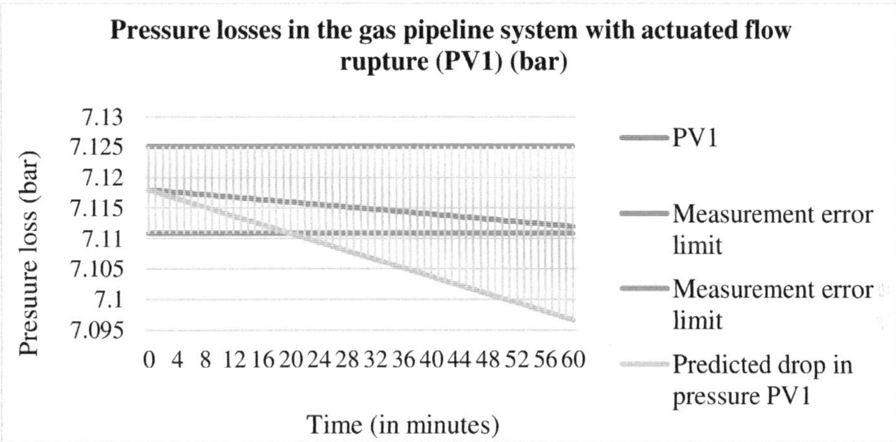

Figure 7. Pressure losses in the gas pipeline system with actuated PV1 bar/1h.

Predicted gas leakage at the site after 24 hours and after all calculations was 0.213 m^3/h (Formula 5) or 6.58 bar at manometer values. The actual gas leak at the object was 151.25 mmHg or 0.178 bar after 24 hours.

By all standards, the allowed pressure drop limit is exceeded, the leakage check has not been passed, the pipeline system cannot be accepted into service.

In Figure 8 you can see an author's calculation schedule for the loss of gas pipeline pressure with a started single flow interruptor (PV1) after 24 hours (bar), where the expected loss of pressure exceeds the limit of a measurement error of 3 hours, while the actual pressure loss of 7 to 8 hours reaches the limit of the measurement error. Gas leakage at each hour was 78 l/h (formula 5), 2.7 times less than expected gas leak – 11.25 l*1 bar/h is the maximum drilling overflow.

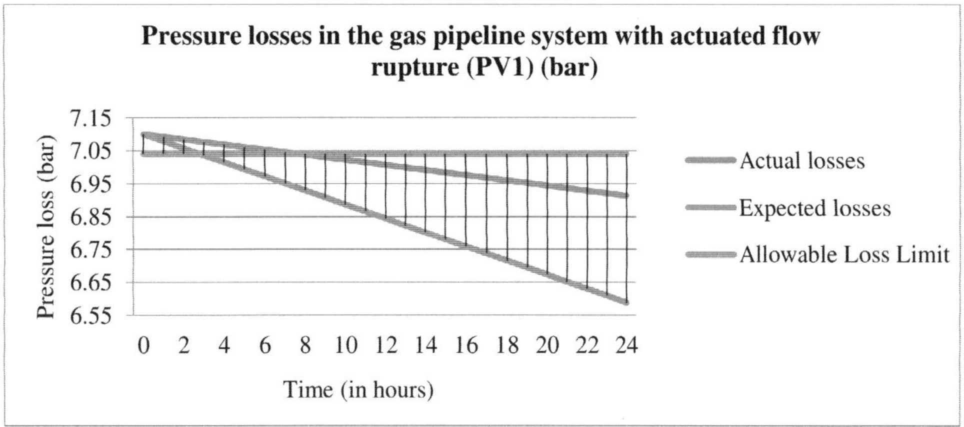

Figure 8. Pressure losses in the gas pipeline system with actuated PV1 bar/24h.

3 - Phase results (steps): The expected gas leak at an installation with two flow interruptors activated after 1 hour and, by all calculations, was 426 l/h (formula 5) or 7,065 bar based on manomerts.
The actual gas leakage at an installation with two flow interruptors activated was 80 l/h or 7.10 bar after 1 hour, according to pressure gauge figures. Gas leakage 5.3 times lower than expected – 11.24 l*1 bar/h and maximum drilling overflow (according to producers' data) was not achieved. If divided into each rupture, the gas leak through one overflow of the well — 5,62 l*1 bar/h.
By all standards, the allowed pressure drop limit is not exceeded, the leakage check is, passed and the pipeline system can be accepted into service.
Author decides to continue the experiment by performing a 3-phase combined leak test with two flow interruptors engaged for 24 hours.
In Figure 9 you can see an author's calculation schedule for the loss of gas pipeline pressure with two flow interruptors (PV2) engaged after 1 hour (bar), where the expected loss of pressure exceeds the limit of the measurement error by 10 minutes, while the actual loss of pressure at 53 per minute.

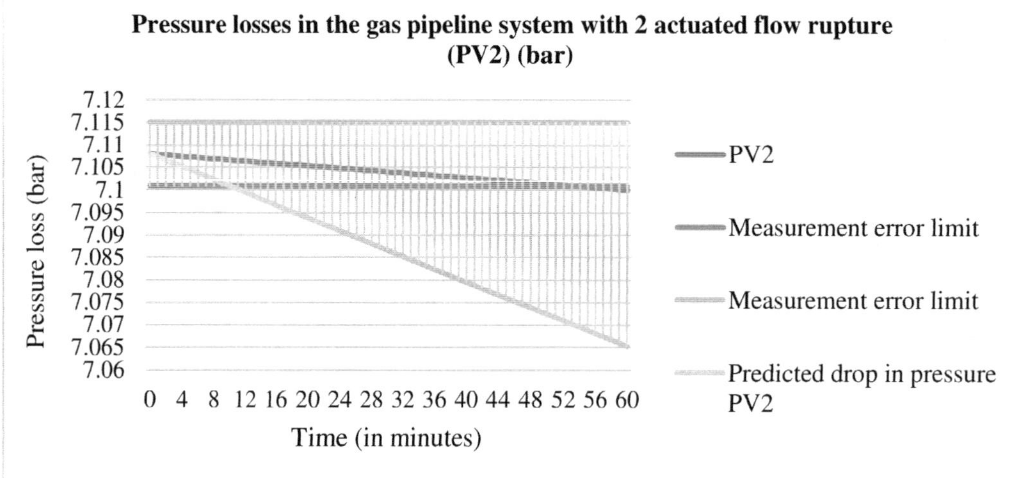

Fugure 9. Pressure losses in the gas pipeline system with actuated PV2 bar/1h.

The expected gas leakage at an installation with two flow interruptors activated after 24 hours and, for all calculations, was 0,415 m^3/h or 5,91 bar based on manomerts. The actual gas leak at an installation with two flow stoppers activated was 174 mmHg or 0.228 bar after 24 hours, according to pressure gauge figures. Taking in, too account all standards, the allowed pressure drop limit is exceeded, the leakage check has not been passed, the pipeline system cannot be accepted into service.
In Figure 10, you can see an author's calculation schedule for the loss of gas pipeline pressure with two flow interruptors (PV2) activated after 24 hours (bar), where the expected loss of pressure exceeds the limit of a measurement error of 1.5 per hour, while the actual loss of pressure of 6.5 per hour reaches the limit of the measurement error.
The gas leak at each hour was 95 l/h, which is 4.4 times less than the expected gas leak – 14,19 l*1 bar/h is the maximum overflow of drills. If divided into each rupture, the gas leakage through one overflow of the well — 7,10 l*1 bar/h.

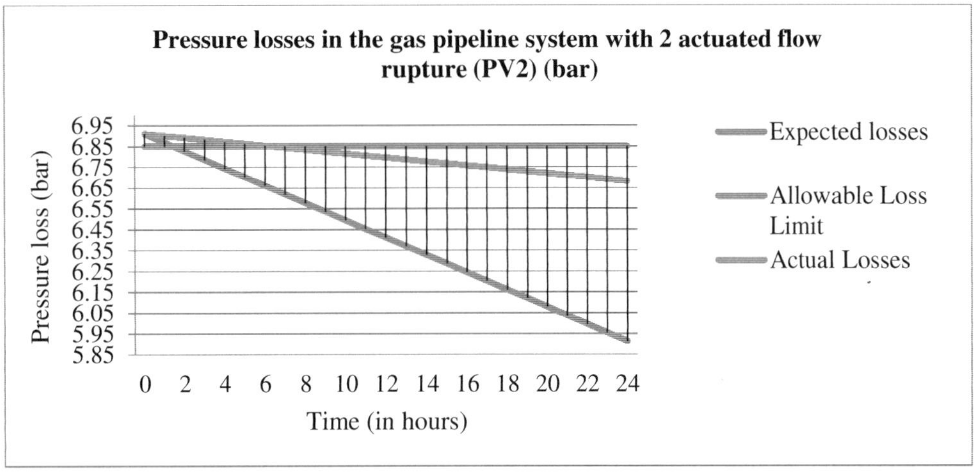

Figure 10. Pressure losses in the gas pipeline system with actuated PV2bar/24.h.

Conclusion

The results of the experiment:

- By starting one flow interruptor for one hour, the pressure losses of the gas pipeline were in line with the regulatory framework (the potential defect tured into a "Hidden defect").
- By starting two flow interruptor pressure losses for one hour exceeded the limit value-"Hidden defect" turned into exposed defect.
- By exposing both one and two interuptor for 24 hours, pressure losses exceeded the limit value "Hidden defect" turned into exposed defect.
- Figure 11 concludet that when sheking the strength of gas pipeline, leakege (combined inspection) and pressure loss calculations were carried out using the given clculation methodology "Site2", pressure losses exceeded the premissible loss limit – the calculation methodology works, defects were found in the pipeline. One the other hand, at "Site 1" (similar to the parameters of the pipeline system), the pressure loses did not exceed the premissible loss limit – no defects were detected in the gas pipeline (erroneous data).

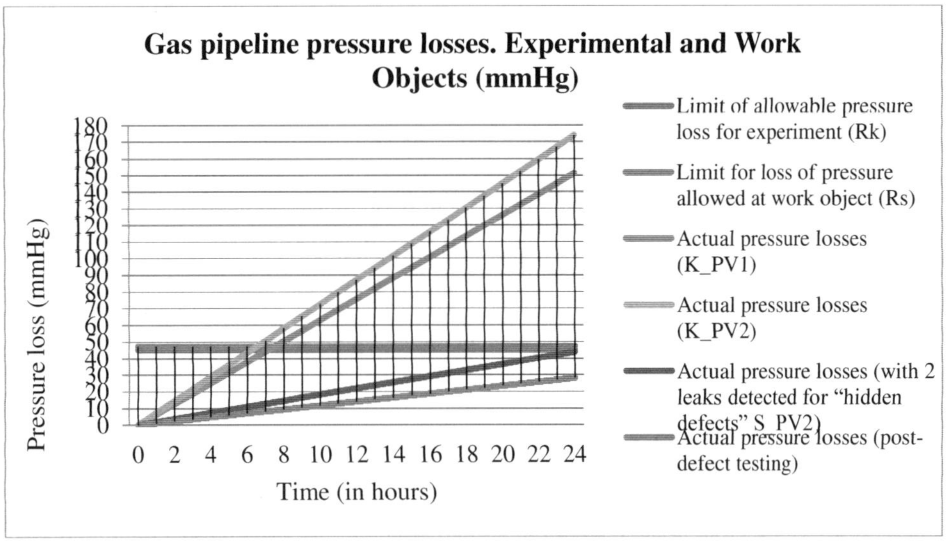

Figure 11. Gas pipeline pressure losses. Object 1(Work) and Objects 2 (Experimental) (mmHg).

- The results of the experiment confirmed the non-compliance of the existing calculation methodology and the regulatory base by testing the gas pipeline systems for strength and tightness of gas pipelines with increased and varying diameters, volumes and lengths.

In the Site 1 "Work" a more significant percentage of gas pipeline length and volume is made by the gas pipeline at the building. Calculating mathematically, this fact significantly reduces the average diameter of the total gas pipelines, which affects the calculation of the permissible gas loss limit for the gas pipeline as whole.
The total volume of gas pipeline drives in the Site 2 "Experimental" is smaller and the calculated average diameter of the gas pipeline is close to the actual average diameter of the gas pipeline.
When designing new gas pipelines with an average diameter close to the minimum diameter of the Dn 50mm gas pipelines, the average geometric volume of 9 - 10m^3 and the length of 2 - 3 kilometers, it is recommended to test the strength and tightness of the gas pipeline in stages to increase safety criteria.
During the period when the building regulations Latvian Building Code, Latvian National Standard are renewed, it is recommended to increase the stress test time by at least 3 hours.
Continuing further research, calculation methodology and standard improvement requirements to determine more accurate gas pipeline test parameters using author methodology.

References

[1] Biršs J 2015 *Errors in Construction and Possible Solutions* (Rīga J.L.V Press) p 239
[2] Deutscher Verein des Gas - und Wasserfaches e.V. Nr. G 469 2010 *Druckprüfverfahren für Leitungen und Anlagen der Gasversorgung* (Bonn 16 S)
[3] Gabrens P 2017 *Overflow valves in natural gas distribution networks* (Master's works at Riga Technical University) p 61
[4] Latvian National Standard Nr. 418 2010 *Gas supply systems, external gas pipelines, construction works* (Riga JSC "Latvijas gāze" Press) p 44
[5] Latvian National Standard Nr. 421 2016 *Gas Distribution and User Systems. Additional requirements for the design, construction and repair of polyethylene gas pipelines* (Riga JSC "Latvijas gāze" Press) p 27
[6] Latvian National Standard Nr. 423 2016 *General requirements for corrosion protection of steel underground gas pipelines* (Riga JSC "Latvijas gāze" Press) p 31
[7] Latvian National Standard No. 445-1 2009 *Operation and maintenance of natural gas distribution and User gas supply systems with maximum operating pressure up to 1.6 Mpa (16 bar). General Requirements* (Riga JSC "Latvijas gāze" Press) p 52
[8] Latvian National Standard No. 445-2 2009 *Operation and maintenance of natural gas distribution and User gas supply systems with maximum operating pressure up to 1.6 Mpa (16 bar). Maintenance Times, Description of Works and Documentation of their Implementation* (Riga JSC " Latvijas gāze " Press) p 118
[9] Latvian National Standard No. EN 12327 2012 *Gas infrastructure. Pressure testing, commissioning and decommissioning procedures. Functional Requirements* (Riga JSC "Latvijas gāze" Press) p 21
[10] Latvian Building Code LBN 242-15 2015 *Natural Gas External Gas Pipeline System* Adopted by the Cabinet of Ministers on June 30, 2015, Latvijas Vēstnesis, website Likumi.lv available at https://likumi.lv/ta/id/275017)
[11] Ohotskis D 2015 *Reliability of Internal Natural Gas Systems* (Master's work at Riga Technical University) p 64
[12] Platais I and Graudiņš P 2008 *Gas Supply Part 2 Development, Installation and Maintenance of Natural Gas Gas Supply Systems* (Riga RTU Publisher) p 219
[13] Romanov A 2018 *Impact of hidden defects on the durability and safety of gas pipelines* (Master's work at Riga Technical University) p 70
[14] Chebbo Z, Vincent M, Boujlal A, Gueugnaut D un Tillier Y 2015 *Numerical and Experimental Study of the Electrofusion Welding Process of Polyethilene Pipes* available www.wileyonlinelibrary.com)

[15] Jianfeng, S., Jinyang, Z., Weican, G., Yongquan, Q. *"Defects Classification and Failure Modes of Electrofusion Joint for Connecting Polyethylene Pipes"* (2011) [tiešsaite]. available: www.wileyonlinelibrary.com

[16] Lithuanian Construction Standard No. 1-162 2016 *Order for Approval of Distribution Pipeline Installation Rules* (adopted by Lietuvos Respublikos Energetikos Minister's decision of 17 May 2016, www.e-tar.lt available: https://www.e-tar.lt/portal / en / legalAct / 7d040a101e8111e69446a4bedc730fe6)

[17] Alenkov D and Tarasenka V 2013 *Analysis of Accidents for Gas Distribution Systems and Gas Consumption* (Polymer Gas No. 2) p 5

[18] Ionin A 1989 *Gas Supply* (Moscow Building Press) p 439

[19] Karnaukh N and other 1998 Polymers in gas reinforcements. Catalog (Moscow Engineering Press) p 856

[20] O'Connor K 2012 *Polyethylene Pipes: How to Avoid Welding Defects?* (Russian Gas No. 3) p 48-58

[21] Stankevich N 1960 *Guidance on the Supply of Gas* (Leningrad Krasnij Pechatnik Press) p 875.

[22] http://plasson.ru/downloads/action/show_file/f/2

This work has been supported by the European Regional Development Fund within the Activity 1.1.1.2 "Post-doctoral Research Aid" of the Specific Aid Objective 1.1.1 "To increase the research and innovative capacity of scientific institutions of Latvia and the ability to attract external financing, investing in human resources and infrastructure" of the Operational Programme "Growth and Employment" ***(No.1.1.1.2/VIAA/ /2/18/259 Efficiency of compact gas hybrid appliance in Latvian climate conditions)***

PostDoc Latvia

NACIONĀLAIS ATTĪSTĪBAS PLĀNS 2020

EIROPAS SAVIENĪBA
Eiropas Reģionālās attīstības fonds

IEGULDĪJUMS TAVĀ NĀKOTNĒ

IOP Conf. Series: Earth and Environmental Science **297** (2019) 012047 doi:10.1088/1755-1315/297/1/012047

Development of a Measuring Tool for Walkability in the Street Scale - the case study of Hamburg

Nasimeh Fallahranjbar, Udo Dietrich, Joerg Pohlan

HafenCity University, Hamburg.

Nasimeh.fallahranjbar@hcu-hamburg.de

Abstract. The humanistic approach is a key principle of any urban development with the aim of urban quality improvement. A healthy environment is fundamentally associated with walkable communities. According to a study in 2016 from the Centre for Diet and Activity Research (CEDAR) and the Medical Research Council Epidemiology Unit at the University of Cambridge, the health benefits of walking and cycling override the negative impacts of air pollution on health even in highly polluted cities. Additionally, planning walkable cities promote low carbon emission developments. Measuring walkability of the streets makes it possible for the planners and experts to address the quality of the built environment. This paper reports a practical assessment tool, developed in a master thesis, based on decisive walkability features by applying a weighting system to score those features in order to achieve a tangible outcome. For this reason, an examined weighting system from a previous study led by the Institute of Urban Planning and the University of Duisburg-Essen is optimized. Afterward, the proposed tool is applied for measuring the walkability in a selection of different streets in Hamburg, in order to answer the questions of how these Hamburg's streets differ in their degree of walkability and how their walkability can be improved.

1. Introduction

As John Gehl once said, cities should be given back to their people [1]. This means considering the human scale and that human needs should become a priority in every urban development. Walking is the primary human transit mode and embeds the interaction between pedestrians and their surrounding built environment. Walkability is not necessarily related to the amount of walking or how vibrant a street is but is a precondition for a positive correlation between walking and the daily pursuits [2]. The activity of walking can be considered as necessary (for shopping and commuting), optional (for sight-seeing or pleasure) and social (combined with sitting, reading, or talking). A livable urban environment is a place that also provides facilities related to activities such as sitting, talking or enjoying public art [3]. The literature of humanizing urban design began in the 60s in the US when large-scaled and car-oriented cities had taken away the urban life from their inhabitants. Livable streets and neighborhoods with safety, cleanness, and security were mentioned to be one of the most important characteristics of a livable urban environment [4]. After decades, livable streets and walkability are still the most important solutions encountering new problems like air pollution, concerns about fossil fuels and public health issues. Obesity has tripled in WHO European Region since the 1980s [5]. Improving walkability can improve both the built environment and public health. This would happen through improving the "sidewalk room", considered as the essential element of urban space. The analysis of the degree of walkability at the street scale is focused on the spatial quality of a sidewalk, defined through its four edges [6] and how it can affect the walkability features. A study on measuring and recording walkability in mixed-use streets of the city of Essen has been carried out by the Institute of Urban Planning and the University of Duisburg-Essen in 2017 ("Essen report", [7]). Its methodology uses indexes, referring to different walkability features, and a weighting system to calculate the final score (see table 2). It is based on previous studies in American cities (see [7]) and was adapted to European cities, which are typically part of an old urban system from middle ages where the system of pedestrian areas, cycling paths and large public transportation has remained over the time despite sprawling [8]. The proposed tool in this paper is developed by further optimizing the methodology of the Essen´s report, in order to achieve a more comprehensive and applicable measuring tool for walkability in the street scale.

2. Walkability features

In previous studies, various physical features and perceptual qualities of walkability have been identified. They differ from being more objective to more subjective (reflected by individual reactions to a place) [9]. Existing methodologies often solely refer to (directly measurable) objective criteria or cover indexes related to perceptual qualities, without presenting and ordering them in a deeper relation. Consequently, there is a risk that the methodology does not describe all aspects of the phenomenon of walkability enough and that correlations between the different used features remain invisible.

For this reason, here it is proposed to relate first to the human needs, like in Max-Neef's human needs chart [10], and how fare they are touched by walking through a city. That allows defining a group of central features describing the context and to bring them into an order where the prior one is a precondition for the following one (see table 1). The features of connectivity and security are the essentials for walkability, operating in parallel. No connectivity means an access is missing. However, one would not choose to use a path if no physical security is guaranteed, even when the access provided. The scale of analysis for connectivity is the neighborhood scale, in which the connection between different nodes can be studied. Regarding this scale, connectivity is not included in this paper. Apart from security, pedestrians also need to feel that themselves and their belongings are safe while walking on the sidewalks, which are additionally stable, well maintained and secure [11]. Therefore, safety becomes another decisive feature. According to Jan Gehl's quality criteria for good design of public spaces [12], other features that refer to protection and comfort come next. Thereupon, the first three essential prerequisites for walkability in street scale are the feeling of security, safety, and comfort. Other available features refer to the urban quality, which most essential ones in the street scale are identity, attractiveness, and invitingness (the state of being inviting to gather people together).

Table 1. Table of features and index classification.

	Feature	Index (extracted from the Essen's report)	
Driven from the human need chart [9] and quality design criteria	Feeling of security	Street hierarchy type / Street traffic type / Parked cars / Buffer	Physical indexes
	Feeling of safety	Sidewalk width and street width ratio / Tree canopy	
	Feeling of comfort	Sidewalk width / Obstacle / Hindrance / Barrier freeness / Accessibility /Urban furniture / Active tram or bus line on the street / Public transportation stations / Other greenery	
	Identity	Landmarks	
	Invitingness	Number of gastronomy units / Number of gastronomy units with outdoor setting / Number of retail stores/ Supermarkets along the sidewalk / Presence of public places / Presence of vacancy	
	Attractiveness	Façade maintenance condition / Street cleanness condition / Street façades complexity / Presence of artwork	Perceptional indexes

Each walkability feature can be identified through special groups of indexes. The value of a feature can be measured through the value of its indexes. The indexes in the table are extracted from the Essen´s report, which is briefly explained in the next chapter.

3. Tool development

Table 1 is the basis of the proposed tool in this study. An appropriate scoring system is required in order to attain a tangible degree of walkability in the street scale. The weighting system from the Essen's report acts as a reference to this tool. In the Essen's report, walkability indexes are investigated in a different classification which is regardless of the walkability features (see table 2). Therefore, in a street analysis, it would be impossible to figure out which features need to be improved or which perform satisfactorily. What makes this study distinguished is a weighting system used to score each category of the mentioned classification. According to this study, each index has a different weight in the total value. The weighting system was achieved by an interdisciplinary expert panel. The experts have weighted independently and then the average of weights was determined (see tables 2 and 3).

Table 2. A segment of physical features index from the Essen's report (see [7]).

Index group	Index	Category	Score	Category weight
1. Sidewalk (criteria weight = 30%)	Sidewalk width	No sidewalk	1	24.40 %
		< 1.5 m	2	
		1.5m -2.5m	3	
		2.5m -5m	4	
		>5m	5	
	Buffer	None / only curb	3	20 %
		Parked cars	3	
		Bicycle path	4	
		Flower boxes	4	
		Green part	5	
	Obstacle / hindrance	Very disturbing	1	17.50 %
		Disturbing	2	
		None / not disturbing	5	
	Barrier freeness / Accessibility	Yes	5	13.10 %
		Partly	3	
		No	1	
	Sidewalk width and street width ratio	>70%	5	25 %
		61-70%	4	
		60%	3	
		50-59%	2	
		<50%	1	
2. Street (criteria weight = 35%)	Street type	Local street	5	21.40 %
		Collector street	4	
		Sub arterial road	3	
		Arterial road	2	
		Other	1	
	One-way street	No, pedestrian zone	5	7.70 %
		No	3	
		Yes	4	

Table 3. A segment of urban quality index from the Essen's report (see [7]).

Index	Category	Score	Weight 0-9
Landmarks	1	5	6.5
	0	3	
Street space condition	Bad condition	1	5.9
	High pollution	2	
	Very high pollution	3	
	Low pollution	4	

The tables in the Essen's report uses step weight factors for each calculation in a set of formulas. However, the tables above have different total weight factors. For table 2 it is 100; while for table 3 with weighting from 0-9, the total is 64.7. By simply multiplying each weight factor to the related score factor and then dividing by the total weight, a single weight factor would be assigned to every single index. This single weight factor is, in fact, the final score calculated for each index. This will make the new arrangement of indexes (see table 1, plus table 1 in the appendix) possible. As an example from the second index group in table 2: the index "Street type" has a weight factor of 21.40%, and belongs to the index group "Street" which has a weight factor of 35%. If a category in "Street type" index has the score factor of 5, then the final score for this category will be:

Category 's final score = (5 x 21.4)/100 x 35/100 = 0.3745

By using this method, a new weighting system can be achieved in which every single category of the index groups has independent weight. A similar method is used to reach the single weight for the categories from urban quality index table in the Essen's report (see table 3). The two finalized tables: Street Physical Index (SPI) plus the Street Quality Index (SQI) (see tables in the appendix), have been provided in a spreadsheet software with the new single weight factors and new index arrangement, in a way that they would be easily applied by clicking on the desired category. After clicking, the final score factor will appear as well as the sum of these score factors will be automatically calculated. To check the validity of the new tables, the sum of all scores from the whole categories should be 5 when selecting the maximum rate in every case. The corresponding maximum score factor for each index is directly shown in the tables. This maximum score factor adapts to other corresponding score factors. For example, in the street hierarchy type index, the maximum score factor is 5 and the final score for an arterial road is 2. This can be calculated with this formula: 2/5 x 0.3745 = 0.1498.

Finally, the relationship between the physical characteristics and the urban qualities, mentioned as the Integrated Walkability Index (IWI) in the Essen's report and as walkability degree in this paper, is calculated with the weight factors shown in the formula below. By implying the final scores into the new tables, the walkability degree will be between 2.1-5:

Walkability Degree = (SPI x 52.85) + (SQI x 47.15) / 100

Furthermore, the description of a few indexes from the Essen's report needed to be adapted in order to improve comprehensibility. For instance, in the index of sidewalk width and street width ratio (see table 2) there is no possibility to choose between 60-61% and 59-60%, therefore the category is changed accordingly to cover all possible numbers. Another example is the index street facades complexity that has the category of heterogeneous with the highest score and homogeneous with the least score. Whilst in the topic of street scale, it is important if a facade is pleasant to watch or not. A façade can be homogeneous and at the same time pleasant to watch. Therefore, the new category is changed to street facades complexity: "a pleasure to watch" or "boring". Additionally, in this tool, the corresponding street length set for every index is 250 meters except for the indexes referring to the number of gastronomy units, which require a different reference length. World's great streets such as Regent Street in London, Les Champs-Elysées and Boulevard St. Michel in Paris have an average of one new public function at least in every 15 meters, which means 6-8 units in every 100 meters [13]. Therefore, more than 5 gastronomy units in 100 meters receive the highest score (while in the Essen's report, more than one unit is considered adequate to receive the maximum score). This tool has been applied to measure walkability in some streets in the city of Hamburg by collecting data through observation. There might be uncertainties when applying this tool. For instance, both sidewalks of a street may not have the same condition. For the scoring purposes, both sidewalks should be checked, and the highest score should be selected. In the following chapters, some of the application results are briefly described.

4. Walkability in Hamburg streets

IOP Conf. Series: Earth and Environmental Science **297** (2019) 012047 doi:10.1088/1755-1315/297/1/012047

Hamburg is a typical European city with the corresponding quite dense street grid. Some streets are always preferred by walking people and some others are not. By applying this tool, it will be possible to understand if a street is as walkable as it looks and what reinforces or deteriorates its walkability. Six streets were selected for this study. Some of them are well known as being walkable and some others are usually desolated. One arterial road was selected in order to obtain low scores and to better comprehend the meaning of the walkability degree, scored between 2.1 - 5. At first sight, two of the six investigated streets seemed to have completely different potential of walkability, but after the assessment was done, their walkability degrees resulted quite alike. These streets are briefly described and compared as follows.

4.1. Muehlenkamp

This street is the main street of a sub-center and perceived as one of the vibrant streets of Hamburg with many gastronomy units, supporting commercial and service units. Muehlenkamp is a two-way collector street with partly more than two lanes. The buffer between the street and the sidewalk is a bicycle path and there is a row of trees in the sidewalk. The sidewalk width varies from 2.5m - 5 m and the street width (building edge to building edge) is 20 m. The sidewalk width ratio (one sidewalk width divided by the total width of the street) is less than 50%. The width of the footway is not fixed through the street and in narrow parts there are some disturbing obstacles like product shelves in front of some shopping unit, while not every part of the sidewalk can be freely accessed. The sidewalks and the facades, with buildings of around 1900, can be scored as with "low dirtiness".

Figure 1. Pictures of Muehlenkamp. Source: author.

4.2. Shanghaiallee

Shanghaiallee is part of the newly built HafenCity neighborhood with new and clean buildings. Less vibrancy and usually empty sidewalks make it distinguishably different from Muehlenkamp. However, the structural characteristics of this street which define the feeling of security are the same as Muehlenkamp. Also, Shanghaiallee has sidewalks free of barriers but with less greenery.

Figure 2. Pictures of Shanghaiallee. Source: author.

4.3. Feature score comparison
After visiting the two mentioned streets, it can be perceived that Muehlenkamp is much more vibrant than Shanghaiallee, since its sidewalks are full of people and sometimes even crowded.

Table 4. Feature score comparison from two investigated streets

Street name	Feeling of security	Feeling of safety	Feeling of comfort	Identity	Attractiveness	Invitingness	Walkability degree
Muehlenkamp	1.15	0.63	1.74	0.3	1.20	2.55	3.77
Shanghaiallee	1.15	0.52	1.90	0.3	1.40	2.05	3.64

However, by calculating the walkability degree, there is a minor difference between these two streets (see table 4). This shows that Shanghaiallee is nearly as walkable as Muehlenkamp and has all preconditions to become vibrant. Shanghaiallee has well-maintained façades, spotless sidewalks, full accessibility, and no hindrance, which prevents a drop in the walkability degree. Still, the weak invitingness leads to empty sidewalks.

Figure 3. Index comparison chart, Muehlenkamp & Shanghaiallee.

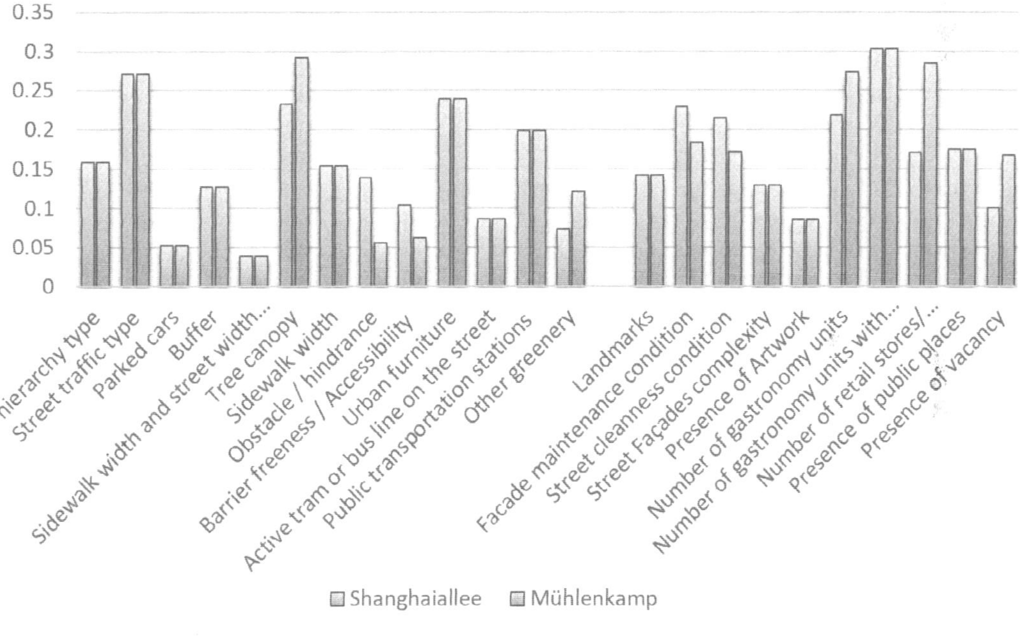

Increasing the gastronomy units and other units with supporting public functions can effectually improve walkability in this street. Street analysis by evaluating each index as well as index comparison among streets are possible through this tool. Figure 3 illustrates the index comparisons of the two mentioned streets.

5. Tool results analysis
Various streets from absolute car-oriented to car-free ones were selected to be assessed by this tool. This permitted to reach a tangible range of walkability's degrees for the final assessment (see figure 5). The proposed tool gives the possibility to analyze streets in several terms: regarding their features, the indexes or the walkability degree. For instance, among six investigated streets, the lowest degree (2.82) applies to an arterial road with car-oriented design, while the two highest degrees belong to relatively car-free streets, with and without an active bus line (see figure 4).

Figure 4. Walkability degree in analyzed streets.

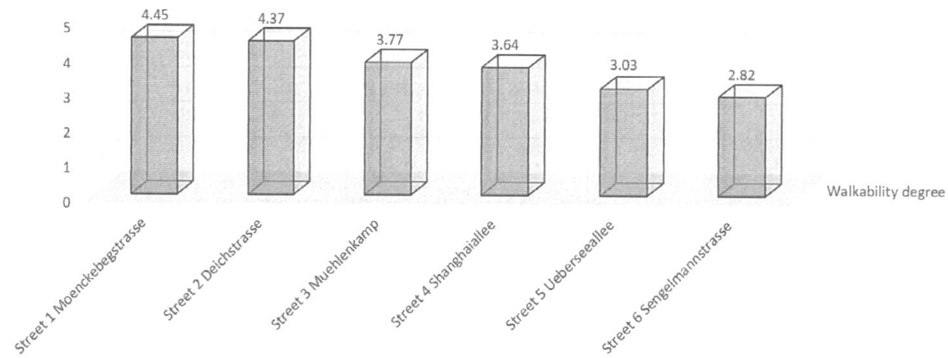

According to the results, the degrees can be interpreted as follows:

Figure 5. Rating of walkability degrees (2.1 – 5).

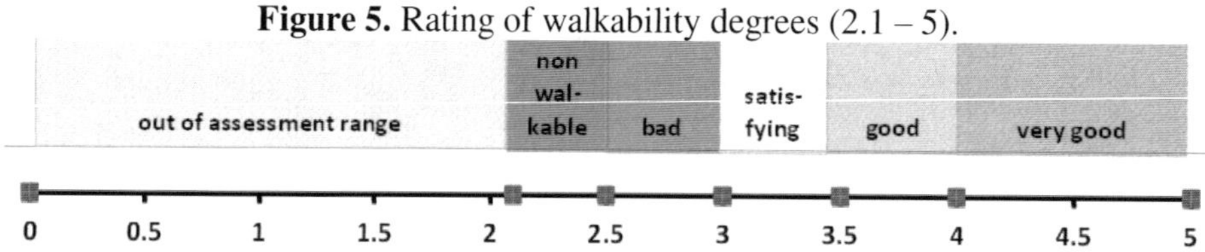

The lowest possible degree that can be achieved by this tool is 2.1 and the highest is 5. The walkability situation of the investigated streets of Hamburg, therefore, can also be interpreted besides the numbers. In other words, the streets no.1 and 2 have very good walkability. Streets no.3 and 4 can be classified as good while no.5 is satisfying, and no.6 is bad. Another way to analyze the driven data is to compare the streets according to their features (see figure 6). In this way, the features that need to be improved would be distinguished. For instance, street no. 5 is a sub-arterial where the walkability needs to be enhanced. As shown in figure 6, the feeling of safety and security is low in this case. Any intervention that strengthens these features, such as adding green areas as a buffer or create a tree canopy, can improve the total score. The street no.4 has an equal level of safety as the street no.5 but higher scores in comfort and invitingness, which contributes to achieving a good grade of walkability.

Figure 6. Feature score comparison in analyzed streets

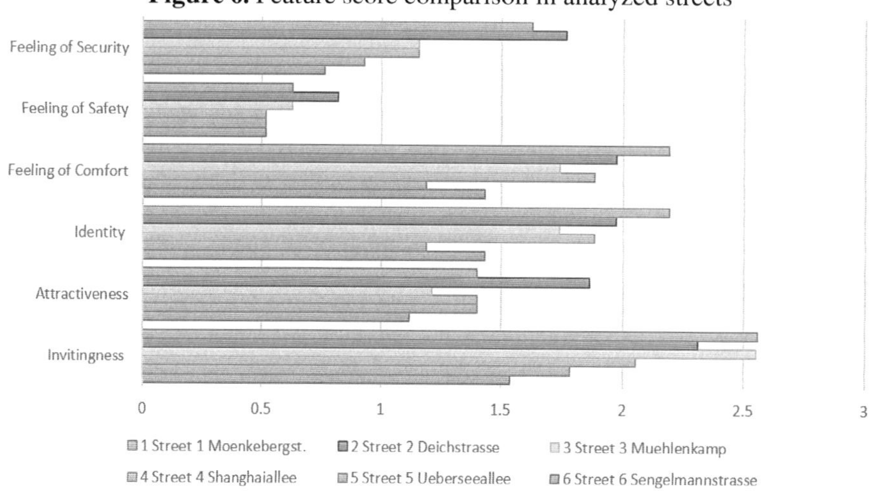

6. Conclusion

Walkability is a vast topic, containing numerous variables and indicators. The proposed tool includes all essential indicators for walkability. It offers various possibilities to analyze a street and delivers tangible results in accordance with expectations. An important data driven from this tool is the walkability degree between 2.1-5. Assessing perceptibly walkable as well as non-walkable streets allowed to define a rated scale for walkability degrees.

This tool can also be applied outside Europe with small adaptations. In the first table of the appendix (SPI) the indexes refer to physical and structural preconditions of a walkable street, thus they are always valid regardless to the culture or climate except for one index referring to public transportation: active tram or bus line can be replaced by any other local public transit mode. In the second table of the appendix (SQI), indexes regarding gastronomy units might need to be tailored. For example, in a country like Malaysia, street food stands characterize the streets and attract people, the reason why the number of food stands should be considered in the index of gastronomy units.

In fact, if a street is certainly walkable or not depends on how a pedestrian perceives it. One can assess the same street differently depending on the person's mood or background. People's age, gender, nationality, being a tourist or not etc. can affect the perception as well. This is nearly unmeasurable and makes the limitation of any measuring tool for walkability. For this reason, it is always recommended to interview different people and inhabitants in diverse timeframes to achieve a higher level of comprehension about walkability.

For a deeper investigation, this tool can be applied separately to each sidewalk of a street. Index scores always refer to the best existing condition even if it is not the same along the sidewalk. Therefore, a checklist recognizing potentials improvement can also be attached to the tool for the cases that sidewalk segments differ in terms of scoring.

Acknowledgments

Authors wishing to acknowledge Clarisa Mazzara for constructive criticism of the manuscript.

References

[1] Gehl J 2010 *Cities for people* (Island press)
[2] Ng S, Lau W, Brown F, Tam E, Lao M and Booth V 2012 *Walkable City, Living Streets* www.civic-exchange.org
[3] Gehl J 2004 *Towards a fine city for people* (Transport for London & Central London Partnership)
[4] Jacobs A & Appleyard D 1987 *Toward an Urban Design Manifesto* APA Journal www.semanticscholar.org
[5] World Health Organization 2018 *Obesity*, http://www.euro.who.int/en/health-topics/noncommunicable-diseases/obesity
[6] NYC (City of New York) 2013 *Active design. Shaping the sidewalk experience* (E-book) retrieved from https://www1.nyc.gov/assets/planning/download/pdf/plans-studies/active-design-sidewalk/active_design.pdf
[7] Institut für Stadtplanung und Städtebau, Universität Duisburg-Essen 2017 *Messung Und Erfassung Der* Fussgängerfreundlichkeit *Von Stadträumen* (PDF) retrieved from https://www.uni-due.de/imperia/md/content/staedtebau
[8] Le Gales P 2007 *Cities are back in town: the US/Europe comparison* (PDF) Sciences Po http://blogs.sciences-po.fr/recherche-villes/files/2010/01/cahierville0606.pdf
[9] Ewing R & Clemente O 2013 *Measuring urban design, metrics for livable places* (Island Press)
[10] Max-Neef M 1991 *Human Scale Development* (New York: The Apex Press) p 32
[11] Department of Transport 2011 *Walkability Audit Tool* (Western Australia: Department of Transport) p 2
[12] https://gehlinstitute.org/wp-content/uploads/2017/08/quality-criteria-final.pdf
[13] Karssenberg H Laven J Glaser M & Hoff M 2015 *The city at eye level, lessons for street plinths-Hans* Karssenberg (PDF) Eburon

Appendix

Table 1. First table of the finalized tool; Street Physical Index (SPI)

Feature	Index	Category			Score	Final score
Feeling of security	Street hierarchy type	Local street			☑ 5	0.3745
		Collector Street			☐ 4	
		Sub Arterial Road			☐ 3	
		Arterial Road			☐ 2	
		Highway			☐ 1	
	Street traffic type	Pedestrian (only bikes, Trams,	Yes		☑ 20	1.04475
			No			
				One-way street		
				Yes	☐ 4	
				No	☐ 3	
				Number of lanes		
				1	☐ 4	
				2	☐ 3	
				> 2	☐ 1	
				Speed limit		
				< 30km/h	☐ 4	
				30-50 km/h	☐ 3	
				> 50km/h	☐ 1	
				Pedestrian crossing type		
				Crosswalk	☐ 4	
				crosswalk	☐ 3	
				Partly crossable	☐ 2	
				No crosswalk	☐ 1	
	Parked cars	None			☑ 5	0.16625
		Beside/ on the street			☐ 3	
		On the sidewalk			☐ 1	
	Buffer	Green part			☑ 5	0.3
		Bicycle path			☐ 4	
		Post/flower boxes			☐ 4	
		Parked cars			☐ 3	
		None / only curb			☐ 3	
Feeling of safety	street width ratio	> 70%			☑ 5	0.375
		63-70%			☐ 4	
		57-63%			☐ 3	
		50-57%			☐ 2	
		<50%			☐ 1	
	Tree canopy	Tree row(s)			☑ 5	0.5515625
		Single trees			☐ 4	
		None			☐ 3	
Feeling of Comfort	Sidewalk width	> 5m			☑ 5	0.366
		2.5m -5 m			☐ 4	
		1.5m - 2.5m			☐ 3	
		< 1.5 m			☐ 2	
		No sidewalk			☐ 1	
	Obstacle / hindrance	None / not disturbing			☑ 5	0.2625
		Disturbing			☐ 2	
		Very disturbing			☐ 1	
	Barrier freeness	Yes			☑ 5	0.1965
		Partly			☐ 3	
		No			☐ 1	
	Urban furniture	Lighting + bench	Yes		☑ 5	0.3506875
		only lighting or no furniture	No		☐ 3	
		Bicycle stand	Yes		☑ 5	0.2421875
			No		☐ 3	
	Active tram or bus line	Yes			☑ 5	0.1645
		No			☐ 3	
	stations	Yes			☑ 5	0.375875
		No			☐ 3	
	Other greenery	Yes			☑ 5	0.2296875
		No			☐ 3	
Sum						5 / 5

Table 2. Second table of the finalized tool; Street Quality Index (SQI)

Feature	Index	Category	Score	Final score
Identity	Landmarks	at least one	5 ☑	0.50231839
		none	3 ☐	
Attractiveness	Facade maintenance condition	Good maintenance	5 ☑	0.48686244
		Low dirtiness	4 ☐	
		Medium dirtiness	3 ☐	
		High dirtiness	2 ☐	
		Bad maintenance	1 ☐	
	Street cleanness condition	Good condition	5 ☑	0.45595054
		Low dirtiness	4 ☐	
		Medium dirtiness	3 ☐	
		High dirtiness	2 ☐	
		Bad condition	1 ☐	
	Street Façades complexity	Pleasure to watch	5 ☑	0.45595054
		Not Boring	3 ☐	
		Boring	1 ☐	
	Presence of Artwork	>1	5 ☑	0.30139104
		1	4 ☐	
		0	3 ☐	
Invitingness	Number of gastronomy units	>5	5 ☑	0.57959815
		1 to 5	4 ☐	
		0	3 ☐	
	Number of gastronomy units with outdoor setting	>1	5 ☑	0.64142195
		1	4 ☐	
		0	3 ☐	
	Number of retail stores/ supermarkets along the sidewalk	>1	5 ☑	0.60278207
		1	4 ☐	
		0	3 ☐	
	Presence of public places	Yes	5 ☑	0.61823802
		No	3 ☐	
	Presence of vacancy	0	5 ☑	0.35548686
		1	3 ☐	
			1 ☐	
		>1		
Sum				5 / 5

The Negative Impact of Solar Reflections Caused by Reflective Buildings' Facades: Case Study of the Nasher Museum in Texas

Sahar Abdelwahab[1], **Mariam Elhussainy**[1], **Rania Labib**[2,3]

[1] Faculty of Engineering, Architecture department, Al Azhar University, Cairo, Egypt
[2] Department of Architecture, Texas A&M University, College Station, Texas, USA
[3] Department of Architecture, Prairie View A&M University, Prairie View, Texas, USA

Abstract. Large reflective facades in buildings can cause severe visual discomfort due to the reflection of sunlight falling on buildings' surfaces. These intense reflections can cause a intolerable glare and contribute to overheating the interior of surrounding buildings.
An example of reflective building materials causing glare is experienced in Dallas, Texas, where a residential tower, which was fitted with a fully glazed façade, has caused intense specular reflections into the Nasher sculpture museum. The problem has caused glare and overheating of the interior spaces in the museum, thus leading to the damage and deterioration of the sculptures on display.

In order to examine the negative impact of the Museum Tower's reflective façade on the visual comfort inside the Nasher museum, glare simulations are carried out at different times of the year. Additionally, to examine a possible solution to the glare problem, a proposed solution of using a less reflective material on the tower's facade is examined via glare simulations. The initial set of simulations confirmed that tower skin contributed to blinding glare particularly at 12:00 and 15:00 on the summer and winter solstice, and the spring equinox. Use of less reflective material on the tower façade concluded that the tower façade still contributed to glare and that is due to its convex geometry.

1- Introduction

Reflective façade is a popular architectural feature that is widely used in buildings, especially tall buildings in condensed urban settlements. In the last few decades, architects have been inspired by the contemporary architectural style, which utilizes full glazed building façades despite the noticeable differences in thermal requirements across diverse climatic regions. The increased preferences to use glazed façades in buildings contributes to severe visual discomfort due to the excessive sunlight internally received and, equally important, external reflection of sunlight falling on buildings' surfaces. The intense external sunlight reflections on buildings can cause an intolerable glare, which can impair the vision of surrounding buildings' occupants thus hindering them from performing their daily tasks. Figure 1 shows various examples of sunlight bounced off glazed façades causing blinding glare.

A 2014 research study that investigated the probability of disability glare in a residential building caused by sun reflections off a nearby tall building façade that is fitted with photovoltaic (PV) panels, confirmed that solar reflections are causing severe glare, that is known as disabling glare for 30 minutes up to 72 days a year and for over an hour for up to 34 days a year [1].

In order to mitigate the problem of glare, building occupants tend to abandon the use of daylighting by permanently closing their blinds and depend on the use of artificial lighting, therefore, increasing the lighting loads which leads to a significant increase in energy consumption that is estimated to account for 25 to 40% of commercial buildings' total energy consumption [2].

Figure 1. Examples of intense solar reflections in urban settings causing severe visual discomfort[1]

Solar reflections caused by high reflective façade materials, such as dark solar glazing, not only cause glare but can also contribute to other undesirable issues; such as the overheating of the areas in the vicinity of the building and negatively impacting the thermal comfort of both outdoors pedestrians and the indoor occupants of surrounding buildings. In turn, this leads to a significant increase in building cooling capacity. This is necessary to overcome the problem of overheated indoor spaces caused by intense sunlight reflections [3–5].

An example of reflective building materials causing disabling glare is an installation of a PV array at a Massachusetts airport. The installation of this array has caused specular solar reflections, leading to intolerable glare, resulting in difficulties viewing computer screens inside the airport control tower [6].

Similarly, when a 42-story high-rise building in Dallas, Texas (the Museum Tower) was fitted with a fully glazed façade, this led to the intense solar reflections that caused visual and thermal discomfort to the Nasher sculpture museum across the street. The problem has caused disabling glare and overheating inside the museum that results in damaging the sculptures in the display. In order to mitigate these problems, museum officials spent over one million, yet, no viable solution has yet been found [4,7,8].

2-The Nasher Sculpture Museum Description

The museum building is nearly 5,000 square meters. It is made up of five identical and parallel rectangular pavilions, whose volumes are delineated by walls that are covered with Italian travertine, which is a type of stone that is used in wall finishing. The stone has been polished inside the pavilions,

[1] Image taken from https://rwdi.com/en_ca/expertise/all-services/glare-and-reflections

giving the walls some reflectivity features. The façades of the museum at each end are made up of large transparent windows.

The museum was constructed on two levels. The three central arcades on the ground floor house the sculptures and paintings that are most sensitive to atmospheric agents, while the two arcades (one on each side of the building) contain the cafeteria, the shop, and the offices. On the lower level, there's a small gallery for light-sensitive works, such as prints and drawings; as well as preservation laboratories, research and teaching areas; and even an auditorium that overlooks a portion of the garden that descends toward the hall in a terraced fashion, thus creating an outdoor theatre.

The ceiling of the museum is comprised of five glass vaults nestled between the seven travertine walls, which are extended above the pavilions. A shielding system, made up of aluminum panels, is positioned above the glass ceiling. These three-dimensional elements, whose design has been patented, are repeated 223,020 times and only allow for the passage of direct light from the north. The Nasher museum is across the street from the Museum Tower. The façade causing glare is facing the south-west direction. (see Figure 2)

Figure 2. Nasher Sculpture museum and the Museum Tower highlighted in red. The tower's façade faces the south-west direction.

To be able to understand how and when the sun reflections that are caused by the museum tower façade impact the visual and thermal conditions inside the Nasher sculpture museum, glare simulations were carried out during different times of the year. To find out if replacing tower façade with a less reflective material can offer a solution to the problem of the intense specular reflections, another set of glare simulations were preformed using a less reflective material.

3-Methodology

3.1. Simulation Tools and Analysis Type
Point-in-time glare analyses were carried out under Texas climate, US. The analyses were conducted on the 21st of March, June, September, and December at 9 am, 12 pm, and 3 pm. The simulations were performed using the graphical algorithm editor Grasshopper, linked to the daylight simulation engine RADIANCE through the DIVA V.4 tool [9]. DIVA calculates glare using RGB-format picture that contains the full luminance distribution of a half hemisphere in the viewing direction using a fish-eye angular lens [10–12]. Each pixel with a luminance value four times higher than the average task-zone luminance is treated as a glare source [13].

Daylight Glare Probability DGP [13] was used as an indicator for the received glare. The evaluation of DGP was based on the DGP rating proposed by Wienold [14]: imperceptible glare= DGP≤0.35; perceptible glare= 0.35<DGP≤0.40; disturbing glare= 0.40<DGP≤0.45; intolerable glare= DGP>0.45. DGP was measured from a perspective of two subjects: visitor 1, and visitor 2 (shown in Figure 3) standing in front of a sculpture (0.6x0.6m cabinet) at a distance of 1.5m. The Museum Tower is in the field of view of visitor 1, while visitor 2 faces the opposite direction. (see Figure 3).

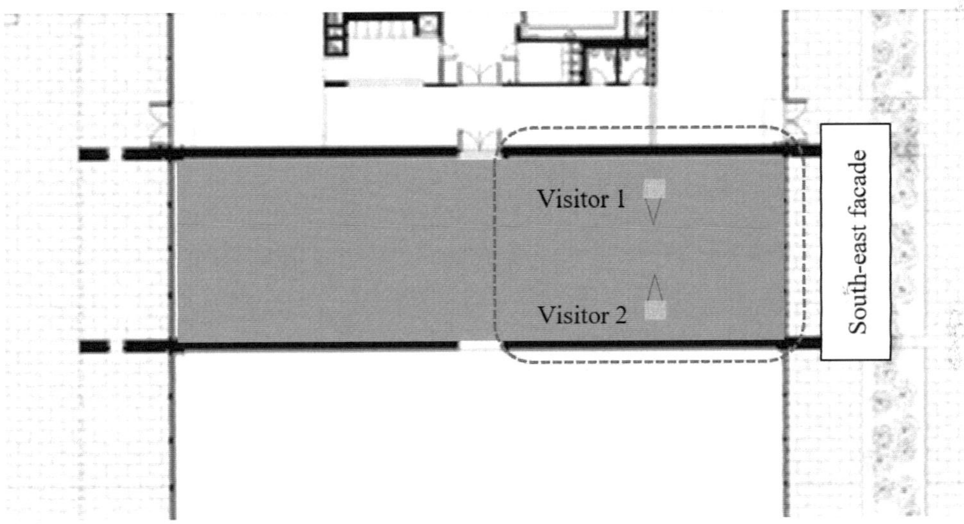

Figure 3. The simulated pavilion and visitor locations.

The two view positions were chosen close to the south-east façade, so the probability of severe glare occurrence from the ceiling and the facade can be clearly tackled (preliminary analysis showed that this particular façade receives excessive direct light compared to the north-west façade).

3.2. Model Parameters
A model of the five-pavilion museum, in addition to the museum tower, was modeled. The pavilion, shown in Figure 3, was tested for glare performance as one of the three middle pavilions that house the sculpture and, more importantly, it is the closest pavilion to the museum tower where severe light reflection from tower's skin is more likely to occur. Figure 4 shows detailed Rhino scenes for the model built for the Museum and the museum tower.

IOP Conf. Series: Earth and Environmental Science **297** (2019) 012048 doi:10.1088/1755-1315/297/1/012048

Material characteristics of the museum and the museum tower were mainly based on the described materials in the museum description section, solar control glazing with low-e coating was used for museum facades, while the interior walls finished with Italian travertine were assumed with a 70% reflectance. The ceiling was modeled with aluminum roof structure (so as façade mullions) and shielding system made of aluminum panels on top of clear glazing. Floor tiles were assumed with 40% reflectance. The external ground was modeled with 20% reflectance. The analysis considered no external abstractions except for the museum tower.

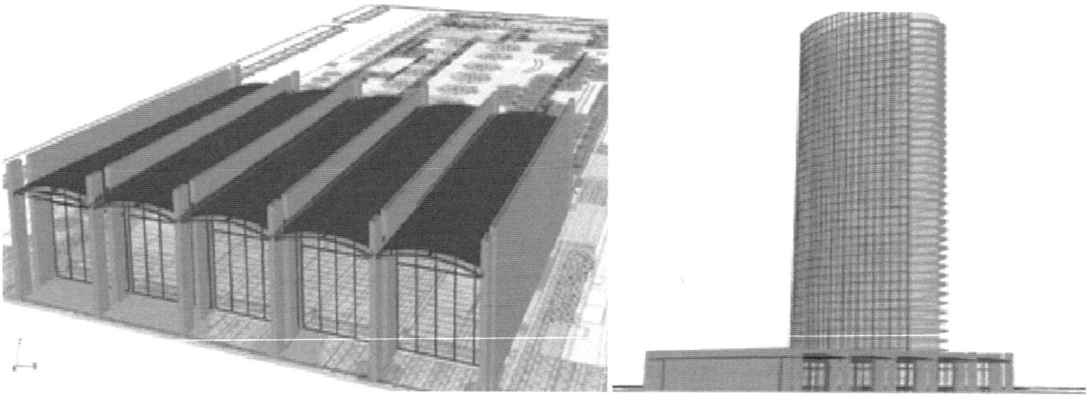

Figure 4. Rhino scene for the modeled Nasher Museum and the museum tower.

3.3 Simulation settings

Compliant with the findings from work by Abdelwahab, S. et al. [15], it suggested using ambient bounces of -ab2 for glare analysis in order to improve simulation accuracy. Simulation settings are illustrated in detail in Table 1. On the other hand, Perez sky model [16] was used for the simulation, as it showed to give the best agreement with interior measured values, particularly when performing discomfort glare analysis [17].

Table 1 Radiance settings used for discomfort glare simulations.

RADIANCE parameters	DGP simulation parameters				
	Ambient bounce -ab	Ambient division -ad	Ambient sampling -as	Ambient resolution -ar	Ambient accuracy -aa
	2	2024	1024	256	0.2

4- Results (Point-in-time glare analysis)

Point-in-time DGP analysis was carried out through DIVA for each time of day and year from the perspective of visitor 1 and 2, separately. DIVA automatically calculates DGP and generates a luminance map for the fish-eye lens scene. It also produces a JPG image that highlights sources of glare within the scene.

The luminance maps generated from the perspective of visitor 1 and 2 showed a high potential of glare occurrence due to excessive light received from the window and, obviously, the ceiling through the shielding system. Table 2 shows a summary of glare analysis conducted at different times of day and year. DGP rating was consistently intolerable (DGP >0.45) particularly at 12 pm and 3 pm on the 21st of March, June, and September. The probability of glare occurrence was low on the 21st of December as DGP was imperceptible (DGP <0.35).

Table 2. Point-in-time glare analysis: DGP received from the perspective of visitor 1 (cam1) and 2 (cam2) at different times of day and year.

	Time of year	Time of day	DGP		Time of year	Time of day	DGP
Camera 1	21st December	9:00 AM	0.21	Camera 2	21st December	9:00 AM	0.26
		12:00 PM	0.32			12:00 PM	0.32
		3:00 PM	0.28			3:00 PM	0.28
	21st March	9:00 AM	0.45		21st March	9:00 AM	0.32
		12:00 PM	0.88			12:00 PM	0.43
		3:00 PM	0.71			3:00 PM	0.38
	21st June	9:00 AM	0.34		21st June	9:00 AM	0.62
		12:00 PM	0.51			12:00 PM	0.68
		3:00 PM	0.55			3:00 PM	1
	21st September	9:00 AM	0.60		21st September	9:00 AM	0.33
		12:00 PM	0.47			12:00 PM	1
		3:00 PM	0.37			3:00 PM	0.33

The examples of luminance maps in Figure 5 show that the museum tower does not provide any shading for the museum's roof. Rather, high reflective envelope reflects excessive sunlight, except for some parts in the tower's south-east-elevation, where low luminance levels from the non-reflective slabs were detected. However, it should be noted that, in general, both light sources (i.e. sun, sky) and the reflected light from the tower may contribute in receiving high DGP at 12 pm because, as expected, the sun has a high angle and therefore sunlight is excessively received through the glazed roof. On the other hand, the tower skin was detected as a potential source of glare (see circular red marks in 4), particularly with low sun angle at 3 pm.

These results, however, were expected since the literature and the observations have a referred glare issue inside the museum to the reflected light from the museum tower's skin. Therefore, it was suggested examining the potential of glare when using less reflective material. The simulations hence were repeated, assuming low-e glazing with Tvis 65% assigned to the Museum Tower's façade, in order to examine its impact on the reflected amount of light and the likely effect on the received glare (from the perspective of visitor 1 who looks at a sculpture facing tower direction). Table 3 shows a comparison between DGP received with the current high reflective tower skin and when using less reflective material

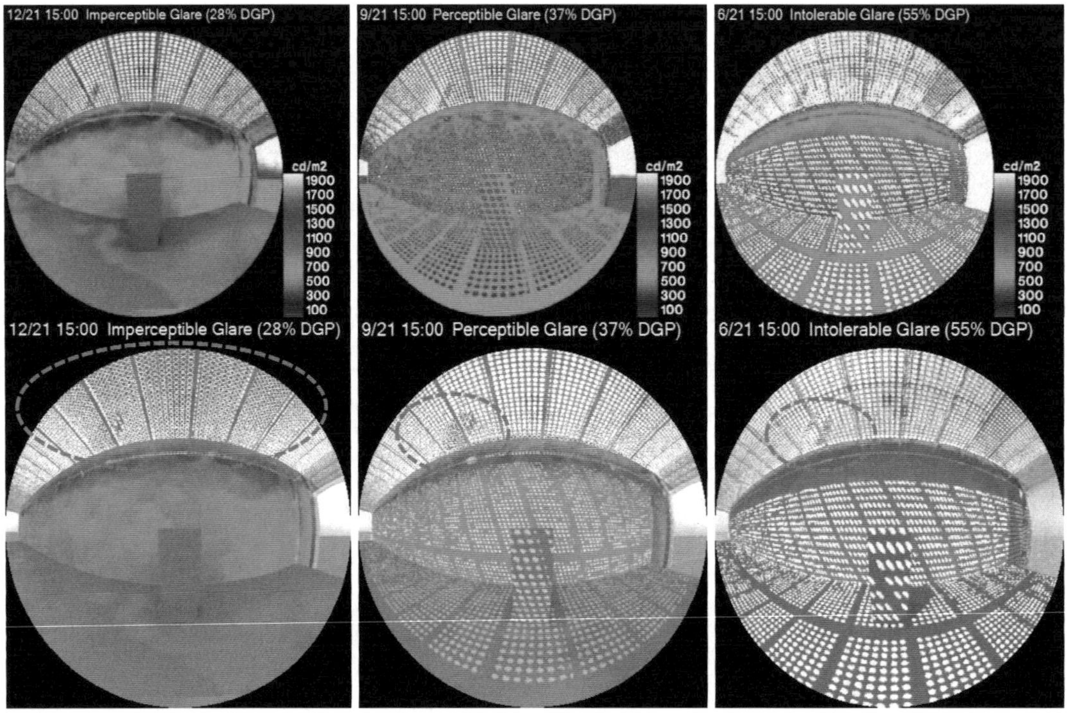

Figure 5 Example for luminance maps generated from the perspective of visitor 1 (facing tower's direction) on the top, and the potential sources of glare detected (blue patches) on the bottom.

Table 3 A comparison between DGP received from the perspective of visitor 1 with the current high reflective tower skin and when using less reflective material

	Time of year	Time of day	**DGP** (High reflective glazing)	**DGP** (Glazing with low-e coating, T_{vis}= 65%)
Camera 1	21st December	9:00 AM	0.21	0.21
		12:00 PM	0.32	0.32
		3:00 PM	0.28	0.28
	21st March	9:00 AM	0.45	0.43
		12:00 PM	0.88	0.50
		3:00 PM	0.71	0.71
	21st June	9:00 AM	0.34	0.34
		12:00 PM	0.50	0.50
		3:00 PM	0.55	0.52
	21st September	9:00 AM	0.60	0.30
		12:00 PM	0.47	0.47
		3:00 PM	0.37	0.36

As seen from the results, DGP values almost remained the same for most of the comparisons. In some cases, however, the estimated DGP slightly decreased when using less-reflective material for the tower's skin, although, this didn't lead to further improvement in the level of DGP received (i.e. DGP rating).

The examples in Figure 6 show lower luminance levels detected on the tower when using less-reflective material compared to the current high reflective glazing (see magenta marks). Nevertheless, the reduction in DGP values resulted only in very few cases and didn't prevent glare. This essentially means that replacing the tower's skin, using less reflective material, may not solve the Museum Tower's negative impacts experienced inside the museum.

In fact, these results speculate that glare probability inside the museum increases, not only because of the material reflection, but it could be the elliptical geometry of the museum tower with multi surfaces that reflect the light. This brings some examples that have similar issues because of its convex outside geometry (e.g. Fenchurch tower in London). Even with low-reflective material, the estimated DGP was generally high. Additionally, in some cases, the analysis showed that not only the tower skin was detected as a potential source of glare, but also direct light from the sun and sky received from the ceiling through the shielding system contributed to glare occurrence (see Figure **5** 6 left-bottom).

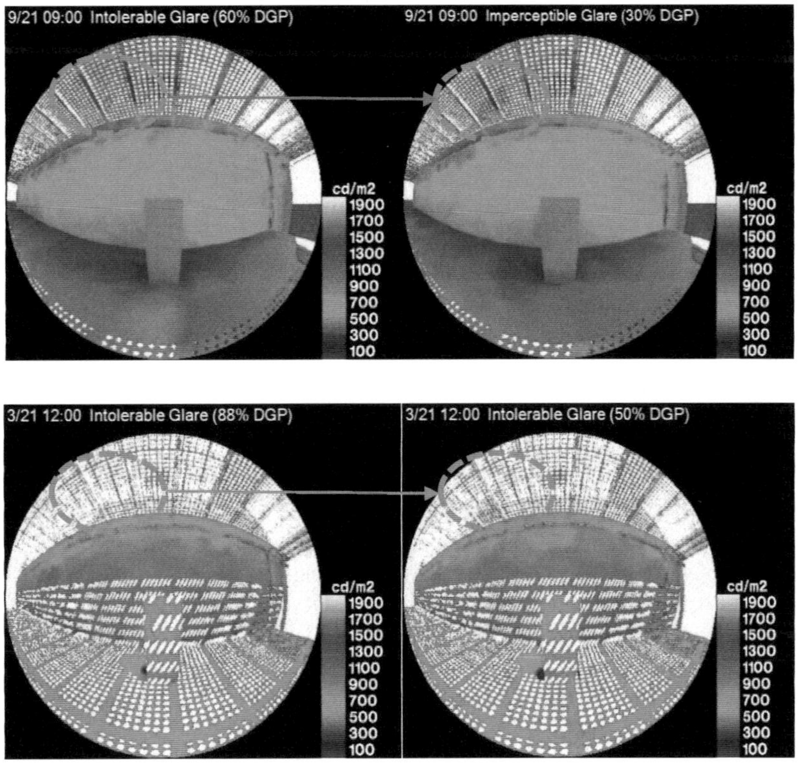

Figure 6 luminance maps generated from the perspective of visitor 1 with the current materials (left) and less reflective glazing assigned on the tower's skin (right)

Since replacing tower's skin didn't lead in further improvements in limiting glare, it is suggested that the solution hence could be through using low Tvis glazing in the museum ceiling, or through adding shading elements on the museum tower facade. Research by Singh et al. [18] showed that glazing with a high visible transmittance can result in higher glare. Other studies in the literature showed that some façade parameters (e.g., shading devices, window size, etc.) have an impact on limiting received glare[19,20]. This can be thoroughly examined in future research.

5- Limitations

Due to the computational limitations associated with modeling the light through complex geometries, the shielding system was simplified, and the number of openings was reduced, keeping the same opening ratio, in order to enable RADIANCE to perform the simulations in a reasonable timeframe.

6- Conclusion

Large glazed facades in urban settings contribute to undesirable issues, such as glare and overheating. To examine the glare problem inside the Nasher museum, which is caused by the Museum Tower's reflective façade, a computer simulation-based approach was carried out. The simulation-based approach examined glare from the point-of-view of two museum visitors; The Museum Tower is in the field of view of visitor 1, while visitor 2 faces the opposite direction. Glare simulations showed a high potential of glare occurrence due to excessive light received from the window, and the perforated ceiling in both views, additionally, the HDR images confirmed that the tower reflective skin is recognized as a potential glare source, especially when the sun is at a low angle at 15:00 in the summer and winter solstice, and the spring equinox. In an effort to examine a solution to the glare problem, a less reflective material was used to carry out another set of glare simulations. Use of less reflective material on the tower façade concluded that the tower façade still contributes to glare and that is due to its convex geometry.

References

[1] W.J. Brotas Luisa, Solar reflected glare affecting visual performance, in: Proceedings of 8th Windsor Conference: Counting the Cost of Comfort in a Changing World Cumberland Lodge, 2014: pp. 10–13.

[2] J.D. Kelso, Buildings Energy Databook, Department of Energy, 2011.

[3] M. Schiler, E. Valmont, Microclimatic Impact: Glare around the Walt Disney Concert Hall, in: Proceedings of the International Solar Energy Society Conference, 2005: pp. 6–12.

[4] Council on Tall Buildings and Urban Habitat, "Oculi Solution" to Museum Tower Glare Rejected, (2012). http://www.ctbuh.org/News/GlobalTallNews/tabid/4810/Article/165/language/en-US/view.aspx (accessed December 10, 2017).

[5] T. Tsoutsos, N. Frantzeskaki, V. Gekas, Environmental Impacts from the Solar Energy Technologies, Energy Policy. 33 (2005) 289–296. doi:10.1016/S0301-4215(03)00241-6.

[6] J.A. Jakubiec, C.F. Reinhart, Assessing Disability Glare Potential Due to Reflections from New Constructions : A Case Study Analysis and Recommendations for the Future, Transportation Research Record. (2014) 1–12. doi:10.3141/2449-13.

[7] M. Granberry, S. Thompson, G. Jacobson, Museum Tower hires technical firm to help solve glare dispute | News | Dallas News, (2014). https://www.dallasnews.com/news/news/2014/04/02/museum-tower-hires-technical-firm-to-help-solve-glare-dispute (accessed December 10, 2017).

[8] R. Stott, Search Ends for Solution to Museum Tower's Glare Problems at Nasher Sculpture Center, (2015). https://www.archdaily.com/773066/search-ends-for-solution-to-museum-towers-glare-problems-at-nasher-sculpture-center (accessed December 10, 2017).

[9] J.A. Jakubiec, C.F. Reinhart, DIVA 2.0: Integrating daylight and thermal simulations using Rhinoceros 3D, DAYSIM and EnergyPlus, Proceedings of Building Simulation 2011: 12th Conference of International Building Performance Simulation Association. (2011) 2202–2209.

[10] M. Hirning, G. Isoardi, S. Coyne, I. Cowling, Applying the use of high dynamic range imaging pipelines to discomfort glare research, in: Proceedings of CIE, 2010: p. 767e75.

[11] J.A. Jakubiec, The use of visual comfort metrics in the design of daylit spaces, (2014).

IOP Conf. Series: Earth and Environmental Science **297** (2019) 012048 doi:10.1088/1755-1315/297/1/012048

[12] I. Konstantzos, A. Tzempelikos, Daylight Glare Probability Measurements And Correlation With Indoor Illuminances In A Full-Scale Office With Dynamic Shading Controls, (2014).
[13] J. Wienold, J. Christoffersen, Evaluation methods and development of a new glare prediction model for daylight environments with the use of CCD cameras, Energy and Buildings. 38 (2006) 743–757.
[14] J. Wienold, Dynamic daylight glare evaluation, in: Proceedings of Building Simulation, 2009.
[15] S. Abdelwahab, M. Kent, S. Altomonte, Sensitivity Analysis for the Daylight Simulation of Complex Façades, in: 2017 Advanced Building Facade, Bern, Switzerland, 2017.
[16] R. Perez, R. Seals, J. Michalsky, All-weather model for sky luminance distribution—preliminary configuration and validation, Solar Energy. 50 (1993) 235–245.
[17] N.L. Jones, C.F. Reinhart, Experimental validation of ray tracing as a means of image-based visual discomfort prediction, Building and Environment. 113 (2017) 131–150.
[18] R. Singh, I.J. Lazarus, V. Kishore, Uncertainty and sensitivity analyses of energy and visual performances of office building with external venetian blind shading in hot-dry climate, Applied Energy. 184 (2016) 155–170.
[19] J. Wienold, F. Frontini, S. Herkel, S. Mende, Climate based simulation of different shading device systems for comfort and energy demand, in: 12th Conference of International Building Performance Simulation Association, 2011: pp. 14–16.
[20] L. Brotas, D. Rusovan, Parametric daylight envelope, PLEA, in: 2013.

IOP Conf. Series: Earth and Environmental Science **297** (2019) 012049 doi:10.1088/1755-1315/297/1/012049

Designing liveable urban open spaces in high density cities

R Thilakaratne

Department of Architecture, Chu Hai College of Higher Education, Hong Kong
ruffinaat@chuhai.edu.hk

Abstract. Hong Kong is one of the densest cities compared to other metropolitan cities with population density in some districts exceeding 57,250 persons per sqkm. Hong Kong's liveability has been affected by factors such as escalating urban density, lack of open spaces, congested living environments, air pollution concentration and high proneness to pandemics.

Due to extremely high land premiums, some residents in Hong Kong live in nano flats small as 15sqm in high-rise towers. Given the congested nature of apartments, open spaces a play a vital role contributing to residents' psychological and physical well-being. Due to peculiar topography and extreme high density, Hong Kong open space per capita is far from the WHO recommended standards of 9 sqm per person. Most urban open spaces are pocket spaces located among developments or adjacent to traffic roads with a high ratio of hardscape compared to vegetation. Due to challenging Summer conditions in Hong Kong with high temperature, humidity and stagnant wind most urban open spaces are underutilized.

This research investigates the factors effecting enjoyment of open spaces and design factors that could contribute to improve their liveability as public amenities. Exploration is based on liveability parameters such as user comfort, user behaviour, and spatial attributes of urban open spaces. Field data was collected on seven urban open spaces within Tsuen Wan district during dry Summer days. Field data analysis was supplemented by simulation methods, photographic survey of users' movement within open spaces and user interviews.

All open spaces are located amidst of high-density urban environment facilitating easy access to residents in the neighbourhood proving sense of security. Findings indicate tendencies in using these open spaces in the afternoon after 4pm when these open spaces cool off from solar radiation. Majority of open spaces benefited from shading effect from surrounding buildings, however the seating arrangements have not benefited from the shading. Satisfactory wind speed could be observed in four open spaces out of the seven that benefited from South and SE wind directions providing desirable thermal comfort to users. The park that consisted of the highest ratio of greenery received the lowest wind speed due walled effect created by surrounding buildings. Ironically open spaces with high percentage of hardscape areas reported better wind performance indicating the possibility of high emissivity material creating low pressure pockets drawing winds in particularly in the afternoons.

1. Urban fabric of Hong Kong

Hong Kong is amongst the densest cities in the world due to its challenging topography where less than 25% of the total land area is developable out of total extent of 1064sqkm. Land population density in some areas exceeds 57,250 persons per sqkm [1]; resulting on high-rise high density-built environment to accommodate 7.3million population within approximately 265sqkm of developable land. People live in high-rise towers tall as 60 storeys and in dwelling units small as 15sqm. Given congested living environments, easily accessible outdoor amenities play a vital role towards residents' well-being.

Urban open spaces not only play a vital role as public amenities contributing to emotional and physical well-being of residents, they are also important in maintaining ecological continuity, creating breathing spaces within the high-density fabric and creating desirable microclimates. Despite the size and nature of open spaces 'between' buildings still provides a frame-work for several activities, planned or spontaneous [2].

A study conducted covering all urban open spaces in Hong Kong reports that average open space per capita is 1.6sqm with a range of 0.89 and 2.82sqm [3] which is far below from open space per capita in many metropolitan cities. Approximately 93% of the population of Hong Kong have access to public open spaces within 400m [3].

1.1. Urban climatic challenges

Hong Kong inherits a sub-tropical climate with hot, humid and wet Summers. Thermal comfort conditions in urban centres are influenced by its peculiar hilly terrain and proximity to sea and high-rise high density built environment. Urban heat island effect, street canyon effect, shading from buildings, stagnant wind, air pollutant concentration and high humidity are unavoidable attributes of Hong Kong living. On average wind speed in most urban areas is approximately 1 m/s. Over the past five years humidity levels from June to September ranged between 75% - 84% with average Summer temperature reaching $34.5^{\circ}C$ [4].

High humidity, hot weather, stagnant wind and air pollution are some of the key challenges towards the enjoyment of outdoor life despite over 70 percent of woodlands and country parks and easily accessible urban open spaces. Urban open spaces are predominantly being used by elderly on a regular basis who are more vulnerable to heat stress and safety related issues.

2. Urban morphology in the case study area

The first 'New Town' 'Tsuen Wan district' developed in early 1960 as an industrial hub which has now been evolved into a mixed-use residential and commercial area was selected as the case study. This 6,000ha district accommodates 302,814 population with a density of 5,300 persons per sqkm and a stock of 103,219 households. About 70,000 people live in public and subsidized housing whilst the majority living in private housing [5]. Out of total population of 302,814 residents, 36.5% represent home makers, retirees, self-employed and others who are not employed who spend more hours of the day within the district. About 21.5% population are above 60 years of age [6].

Five to six storey Chinese tenement housing is the predominant typology with a commercial element at the road level. Public housing in the form of residential towers were developed in late 1970s to accommodate influx of population. Street grids are aligned to South East and North West directions without facilitating adequate wind flow along the roads in Summer where predominant Summer wind directions are from the East, South and South-East quarters [7].

Being an old district with many old housing estates, these open spaces are often being used by senior citizens, home makers, parents and domestic helpers who are waiting to pick up their children from schools. Tsuen Wan inherits a combination of parks, public open spaces and sports grounds of varying scales ranging from > 1ha to 5hs<, out of which 87 percent of residents are within 400m to public open spaces and 94 percent residents are within 800m to public open spaces [3]. In terms of accessibility public open spaces within 400m to residential areas, Tsuen wan ranks 16th position out of 18 districts in Hong Kong [3]. Six out of seven open spaces considered for this study are less than 1ha.

2.1. Open spaces planning standards and guidelines
Draft Outline Zoning Plan approved for Tsuen Wan district in 2017 includes land allocations for "Open Space and Green Belt" land uses [8]. 131.84ha out of 6,000ha is land is zoned as "Open Space" to provide active and passive recreational facilities for the community [9]. Parks in Tsuen Wan district have been planned prior to the release of the Draft Outline Zoning Plan which has provided several recommendations for creating sustainable and liveable open spaces. Therefore, findings from this study may be beneficial in redesigning and improving liveability and functionality of open spaces in Tsuen Wan and similar districts. Majority of these open spaces are pocket open spaces that are located among commercial and residential land uses. Local open spaces are primarily intended for passive uses, and hence the 3:2 active to passive ratio does not apply as for district level open spaces [10].

Planning Department [10] recommends that open spaces should be safe, functional, accessible and usable spaces for the community; not merely fulfilling regulations. Safety is a major consideration in open space design with regards to location, identifiable entrances, surface texture and facilities provided. Open spaces should cater to elderly, persons with disability, children and adults in an integrated manner with attention to surface finishes and layout providing adequate outdoor furniture under shading and away from the pedestrian paths with slip resistant hardscape finishes. Where possible parks are required to have a nature buffer to reduce pollutants infiltration. Longitudinal span of 90 percent of the open spaces considered for this study are aligned to North-East and South-West direction only allowing that are abutting streets to receive Summer wind from East, South and South-East quarters prevailing wind through their shorter span.

3. Methodology
The research engaged a multi-dimensional approach for data collection and analysis; field data on microclimatic conditions were collected during Summer months using Extech HT200 Heat Stress WGBT Meter and Kestrel 5000 Environmental Meter. Microclimatic data were supplemented by contextual analysis of these open space, photographic survey of user behaviour and interviews with open space users. Behavioral indicators contributing heat stress such metabolic rate, clothing insulation and moisture permeability were excluded as the WBGT index. Epstein and Moran cite Fanger to explain the scope of WBGT index as an empirical index that indicates heat stress under direct sunlight measuring temperature, humidity, wind speed, sun angle and solar radiation without counting users' physiological information [11] [12]. Field data were verified with the nearest Observatory weather station data. Shadow analysis was conducted using Grasshopper Ladybug plugin; results were correlated with field measurements and the photographic survey.

Field data were collected at 1130hrs and 1530hrs based on initial observations on peak usage trends by the non-working residents in the surrounding housing estates. Users were interviewed based on random sampling supplemented by a structured questionnaire to gather their perception on the factors that affect liveability during summer and their suggestions for improvements. Number of interviews and questionnaire surveys were limited when the responses followed similar trends. Photographic survey was conducted for recording user behavior, park layout, landscape and other features.

The study focused on seven urban pocket open spaces that are less than 1ha in extent that represent different geometries, contextual attributes and land use compositions to investigate their success in functioning as public amenities providing user comfort.

3.1. Developing liveability analysis matrix
The term liveability is mostly researched and discussed at cities' scale although there is no unified matrix for analysis. Even the most widely adopted indices such as Mercer Quality of Living ranking and The Global Liveability Index by The Economist Intelligence Unit vary in their parameters and hence the results.

Gehl [2] & [13] has contributed to knowledge immensely by comprehensive research on quality of public spaces. Twelve quality criteria by Gehl [2] focuses on human senses; protection, comfort and enjoyment, than on physical aspects of a public space. However, since Gehl's studies offers key insights for designing public places, this research extended Gehl's index to incorporate physical design parameters, microclimatic aspects, operational aspects and impacts from the surrounding and mapping them to sensory aspects of Gehl's matrix (Table 1).

Besides human sensation related factors, physical comfort conditions such as thermal comfort and design quality of open spaces play an important role when defining the liveability and functionality. ASHRAE 55 defines thermal comfort as a sensation that indicates the satisfaction with the thermal environment.

Table 1: mapping of Gehl's twelve quality criteria on to the extended matrix

Extended index		**Gehl's twelve quality criteria**
Planning aspects	Surrounding environment	Protection - against harm by others
	Location and accessibility	Protection - against traffic and accidents
	Clear separation of pedestrian traffic from vehicular traffic (automobiles)	
Spatial design and quality	Easy navigation within	Comfort - options for mobility
	Clear separation of relaxing areas from walkways	Comfort - options for mobility
	Amenities –active recreational facilities for all ages	Comfort - options for play, exercise, and activities
	Amenities – access to sanitary facilities, water fountains	
	Spatial arrangement – defined seating, shelters, choice of seating for few and a crowd, resting areas, pedestrian paths	Comfort - options to stand and linger Comfort - options for sitting Comfort - options for talking and listening/ hearing
	Sense of scale	Enjoyment – scale
	Aesthetics – landscapes, waterscapes	Enjoyment - Experience of aesthetic qualities and positive sensory experiences
	Inclusive design	Comfort - options for mobility
	Texture of surfaces	
	Views	Comfort - options for seeing
Operation and maintenance	Cleanliness / hygiene	Protection against unpleasant sensory experience
	Odor	
Microclimatic aspects	Noise	Protection against unpleasant sensory experience
	Temperature	Enjoyment - opportunities to enjoy the positive aspects of climate
	Humidity	
	Wind	

4. Findings

4.1. Planning aspects

As illustrated in Figure 1, the seven urban open spaces considered for this study are well integrated into the high density residential and commercial urban fabric of Tsuen Wan providing a sense of protection, convenience and accessibility. All open spaces are located within 400m walking distance to residential estates and nearby schools and are connected through a network of routes. 400m is considered as a healthy walkable distance according to LEED green building rating tool. Other than three, others are located inland away from major traffic roads and majority are located within commercial or residential developments providing sense of safety for the users and sense of enjoyment due to lack of traffic noise and air pollution.

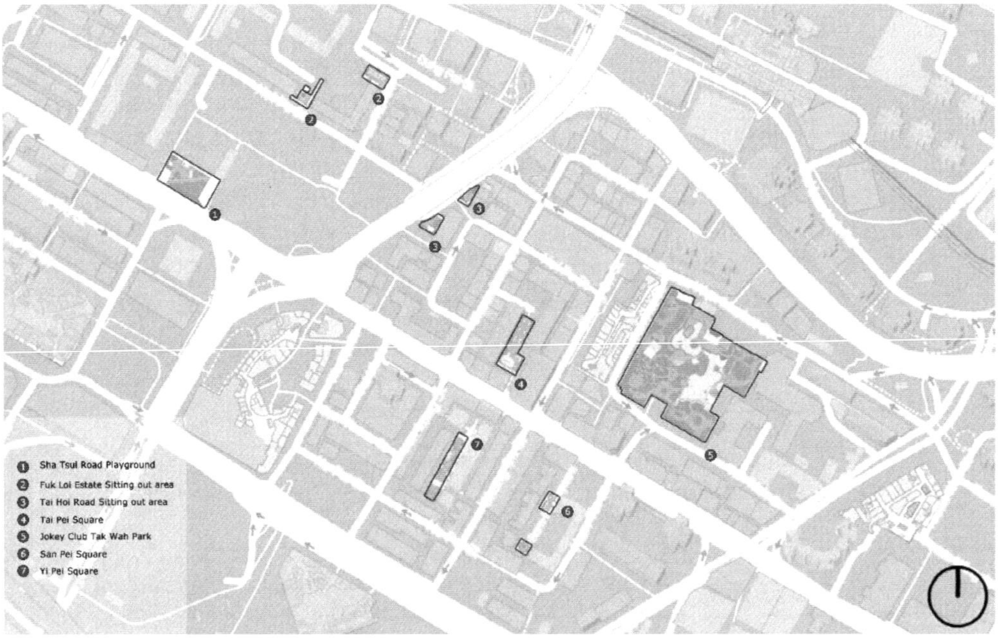

Figure 1. Park network in the case study area (Author, 2018)

4.2. Spatial design and quality

Out of the seven cases, only Jockey Club Tak Wah park is designed according to classical Chinese garden theme whilst other open spaces can be classified as urban pockets. This park is approximately 1.6 ha in extent and about ten times larger compared to all other pocket spaces.

All open spaces facilitated easy navigation within because of their simple configurations. However, the separation of sitting out areas from pedestrian walkways are not adequate to provide sense of relaxation to users as seats are often positioned along the walkway. Out of the seven open spaces only one consists of recreational equipment for elderly and two consisted of simple play elements for children. Seating arrangements and shelters provided in majority of these open spaces and the usage patterns confirm that these open spaces are primarily for passive recreation. Out of the seven, only the Jockey Clun Tak Wah park had sanitary facilities. All these open spaces have followed inclusive design principles making them elderly and disable friendly. None of them however had any views as all of them are located within high density built environment.

Table 2 presents an analysis of spatial attributes indicating hardscape to greenery ratio and orientation in relation to wind direction. Open space sizes vary from 200 sqm to 10,600 sqm with only three open spaces with over 1000sqm extent. These open spaces predominantly comprise of hard surfaces and large canopy trees with limited green cover.

Table 2. Analysis of spatial attributes and response to prevailing wind directions (E, SE & S) (Author, 2018)

<table>
<tr><td colspan="2">Sha Tsui road playground – located adjoining a major traffic road
Orientation in relation to wind direction: A playground on NE, E and SE; a traffic road on S, SW; tall buildings on NW
Benefit from E, SE and S wind directions (average wind speed 1.79m/s)
Hardscape: Greenery = 774:1202 sqm = 1:1.55 (61% greenery in the park)</td></tr>
<tr><td>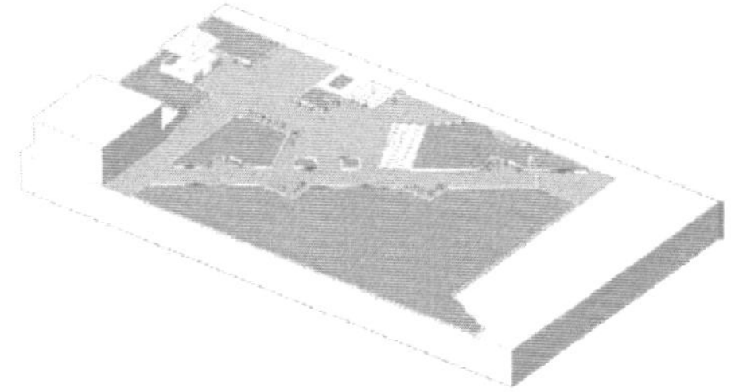</td><td></td></tr>
<tr><td>Fuk Loi Estate urban pocket park – located adjoining a secondary road and surrounded by a residential estate
Orientation in relation to wind direction: N, NE, E, SE facing pedestrian roads; S, W facing school buildings; SW facing school ball court
Benefit from E and SE wind directions
(average wind speed 1.34m/s)
Hardscape: Greenery = 170:34 sqm = 1:02 (17% greenery in the park)</td><td>Fuk Loi Estate urban pocket park – located adjoining a secondary road and surrounded by a residential estate
Orientation in relation to wind direction: E, SE abutting a residential block; S, SW facing a road; NE facing housing estate courtyard surrounded by residential blocks
Benefit from S wind directions
Hardscape: Greenery = 169:61 sqm = 1:04 (27% greenery in the park)</td></tr>
<tr><td>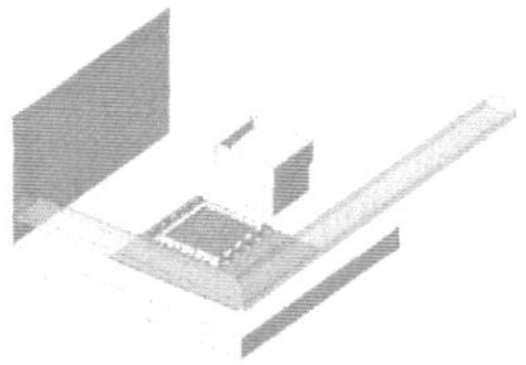</td><td>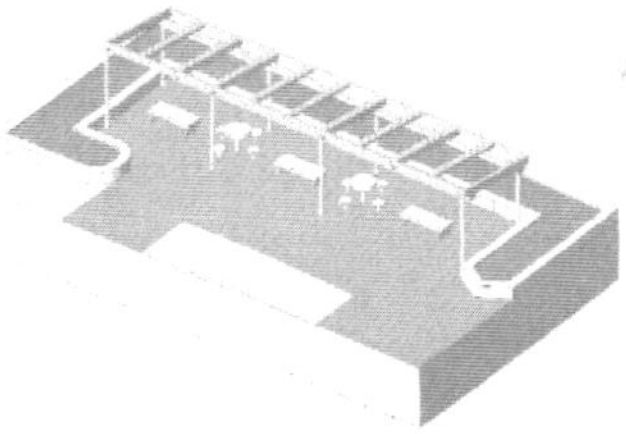</td></tr>
<tr><td></td><td></td></tr>
<tr><td>Tai Hoi road urban pocket park – located adjoining a major traffic road
Orientation in relation to wind direction: N, NW facing a major road; W, SE, S blocked by buildings; SSE facing a road
Benefit from SSE wind directions (average wind speed 1.34m/s)
Hardscape: Greenery = 174:172 sqm = 1:0.99</td><td>Tai Hoi road urban pocket park – located adjoining a major traffic road
Orientation in relation to wind direction: N, NW facing a major road; W, SW, SE, S blocked by buildings; SSE facing a road
Benefit from SSE wind directions
Hardscape: Greenery = 250:180 sqm = 1:0.72 (42% greenery in the park)</td></tr>
</table>

IOP Conf. Series: Earth and Environmental Science **297** (2019) 012049 doi:10.1088/1755-1315/297/1/012049

(50% greenery in the park)	
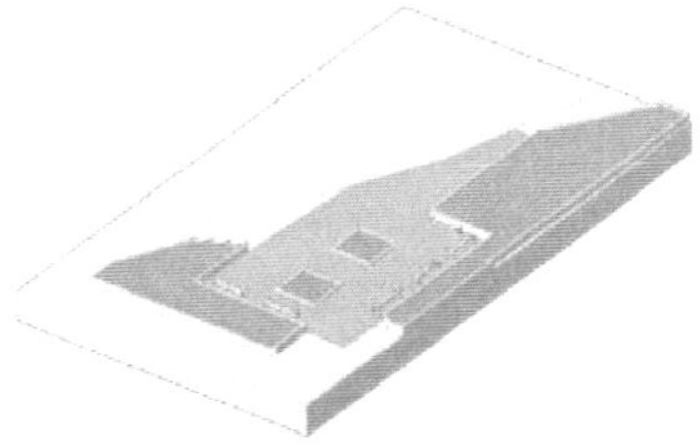	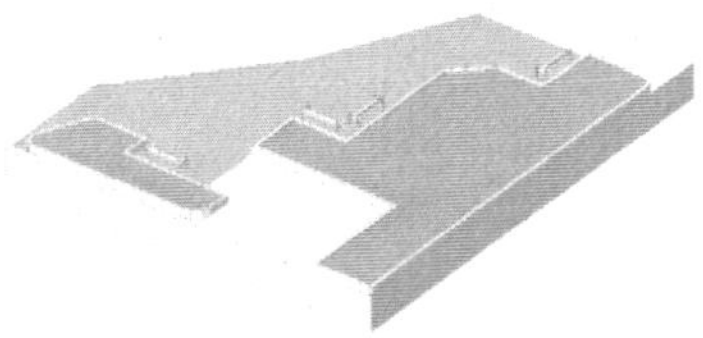

Tai Pei square – located away from traffic routes Orientation in relation to wind direction: surrounded by tall buildings with NE and SW connecting to pedestrian walkways and a small square on NE; S, SW facing a major road Benefit from S wind direction ((average wind speed 1.79m/s) Hardscape: Greenery = 784:279 sqm = 1:0.36 (26% greenery in the park)	
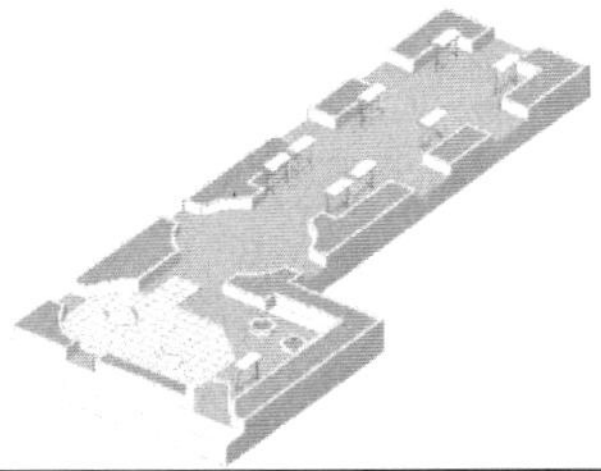	
Jockey Club Tak Wah park – located in the middle of institutional and residential areas. Accessible through many pedestrian walkways Orientation in relation to wind direction: Square park. E, SE facing a hospital; all other sides are facing pedestrian route and buildings Not directly exposed to any wind direction (average wind speed 0.58m/s) Hardscape: Greenery = 4903:4904 sqm = 1:1(42% greenery in the park) Waterscape = 1745 (17%)	

Sam Pei Square – located away from traffic roads Orientation in relation to wind direction: E, SE, facing a school block; S facing school ball court Benefit from S wind directions (average wind speed 2.32m/s) Hardscape: Greenery = 271:69 sqm = 1:0.26 (20% greenery in the park)	

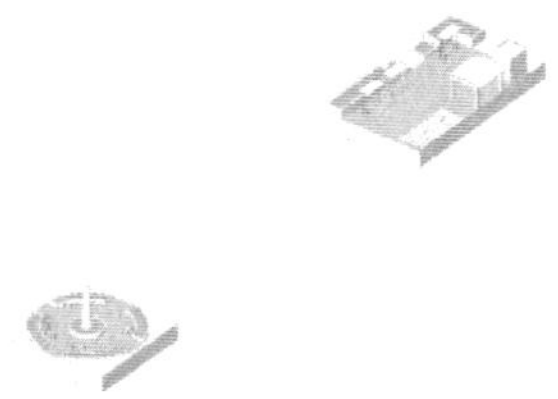

Yi Pei Square – located away from traffic roads

Orientation in relation to wind direction: All sides surrounded by buildings with access from SW, NW, NE, SE

Benefit from SE wind direction (average wind speed 2.46m/s)

Hardscape: Greenery = 621:115 sqm = 1:0.18 (16% greenery in the park)

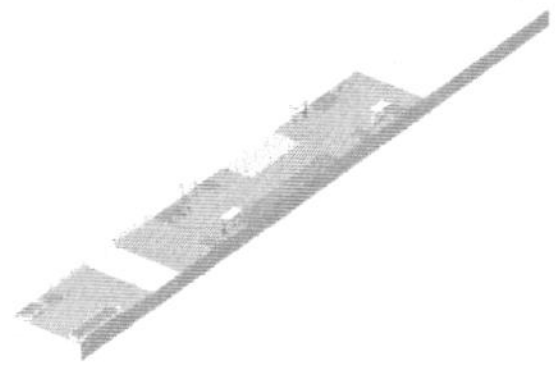

4.3. Impacts on microclimate

Planning Department stipulates greenery requirements for passive recreational areas; 70 percent soft landscape inclusive of 60 percent trees [10]. None of the open spaces in the study area achieved this greenery requirement; only two consisted of more than 50 percent landscape indicating the predominance of hardscape.

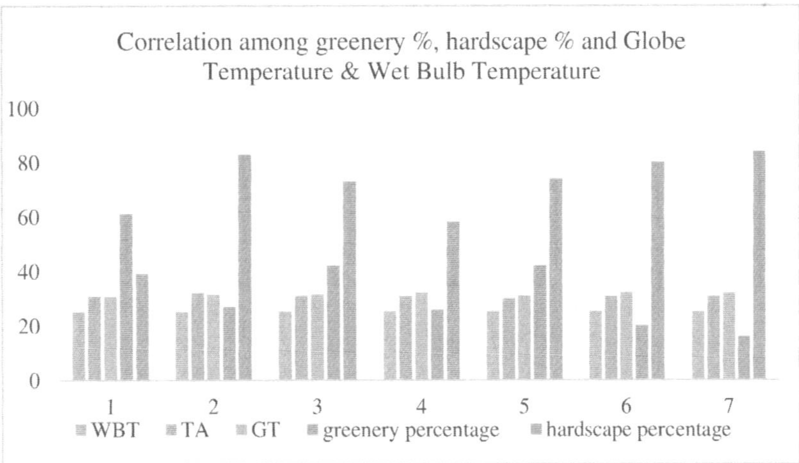

Figure 2. Greenery percentage, hardscape percentage and Globe Temperature in case study open spaces (Author, 2018)

Five out of seven open spaces consist of cement block paving and two consisted of orange color eco blocks; with two having dark color rubber mats in the activity areas. Although light color high albedo materials have been used on hardscape areas, compared to greenery hard surfaces contribute to high emissivity and therefore elevated mean radiant temperature increasing Urban Heat Island (UHI) Effect. Hong Kong is notorious for UHI effect with an average increase of about 2°-3.5°C in air temperature in urban areas [14]. A study on Taipei with a similar climate suggests how solar radiation absorption within unshaded hardscape areas effected the thermal environment within a park during daytime elevating temperature in the surrounding areas [15]. The same study recommends reducing unshaded paved areas to less than 50 percent to alleviate negative effects. Supporting the above argument, Wong and Chen report positive contributions from greenery on microclimate within the parks and the surrounding [16]. Our analysis between hardscape percentage, greenery percentage and Globe Temperature (GT) presented in Figure 2 indicates the lowest GT in the park with the lowest percentage of hardscape and highest percentage of greenery. Parks with the lower percentage of greenery and relatively a high percentage of hardscape indicated a higher GT and elevated air temperature (TA) values. A study that compared thermal properties of material in a desert climate concluded that unlike concrete and asphalt, grass and poly material contribute to the least amount of thermal energy during a 24-hour average, thereby reducing UHI [17]. EPA US reports the importance of vegetation claiming that evapotranspiration from greenery could reduce peak summer temperatures by 1–5°C [18]. Shahmohamadi et al. opined that replacement of soil and vegetation by concrete and asphalt increases ambient temperature. These studies support the role of urban greening on thermal comfort [19]. Wet Bulb Temperature did not vary across the open spaces although the relative humidity ranged between 60.5 to 66.6 percent.

4.4. Outdoor thermal comfort

With 15.9 percent elderly in Hong Kong with an increasing trend and 91.9 percent of them living in compact domestic households, provision of active and passive public recreational facilities has become an agenda towards facilitating healthy communities. Given the hot humid climate in Hong Kong and challenging weather conditions in Summer it is important that these outdoor spaces don't create heat stress in vulnerable population.

WBGT Index was used to assess outdoor thermal comfort as it combines different temperature readings, relative humidity, solar radiation and wind. Although the WBGT Index has its limitations excluding physiological changes. However, given that the majority of users are elderly with similar physiological state WBGT Index was considered adequate for this analysis. Thermal comfort is a state of mind than what reads on the measuring devices; park users was interviewed to accommodate these responses.

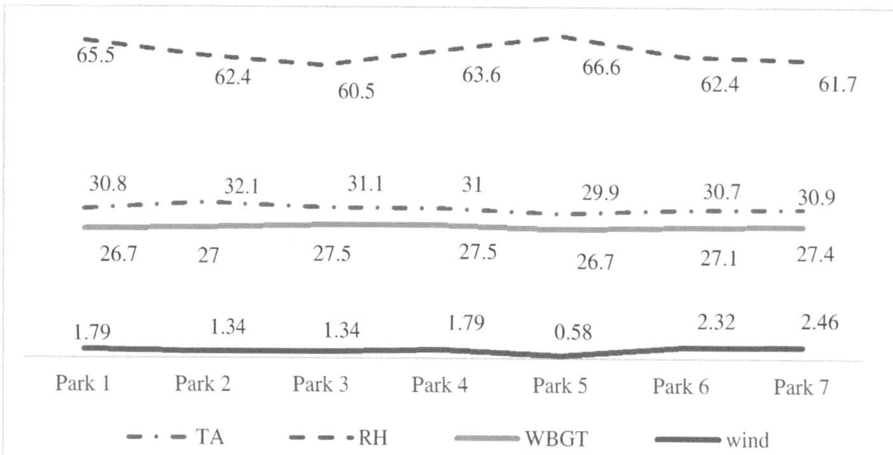

Figure 3. Microclimatic conditions in the seven open spaces collected in Summer (Author, 2018)

Figure 3 presents a comparison of TA, WBGT, RH and wind speed for discussing their effect on thermal comfort. Based on CBE Thermal Comfort Tool analysis none of the open spaces meets the thermal comfort standards required in ASHRAE 55 standards (CBE). To meet ASHRAE 55 standards, at 60% RH, the TA should be between 24-29°C. Ambient Temperature varied from 29.9 to 32.1°C in park 2 and park 5 respectively with a median of 30.9°C. Whereas WBGT results varied from 26.7 to 27.5°C within a median of 27.1°C maintaining a narrow range. Humidity varied between 60.5% to 65.5% with a median of 62.4%. A study on neutral physiological equivalent temperature (PET) reports that in summer in Hong Kong under shaded conditions when the temperature is 28°C, a wind speed of 0.9-1.3m/s is required to provide neutral thermal comfort for a person in light clothing [20]. All open spaces reported air temperature above 29.9°C; however, WBGT levels were between 26.7-27.5°C with lower humidity levels and wind speed in majority of open spaces ranged between 1.34-2.46m/s providing users with comfortable thermal environments. Open spaces (3B, 4, 5, 6 & 7) that are abutting to major traffic roads or surrounded by streets or located amidst of buildings reported elevated Ambient Temperature around noon. Open spaces 5, 6 and 7 cooled off around 4pm perhaps because of high wind speed reported in open space 6 and 7 and high ratio of greenery in open space 5. Hong Kong Observatory station located on a higher plane surrounded by vegetation and about 1.5km direct distance away from the study area reported, 26.8 to 32.2oC range of Ambient Temperature alluding to the effect of greenery on outdoor temperature.

Due to high rise high density built environment, most urban areas in Hong Kong experience stagnant wind. Average Summer prevailing wind speed at 200m elevation for Tsuen Wan is approximately 3.3 m/s [21]. Attributed by microclimatic conditions such as orientation to wind direction, vegetation ratio and surrounding density etc., these open spaces demonstrated varying wind speeds. As reported in Table 2 & Figure 3 satisfactory wind speed could be observed in open spaces 1, 4, 6 and 7 that benefit from South and SE wind directions. Although having a high greenery ratio amongst all, open space 5 reported the lowest wind speed because of the surrounding buildings and perhaps because of the friction created by the tree crowns. Ironically open spaces with high percentage of hardscape areas reported better wind performance indicating the possibility of high emissivity material creating low pressure pockets drawing winds in particularly in the afternoons.

4.5. Impacts from surrounding built environment on the open spaces

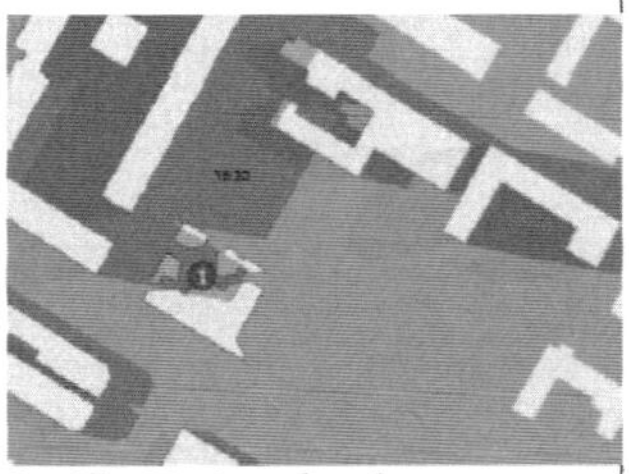 1 – Sha Tsui road park 10% shaded at 11.30 30% shaded at 15.30	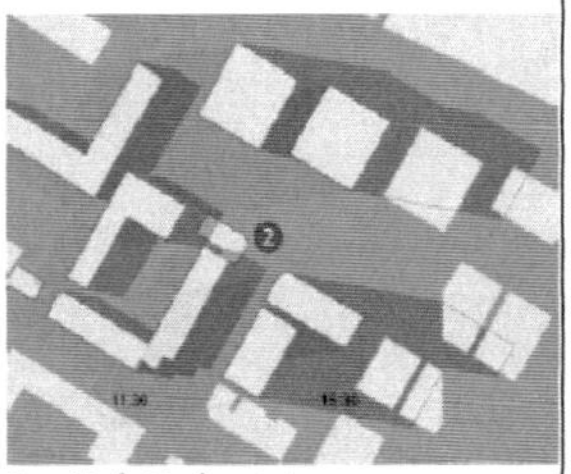 2 – Fuk Loi estate open space 40% shaded at 11:30 20% shaded at 15:30	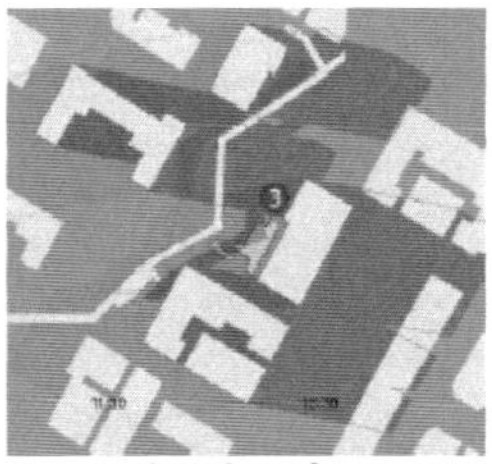 3A – Tai Hoi road open space 50% shaded at 11:30 50% shaded at 15:30
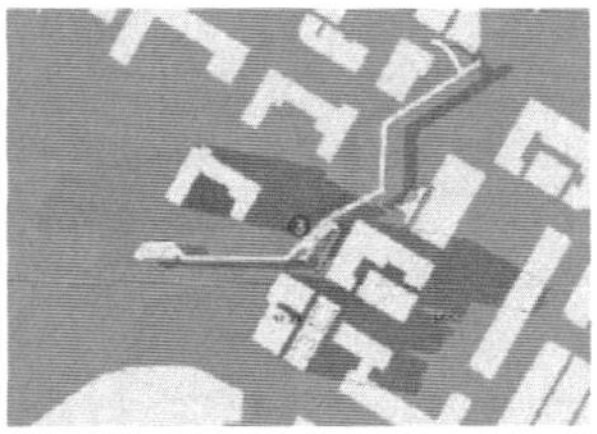 3B – Tai Hoi road open space 50% shaded at 11:30 50% shaded at 15:30	 4- Tai Pei square 70% shaded at 11:30 60% shaded at 15:30	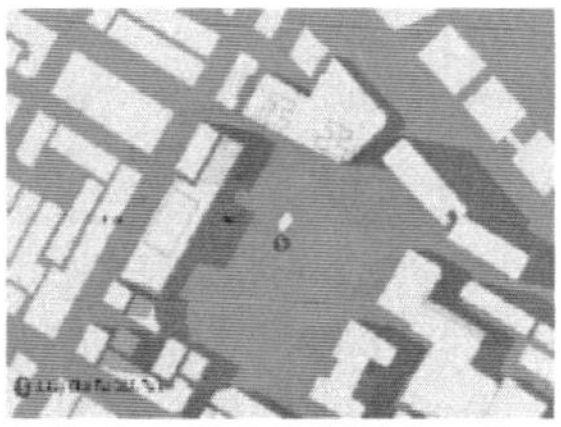 5- Jockey Club Tak Wah park Shaded closer to building edges at 11:30 & 15:30 Centre of the park is shaded by mature trees and small pavilions

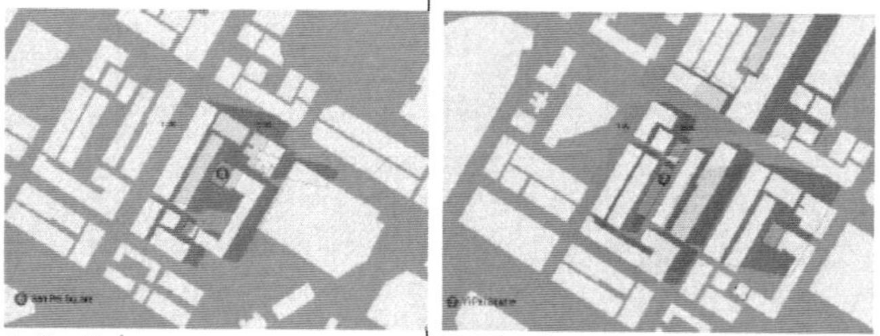

6- San Pei square
100% shaded at 15:30

7- Yi Pei square
100% shaded at 11:30 & 15:30

Figure 4. Shadow analysis that compare 11:30 & 15:30 hrs. on a sunny day (Author, 2018)

Open spaces are surrounded by the residential, commercial and institutional buildings around providing convenience to park users. Majority of these buildings are above six stories high casting shadows on the open spaces. Figure 4 shadow analysis indicates a positive effect by creating partial shading on the exposed areas at different times of the day contributing to lower solar radiation caused by the hardscapes. Most open spaces seem to benefit from shadows in the afternoon mitigating radiant heat. Shadow analysis correlated with people's preferred locations within the open spaces which, could be used as a basis for positioning shelters and seating for better integration of amenities.

4.6. Users' perceptions

Protection, comfort and enjoyment criteria from Gehl's study was adopted in the user survey [2]. Since these urban spaces are primarily intended for passive recreation, around 80 percent of the users are elderly from the vicinity who enjoy these open spaces 2-7 days per week between 3-5pm for socializing with friends and for relaxation. Not all seating places are provided under shades by trees or shelters, photographic survey confirmed users' preference for these spots. Elderly expressed the need for shade and young responded expressed the need for more seating. There was a consensus about the proximity of these open spaces to their residences and the emotional comfort they receive from being outdoors. Majority of respondent reported the benefit of large mature trees and nature on these parks providing them breeze and comfort. Although ambient temperature is above the comfortable range as per Ng & Cheng study [19], respondent didn't express any discomfort perhaps because of relatively lower humidity in these open spaces compared to average humidity level in Hong Kong.

For future development of parks, users suggested the need for more amenities such as more seating and toilets and proper drainage systems to reduce storm water stagnation. Compared to small open spaces they expressed the preference for large parks with big squares, more trees, seats and more natural elements such as water and rocks.

5. Conclusions & recommendations

Open spaces are vital in Hong Kong given the nature of high-density high-rise living and compact apartments. Although the parks are rather small in extent with limited amenities, facilities, and greenery, their location in close proximity to residential and commercial areas, provided convenience and sense of security and protection for elderly and children. Other than two open spaces, all others are located away from traffic roads providing protection against vehicular traffic and screening from noise and air pollution. Other than Jockey Club Tak Wah park, other parks layouts provide user-friendly navigation within. Due to compact nature of majority of open spaces, they have not been able to provide adequate separation between relaxing areas and walkways enhancing users' comfort and enjoyment. Majority of open spaces benefit from shadow casting in some areas from surrounding buildings most of the day; however, seating arrangements have not followed these natural patterns. These parks are often been used in the late afternoon when the parks cool off. Due to non-alignment with prevailing wind directions, only two parks received above 2m/s wind speed; planning and design measures such as proper positing of vegetation and wind deflectors may be considered for improving wind penetration through these open spaces.

Although these open spaces are connected forming a network, they don't appear to be successful in forming ecological continuity due to limited amount of greenery compared to predominant hardscapes. Regardless of high Urban Heat Island effect in Hong Kong urban areas, these parks consist of over 66 percent of hardscape on average. Users expressed concerns with storm water drainage on these open spaces that could be attributed to lack of ground cover vegetation to absorb water. Considering the increasing amount of rainfall and global warming in the recent years, it would be sensible to replace impervious paving with vegetation cover. Instead of adopting engineering approaches, these parks should consider adopting 'nature based solutions' to promote ecological continuity and to mitigate adverse impacts from Urban Heat Island effects and storm water stagnation. Native plants and local material could be sensible options to consider when implementing 'nature-based solutions' to reduce maintenance and water efficient. However, feasibility studies and proper planning is required in order to reap best benefits out of 'nature-based solutions'. Considering the rising ageing population in Hong Kong, public amenities should adopt an integrated approach focusing on user comfort and safety, careful choice of material, natural shading patterns, providing adequate amenities and improving ecological value, and microclimatic conditions of the open spaces and the surrounding.

6. Acknowledgements
The author is grateful to Chu Hai College of Higher Education for research funding. Author would like to thank research assistants, Mr. Lam Ho and Mr. Yiksun Lai for their contribution in data collection and simulation studies and Prof. Paul Chu and Mr. Roger Tan for advice and thought-provoking ideas.

References

[1] HKSAR Information Services Department 2015 Hong Kong – The Facts [Available from: https://www.gov.hk/en/about/abouthk/factsheets/docs/population.pdf

[2] Gehl J 2011 Life between buildings: using public space. 6 ed. Washington DC: Island Press

[3] Chow J 2018 Public open space accessibility in Hong Kong, a geospatial analysis Reconnecting open space programme

[4] Observatory HK 2018 Daily Weather Summary and Radiation Level on 26/06/2018 [updated 27 June. Available from: http://www.hko.gov.hk/cgi-bin/hko/yes.pl?year=2018&month=06&day=26&language=english&B1=Confirm

[5] HKSAR Home Affairs Department 2019 My Districts - Tsuen Wan District [Available from: https://www.had.gov.hk/en/18_districts/t_my_map_16.htm

[6] HKSAR Census and statistics Department 2016 Population by Census district profiles [updated 8 December 2017. Available from: https://www.bycensus2016.gov.hk/en/bc-dp-new-town.html

[7] Ng E 2011 Cat. A1– Term consultancy for expert evaluation and advisory services on air ventilation assessment Contract No.: PLNQ 35/2009

[8] Town Planning Board 2017 Consideration of Representations in respect of the Draft Tsuen Wan Outline Zoning Plan No. S/TW/32. Hong Kong: Town Planning Board

[9] HKSAR Planning Department 2017 Recreation, Open Space and Greening

[10] HKSAR Planning Department 2015 Chapter 4: Recreation, Open Space and Greening [updated October 2015 Available from: https://www.pland.gov.hk/pland_en/tech_doc/hkpsg/full/ch4/ch4_text.htm#1.9

[11] Epstein Y and Moran D 2006 Thermal comfort and the heat stress indices *Industrial Health* **44**: 388–98

[12] Fanger P 1970 Thermal comfort Copenhagen: Danish Technical Press

[13] Gehl J and Svarre B 2013 How to study public life. 2 ed. Washington DC: Island press

[14] Fung W, Lam K, Nichol J and Wong M 2009 Heat Island Study - Satellite Derived Air Temperature. *J. Applied Meteorology and Climatology* **48**(4):863-72
[15] Chang C and Li M 2014 Effects of urban parks on the local urban thermal environment *Urban Forestry & Urban Greening* **13**(4)
[16] Wong N and Chen Y 2006 Thermal benefits of city parks *Energy and Buildings* **38**(2) :105-120
[17] Santillán-Soto N, Garcia R, Haro Z, Ojeda-Benitez S, Quintero-Nuñez M and Velázquez N 2015 Radiation balance of urban materials and their thermal impact in semi-desert region: Mexicali, México study case *Atmosphere* **6**: 1578-1589
[18] USA Environmental Protection Agency Reducing Urban Heat Islands: Compendium of strategies trees and vegetation Available from: https://www.epa.gov/sites/production/files/2014-06/documents/treesandvegcompendium.pdf
[19] Shahmohamadi P, Ani A, Ramly A, Maulud K and Nor M 2010 Reducing urban heat island effects: A systematic review to achieve energy consumption balance *International J. Physical Sciences* **5**(6):626-36
[20] Ng E and Cheng V 2012 Urban human thermal comfort in hot and humid Hong Kong. *Energy and Buildings* **55**
[21] HKSAR Planning Department Site wind availability Available from: https://www.pland.gov.hk/pland_en/info_serv/site_wind/site_wind/067054.html

IOP Conf. Series: Earth and Environmental Science **297** (2019) 012050 doi:10.1088/1755-1315/297/1/012050

Evaluation of hybrid heating systems with a combination of fossil and renewable energy sources

Anatolijs Borodinecs[1], Raimonds Bogdanovics[1], Aleksejs Prozuments[1], Jelena Tihana[1], Baiba Gaujena[2]

[1] Institute of Heat, Gas and Water Technology, Riga Technical University
[2] Department of Building Production, Riga Technical University

Abstract. Natural gas heating systems are traditionally very popular in private sector in Latvia because of their numerous advantages, such as high efficiency, relatively low CO_2 emissions, automation possibilities and convenient fuel supply, as well as widely available natural gas distribution network. Household natural gas consumption has remained fairly stable over the last decade and accounted for 11.3% of total natural gas consumption in 2017, however, natural gas consumption may decline in the near future, because of installation of renewable energy powered systems which improve systems' resilience and reduce expenses on heating and hot water preparation. Curently hybrid gas boilers in combination with solar thermal panels are promoted on the market as the most efficient and environmentally friendly solutions. The main aim of this study is to perform a detailed simulation on hybrid gas systems energy production in Latvian climate and to evaluate potential energy savings. The performed simulation results show that solar thermal panels do not actually reduce the consumption of natural gas in November, December and January, whereas in the summer months solar thermal panels almost entirely provide the needed energy to prepare hot water. Nevertheless, the annual savings for a single family house are minor and the payback time for the installation of hybrid gas/solar systems could exceed 20 years.

1. Introduction

Complying with the European Union goals of reducing greenhouse gas emissions by at least 20% compared to 1990, a 20% increase of energy production from renewable sources and a 20% improvement in energy efficiency by 2020 [1], a choice of suitable and modern heating system in private housing sector becomes very reasonable. According to Eurostat, household sector consumes approximately quarter of all energy used in Europe (EU-28) [2].

In Latvia natural gas boilers traditionally are very common in private housing sector due to a number of advantages, such as high efficiency, relatively low CO_2 emissions, automation possibilities and convenience of fuel supply, as well as widely available natural gas distribution network. Nevertheless, over the last decade the total natural gas consumption in Latvia has been falling steadily; according to the information available in Central Statistics Center's database between the years of 2010 and 2017 natural gas consumption in Latvia has decreased by 33% [3] (Figure 1).

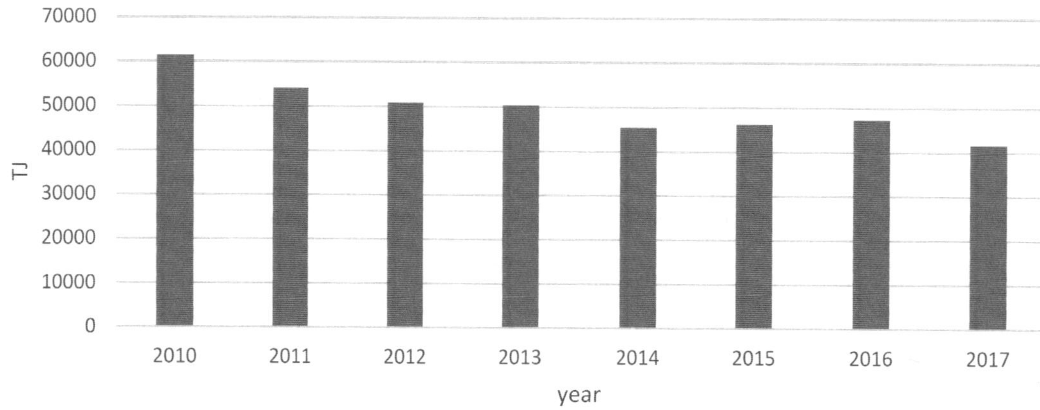

Figure 1 Natural gas consumption in Latvia [3]

Household natural gas consumption has remained fairly stable over the last 10 years and in 2017 it constituted 11.3% of the total natural gas consumption. It is very likely that in the nearest future natural gas consumption may continue to decline, because of the following factors:

- the end users more often tend to choose modern technological solutions for heating systems;
- national and EU legislation requires construction of energy efficient buildings.

The bulk of natural gas consumption in households is attributed to meeting heating and hot water demands. The possibilities of gas boiler modernizations and a decrease in natural gas consumption, while the efficiency of the system is increased, are explored in the following sections.

Nowadays convection-type gas boilers are outdated and are barely used. The exhaust gas temperature in such boilers reaches 150-200 °C in order to prevent water condensation in the boiler and, thus, to avoid corrosion [13]. Thereby, huge amount of energy is lost through flue gas. Nowadays condensing – type boilers are used in newly constructed buildings. The dependence of gas boiler efficiency on air surge α and flue gas temperature is shown in Figure 2.

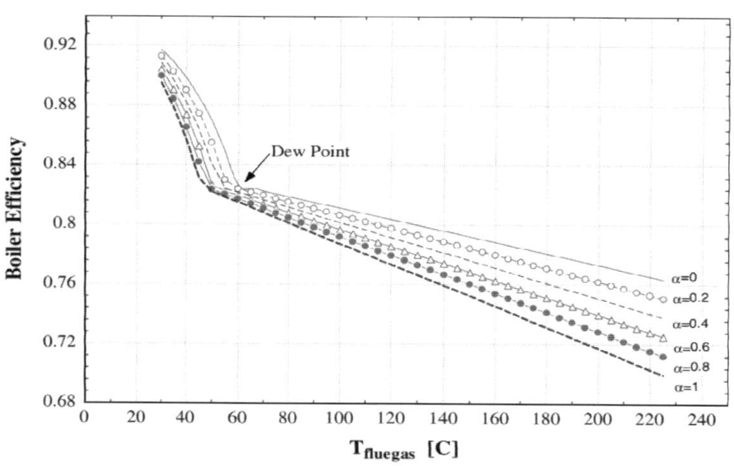

Figure 2 Dependence of gas boiler efficiency on air surge α and flue gas temperature [13].

Based on [5] article, it can be concluded that condensing-type gas boilers consume less fuel and emit less CO2 into the atmosphere (16.8 g - 18.5 g CO2eq per 1 MJ of energy produced) under similar climatic conditions compared to the conventional gas boilers (19.8 g - 21.3 g CO_{2eq} to 1 MJ).

2. Review of hybrid heating systems and their performance

2.1. Gas boiler and heat pump

In order to increase the efficiency of the building's heating system and reduce the operating costs, it is possible to combine a natural gas boiler with a heat pump in a single hybrid system. With the help of sensors and automation, the hybrid system itself chooses the most efficient heating mode based on the operating conditions [8].

Air-to-water heat pumps are quite common in Latvian market, as they can be easily and quickly installed in almost any building type. These pumps also require minimal maintenance, do not create exhaust gases and therefore do not pollute the environment, and moreover, they are cheaper than ground heat pumps. The highest drawback and remaining concern of the air-water heat pumps is the efficiency drop, when the outdoor temperature decreases. [9]. According to the local standards the heating systems have to be designed for the coldest five-day average air temperature, at which the air-to-water heat pumps' efficiency is relatively low. If a heating system consists of just an air-to-water heat pump, most part of the heating period the heat pump will operate in low capacity mode. [11] performed a study with TRNSYS software, where the effectivity of hybrid system (air heat pump and condensing gas boiler) was analyzed. It was shown in the study that the total efficiency of hybrid system reaches 125.2% which is more efficient in comparison to heat pump (96.6%), gas condensing boiler (104.1%) and microCHP system (119.8%).

2.2. Gas boiler and solar thermal collectors

The main downside of a single solar thermal collector system is the unpredictability of the produced thermal energy as well as an insufficient amount of energy production in winter time.

Combining the use of solar energy with gas boilers in a hybrid system can simultaneously provide stability of energy supply, significantly reduce the consumption of natural gas and minimize negative environmental impact. When the amount of incident solar energy is sufficient, i.e., in the summer or spring, the system can operate solely on the solar energy, while in the cold season the gas boiler serves as a heat source. Natural gas savings were examined in [12] study, in which the comparison of a hybrid system and a gas convection boiler under similar climatic conditions was described. The results of the study can be seen in Figure 3. The total annual gas consumption in a hybrid system was only half the gas consumption compared with the system employing a convection-type gas boiler.

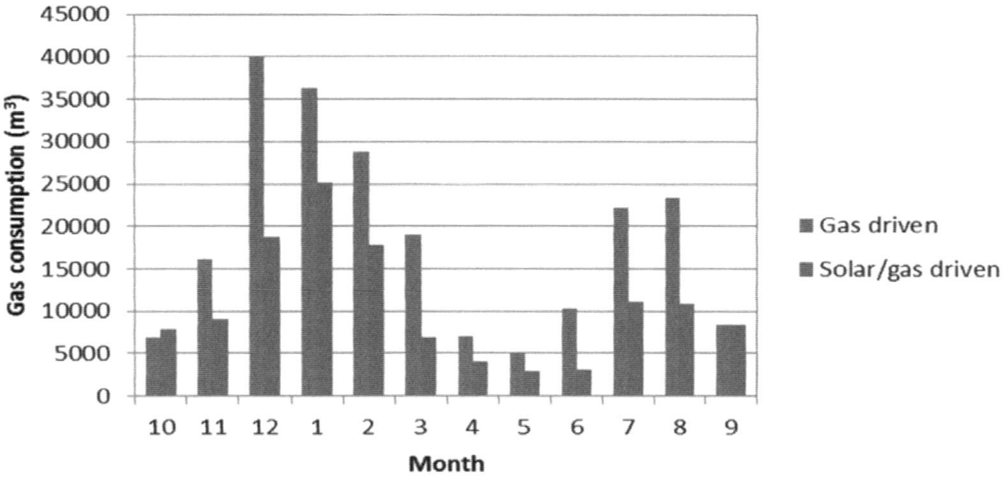

Figure 3 Natural gas consumption in hybrid (solar / gas) and convection-type gas boiler heating systems [12]

It should be noted, that the above-mentioned study focuses on the application of hybrid systems in commercial buildings. Such buildings have a profoundly different energy usage profile. Thus, the results from that study cannot not be effectively applied to family houses.

3. Evaluation of gas/solar system energy efficiency for a single family house in Latvian climate conditions

In order to analyze the consumption of natural gas in households, a dynamic simulation model with the use of the TRNSYS program was developed.

A 2-floor single family house with geometric parameters of 10x10x6 m was selected as the most viable customer type model for future mass installation. Heat transfer coefficients of the external building elements (walls, floors, ceilings, windows and thermal bridges) were defined according to LBN 002-15 [14] "Thermal performance of building envelope". Some additional input parameters for the simulation model are shown in Table 1.

Table 1. Additional input parameters for the simulation model.

Parameter	Value
Average heat transfer coefficient, W/(m^2·K)	0.25
Indoor air temperature, °C	+21
Total enclosure area, m^2	460
Total building volume, m^3	550

Ventilation system losses were assumed to be compensated by heat gains. Meteorological data (except for the outside temperature) was derived from TRNSYS library collections. The hourly outdoor temperature was calculated as the average specific hourly temperature for the 5 years period (2013-2017) based on meteo.lv available data [15]. Heating would be switched on during the heating season, which lasts for 207 days [16]. It is assumed that the building is occupied by 4 people and that one person per day consumes 36 liters of hot water (according to Table A.1 of Annex A, table LVS 15316-3-1). According to the recommendations of solar collector manufacturers, 4 flat type solar collectors with a total surface area of 10 m^2 were added to the model. To use solar energy for both heating and hot water, a storage tank with a capacity of 750 liters was added. The efficiency of the condensing boiler (calculated from the highest calorific value) was defined as 90%.

City of Riga was selected as the location of the analyzed building, the most populated and growing area in Latvia. Also, the City of Riga as well as its suburban areas have a wide network of natural gas pipelines. It should be mentioned that the number of sunny days doesn't vary extensively across the country of Latvia due to its relatively small area and geographical location.

4 Results

The performed simulation has shown that solar collectors do not reduce natural gas consumption in November, December and January, while in the summer months they almost entirely provide all the energy needed for hot water preparation. Consequently, the installation of solar collectors in private homes increases the imbalance of gas consumption during the year.

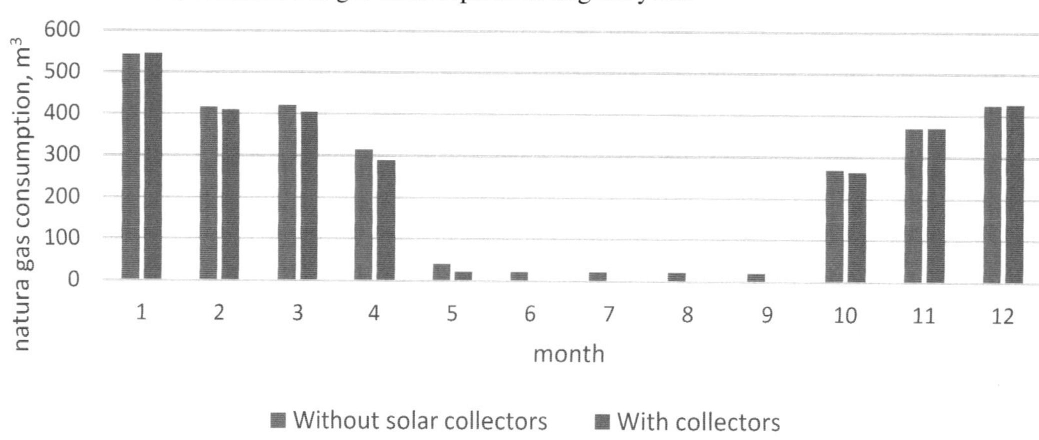

Figure 4 Consumption of natural gas in a hybrid system

According to the simulation results, just with condensing boiler heating and hot water needs 15143.6 kWh of energy per year, or 1437.7 m^3 of natural gas (medium highest calorific value of natural gas during the year was 10,589 kWh/m³, [17]). The installation of solar thermal collectors reduces the energy which need to be supplied from gas boiler, to 12889.1 kWh, or 1,225.3 m^2 of natural gas. The annual calculated saving is 2254.5 kWh or 212.4 m3 of natural gas. If it is assumed that the natural gas tariff is EUR 0.04524 / kWh (effective from 1 July 2018) [18], the approximate annual saving is € 102.
Considering the design, construction and operating costs, the payback period may exceed 20 years and therefore the matter of economic profitability and overall reasonability of implementing such system becomes uncertain.

5 Conclusions

Installing solar collectors in a conventional private house in Riga as an additional source of energy for a condensing gas boiler for hot water preparation and heating systems allows to reduce natural gas consumption by up to 15% per year.

According to the scientific articles reviewed, a similar economy (5-20%) can be provided by increasing the efficiency of the system by cooling the flue gases with an integrated absorption-type heat pumps or by recirculation-condensation water recirculation-heat loss recovery condensing boiler, or installing hybrid heating systems with air-to-water heat pump.

A the current cost level the hybrid gas/solar system allows to save 102 euros per year, which, given the high initial cost of system modernization, does not provide a clear basis for the economic justification for such modernization. Therefore, there is no reason to expect a rapid increase in popularity of alternative energy sources in the Latvian market because they are not able to compete with natural gas at current prices. This, in turn, shows that the consumption of natural gas in households will not continue to decrease further in the near future.

It can be concluded that the modernization plan of the existing heating system with natural gas as a heating source has to take into account the price of the new equipment and the relatively long payback period, as only optimal combination of the initial price and the running costs may ultimately lead to the competitiveness of the particular solution.

However, under fluctuating natural gas prices the long-term efficiency of hybrid natural gas/solar thermal collectors should be evaluated in a more detailed manner. It should be noted that Latvian energy sector depends on natural gas imports quite aggressively. It means that the hybrid natural gas systems shall be promoted by government and local incentives in order to minimize Latvian dependency on imported fuels. This also allows for minimal systems' operation in the event of full shut down of gas supply, thus ensuring some degree of resilience of the energy system.

Aknowledgements

This study was supported by European Regional Development Fund project Nr.1.1.1.1/16/A/048 "Nearly zero energy solutions for unclassified buildings".

Reference

[1] European Commission. 2020 climate & energy package [online]. [viewed 15.09.2018]. Available from: https://ec.europa.eu/clima/policies/strategies/2020_en
[2] Eurostat. Final energy consumption by sector [online]. [viewed 15.09.2018]. Available from: http://ec.europa.eu/eurostat/web/products-datasets/-/tsdpc320
[3] Central Statistical Bureau database. ENG020. Energybalab, TJ, thausn .toe (NACE 2. red.) in Latvian.
[4] M.C. Barma, R. Saidur, S.M.A. Rahman, A. Allouhi, B.A. Akash, Sadiq M. Sait: A review on boilers energy use, energy savings, and emissions reductions. Renewable and Sustainable Energy Reviews. 79, pp. 970-983 (2017).

[5] Vignali, G.: Environmental assessment of domestic boilers: A comparison of condensing and traditional technology using life cycle assessment methodology. Journal of Cleaner Production. 142, pp. 2493-2508 (2017).

[6] Zhu, K., Xia, J., Xie, X., Jiang Y.: Total heat recovery of gas boiler by absorption heat pump and directcontact heat exchanger. Applied Thermal Engineering. 71, pp. 213-218 (2014).

[7] Lee, C.-E., Yu, B., Lee, S.: An analysis of the thermodynamic efficiency for exhaust gas recirculation-condensed water recirculation-waste heat recovery condensing boilers (EGR-CWR-WHR CB). Energy. 86, pp. 267-275 (2015).

[8] Rethink the future of heating Daikin Altherma Hybrid heat pump. 4 p.

[9] Asaee, S., Ugursal, V., Beausoleil-Morrison, I.: Techno-economic feasibility evaluation of air to water heat pump retrofit in the Canadian housing stock Applied Thermal Engineering. 111, pp. 936-949 (2017).

[10] Klein, K., Huchtemann, K., Müller, D.: Numerical study on hybrid heat pump systems in existing buildings. Energy and Buildings. 69, pp. 193-201 (2014).

[11] Di Perna, C., Magri. G., Giuliani, G., Serenelli, G.: Experimental assessment and dynamic analysis of a hybrid generator composed of an air source heat pump coupled with a condensing gas boiler in a residential building. Applied Thermal Engineering. 76, pp. 86-97 (2015).

[12] Sun, H.Q., Xu, Z.Y., Wang, H.B., Wang, R.Z.: A solar/gas fired absorption system for cooling and heating in a commercial building. Energy Procedia. 70, pp. 518-528 (2015).

[13] Qu, M., Abdelaziz, O., Yin, H.: New configurations of a heat recovery absorption heat pump integrated with a natural gas boiler for boiler efficiency improvement. Energy Conversion and Management. 87, pp. 175-184 (2014).

[14] Latvian Buildings Code LBN 002-15 "Thermal performance of building envelope"

[15] Latvian Environment, Geology and Meteorology Centre climatic data: https://www.meteo.lv/meteorologija-datu-pieejamiba/?&nid=462

[16] Data on length of heating season in Riga (in Latvian): http://www.rs.lv/lv/aktualitates/noslegusies-apkures-sezona

[17] Data on natural gas properties in Latvia (in Latvian). Pieejams: https://mans.gaso.lv/Caloricity.aspx

[18] Natural gas price calculator: https://lg.lv/majai/tarifi-un-kalkulators

[19] Shipkovs, P., Kashkarova, G., Lebedeva, K., Shipkovs, J. Perspectives for solar thermal energy in the baltic states. In D. Goswami (Ed.), Proceedings of the Solar World Congress 2005: Bringing Water to the World, Including Proceedings of 34th ASES Annual Conference and Proceedings of 30th National Passive Solar Conference 2. 2, pp. 1378-1383. Orlando: Curran Associates, Inc (2005).

AUTHOR INDEX

AUTHOR INDEX